# Applications for Earth-Abundant Transition Metals

## Edited by

**Inamuddin[1], Arwa Alrooqi[2], Hind Alluqmani[2], Mohammad Abu Jafar Mazumder[3,4]**

[1]Department of Applied Chemistry, Zakir Husain College of Engineering and Technology, Faculty of Engineering and Technology, Aligarh Muslim University, Aligarh-202002, India

[2]Department of Chemistry, Faculty of Science, Al Baha University, Al Baha, Saudi Arabia

[3]Chemistry Department, King Fahd University of Petroleum & Minerals, Dhahran 31261, Saudi Arabia

[4]Interdisciplinary Research Center for Advanced Materials, King Fahd University of Petroleum & Minerals, Dhahran 31261, Saudi Arabia

Published by **Materials Research Forum LLC**
Millersville, PA 17551, USA

Published as part of the book series
**Materials Research Foundations**
Volume 179 (2025)
ISSN 2471-8890 (Print)
ISSN 2471-8904 (Online)

Print ISBN 978-1-64490-370-4
eBook ISBN 978-1-64490-371-1

This book contains information obtained from authentic and highly regarded sources. Reasonable efforts have been made to publish reliable data and information, but the author and publisher cannot assume responsibility for the validity of all materials or the consequences of their use. The authors and publishers have attempted to trace the copyright holders of all material reproduced in this publication and apologize to copyright holders if permission to publish in this form has not been obtained. If any copyright material has not been acknowledged please write and let us know so we may rectify this in any future reprints.

Distributed worldwide by

**Materials Research Forum LLC**
105 Springdale Lane
Millersville, PA 17551
USA
https://mrforum.com

Manufactured in the United States of America
10 9 8 7 6 5 4 3 2 1

# Table of Contents

# Preface

The increasing global demand for sustainable technologies, renewable energy, and cost-effective industrial processes has drawn significant attention to the use of earth-abundant transition metals. Metals such as iron, copper, manganese, cobalt, and nickel offer viable alternatives to expensive noble metals commonly used in catalytic, industrial, and medicinal applications. Their natural abundance, lower cost, and diverse chemical behavior position them as crucial contributors to the advancement of greener technologies and more accessible scientific solutions.

The book *Applications of Earth-Abundant Transition Metals* is a comprehensive compilation of recent studies and emerging applications that leverage the potential of these metals across various industries. It draws from interdisciplinary research to provide a broad and insightful overview of how earth-abundant transition metals are transforming modern technology. This book aims to serve as a valuable resource for researchers, academicians, and industry professionals working in the fields of catalysis, environmental science, materials chemistry, pharmaceuticals, and sustainable energy. By showcasing the versatility and promise of earth-abundant transition metals, it contributes to the global pursuit of environmentally responsible and economically viable technological solutions. The chapters are summarized below:

**Chapter 1** presents an in-depth analysis of the role of earth-abundant metals (EAMs) in enhancing the performance of solid propellants. The chapter explores critical catalytic mechanisms—such as radiation, exothermic decomposition, and physical transitions—while addressing practical challenges in nozzle design, combustion dynamics, and propellant stability. Environmental and sustainability benefits of EAM-based catalysts in propulsion technologies are also discussed.

**Chapter 2** explores Earth-Abundant Transition Metal Oxide-based catalysts for composite solid propellants. It examines the use of single, binary, and composite metal oxides and provides insight into their catalytic mechanisms.

**Chapter 3** reviews the applications of earth-abundant metals in pharmaceuticals. It covers both bulk and nanosized particles and discusses their therapeutic roles. Since these metals are trace elements in biological systems, the chapter emphasizes the importance of sustained, targeted, and controlled delivery to minimize cytotoxicity and maximize pharmaceutical potential.

**Chapter 4** discusses the medicinal use of metal-containing compounds. Currently, 76 elements have been incorporated into clinically approved or experimental compounds, including both stable and radioactive isotopes. The chapter explains how these metals are utilized in the diagnosis and treatment of various diseases.

**Chapter 5** introduces transition metal chalcogenides (TMCs) and their application in dye-sensitized solar cells (DSSCs). It highlights TMCs such as tungsten disulfide ($WS_2$), molybdenum disulfide ($MoS_2$), and titanium disulfide ($TiS_2$), which enhance solar conversion efficiency by improving charge transfer and reducing recombination losses.

**Chapter 6** discusses the application of transition metal composites as high-performance supercapacitor electrodes. It also provides a detailed explanation of energy storage mechanisms. Emphasis is placed on recent advancements, performance evaluations, and the role of ruthenium, manganese, cobalt, nickel, and other metal-based composites in enhancing the efficiency of next-generation energy storage systems.

**Chapter 7** provides a focused overview of the mechanisms and components of DSSCs using TMCs. It underscores the role of innovation and affordability in promoting sustainable energy technologies and encourages further research in photovoltaic materials.

**Chapter 8** explores the application of earth-abundant transition metal-based materials in the development of chemical sensors. It presents various nanomaterials derived from these metals and illustrates their use in diverse sensor formats. The advantages of chemical sensors over conventional analytical techniques are also discussed.

**Chapter 9** emphasizes the urgent need for sustainable water treatment solutions. It explores the unique redox and catalytic properties of EATMs that make them suitable for applications like adsorption, ion exchange, and membrane-based processes. The chapter offers an extensive evaluation of their potential in contaminant removal and antimicrobial functions.

**Chapter 10** reviews the recent progress in using single elements, compounds, and complexes of earth-abundant transition metals as catalysts for water electrolysis and photolysis. It also outlines existing challenges and future directions for improving catalytic efficiency.

**Chapter 11** focuses on the role of earth-abundant metals as cost-effective alternatives to noble metals in hydrogen evolution reaction (HER) catalysis. It describes various synthesis methods, such as top-down and bottom-up approaches, and examines HER mechanisms under electrochemical and photochemical conditions, emphasizing sustainable hydrogen production.

**Chapter 12** details the conventional extraction methods and applications of transition metals in civil engineering, chemical manufacturing, and energy sectors. It highlights recent innovations in catalysis, as well as modifications and transformations that expand their utility in areas such as energy storage, biosensing, and hydrogenation.

**Chapter 13** discusses the use of transition metal-based catalysts in petrochemistry. While traditional catalysts often rely on expensive metals like platinum, palladium, and rhodium, this chapter examines lower-cost alternatives using earth-abundant metals. It presents research findings and evaluates the economic and environmental advantages of adopting these alternatives.

Applications for Earth-Abundant Transition Metals
Materials Research Foundations 179 (2025) 1-19

Materials Research Forum LLC
https://doi.org/10.21741/9781644903711-1

# Chapter 1

# Earth-Abundant Metal-Based Catalysts for Composite Solid Propellants

Driss Soubane[1,2] *

[1]Institut National de la Recherche Scientifique - Centre Énergie, Matériaux et Télécommunications (INRS-EMT), 1650, boul. Lionel-Boulet, Varennes, Québec    J3X 1P7 Canada

[2]Department of Physics, Poly-disciplinary Faculty of Safi, Cadi Ayyad University, Sidi Bouzid, B.P. 4162, Safi 46 000 Morocco

*driss.soubane@inrs.ca, https://orcid.org/0000-0002-8873-3799

## Abstract

This chapter offers a comprehensive review of earth-abundant metal-based (EAM) catalysts in solid propellants, tracing the evolution of combustion laws and rocket propulsion principles, including Tsiolkovsky's logarithmic equation. We explore the historical development and scientific advancements in solid propellant technologies, with a focus on how metals such as iron, manganese, and copper improve burn rates, enhance combustion efficiency, and lower decomposition temperatures. The chapter also examines catalytic mechanisms, addressing the effects of radiation, exothermic catalytic decomposition, and physical transitions, along with the efficiency and durability of EAMs under extreme conditions. Additionally, we assess the impact of metal-organic frameworks on propellant performance and discuss current technologies and future challenges. The chapter concludes by outlining future research directions, including optimizing catalyst compositions, utilizing nanoscale materials, and developing sustainable, high-performance catalysts for long-duration missions.

## Keywords

Solid Propellants, Combustion Efficiency, Burn Rate, Ammonium Perchlorate, Earth-Abundant Metal-Based Catalysts, Sustainability, Future Research Directions

## Contents

## 1.      Introduction

The word '*propellant*' originates from the Latin terms '*pro*' and '*pellere*', meaning '*forward*' and '*to push*', respectively. Together, they describe the motion of thrust or propulsion, pushing an object forward [1]. In this context, a propellant refers to the force responsible for such motion. The lack of comprehensive public data in this sensitive area of research, particularly regarding the full details of chemical compositions, is evident in the predominance of patents and, occasionally, publications. Major advancements in this field are often driven by industrial and military developments, which are frequently kept confidential for security reasons. Such data may remain unpublished or disclosed only after a significant delay. Despite this, we aim to present an up-to-date overview of the key developments, while relating them to basic physical laws and principles, offering a broad perspective on the recent advancements in the field.

In this chapter, we delve into the catalytic role of earth-abundant metals in solid propellants, starting with the foundational principles of rocket propellants and progressing to the historical context of Tsiolkovsky's logarithmic equation. We will explore the laws of combustion and the catalytic mechanisms that enhance the performance of solid propellants, with particular emphasis on the radiative, exothermic catalytic decomposition, and physical transitions occurring during combustion.

This discussion is followed by an examination of the current state and challenges of propulsion system technologies, focusing on recent advances in the catalytic properties of earth-abundant metals. We will address the efficiency and longevity of these catalysts in improving burn rates, lowering decomposition temperatures, and enhancing heat release. The chapter will first review the effects of EAMs on propellant performance, followed by an analysis of future research directions in this field. Finally, we summarize key findings and explore potential future developments in catalyst technology aimed at creating more efficient and environmentally friendly solid propellants.

Propulsive rockets can be categorized into two types: those that use air as an oxidizer, such as jet engines for applications within Earth's atmosphere, and those designed for space, where a vacuum exists. Jet engines typically operate on the Brayton cycle [2], with sub-categories functioning under constant pressure or constant volume conditions. In contrast, non-air-based propulsion systems, such as those used in space, are designed to operate in a vacuum without the need for atmospheric oxygen.

The core of rocket technology lies in the conversion of nuclear, electrical, photovoltaic [3] [4], or chemical energy into kinetic energy.

The physical state of chemical propellants—gas, liquid, solid, or a combination of solid and gas in hybrid motors—determines the type of rocket motor. Propellants can thus be categorized into solid, liquid, gas, or hybrid types, as shown in Figure 1. 'Hybrid' refers to a combination of two of these physical states [5].

Solid propellants are composed of organized atoms forming a solid material commonly used in rocket engines. Their operation relies on burning the solid material to generate thrust once ignited. For instance, nitrocellulose, derived from cotton, is a simple propellant used in fireworks and certain military applications, classified as a single-base propellant. When nitroglycerin is added, it becomes a double-base propellant. The higher energy output of these double-base propellants has made them suitable for use in rockets and missiles.

Composite solid propellants (CSP) typically consist of an oxidizer, usually ammonium perchlorate (AP), combined with a fuel, such as aluminum powder, and a binder to hold these components together. The most efficient and modern binder used is hydroxyl-terminated polybutadiene. The oxidizer can account for up to 86% of the total mass of the CSP [6].

While liquid propellants offer superior specific impulse compared to solid and gas propellants, they are less efficient in terms of storage. Solid propellants, with their lower entropy, justify their stability and compact storage advantages. In contrast, liquid propellants require either a cooling system or pressurized tanks for proper storage.

The exploration of space, the maintenance of space shuttles, and the operation of the International Space Station (ISS), including supporting the survival needs of astronauts aboard the ISS, heavily rely on advanced rocket and missile propulsion systems [8]. To enable these operations, it is essential to study and develop materials that power rocket motors or engines. these two terms can be used depending on the physical state of the propellant either solid or liquid material, respectively. For instance, solid propellants, comprising both fuel and oxidizer, generate sufficient thrust to launch satellites into space.

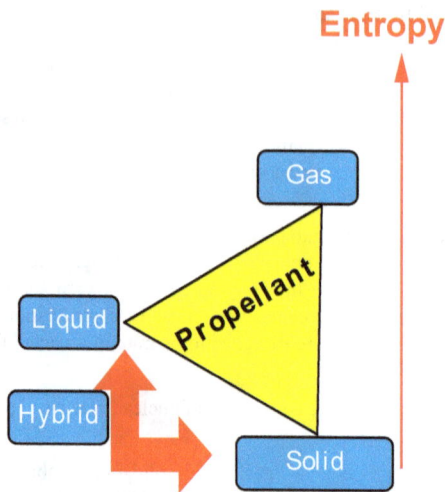

*Figure 1: Different propellant classes versus entropy evolution [7]*

Thrust is produced through a powerful exothermic oxidation-reduction reaction [9], creating high pressure within the combustion chamber and nozzle. This reaction initiates a pyrolytic effect, which sustains the process as long as the gas phase remains in thermodynamic non-equilibrium state. According to Newton's third law of motion, the high-speed exhaust gases expelled from combustion are thrown rearwards, propelling the spacecraft forward [10]. One of the major tasks of the nozzle is to accelerate the subsonic gases within the combustion chamber to supersonic speeds in the exhaust. The term supersonic, derived from the Latin '*super-sonus*' (meaning 'beyond sound'), refers to any object traveling faster than the speed of sound, or Mach 1, named after Austrian scientist Ernst Mach. This implies that the throat of the nozzle is narrower, and according to Bernoulli's principle, the pressure of the gas decreases while its velocity increases [11] [12].

## 2.  Brief History

The discovery of black powder, created by combining potassium nitrate, sulfur, and charcoal by ancient Chinese civilization around 200 BC, marked the inception of the first solid-propellant rockets, initially used for fireworks. This knowledge was later advanced by Muslim civilizations [13], which facilitated its transfer to Europe, particularly from the southern region of Spain, Andalusia. This evolution transformed the use of rockets from mere displays of color in the night sky to significant military applications [14] [15]. In the 18th century, Sir William, a military leader from the United Kingdom, utilized and further developed rockets to combat French troops [16].

Later, Tsiolkovsky proposed the concept of a liquid propellant engine, a vision that was shared by American scientist Robert Goddard and German engineer Wernher von Braun, who is often

Applications for Earth-Abundant Transition Metals                    Materials Research Forum LLC
Materials Research Foundations 179 (2025) 1-19            https://doi.org/10.21741/9781644903711-1

referred to as the father of modern space flight [17]. In 1919, Americans and peculiarly Goddard [18] began to devote significant efforts to enhancing their rocket technology, laying the groundwork for the ambitious dream of conquering and colonizing the moon.

Several breakthroughs in rocket technology led to the development of the first long-range ballistic missile, credited to Wernher Von Braun. Following World War II, his foundational work continued in Germany, eventually flourishing in the Soviet Union, where significant achievements included the Vostok, Voshod, and Soyuz programs [19]. These advancements positioned Soviet cosmonauts at the forefront of space exploration. Subsequently, NASA pushed the boundaries of space travel further, achieving an unprecedented and non-reproducible milestone with the Apollo 11 mission in 1969, which successfully landed humans on the moon.

Solid propellant motors are composed of minimal components, typically an oxidizer and a fuel that react to produce a fume-less gas phase of significant volume [20]. Depending on their intended application, they can be further categorized as space launch booster propellants or tactical missile propellants. Among these, composite solid propellants are the most widely used today. Propellants may be classified as either non-detonable or detonable. While aluminum powder is the most common fuel used, alternatives such as boron or beryllium can also be employed. Additionally, additives are often incorporated to modify the curing time of the propellant, with catalysts utilized to either accelerate or slow down this process. The geometry of the propellant particles can also play a critical role in determining performance.

Solid propulsion technology holds significant interest for both military engineers and meteorologists. Unlike stable liquid propellant systems, the consumption rate of solid propellants cannot be precisely controlled or interrupted once ignited. Their lower entropy contributes to easier storage, faster start-up times, in addition to the absence of moving parts. Currently, solid propellants are employed for launching satellites and in various military applications, including ballistic missiles and anti-tank rockets. They can generate thrust ranging from 2 to over $10^7$ N [5].

Additional advantages of solid propellant include its lower hazard level and higher reliability [20]. Its simpler design compared to liquid propellants positively influences fabrication costs and allows its rapid combustion. Moreover, the specific power of solid propellants—measured as energy per unit volume—significantly exceeds that of their liquid counterparts, primarily due to differences in entropy [7].

This section focuses on solid material propulsion, specifically its integration into rocket motors. The propellant consists of a grain housed within the combustion chamber, which is composed of both fuel and oxidizer. As illustrated in Figure 2, the motor includes key components such as the grain, an igniter responsible for initiating combustion, and a throat connected to a nozzle. The rear part of the motor features the throat, which directs the extremely hot gas flow at high speed. For this purpose, refractory materials are preferred [21].

From a chemical composition perspective, solid propellants can be categorized into three types: colloidal or powder propellants, composite solid propellants, and mixed propellants [22]. Colloidal propellants primarily consist of two auto-oxidizing entities: nitrocellulose and liquid polyhydroxy alcohols. CSP, on the other hand, mainly contains an oxidizer, organic materials, and binding components. The first ballistite, a type of colloidal propellant made from nitrocellulose and nitroglycerin, was fabricated by Alfred Nobel in 1887 [23]. In this chapter, we will focus exclusively on the composite solid propellant class.

Adding fuel to an oxidizer generates a substantial volume of gas, creating a flame in an environment with low air pressure, which is characteristic of solid propellants [20] [24]. Composite solid propellant is a complex mixture of oxidizing agents that stores energy [25].

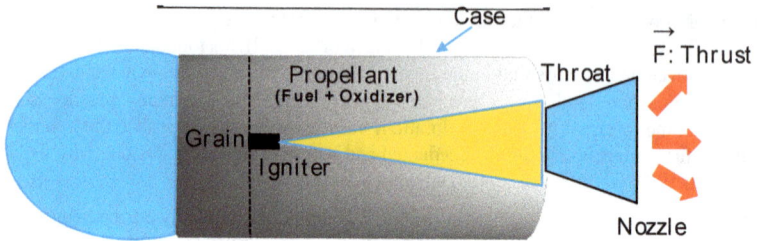

*Figure 2: Sketch representing a solid rocket motor [25]*

To ignite the solid material, an igniter is essential for initiating the combustion process. In this context, solid propellant plays a crucial role [26] [27]. In liquid-propelled engines, the combustion chamber typically burns liquid propellants, while in solid motors, the chamber contains a static, single-use solid propellant. Recent advancements in research, particularly the use of energetic metal-organic frameworks, have significantly enhanced the energy output and impulse of solid propellants [28] [29] [30] by facilitating the production of metal oxides [26] and incorporating silicon [31] [32] and nanostructured materials [33] [34] [35].

The combustion process begins with the ignition of the solid material, generating extremely hot gases that rapidly accelerate as they pass through the throat and nozzle, producing thrust. As the solid mass gradually burns away, the velocity of the exhaust gases increases, propelling the rocket forward. The thrust in this context can be derived from the fundamental principles of dynamics [36] [37]:

$$F = F_m + F_p = \frac{dm}{dt} v_e + (P_e - P_0)A_e \qquad \text{Equation 1}$$

Where

- $F_m$ and $F_p$ : are respectively momentum thrust and pressure thrust
- $\frac{dm}{dt}$ : mass flow rate that measures the solid propellant converted and then expelled as gases.
- $V_e$: The exhaust velocity that is intrinsically linked to the nozzle design and propellant properties.
- $P_e$ and $P_0$: The exit pressure and the ambient pressure either atmospheric pressure or vacuum in space.
- $A_e$: The nozzle exit orifice.

Peculiarly, thrust is governed by the exhaust gases as they pass through the nozzle, which is influenced by the geometric shape and size of the flame. Additionally, the throat and nozzle are designed to ensure that the Mach number reaches a value of 1, preventing the hot exhaust stream from exceeding supersonic speeds within the combustion chamber.

Applications for Earth-Abundant Transition Metals           Materials Research Forum LLC
Materials Research Foundations 179 (2025) 1-19      https://doi.org/10.21741/9781644903711-1

### 3. Tsiolkovsky's Logarithmic Equation

Historically, in 1897, Konstantin Tsiolkovsky, a physicist from the Soviet Union, formulated the equation that describes the logarithmic motion of a rocket. This equation tracks the variation in rocket velocity $\Delta V$ over time as a function of the rocket's empty mass M, the fuel mass m, and the exhaust velocity $V_e$. As the rocket travels, the propellant is gradually consumed [38]:

$$\Delta V = V_e ln(1 + \frac{m}{M}) \qquad\qquad\qquad\qquad\qquad\qquad\qquad\qquad \text{Equation 2}$$

To derive this formula, we consider an observer on Earth using it as an absolute reference frame. The isolated system in space consists of the rocket and the exhaust gas particles produced during fuel combustion, as illustrated in Figure 3.

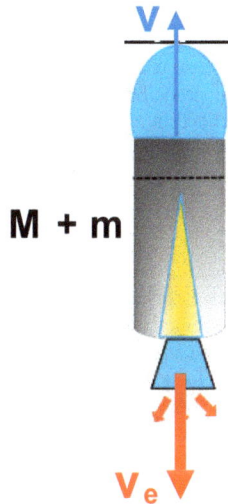

*Figure 3: Graphical representation of Newton's third law*

First, the rocket's motion is analyzed within the framework of classical mechanics, as the rocket's speed is significantly less than the speed of light, and its dimensions are considerably larger than those of subatomic particles. Therefore, the formalism of special relativity or quantum mechanics is not required; the reduced Planck constant $\hbar$ is negligible in this context compared to the rocket's action.

Applying Newton's second law, we define force as the time derivative of the system's momentum. Since the system is isolated with no external forces acting upon it, we can invoke Newton's first law. Thus, the equation can be expressed as follows:

$$F = \frac{dP}{dt} = \frac{d(mv)}{dt} \rightarrow F = m\frac{dv}{dt} + v\frac{dm}{dt}$$

Equation 3

Since the system is isolated with no external forces acting on it, we can apply Newton's first law. Consequently, the equation above can be written as:

$$m\frac{dv}{dt} + v\frac{dm}{dt} = 0$$

Equation 4

Newton's first principle implies that the exhaust velocity is constant, denoted as $V_e$. By integrating the equation above, we can straightforwardly obtain Tsiolkovsky's equation. It is important to note that $V_e = I_{sp} \cdot g$, where g is the gravitational acceleration and $I_{sp}$ is the specific impulse of the rocket motor.

The basic components of a solid propellant motor include the case, which fully encases the system. This case is typically made of glass composites or metals such as steel. A key criterion for the case is to balance reliability with lightweight construction.

To withstand high pressures and extreme temperatures, the choice of propellant grain geometry must be optimal. Commonly, solid propellant consists of an oxidizer and fuel; ammonium perchlorate particles are often suspended in a plastic binder. The efficiency of the propellant is enhanced through metallization with earth-abundant metals, which will be discussed in detail in the next section. The igniter, responsible for initiating combustion, employs an electric spike to ignite the grain.

## 4. Law of Combustion

The burning rate of composite solid propellants $\dot{r}$ is primarily influenced by temperature and pressure. The Benhaim group has developed an experimental approach to measure this burning rate [39]. Summerfield et al. formulated an approximate mathematical equation that describes the burning rate as a function of pressure [40]:

$$\frac{1}{\dot{r}} = \frac{\alpha}{p} + \frac{\beta}{\sqrt[3]{P}}$$

Equation 6

In this equation, $\alpha$ and $\beta$ are constants specific to a particular fuel [41] [42].

The efficiency of solid propellants, represented by its power function, is evident in the variation of the burning rate $\dot{r}$ as a function of pressure P. More broadly, an empirical law that tracks the burning rate is frequently employed due to its simplicity and applicability to colloidal propellants as well [43]:

$$\dot{r} = \zeta p^n$$

Equation 7

In Vieille's law [44], also known as Piobert's law [45], the combustion index n (or pressure exponent) quantifies the sensitivity of the burning rate to changes in chamber pressure. The temperature effect is intrinsically embedded in the constant $\zeta$. Various catalysts, including organic materials, have been tested as additives; depending on the nature of the fuel, n can vary, reaching an upper limit of 0.9. To maintain stable burning, n should not exceed 1; however, a plausible value for n is around 0.65 [25].

Based on the burning rate, we can categorize grains into two types: neutral grains, which exhibit a steady burning rate, and regressive grains, where the burning rate decreases over time.

## 5.    Catalyst Mechanism

The performance of solid propellants is significantly enhanced by the incorporation of catalysts, leading to a marked change in the propellant's burning rate. The catalyst combustion process can manifest through various mechanisms, either individually or in combination:

### 5.1    Radiation

Thermal energy (KT) can be understood from a thermodynamic perspective as kinetic energy. The dynamics of electrons and their transitions generate electromagnetic quanta, such as photons, which travel and collide with the fuel. The absorption of these photons directly influences the burning rate of metallic atoms, including iron (Fe), copper (Cu), cobalt (Co), nickel (Ni), potassium (K), and lead (Pb). These metals have been shown to significantly enhance the combustion rate [40].

### 5.2    Exothermic and Catalytic Decomposition

While we previously discussed the role of catalyst radiation involving various metals, it is important to note that not all metals exhibit chemiluminescence characteristics that positively influence the burning rate. For instance, compounds based on metals such as chromium do not display chemiluminescence. However, chromates and dichromates can absorb radiation and possess sufficient thermal dissociation energy. This property has been evidenced by a significant increase in the thermal dissociation energy of ammonium perchlorate—up to 500%—when utilizing metal oxides, specifically ammonium dichromate [40].

Certain metal oxides, including copper oxide (CuO) [46], zinc oxide (ZnO) [41] , and chromium trioxide ($Cr_2O_3$) [48], have shown high efficiency in accelerating the thermal decomposition of ammonium perchlorate.

### 5.3    Physical Transition

During combustion, as the temperature rises, a physical transformation occurs, transitioning from a low-entropy crystallographic solid structure to increasingly higher entropy states, including liquid and gas. This thermal expansion leads to surface distortion, which facilitates faster combustion [40] [49] .

## 6.    Actual Technology and Challenges

Intensive research and development efforts are underway to innovate new formulations of solid propellants, aiming to enhance energy output and overall performance. Utilizing abundant metals is one promising approach to increase propellant energy density. Key directions for exploration include improving and innovating propellant constituents to achieve optimal burning rates, with the ultimate goal of enhancing the performance of solid propellant devices.

Simultaneously, system reliability, efficiency, and safety remain not only priorities but also significant concerns. The longevity of solid propellants is affected by various factors, including temperature fluctuations and humidity, which can lead to degradation over time. Ensuring long-term reliability and stability is crucial. Additionally, it is important to consider the environmental impact of solid propellants, particularly regarding radiation emissions during operation, which pose serious challenges.

## 7.    Earth-Abundant Metals

Earth-abundant metals, including aluminum (Al), iron (Fe), magnesium (Mg), calcium (Ca), sodium (Na), potassium (K), titanium (Ti), manganese (Mn), nickel (Ni), zinc (Zn), and copper (Cu) [50], are widely available and more affordable compared to rare precious metals. Their lower price point makes iron, manganese, cobalt, nickel, and copper viable alternatives to expensive metals typically used in catalysis, such as platinum, palladium, and rhodium [51] [52] [53]. The exploration of earth-abundant metals (EAMs) is driven by the high costs and limited availability of precious metals like platinum, palladium, and rhodium [54] [55] [56].

Metal oxides, in particular, have garnered significant interest due to their ability to adjust both structure and chemical composition, thereby enhancing their catalytic activities. The catalytic properties of transition metals can be attributed to their d-electronic structure [57] [58]. For instance, the production of 1 kg of platinum generates approximately $1.2 \times 10^4$ kg of carbon dioxide as a byproduct, underscoring the environmental impact associated with the use of precious metals [59].

## 8.    Earth-Abundant Metals Catalytic Properties

### 8.1    Catalyst Mechanism

Catalysts primarily serve to accelerate chemical reactions by lowering the activation energy, or energy barrier, required to initiate these reactions. To achieve this, various reaction pathways are explored to identify the most favorable route for the process.

### 8.2    Catalyst Efficiency

Similar to expensive metals such as platinum, earth-abundant metals (EAMs) can efficiently drive a variety of chemical reactions, achieving comparable success rates at a significantly lower cost. This efficiency makes EAMs particularly advantageous for large-scale industrial applications.

Applications for Earth-Abundant Transition Metals                    Materials Research Forum LLC
Materials Research Foundations 179 (2025) 1-19              https://doi.org/10.21741/9781644903711-1

## 8.3 Longevity of the Catalyst

EAMs demonstrate resilience, maintaining their efficiency even under extreme conditions. In an era where sustainability and environmental impact are paramount, these metals also confirm their recyclability, making them a practical choice for environmentally friendly catalytic processes.

## 9. Earth-Abundant Metals in Solid Propellants

## 9.1 Performance

The inherent catalytic properties of earth-abundant metals (EAMs) have been extensively examined in various research and development initiatives. Studies have demonstrated that incorporating EAMs can significantly enhance efficiency and energy release during the combustion process. As a result, the addition of EAMs not only increases thrust but also boosts power output.

## 9.2 Stability Improvement

*EAM catalysts can enhance the stability of solid propellants, extending their lifespan while maintaining superior performance. Research indicates that metal oxides play a critical role in stabilizing these propellants, thereby improving overall reliability.*

## 9.3 Energy Output

The catalytic effects of EAMs in combustion reactions lead to increased thrust and a rise in energy output, making them a valuable addition to solid propellant formulations.

## 10. Brief Review on Earth-Abundant Metals-based Catalysts Impact on Propellant Performance

To achieve prompt propulsion in rockets, a critical criterion for composite solid propellants is the sublimation of their constituents at high temperatures, with boiling point temperatures typically ranging from 2500 to 3900 K [60]. Earth-abundant metals are often added to oxidizers, such as ammonium perchlorate, and uniformly dispersed in the fuel binder. Incorporating catalyst-based EAMs has been shown to significantly enhance the burning rate of solid propellants by facilitating chemical decomposition reactions.

Various oxidized metals, including but not limited to Fe, Cr, and Cu, have effectively increased the burning rate [61] [62]. Furthermore, research into downscaling these catalyst materials has unraveled their geometric properties, size, and surface-to-volume ratio, revealing their positive impact on solid propellant efficiency [62]. Downscaled catalysts exhibit quantum phenomena that can reduce activation energy, decrease reaction duration, and enable faster ignition [63] [64]. Notably, the addition of less than 3% of EAM-based catalysts facilitates the thermal decomposition of oxidizers [6].

Yadav et al. [6] investigated the effects of various metal oxides, powders, and complexes on ammonium perchlorate composite solid propellants. Their overall findings indicated that nanoscale earth-abundant metals improved the exhausted energy of ammonium perchlorate by 30-40%. Similarly, researchers tested copper, nickel, and aluminum at different nanoscale sizes,

Materials Research Forum LLC
https://doi.org/10.21741/9781644903711-1

observing a notable increase in heat release [65]. Additionally, NiCu bimetallic nanoparticles have been shown to lower the decomposition temperature of composite solid propellants.

Other studies explored catalysts such as Zn-Cu, Zn-Ni, and Zn-Fe, which exhibited similar effects on ammonium perchlorate [66]. Incorporating 1% and 3% $Fe_2O_3$ into ammonium perchlorate revealed promising results [63]. Research focusing on iron and manganese as earth-abundant metal catalysts demonstrated their effectiveness in accelerating the burn rate of solid propellants [64] [67]. Notably, 1% $Fe_2O_3$ nanoparticles increased the burn rate of ammonium perchlorate-based propellants by approximately 20% under high-pressure conditions [68] [63] [69] [70].

Furthermore, studies have shown that copper oxide and iron oxide significantly reduce the high-temperature decomposition (HTD) peaks of ammonium perchlorate, achieving reductions of approximately 40-60°C when materials such as $CuCr_2O_4$ [71] [72] [73] [74] or $Fe_2O_3$ are incorporated. This enhancement not only facilitates better combustion but also lowers the decomposition temperature of the propellant.

Investigations into heat release have demonstrated that EAM-based catalysts significantly enhance thermal energy release during combustion. The addition of metallic catalysts such as Fe, Cu, and Mn to solid propellants resulted in an increased heat release of 30-40%, substantially improving their performance [6] . Furthermore, quantum confinement effects, which lead to increased surface-to-volume ratios, play a critical role in influencing chemical reactions [75] [6]. Nanoscale metal catalysts have been shown to decrease the activation energy required for combustion reactions in solid propellants.

EAM-based catalysts are not only cost-effective but also significantly enhance the burning rate of solid propellants, paving the way for applications in space exploration and the military sector [76] [77]. Research has shown that incorporating $Fe_2O_3$ and CuO quantum dots can lead to remarkable improvements; for instance, a 3% addition of CuO/graphene foam has been found to increase the burn rate of ammonium perchlorate-nitrocellulose by seven times [72]. Additionally, the integration of EAM-based catalysts can reduce the activation energy for solid propellant decomposition by 22%, facilitating rapid combustion and lowering the thermal decomposition temperature of ammonium perchlorate for efficient ignition [78].

A similar study on the use of iron oxide as a catalyst for solid propellant combustion found that hematite ($Fe_2O_3$) significantly enhanced the burning rate of ammonium perchlorate, increasing it by 20-30% with just 0.28 wt.% of the iron oxide additive [79]. This research demonstrated that the iron oxide catalyst lowered the onset decomposition temperature of ammonium perchlorate by 13°C, shifting it from 360.8°C to 347.4°C. Additionally, the incorporation of iron oxide increased heat release, enabling more efficient combustion. Notably, the first exothermic step saw an increase in enthalpy ($\Delta H$) from 1.51 kJ/g for the baseline to 2.13 kJ/g [79].

## 11.  Future Trends and Directions of Research

In the short term, research and development in the field of earth-abundant metal (EAM) catalysts should prioritize enhancing their performance through advancements in material science and engineering. Innovations in fabrication and characterization techniques are expected to significantly impact the activity, stability, and cost-effectiveness of EAM-based catalysts, while also facilitating the development of novel EAMs and sustainable solid propellant formulations that deliver improved performance.

Researchers can adopt various strategies, including fine-tuning the crystallographic structures, compositions, and quantum confinement of EAM-based catalysts to enhance their efficiency and reduce decomposition temperatures. Downscaling these catalysts to increase the surface area relative to bulk volume can promote better interactions with solid propellant constituents, thereby accelerating ignition times and increasing burning rates. Additionally, exploring the effects of mixing metal oxides with bimetallic catalysts may yield valuable insights into how these combinations influence burn rates and overall propellant efficiency.

A key focus for EAM-based catalysts is to improve their longevity and stability under extreme temperature and pressure conditions, which is crucial for extending their applicability. This can be achieved by developing novel synthesis techniques that enable the fabrication of uniform catalyst particles with controlled morphologies. Importantly, environmental considerations should be prioritized to ensure the recyclability of earth-abundant metal catalysts, contributing to more sustainable propellant technologies.

Future research should aim to enhance the efficiency of EAM catalysts in solid propellants, optimizing energy output while balancing performance and sustainability for long-term viability. This includes investigating various compositions of EAM catalysts and experimenting with different ratios and combinations to achieve higher burn rates and improved combustion efficiency. The development of nanoscale catalysts can facilitate better interactions with propellant components, ensuring faster ignition. Additionally, the potential synergistic effects of mixed-metal and bimetallic catalysts should be explored to further enhance performance.

The stability of these catalysts under extreme conditions will be crucial for long-duration missions, and synthesis techniques that produce uniform particles with controlled morphologies will contribute to consistent catalyst performance. From an environmental perspective, evaluating the impact and recyclability of earth-abundant metals is essential for promoting sustainable propellant technologies.

Advanced characterization techniques and computational modeling will be instrumental in understanding catalytic mechanisms and predicting performance outcomes; however, field testing will be necessary to validate laboratory results. Collaborative efforts between material scientists, chemists, and aerospace engineers will be vital to bridge the knowledge gap. Such interdisciplinary work can lead to the development of propellant technologies that maximize the advantages of earth-abundant metals, achieving performance thresholds while ensuring sustainability in solid propellant development.

## Conclusion

The incorporation of earth-abundant metals into solid propellant formulations offers significant potential for enhancing combustion characteristics and overall performance. These metals serve effectively as catalysts, reducing decomposition temperatures and increasing burning rates, thereby contributing to a greener and more cost-effective propellant development process. Future research should focus on optimizing catalyst structures, investigating mixed-metal systems, and assessing the environmental impact of these materials. Ultimately, advancements in earth-abundant metal-based catalysts will lead to improved performance and more reliable solid propellants for aerospace and military applications.

Applications for Earth-Abundant Transition Metals                    Materials Research Forum LLC
Materials Research Foundations 179 (2025) 1-19                    https://doi.org/10.21741/9781644903711-1

Acknowledgments

- This work was partially supported by the MRIF (Ministère des Relations Internationales et de la Francophonie), Quebec – Canada – Morocco Project: APQRM 2021–2022, through the INRS-EMT/UCA-FPS Grant.

- We acknowledge Yassine Soubane for his valuable assistance in the development and refinement of this chapter.

**References**

[1] C. T. Lewis, *A Latin dictionary: founded on Andrew's edition of Freund's Latin dictionary*, Reproduction en fac-Similé. Oxford: Clarendon Press, 1980.

[2] L. C. Lichty, *Combustion engine processes*, 7th ed. New York: McGraw-Hill, 1967.

[3] D. Soubane, "Renewable energy resources in a nutshell: Present and future," *International Journal of Renewable Energy Resources*, vol. 7, no. 2, pp. 37–42, 2017.

[4] D. Soubane, A. Ihlal, and G. Nouet, "The Role Of Cadmium Oxide Within The Thin Films Of The Buffer Cds Aimed At Solar Cells Based Upon CIGS Films Fabrication," *Revue marocaine de la matière condensée*, vol. 9, no. Revue marocaine de la matière condensée, Vol. 9 (2007), Mar. 2011, doi: 10.34874/PRSM.MJCM-VOL9ISS0.195.

[5] D. P. Mishra, *Fundamentals of rocket propulsion*. Boca Raton: CRC Press, Taylor & Francis Group, CRC Press is an imprint of the Taylor & Francis Group, an informa business, 2017.

[6] N. Yadav, P. K. Srivastava, and M. Varma, "Recent advances in catalytic combustion of AP-based composite solid propellants," *Defence Technology*, vol. 17, no. 3, pp. 1013–1031, Jun. 2021, doi: 10.1016/j.dt.2020.06.007.

[7] D. Soubane, M. E. Garah, M. Bouhassoune, A. Tirbiyine, A. Ramzi, and S. Laasri, "Hidden Information, Energy Dispersion and Disorder: Does Entropy Really Measure Disorder?," *WJCMP*, vol. 08, no. 04, pp. 197–202, 2018, doi: 10.4236/wjcmp.2018.84014.

[8] H. J. Kramer, "Space Stations," in *Observation of the Earth and Its Environment*, Berlin, Heidelberg: Springer Berlin Heidelberg, 2002, pp. 983–1004. doi: 10.1007/978-3-642-56294-5_13.

[9] K. J. Laidler, "A glossary of terms used in chemical kinetics, including reaction dynamics (IUPAC Recommendations 1996)," *Pure and Applied Chemistry*, vol. 68, no. 1, pp. 149–192, Jan. 1996, doi: 10.1351/pac199668010149.

[10] S. T. Thornton and J. B. Marion, *Classical dynamics of particles and systems*, 5. ed., International student ed., [Nachdr.]. Belmont, Calif.: Thomson, Brooks-Cole, 2008.

[11] O. Darrigol and U. Frisch, "From Newton's mechanics to Euler's equations," *Physica D: Nonlinear Phenomena*, vol. 237, no. 14–17, pp. 1855–1869, Aug. 2008, doi: 10.1016/j.physd.2007.08.003.

[12] R. W. Johnson, Ed., *Handbook of fluid dynamics*, Second edition. Boca Raton London New York: CRC Press Taylor & Francis Group, 2016.

[13] G. Ágoston, *Guns for the sultan: military power and the weapons industry in the Ottoman Empire*. in Cambridge studies in Islamic civilization. New York: Cambridge University Press, 2005.

[14] R. Irons, *Hitler's terror weapons: the price of vengeance*. London: Collins, 2003.

[15] K.-H. Wellmann, Ed., *Warum ist es nachts dunkel? was wir vom Weltall wirklich wissen; in Zusammenarbeit mit dem neuen Funkkolleg*. Stuttgart: Kosmos, 2006.

[16] D. W. Wragg, Ed., *A dictionary of aviation*. Reading: Osprey, 1973.

[17] E. Bergaust, *Wernher von Braun: the authoritative and definitive biographical profile of the father of modern space flight*. Washington: National Space Institute, 1976.

[18] J. D. Hunley, "The Enigma of Robert H. Goddard," *Technology and Culture*, vol. 36, no. 2, p. 327, Apr. 1995, doi: 10.2307/3106375.

[19] R. Hall, R. D. Hall, and D. J. Shayler, *Soyuz: a universal spacecraft*. in Springer-Praxis books in astronomy and space sciences. London Berlin Heidelberg: Springer [u.a.], 2003.

[20] M. C. Vilela Salgado, C. M. N. Belderrain, and T. C. Devezas, "Space Propulsion: a Survey Study About Current and Future Technologies," *J.Aerosp. Technol. Manag.*, vol. 10, Feb. 2018, doi: 10.5028/jatm.v10.829.

[21] M. El Garah, D. Soubane, and F. Sanchette, "Review on mechanical and functional properties of refractory high-entropy alloy films by magnetron sputtering," *emergent mater.*, vol. 7, no. 1, pp. 77–101, Feb. 2024, doi: 10.1007/s42247-023-00607-8.

[22] A. Okninski, W. Kopacz, D. Kaniewski, and K. Sobczak, "Hybrid rocket propulsion technology for space transportation revisited - propellant solutions and challenges," *FirePhysChem*, vol. 1, no. 4, pp. 260–271, Dec. 2021, doi: 10.1016/j.fpc.2021.11.015.

[23] M. W. Beckstead, "Model for double-base propellant combustion," *AIAA Journal*, vol. 18, no. 8, pp. 980–985, Aug. 1980, doi: 10.2514/3.7701.

[24] I. E. Smith, "Chemical Rockets and Flame Explosives Technology. Richard T. Holzmann. Marcel Dekker. New York, 1969. Illustrated. 499 pp. £12.25.," *Aeronaut. j.*, vol. 75, no. 725, pp. 354–354, May 1971, doi: 10.1017/S0001924000045504.

[25] G. T. Healey, "Fundamental Aspects of Solid Propellant Rockets. F. A. Williams, M. Barrere and M. C. Huang. Technivision Services, Slough. 791 pp. Illustrated.," *Aeronaut. j.*, vol. 74, no. 717, pp. 764–764, Sep. 1970, doi: 10.1017/S0001924000048296.

[26] B. Tan, X. Yang, J. Dou, B. Duan, X. Lu, and N. Liu, "Research progress of EMOFs-based burning rate catalysts for solid propellants," *Front. Chem.*, vol. 10, p. 1032163, Oct. 2022, doi: 10.3389/fchem.2022.1032163.

[27] L. R. Warren, Z. Wang, C. R. Pulham, and C. A. Morrison, "A Review of the Catalytic Effects of Lead-Based Ballistic Modifiers on the Combustion Chemistry of Double Base Propellants," *Propellants Explo Pyrotec*, vol. 46, no. 1, pp. 13–25, Jan. 2021, doi: 10.1002/prep.202000167.

[28] J. Zhang and J. M. Shreeve, "3D Nitrogen-rich metal–organic frameworks: opportunities for safer energetics," *Dalton Trans.*, vol. 45, no. 6, pp. 2363–2368, 2016, doi: 10.1039/C5DT04456A.

[29] I. L. Dalinger *et al.*, "Novel energetic CNO oxidizer: Pernitro-substituted pyrazolyl-furazan framework," *FirePhysChem*, vol. 1, no. 2, pp. 83–89, Jun. 2021, doi: 10.1016/j.fpc.2021.04.005.

[30] Z. Qin *et al.*, "Effect of spherical Al-Mg-Zr on the combustion characteristics of composite propellants," *FirePhysChem*, vol. 2, no. 1, pp. 14–19, Mar. 2022, doi: 10.1016/j.fpc.2022.03.004.

[31] C. Rossi *et al.*, "Solid Propellant Microthrusters on Silicon: Design, Modeling, Fabrication, and Testing," *J. Microelectromech. Syst.*, vol. 15, no. 6, pp. 1805–1815, Dec. 2006, doi: 10.1109/JMEMS.2006.880232.

[32] Driss Soubane, Tsuneyuki Ozaki, Nathaniel Quitoriano, "Metal-Oxide-Insulator-Semiconductor Memristor based on Low Thermal Budget Approach Silicon Nano-crystals embedded in a silica Matrix," 40707, 2019

[33] J. Koo, H. Stretz, A. Bray, J. Weispfenning, Z. P. Luo, and R. Blanski, "Nanostructured Materials for Rocket Propulsion System: Recent Progress," in *44th AIAA/ASME/ASCE/AHS/ASC Structures, Structural Dynamics, and Materials Conference*, Norfolk, Virginia: American Institute of Aeronautics and Astronautics, Apr. 2003. doi: 10.2514/6.2003-1769.

[34] D. Soubane and N. J. Quitoriano, "Photoluminescence from low thermal budget silicon nano-crystals in silica," *Nanotechnology*, vol. 26, no. 29, p. 295201, Jul. 2015, doi: 10.1088/0957-4484/26/29/295201.

[35] F. Dnaya *et al.*, "A holistic understanding of optical properties in amorphous H-terminated Si-nanostructures: Combining TD-DFT with AIMD," *Results in Optics*, vol. 16, p. 100694, Jul. 2024, doi: 10.1016/j.rio.2024.100694.

[36] P. G. Hill and C. R. Peterson, *Mechanics and thermodynamics of propulsion*, 2nd ed. Reading, Mass: Addison-Wesley, 1992.

[37] G. P. Sutton and O. Biblarz, *Rocket propulsion elements*, Ninth edition. Hoboken, New Jersey: John Wiley & Sons Inc, 2017.

[38] D. Koks, "The Rocket Equation and that Minus Sign: An Avoidable Pitfall in the Language of Physics," *The Phys. Educat.*, vol. 04, no. 02, p. 2250009, Jun. 2022, doi: 10.1142/S2661339522500093.

[39] D. Grune, "Studies of the Combustion of Solid Propellants at High Pressures," *Propellants Explo Pyrotec*, vol. 1, no. 2, pp. 27–28, Jun. 1976, doi: 10.1002/prep.19760010203.

[40] K. Krowicki and M. Syczewski, "SOLID ROCKET PROPELLANTS".

[41] M. Summerfield, Ed., *Solid Propellant Rocket Research*. New York: American Institute of Aeronautics and Astronautics, 1960. doi: 10.2514/4.864766.

[42] J. C. SCHUMACHER, *Perchlorates: Their Properties, Manufacture and Uses.* in American Chemical Society Monograph, no. no. 146. New York, London: Reinhold Publishing Corp. ; Chapman & Hall, 1960.

[43] R. Fry *et al.*, "Solid propellant burning rate measurement methods used within the NATO propulsion community," in *37th Joint Propulsion Conference and Exhibit*, Salt Lake City,UT,U.S.A.: American Institute of Aeronautics and Astronautics, Jul. 2001. doi: 10.2514/6.2001-3948.

[44] L. Medard, "L'œuvre scientifique de Paul Vieille (1854-1934)/The scientific work of Paul Vieille (1854-1934)," *rhs*, vol. 47, no. 3, pp. 381–404, 1994, doi: 10.3406/rhs.1994.1211.

[45] M. S. Russell, *The chemistry of fireworks*, 2nd ed. Cambridge, UK: RSC Pub, 2009.

[46] R. Poreddy, C. Engelbrekt, and A. Riisager, "Copper oxide as efficient catalyst for oxidative dehydrogenation of alcohols with air," *Catal. Sci. Technol.*, vol. 5, no. 4, pp. 2467–2477, 2015, doi: 10.1039/C4CY01622J.

[47] D. Soubane, A. Tirbiyine, M. Bellioua, S. Laasri, and A. Hajjaji, "Optical Features of Catalyst-Free Zinc Oxide Nanostructures Confined to One Dimension," *Opt. Spectrosc.*, vol. 127, no. 3, pp. 522–526, Sep. 2019, doi: 10.1134/S0030400X19090261.

[48] A. Bumajdad, S. Al-Ghareeb, M. Madkour, and F. A. Sagheer, "Non-noble, efficient catalyst of unsupported α-$Cr_2O_3$ nanoparticles for low temperature CO Oxidation," *Sci Rep*, vol. 7, no. 1, p. 14788, Nov. 2017, doi: 10.1038/s41598-017-14779-x.

[49] V. R. Gutman, "Solid Propellent Burning Rate Theory: Recently Advanced Theories, Their Limitations and Possible Areas for Future Investigation," *Aircraft Engineering and Aerospace Technology*, vol. 32, no. 9, pp. 255–260, Sep. 1960, doi: 10.1108/eb033297.

[50] K. M. P. Wheelhouse, R. L. Webster, and G. L. Beutner, "Advances and Applications in Catalysis with Earth-Abundant Metals," *Org. Process Res. Dev.*, vol. 27, no. 7, pp. 1157–1159, Jul. 2023, doi: 10.1021/acs.oprd.3c00207.

[51] G. A. Filonenko, R. Van Putten, E. J. M. Hensen, and E. A. Pidko, "Catalytic (de)hydrogenation promoted by non-precious metals – Co, Fe and Mn: recent advances in an emerging field," *Chem. Soc. Rev.*, vol. 47, no. 4, pp. 1459–1483, 2018, doi: 10.1039/C7CS00334J.

[52] S. Abelló, E. Bolshak, and D. Montané, "Ni–Fe catalysts derived from hydrotalcite-like precursors for hydrogen production by ethanol steam reforming," *Applied Catalysis A: General*, vol. 450, pp. 261–274, Jan. 2013, doi: 10.1016/j.apcata.2012.10.035.

[53] G. Busca, U. Costantino, T. Montanari, G. Ramis, C. Resini, and M. Sisani, "Nickel versus cobalt catalysts for hydrogen production by ethanol steam reforming: Ni–Co–Zn–Al catalysts from hydrotalcite-like precursors," *International Journal of Hydrogen Energy*, vol. 35, no. 11, pp. 5356–5366, Jun. 2010, doi: 10.1016/j.ijhydene.2010.02.124.

[54] A. K. Neyestanaki, F. Klingstedt, T. Salmi, and D. Y. Murzin, "Deactivation of postcombustion catalysts, a review," *Fuel*, vol. 83, no. 4–5, pp. 395–408, Mar. 2004, doi: 10.1016/j.fuel.2003.09.002.

[55] C.-J. Yang, "An impending platinum crisis and its implications for the future of the automobile," *Energy Policy*, vol. 37, no. 5, pp. 1805–1808, May 2009, doi: 10.1016/j.enpol.2009.01.019.

[56] M. L. Firmansyah, F. Kubota, and M. Goto, "Solvent extraction of Pt(IV), Pd(II), and Rh(III) with the ionic liquid trioctyl(dodecyl) phosphonium chloride," *J of Chemical Tech & Biotech*, vol. 93, no. 6, pp. 1714–1721, Jun. 2018, doi: 10.1002/jctb.5544.

[57] W. M. Haynes, D. R. Lide, and T. J. Bruno, Eds., *CRC Handbook of Chemistry and Physics*, 97th ed. CRC Press, 2016. doi: 10.1201/9781315380476.

[58] R. M. Bullock *et al.*, "Using nature's blueprint to expand catalysis with Earth-abundant metals," *Science*, vol. 369, no. 6505, p. eabc3183, Aug. 2020, doi: 10.1126/science.abc3183.

[59] P. Nuss and M. J. Eckelman, "Life Cycle Assessment of Metals: A Scientific Synthesis," *PLoS ONE*, vol. 9, no. 7, p. e101298, Jul. 2014, doi: 10.1371/journal.pone.0101298.

[60] K. K. Kuo and R. Acharya, *Fundamentals of Turbulent and Multiphase Combustion*, 1st ed. Wiley, 2012. doi: 10.1002/9781118107683.

[61] K. Kishore and M. R. Sunitha, "Effect of Transition Metal Oxides on Decomposition and Deflagration of Composite Solid Propellant Systems: A Survey," *AIAA Journal*, vol. 17, no. 10, pp. 1118–1125, Oct. 1979, doi: 10.2514/3.61286.

[62] R. A. Yetter, G. A. Risha, and S. F. Son, "Metal particle combustion and nanotechnology," *Proceedings of the Combustion Institute*, vol. 32, no. 2, pp. 1819–1838, 2009, doi: 10.1016/j.proci.2008.08.013.

[63] M. B. Padwal and M. Varma, "Thermal decomposition and combustion characteristics of HTPB-coarse AP composite solid propellants catalyzed with $Fe_2O_3$," *Combustion Science and Technology*, vol. 190, no. 9, pp. 1614–1629, Sep. 2018, doi: 10.1080/00102202.2018.1460599.

[64] S. Singh, M. Chawla, P. F. Siril, and G. Singh, "Manganese oxalate nanorods as ballistic modifier for composite solid propellants," *Thermochimica Acta*, vol. 597, pp. 85–92, Dec. 2014, doi: 10.1016/j.tca.2014.10.016.

[65] L. Liu, F. Li, L. Tan, L. Ming, and Y. Yi, "Effects of Nanometer Ni, Cu, Al and NiCu Powders on the Thermal Decomposition of Ammonium Perchlorate," *Propellants Explo Pyrotec*, vol. 29, no. 1, pp. 34–38, Feb. 2004, doi: 10.1002/prep.200400026.

[66] L. Liu, F. Li, L. Tan, L. Ming, and Y. Yi, "Effects of Nanometer Ni, Cu, Al and NiCu Powders on the Thermal Decomposition of Ammonium Perchlorate," *Propellants Explo Pyrotec*, vol. 29, no. 1, pp. 34–38, Feb. 2004, doi: 10.1002/prep.200400026.

[67] Lalith V. Kakumanu, Narendra Yadav, Srinibas Karmakar, "Composite Solid Propellant, Burn Rate, Combustion Characteristics, Metal Catalysts, Fuel-Binder," vol. 3, no. 2, pp. 31–36, 2014.

[68] H. Xu, X. Wang, and L. Zhang, "Selective preparation of nanorods and micro-octahedrons of Fe2O3 and their catalytic performances for thermal decomposition of ammonium perchlorate," *Powder Technology*, vol. 185, no. 2, pp. 176–180, Jul. 2008, doi: 10.1016/j.powtec.2007.10.011.

[69] K.-T. Lu, T.-M. Yang, J.-S. Li, and T.-F. Yeh, "Study on the Burning Characteristics of AP/Al/HTPB Composite Solid Propellant Containing Nano-Sized Ferric Oxide Powder," *Combustion Science and Technology*, vol. 184, no. 12, pp. 2100–2116, Dec. 2012, doi: 10.1080/00102202.2012.703271.

[70] P. R. Patil, V. N. Krishnamurthy, and S. S. Joshi, "Differential Scanning Calorimetric Study of HTPB based Composite Propellants in Presence of Nano Ferric Oxide," *Propellants Explo Pyrotec*, vol. 31, no. 6, pp. 442–446, Dec. 2006, doi: 10.1002/prep.200600059.

[71] A. M. Kawamoto, L. C. Pardini, and L. C. Rezende, "Synthesis of copper chromite catalyst," *Aerospace Science and Technology*, vol. 8, no. 7, pp. 591–598, Oct. 2004, doi: 10.1016/j.ast.2004.06.010.

[72] D. Kshirsagar *et al.*, "Effect of Nano Cr2O3 in HTPB/AP/Al Based Composite Propellant Formulations," *Def. Sc. Jl.*, vol. 66, no. 2, p. 100, Mar. 2016, doi: 10.14429/dsj.66.9250.

[73] A. P. Sanoop, R. Rajeev, and B. K. George, "Synthesis and characterization of a novel copper chromite catalyst for the thermal decomposition of ammonium perchlorate," *Thermochimica Acta*, vol. 606, pp. 34–40, Apr. 2015, doi: 10.1016/j.tca.2015.03.006.

[74] S. G. Hosseini, R. Abazari, and A. Gavi, "Pure CuCr 2 O 4 nanoparticles: Synthesis, characterization and their morphological and size effects on the catalytic thermal decomposition of ammonium perchlorate," *Solid State Sciences*, vol. 37, pp. 72–79, Nov. 2014, doi: 10.1016/j.solidstatesciences.2014.08.014.

[75] Z. Zhou, S. Tian, D. Zeng, G. Tang, and C. Xie, "MOX (M=Zn, Co, Fe)/AP shell–core nanocomposites for self-catalytical decomposition of ammonium perchlorate," *Journal of Alloys and Compounds*, vol. 513, pp. 213–219, Feb. 2012, doi: 10.1016/j.jallcom.2011.10.021.

[76] C. Dennis and B. Bojko, "On the combustion of heterogeneous AP/HTPB composite propellants: A review," *Fuel*, vol. 254, p. 115646, Oct. 2019, doi: 10.1016/j.fuel.2019.115646.

[77] A. Davenas, *Solid rocket propulsion technology*, 1st English ed. Oxford New York: Pergamon Press, 1993.

[78] A. Adharsh Unni, R. Kulkarni, C. Singh, V. Singh, V. M. Priya Varshini, and G. Shanmugaraj, "Effects of adding powdered metals with the solid propellants – A review," *J. Phys.: Conf. Ser.*, vol. 1473, no. 1, p. 012048, Feb. 2020, doi: 10.1088/1742-6596/1473/1/012048.

[79] F. Maggi *et al.*, "Iron oxide as solid propellant catalyst: A detailed characterization," *Acta Astronautica*, vol. 158, pp. 416–424, May 2019, doi: 10.1016/j.actaastro.2018.07.037.

Applications for Earth-Abundant Transition Metals          Materials Research Forum LLC
Materials Research Foundations 179 (2025) 20-36          https://doi.org/10.21741/9781644903711-2

Chapter 2

# Earth-Abundant Transition Metal Oxide-based Catalysts for Composite Solid Propellants

Gladiya Mani[1], Suresh Mathew[1]*

[1]School of Chemical Sciences, Mahatma Gandhi University, Kottayam, Kerala, India- 686560

* sureshmathewmgu@gmail.com

**Abstract**

Earth-abundant transition metal oxides are well known for their catalytic potency for ammonium perchlorate (AP) thermal decomposition, the widely used oxidiser in composite solid propellants. AP constitutes approximately 60–75% of propellant mass. The extreme sensitivity of AP to additives influences its thermal decomposition pattern, hence the propellant's performance. Earth-abundant transition metal oxides are garnering great attention as ideal catalyst candidates owing to their cost-effectiveness, abundance, low toxicity, and sustainability. Additionally, the structural flexibility and tunability of transition metal oxides make them ideal for mixed metal oxide and composite formation, which helps in exploring new combinations with superior catalytic efficacy for harsh conditions as well as future applications.

**Keywords**

Earth-Abundant Transition Metal Oxides, Ammonium Perchlorate, Composite Solid Propellants, Iron Oxide, Copper Oxide, Nickel Oxide

## Contents

## 1.    Introduction

Composite solid propellants (CSPs) are commonly employed in propeller systems to propel vehicles such as rockets, missiles, and projectiles. Generally, CSP performances were evaluated based on the burning rate, thermal stability and sensitivity towards external stimuli. Of these attributes, the CSPs' burning rate is one of the most important ones [1]. Composite solid rocket propellants are generally heterogeneous mixtures made up of a polymeric matrix with solid oxides, energetic fuels, adhesives, burn rate modifiers, plasticizers and other auxiliary components. In comparison to homogeneous propellants, this type of propellant has achieved greater stability, a higher specific impulse, a lower burning rate, and better mechanical qualities [2]. The oxidiser constitutes a significant component of the propellant's makeup. It primarily supplies the oxygen required for the energy release of the fuel, while also significantly enhancing the energy of solid propellant [3]. The essential characteristics determining the efficacy of a propellant oxidiser are high oxygen content, little hygroscopicity, low heat of formation, high density, robust mechanical qualities, significant thermal stability, extended shelf life, interoperability with other elements, and safe handling [4]. Ammonium perchlorate has served as the primary oxidiser in composite solid propellants for several decades. Ammonium nitrate, hydrazinium nitroformate and, ammonium dinitramide are also currently explored as potential oxidizer for upcoming propellant applications.

AP- the workhorse oxidiser for propellants, received significant consideration in research. Ammonium perchlorate constitutes approximately 60–75% of the entire propellant mass, and its unique thermal conductivity significantly influences the CSP performance [5]. AP is very responsive to additions, which modify its ballistic characteristics, particularly the burning rate, hence influencing propellant efficacy [6]. AP possesses several advantages, including enhanced oxygen balance, combustion rate, specific impulse, mechanical characteristics, density, long-term stability, and relative simplicity of handling. It is extensively utilised in space shuttles, heavy-lift launch vehicles, and military applications [4]. The burning rate and energy characteristics of CSPs are intricately linked to AP thermolysis. The decrease in the decomposition temperature of AP implies lowered ignition delay time and enhanced burn rate [7]. As mentioned before, the sensitivity of ammonium perchlorate to additives makes it the most effective approach for modifying AP thermal decomposition behaviour, leading to an accelerated burning rate and an enhanced propellant performance.

A variety of physicochemical methodologies are devised to augment the propellant combustion rate, a fundamental characteristic that dictates the ballistic efficacy of missiles and rockets. Among these, the frequently employed approaches involve [8],

- AP particle size reduction
- Integration of metal powder fuel additives
- Use of high energy polymer binders
- Addition of burn rate modifiers.

The use of superfine AP or the addition of combustion catalysts are the commonly employed methods to accelerate the propellant burn rate. The incorporation of fine oxidiser particles beyond an optimal threshold to increase the burn rate might elevate the viscosity of the slurry, leading to a diminished period of retention. The molding of such slurries is exceedingly challenging. Moreover, the ultrafine particles of AP tend to aggregate, making homogeneous mixing with the binder challenging, which in turn complicates the attainment of a consistent burn rate. Similarly, use of energetic binders will affect the propellant mechanical properties and make the CPS sensitive to shock and impact [8]. Therefore, the use of catalysts such as transition metal oxides (TMOs) is considered the competent strategy to enhance the combustion behavior of solid propellants. The distinctive characteristics of these catalysts, including cost-effectiveness, ease of application, readily accessible, facile application, tuneable size and structure, ability to form complexes, absence of processing or strength issues, and minimal quantity requirement, render it the optimal choice.

Transition metal catalysts have emerged as revolutionary materials in chemical synthesis, with the exploration and use of novel metal complexes prevalent in both academia and industry. Though the origin of transition metal chemistry can be traced back decades, the application of first-row transition metals in catalysis remained way behind the incredible advancements achieved with precious metals [9]. Although the scientific community has acknowledged the intellectual contributions to transition-metal catalysis several times (Nobel Prizes), the research circle cannot rely much more on precious metals despite their well-established applications in catalysis. Given the limited natural availability and environmental repercussions of mining these precious metals, concerns such as sustainability, cost, and price volatility are crucial motivators for investigating and advancing catalysts utilising other materials. Recognising and utilising catalysis involving easily accessible earth-abundant metals enables chemists to tackle the sustainability issues associated with precious metals directly [10]. The essential scientific inquiry must prioritise not just the current demands of industry but also maintain a vision for the future. Therefore, propellant research must adopt sustainable practices to ensure a future of responsible development, safeguarding the rights of future generations.

## 2. Transition metals: An overview

By classification, the three-transition *series* forming short group of elements, placed between the longer main groups. The remarkably varied chemical and physical properties of transition elements are due to the *progressive filling of d orbitals* across each series [11].

Catalysis has been a significant driver of global economic growth for more than a century. Due to the presence of *incompletely filled d orbitals* and *capacity to donate or accept electrons*, transition metals serve as efficient catalysts for several chemical processes. Transition metals, whether in atomic or ionic form, can effectively catalyse electron transfer reactions by oscillating between various oxidation states. They can also catalyse other processes by enabling bond cleavage, accelerated reactant molecule disintegration, and by stabilisation of reaction intermediates. In homogeneous catalysis, transition metal coordination complexes are employed to operate in the same phase as that of the reactants [12]. Similarly, in heterogeneous catalysis, transition metals are generally employed as nanoparticles because of their resilience and ease of separation or recovery from the reaction mixture. Furthermore, due to the presence of unfilled or empty orbitals, metal ions ($M^+$) can coordinate with ligands to establish *metal coordination bonds*, necessitating electron pair donation from the ligand (Lewis base) to the $M^+$ (Lewis acid). Consequently, in comparison to conventional covalent bonds, these non-covalent coordination bonds exhibit enhanced kinetic lability or the ability to dissociate and reassemble, though at the expense of reduced thermodynamic stability [13].

Transition metals have special electronic properties that come from two things: the d shells not being filled up and the d orbitals overlapping with the anion and cation orbitals next to them. In the case of 3d series elements, during compound formation, the metal atom will transfer the 4s electrons to the anion's p orbital, resulting in the partially filled 3d orbital being the outermost. The collaboration with the negatively charged anion ligands will therefore influence the degeneracy of such 3d orbitals [14]. Despite the remarkable advancements observed in the scientific community and industry in the last few years by the use of transition metals, particularly those from the 4d and 5d series and their oxides, the intrinsic limitations such as high cost, toxicity, restricted availability and low Earth abundance of these metals impede their widespread adoption as future materials. In this regard, the use of earth-abundant first-row transition metals as a sustainable alternative has garnered increasing attention.

## 2.1    Earth-abundant transition metals

Earth-abundant metals are currently the subject of extensive research for a variety of catalytic applications due to their numerous benefits, including sustainability, inferior environmental implications, and cost-effectiveness. In contrast to the significant progress found with Pt, Pd, Ag, Au, and Ru; the Earth-abundant first-row transition metals, including Co, Fe, Ni, Cu, and Mn—often referred to as "biometals"—have been largely unexplored until recent years. This underutilisation can be attributed to their labile characteristics, intricate mechanistic pathways, and less apparent catalytic properties [15]. The extensive presence of earth-abundant metals in the Earth's crust renders them significantly more economical than conventionally utilised precious metals in catalysis. The reduced expense of these plentiful metals diminishes the total cost of catalytic processes, enhancing the economic feasibility of industrial chemical operations, particularly for large-scale applications. Furthermore, the use of abundant earth metals minimises the reliance on crucial or scarce raw materials. Several earth-abundant metals exhibit significant tunability for catalytic reactions. Appropriate ligand design or modification of reaction conditions can adjust reactivity and selectivity to fulfil specific requirements, occasionally surpassing the performance of precious metals in certain reactions. For example, earth-abundant metals are being extensively investigated and explored for various renewable energy technologies like water splitting for sustainable applications [16, 17], $CO_2$ reduction [18, 19] and many more. The natural prevalence of these metals reduces the potential for human toxicity and environmental contamination. This also addresses the issue of bioaccumulation and long-term ecological impacts, particularly when catalysts enter biological systems via industrial waste disposal. The natural availability of these resources also diminishes the environmental footprints associated with their mining and processing. Earth-abundant metals facilitate recycling, allowing for more efficient recovery and reuse of catalysts. Reusing and recycling materials reduces waste and aligns with the principles of a circular economy, thereby reducing resource extraction. This facilitates the adoption of sustainable practices within the industry. In line with the call for green chemistry and sustainable development goals, research in propellants and aerospace has increasingly adopted the utilisation of readily available metals, particularly in the thermal degradation investigations of AP. In light of the practical requirements for substantial quantities of chemicals in a single launch, the incorporation of accessible biometals is poised to significantly improve both the economic feasibility and the environmental sustainability of propellant launches.

## 3.    Earth-abundant transition metal catalysts for AP decomposition

AP-based CSPs necessitate combustion catalysts to attain elevated burn rates, with TMOs serving as traditional catalysts. The thermolysis behaviour of AP changes significantly with the introduction of a minimal quantity of metal catalyst. Recently, earth-abundant transition metal-based catalysts have garnered significant attention, as previously noted. We have categorised TMO catalysts into single, binary, and composite metal oxides, including their composites with carbonaceous materials such as graphene and polymeric carbon nitride.

### 3.1    Iron-based catalysts

### 3.1.1    Single metal oxides as catalysts

Iron oxide's advantageous stability, favourable catalytic action, and inexpensive cost make it the most promising CSP combustion catalyst. It was found that for ferric oxide, the catalytic efficacy is intricately linked to the morphology and particle dimensions. With the decreasing particle size (to the nanoscale), the catalytic activity gets enhanced significantly. Budhwar et al. [20] utilised ferric oxide ($Fe_2O_3$) with a particle size of 0.25-0.65 microns as a catalyst for composite solid propellants, favouring it over other metal oxides due to its recognised capability to decompose ammonium perchlorate at elevated temperatures. It has been extensively reported that as the particle size reduces from the micro size regime to the nanometer range, metal oxides show exceptional catalytic efficacy owing to their higher surface area, and excess oxygen vacancies. TG DSC analyses were done for standard propellant formulation and other propellants with 0.25 wt% catalysts. DSC results showed that the high-temperature decomposition (HTD) value of AP appeared at 356°C shifted to 354 and 351°C for micron and nano-sized catalysts respectively. Similar observations were made by Arun Chandru et al. [21] in their study using nano-structured $Fe_2O_3$ for solid rocket propellants. Here, nano iron oxides were prepared via two synthesis routes denoted as SSP and TDP. From the thermal analyses, it was observed that both synthesized samples decreased the decomposition temperature with increased apparent heat evolution to varying levels. With the addition of 2% w/w TDP, the HTD value (AP) lowers from 420°C to 376°C while SSP to 314°C. The apparent heat release was also enhanced by 210% and 350% respectively for TDP and SSP. This enhanced catalytic efficiency of SSP-prepared $Fe_2O_3$ accounts for its higher surface area. The higher surface area enables faster adsorption and desorption of reactive molecules during the decomposition process (HTD) resulting in faster decomposition of AP and thus. Lower decomposition temperature. The efficacy of prepared iron oxide catalysts as ballistic modifiers was evaluated through the deflagration of an AP-HTPB-catalyst composite propellant at atmospheric pressure. With the addition of SSP, the burning rate was enhanced by 88 % compared to TDP with a 50% increment. Similar observations were made by Joshi et al. [22]

Applications for Earth-Abundant Transition Metals                Materials Research Forum LLC
Materials Research Foundations 179 (2025) 20-36        https://doi.org/10.21741/9781644903711-2

and Cao et al. [7]. With the addition of nano-sized $\alpha$- $Fe_2O_3$, both the lower temperature decomposition (LTD) and HTD process showed a considerable decrement in temperature with enhanced heat release from 864 J/g (pure AP) to 1235 J/g. The decelerated in activation energy and accelerated rate of reaction further demonstrates the significant catalytic efficiency of the synthesised $\alpha$-$Fe_2O_3$. Likewise, the role of iron oxide as a catalyst in AP decomposition followed by the catalytic effect of the amount of $Fe_2O_3$ used in decomposition studies as well as solid propellant was studied by Kohga and his co-workers [23, 24]. The studies showed that owing to the addition of $Fe_2O_3$, the decomposition behaviour of AP changed significantly. Though, the degradation behaviour remained independent of the amount of catalyst added. While the burning characteristics are dependent on the catalyst amount used.

### 3.1.2 Composites of metal oxides as catalysts

Despite the confirmed exemplary catalytic behaviour of $Fe_2O_3$ nanoparticles in AP thermolysis, the agglomeration proneness of pure nanoparticles leads to fewer active site exposures, potentially inhibiting their catalytic activity [25]. A unfailing way to address this issue is to anchor the metal oxide nanoparticles onto a suitable substrate such as graphene oxide [25], polymeric carbon nitride [26], carbon nanotubes [27]etc. Studies show that graphene oxide (GO) is an ideal material for anchoring nanoparticles owing to its special surface features, layered structure and presence of oxygen-containing functional groups. These groups make GO chemically active to adsorb polar molecules onto its surface [28].

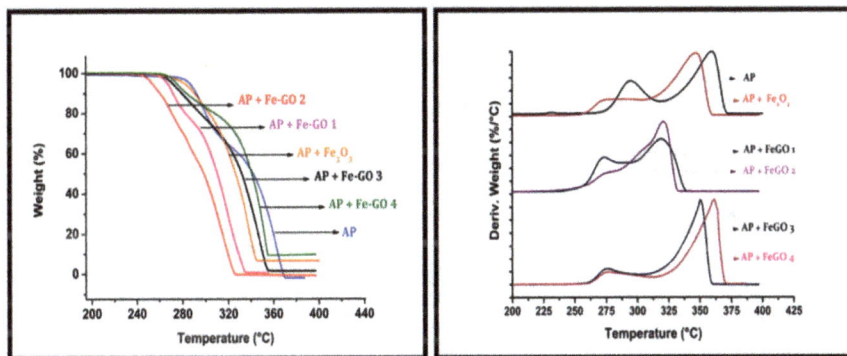

Fig. 1: TG and DTG curves of AP and AP + iron oxide/GO composite catalysts [28].

Literature indicates that, in GO, the intercalation process exhibits both chemical and physical characteristics, involving the adsorption of guest species and the exfoliation of GO sheets. In comparison to pure $Fe_2O_3$ nanoparticles, the $Fe_2O_3$-GO composite exhibits superior catalytic activity by decreasing both LTD-HTD values (Fig. 1) and activation energies (Table 1). The improved effectiveness is due to the even distribution of iron oxide within and on the graphene layer (the composite showing the best activity is named FeGO2) [28]. During the decomposition process, graphene will serve as the supporting material for electron flow, hence enhancing the reaction rate [28, 29].

*Table 1 Kinetic parameters for the thermal decomposition of AP with and without catalysts Fe₂O₃ and FeGO2 [28].*

| Sample code | Kinetic parameters calculated | |
|---|---|---|
| | Activation energy, $E_a$ [kJ mol$^{-1}$] | Rate constant, k [min$^{-1}$] |
| Pure AP | 90 | $5.5 \times 10^3$ |
| AP+Fe₂O₃ | 60 | $8.2 \times 10^3$ |
| AP+FeGO2 | 55 | $9.8 \times 10^3$ |

Moreover, iron oxide incorporated graphene (GINC) was also employed as a ballistic modifier in composite propellant by Dey et al. [30]. With the addition of the composite GINC, a significant rise in burn rate was noticed, i.e. an increment of 52%. A schematic representation of the preparation procedure and application of rGO- Fe₂O₃ nanocomposite in ammonium perchlorate thermal decomposition study is given in Fig.2 [31].

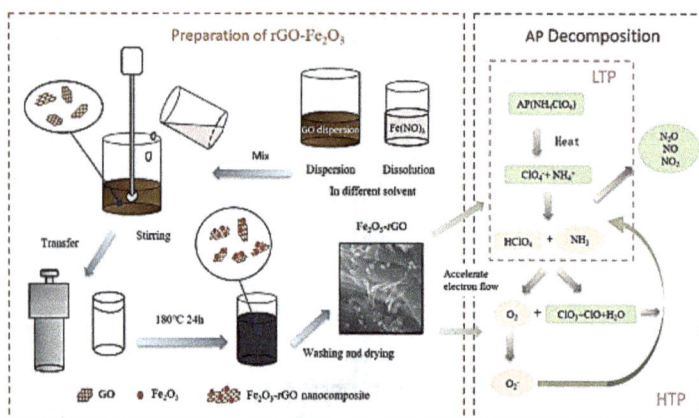

*Fig. 2: A schematic illustration and catalytic application of rGO–Fe₂O₃ composites in AP thermolysis [31].*

### 3.1.3 Binary metal oxides as catalysts

Apart from single metal oxide catalysts, spinel-structured binary metal oxides also have gathered significant interest for catalysing ammonium perchlorate thermal decomposition, on the grounds of exceptional catalytic efficacy resulting from the synergetic interaction between constituting elements [32]. Spinel structured metal oxides are typically represented by the formula $AB_2O_4$, with

A and B denoting divalent and trivalent metal cations, respectively. Zhang et al. [33] have prepared three types of ferrates namely, $NiFe_2O_4$, $ZnFe_2O_4$ and $CoFe_2O_4$. The catalytic performance of all ferrates was studied using TG and DSC and, the catalytic behaviour of each ferrate was significantly different from each other. Though nickel ferrate exhibited a higher BET specific surface area of 105 $m^2$ $g^{-1}$, the best catalytic efficacy towards AP thermolysis was observed for cobalt ferrate with a decrement of HTD value by 109°C and $E_a$ value by 37 kJ $mol^{-1}$. The gas phase MS-FTIR technique revealed that by the usage of ferrate as catalyst for AP thermal decomposition, the decomposition pathway did not change, but influences the energy barrier of high-temperature AP decomposition. Similar ferrate-based studies found in the literature [34, 35] and, all the ferrates have excellent catalytic efficacy towards AP thermolysis. This could be ascribed to the strong interaction between individual metal atoms.

## 3.2 Copper-based catalysts

### 3.2.1 Single metal oxides as catalysts

Copper oxides, as significant transition metal oxides, have been thoroughly investigated for the thermal breakdown of ammonium perchlorate due to their exceptional catalytic efficacy. Elbasuney and Yehia [36] have studied the environmentally benign and cost effective synthesis of colloidal CuO nanoparticles as an innovative catalyst for AP decomposition. Using a rapid solvent-antisolvent technique, they effectively coated the synthesised colloidal CuO nanoparticles with AP particles through co-precipitation. Later, the catalytic effect of CuO particles on AP thermal decomposition were studied using the thermal techniques TGA and DSC. By the usage of 1wt% of CuO, a 30% reduction in the LTD temperature was observed. Further, the merging of a two-stage decomposition pattern into a solo peak, accompanied by a 53% increment in heat release affirms the catalytic efficacy of synthesised CuO. In another study, Nourine et al. [37] coated nano copper oxide (nCuO) with nitrocellulose (NC) to further enhance the thermocatalytic efficiency of nCuO. The findings pointed out that; NC had a distinct effect only during the LTD stage and no significant influence on the HTD stage; though there was a significant increase in the heat release. Nano CuO burn rate catalysts were synthesised through a facile hydrothermal route, deliberately omitting alkali or surfactants. The incorporation of a modest quantity of cotton functioned as a pH-regulating agent via its hydrothermal carbonisation, utilising the $HNO_3$ generated from copper nitrate. The TG-MS analysis of the volatile byproducts of AP decomposition indicated an elevated amount of nitrogen, chlorine, and oxygen, alongside a diminished release of ammonia. This signifies a transition in the decomposition pathway of AP, shifting from a proton transfer (applicable to AP) to an electron transfer mechanism facilitated by Cu(II)O catalysed AP. This study elucidated the catalytic pathway of CuO mediated thermal decomposition of ammonium perchlorate [38].

### 3.2.2 Composites of metal oxides as catalysts

As previously noted, to mitigate the tendency for agglomeration in nanosized CuO, the utilisation of an appropriate substrate or support is essential. Dispersion of nano CuO on graphitic carbon nitride has received ample attention for versatile applications, especially for AP thermolysis studies.

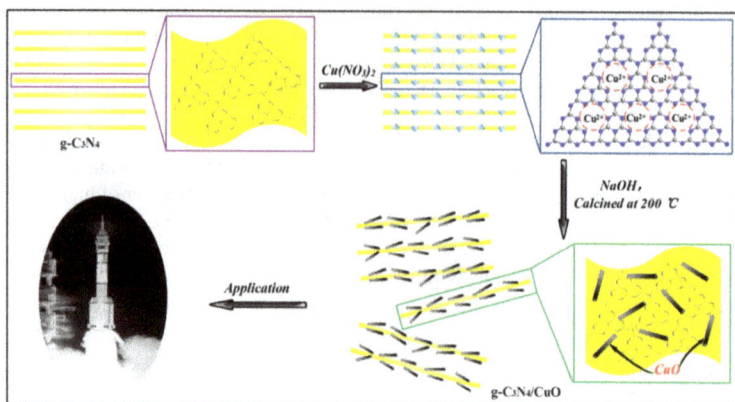

*Fig. 3: A schematic representation of synthesis and application of the g-C₃N₄/CuO*
*nanocomposite in AP decomposition [39].*

According to Tan et al., in the presence of the composite g-C$_3$N$_4$/CuO, the rate of AP thermal decomposition was enhanced significantly. The introduction of g-C$_3$N$_4$/CuO resulted in a reduction of HTD value of AP by 105.5°C, with the observation of a singular decomposition step, illustrating the strong synergy between g-C$_3$N$_4$ and CuO. Similarly, a mesoporous graphitic carbon nitride/CuO (mpg-C$_3$N$_4$/CuO) composite was studied for its catalytic efficacy for AP decomposition. The high specific surface area of the composite 111.33 m$^2$ g$^{-1}$ along with good synergy between mpg-C$_3$N$_4$ and CuO was highly beneficial for the mpg-C$_3$N$_4$/CuO catalysed AP thermolysis [40]. Later, Mani et al. [6] synthesised g-C$_3$N$_4$/CuO composites with varying metal oxides via an ex-situ method and investigated their respective catalytic efficacies. The thermal analyses show that the composite with g-C$_3$N$_4$ -CuO in a 1:1 ratio exhibited the best catalytic performance.

### 3.2.3 Binary metal oxides as catalysts

Aside from super fine iron oxide, copper chromite (CC) stands out as the most favoured catalyst for enhancing the burn rate of CSPs. Literature indicates that CuCr$_2$O$_4$ exerts minimal influence on the thermal degradation of AP at lower temperature range, specifically up till 270 °C. Nevertheless, its impact develops increasingly evident at elevated temperature range spanning from 280 to 340 °C [41]. The physicochemical features of CC differ with the methodology of preparation. Sanoop et al. have synthesised CC using ethylamine as the precipitating agent. This has followed by the calcination of precursors copper ethylamine chromate and copper ammonium chromate at 150-200°C and finally heating at 300-350°C in a muffle furnace [42]. Likewise, Zhang and co-workers [43] synthesised AP-CC composites utilising both ultrasonic dispersion techniques and mechanical grinding methods. A sequence of CuCr$_2$O$_4$ - AP composites with varying dispersions were synthesised to investigate the optimal catalytic role of nano sized CuCr$_2$O$_4$ on AP (ultrafine). In the presence of copper chromite catalyst, AP thermal decomposition temperature was decreased considerably.

### 3.3    Nickel-based catalysts

Like other metal oxides, nickel oxide (NiO) also exhibits inherent catalytic efficiency for a range of applications. NiO nanorods and micro-flowers were prepared via hydrothermal method without using surfactants and post-calcination. The reduced decomposition temperature and high burning rate with NiO micro-flowers make it an outstanding catalyst [44]. With $TiO_2$ and Mxene, NiO@TiO2/MXene nanocomposite was synthesised in which the 2-D MXene will act as the substrate for catalyst dispersion. As the $Ni^{2+}$ concentration rises in the composites, the catalytic capability towards AP also improves consequently. This led to the decrement of HTD value from 437.3 °C to 313.3 °C, a reduction in $E_a$ values from 216.1 kJ $mol^{-1}$ to 114.6 kJ $mol^{-1}$, and enhanced heat release by 118 % [45]. Apart from NiO, nano Ni powder has also been used in propellants for catalytic purposes. Compared to nano-metal oxides, metal nanoparticles seem to be highly sensitive to the presence of oxygen and can enhance the decomposition procedure of AP. Also, Ni nano powders can be easily oxidized to metal oxide and can release a significant amount of apparent heat during AP decomposition [47, 48]. Graphene/Ni aerogel was prepared by Lan et al. [46] by sol-gel and supercritical $CO_2$ drying technique. Due to the high surface area and mesoporosity of the prepared G/Ni aerogel, the nanocomposite exhibited outstanding catalytic performance towards AP thermolysis by reducing the decomposition temperature by 122°C. The catalysed thermal decomposition behaviour AP is given Fig. 4.

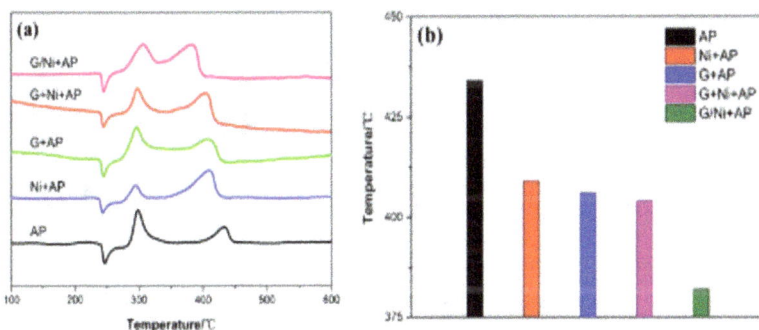

*Fig. 4: DSC curves of AP with and w the mixture of AP with Ni, G, G+Ni, G/Ni aerogels [46].*

## 4.    Mechanism of catalytic action

In recent decades, extensive research has focused on the decomposition mechanisms of ammonium perchlorate, with several potential mechanisms postulated; however, the exact decomposition mechanism remains a contentious topic. This chapter provides a concise explanation of two potential mechanisms underlying the thermal decomposition pathway of AP.

### 4.1    Electron transfer mechanism

As per the mechanism proposed by Bicromshaw and Newman [49], the thermal decomposition proceeds through electron transfer from anion to cation.

$$ClO_4^- + NH_4^+ \longrightarrow ClO_4^0 + NH_4^0$$

The likelihood of electron release is heightened by the shorter distance between the ions. Consequently, an effective electron receiver is deemed to be just those ammonium ions situated in interstices. Upon receiving an electron, the ammonium radical entity disintegrated into $NH_3$ molecule and H atom.

$$NH_4^0 \longrightarrow NH_3 + H$$

H atom transmissions happen across the lattice. The electrons flow precisely in the same manner across the anion sublattice.

$$ClO_4^0 + ClO_4^- = ClO_4^- + ClO_4^0$$

$HClO_4$ is generated by the reaction between the perchlorate ($ClO_4$) entity and hydrogen. The resulting perchloric acid may further react with hydrogen, releasing water molecule and chlorate ($ClO_3$) radical.

The chlorate radical functions as an electron acceptor capable of capturing electrons. Upon electron acquisition, the chlorate radical is converted into the $ClO_3$ anion. Subsequently, the chlorite ($ClO_2^-$) ion and perchlorate radical undergo decomposition, resulting in compounds that can combine with $NH_4^+$ ions. Consequently, by-products such as chlorine, nitrogen dioxide, and water are generated. Generally, transition metal oxide or metal oxide-catalysed AP decompositions are assumed to proceed via an electron transfer mechanism [28, 38, 47]. The presence of GO as a support will facilitate enhanced electron flow and thereby faster decomposition rate.

An objection to the electron mechanism is that AP is a standard dielectric. Because of its low probability, electron transfer is unable to sustain the process at low temperatures of decomposition. Subsequently, all efforts to identify the complex molecular structure proposed by Galwey and Jacobs were unsuccessful. Conversely, the major products identified in the trials conducted by several groups were $NH_3$ and $HClO_4$. This made it possible to assume that H+ transport plays a role in the first stage of the thermal decomposition of AP [50].

## 4.2    Proton transfer mechanism

Arguments in favour of the proton transfer mechanism are alike $E_a$ values of sublimation and thermolysis, undistinguishable product composition of sublimation and thermal degradation, inhibition of process in $NH_3$ vapour and speeding up of reaction presence of $HClO_4$, sensitivity towards dopants and so on [50].

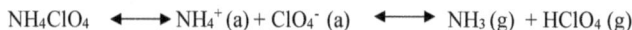

$$NH_4ClO_4 \longleftrightarrow NH_4^+(a) + ClO_4^-(a) \longleftrightarrow NH_3(g) + HClO_4(g)$$

The proton transfer mechanism comprises several stages. Stage I: formation of an ion pair in the perchlorate ammonium lattice. Stage II: the decomposition process begins with a proton transfer from ammonium cation to $ClO_4^-$ anion through the molecular complex. Stage III: the molecular complex dissociates into $NH_3$ and $HClO_4$. The decomposition products either undergo reactions within the absorbed layer on the surface of AP or engage in interactions through desorption and sublimation in the gas phase. These gaseous molecules of $NH_3$ and $HClO_4$ undergo rapid reaction leaving $H_2O$, NO, $N_2O$, $O_2$, and $Cl_2$ as by-products.

Further, inside the absorbed layer, $HClO_4$ desorbs faster than $NH_3$, leading to the incomplete $NH_3$ oxidation resulting in a $NH_3$ saturated environment. Consequently, the HTD process will slow

Materials Research Forum LLC
https://doi.org/10.21741/9781644903711-2

down and undergo incomplete transformation. Though both lower temperature and higher temperature decomposition of AP starts with $H^+$ transfer from ammonia ion to perchlorate ion, the distinction lies in the fact that LTD occurs at the defects of the AP crystal, while the HTD stage transpires within the lattice of the unreacted AP.

In the case of $gC_3N_4$-based metal oxide catalysts, the decomposition proceeds via a proton transfer mechanism [26, 39, 40]. Owing to the low band gap of graphitic carbon nitride, thermal excitation results in electrons and holes formation on the carbon nitride surface. The doped metal oxides along with $HClO_4$ will act as electron traps to avoid the electron-hole pair recombination. Thus, the perchloric acid molecules get reduced to form superoxide anion radicals. These superoxide anions along with holes will oxidize $NH_3$ into corresponding low molecular weight oxides of nitrogen. An example of the $H^+$ transfer mechanism in iron oxide/carbon nitride composite is shown in Fig. 5. Despite extensive studies on catalysed AP thermal decomposition mechanisms, the fundamental mechanism remains inadequately elucidated, indicating a significant need for further investigation through diverse technological approaches.

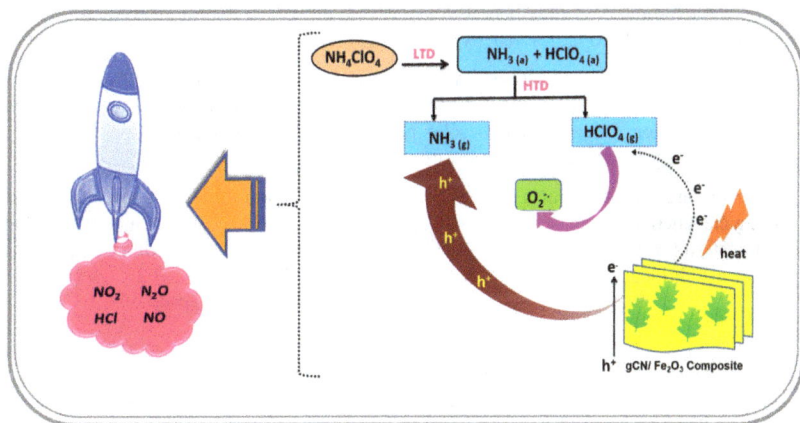

*Fig. 5: An illustration of iron oxide/carbon nitride composite catalysed AP thermal decomposition via proton transfer mechanism. Reproduced with permission from ref. [26]. Copyright 2022 American Chemical Society*

## Conclusions and Future Outlook

Transition metal oxides, especially those that are abundant in the earth's crust, have demonstrated considerable potential as catalysts for AP thermal decomposition. AP thermolysis is crucial for accelerating the efficacy of CSPs, positioning it as a significant domain of inquiry within propulsion technologies. Earth-abundant transition metal oxides are economically advantageous due to their greater availability and lower cost compared to rare or precious metals. This renders them appealing for extensive applications in the defence and aerospace sectors. TMOs demonstrate notable catalytic potential by decreasing the AP decomposition temperature and enhancing the

Materials Research Forum LLC
https://doi.org/10.21741/9781644903711-2

decomposition rate. This leads to enhanced combustion efficiency in solid propellants, minimising undesirable by-products and augmenting thrust. The structural flexibility and tunability of TMOs allow them to form mixed metal oxides and composites with carbonaceous materials making them an ideal choice for propellant applications.

- Although prevalent TMOs specifically as NiO, CuO, and $Fe_2O_3$ have been extensively researched, future investigations may concentrate on examining novel combinations and compositions that could provide greater catalytic activity and improved stability under extreme temperatures.

- Although advancements have been made, a more comprehensive understanding of the catalytic mechanisms involved in AP decomposition on TMO surfaces remains necessary. In situ spectroscopy and computational modelling can provide insights into reaction pathways and active sites, facilitating the design of more efficient catalysts.

- The integration of TMOs with carbon-based materials, polymers, or alternative metal oxides may yield synergistic effects that enhance catalytic performance. Next-generation propellant formulations may benefit from the exploration of hybrid catalysts that exhibit enhanced thermal properties, reactivity, and stability.

- Scaling up the synthesis of TMO-based catalysts is essential for transitioning from laboratory settings to real-world applications. Research must focus on the challenges associated with mass production, integration into solid propellants, and the consistent performance of catalysts under diverse environmental conditions.

- Future research must evaluate the environmental implications of TMOs throughout their lifecycle, encompassing synthesis to disposal. The safety implications of integrating TMO catalysts into large-scale rocket propellants, especially regarding handling and storage, require additional examination.

**References**

[1] P.N. Dave, R. Sirach, Investigating the catalytic effect of nanoferrites additives on the thermal decomposition of ammonium perchlorate and burning rate modification of the composite solid propellants, Mater. Sci. Eng. B 301 (2024) 117136. https://doi.org/10.1016/j.mseb.2023.117136

[2] F.-M. Dîrloman, A.-N. Rotariu, T. Rotariu, G.-F. Noja, R.-E. Ginghină, N.-D. Zvîncu, Ballistic and thermal characterisation of greener composite solid propellants based on phase-stabilized ammonium nitrate, Case Stud. Therm. Eng. 54 (2024) 103987. https://doi.org/10.1016/j.csite.2024.103987

[3] F.Y. Chen, C.L. Xuan, Q.Q. Lu, L. Xiao, J.Q. Yang, Y.B. Hu, G.P. Zhang, Y.L. Wang, F.Q. Zhao, G.Z. Hao, W. Jiang, A review on the high energy oxidizer ammonium dinitramide: Its synthesis, thermal decomposition, hygroscopicity, and application in energetic materials, Def. Technol. 19 (2023) 163-195. https://doi.org/10.1016/j.dt.2022.04.006

[4] J. Jos, S. Mathew, Ammonium nitrate as an eco-friendly oxidizer for composite solid propellants: promises and challenges, Crit. Rev. Solid State Mater. Sci. 42(6) (2017) 470-498. https://doi.org/10.1080/10408436.2016.1244642

[5] V.V. Boldyrev, Thermal decomposition of ammonium perchlorate, Thermochimica Acta 443(1) (2006) 1-36. https://doi.org/10.1016/j.tca.2005.11.038

[6] G. Mani, J. Jos, P.R. Nair, S. Mathew, Investigation of kinetic parameters for ammonium perchlorate thermal decomposition in presence of gCN/CuO by TG-MS analysis and kinetic compensation correction, J. Solid State Chem. 301 (2021) 122301. https://doi.org/10.1016/j.jssc.2021.122301

[7] S.-b. Cao, X.-g. Han, L.-l. Zhang, J.-x. Wang, Y. Luo, H.-k. Zou, J.-f. Chen, Facile and scalable preparation of α-Fe2O3 nanoparticle by high-gravity reactive precipitation method for catalysis of solid propellants combustion, Powder Technol. 353 (2019) 444-449. https://doi.org/10.1016/j.powtec.2019.05.062

[8] N. Yadav, P.K. Srivastava, M. Varma, Recent advances in catalytic combustion of AP-based composite solid propellants, Def. Technol. 17(3) (2021) 1013-1031. https://doi.org/10.1016/j.dt.2020.06.007

[9] R.M. Paul Chirik, Getting Down to Earth: The Renaissance of Catalysis with Abundant Metals, Accounts of Chemical Research 48(9) (2015) 2495-2495. https://doi.org/10.1021/acs.accounts.5b00385

[10] K.M.P. Wheelhouse, R.L. Webster, G.L. Beutner, Advances and Applications in Catalysis with Earth-Abundant Metals, Organometallics 42(14) (2023) 1677-1679. https://doi.org/10.1021/acs.organomet.3c00292

[11] P.A. Cox, Transition Metal Oxides: An Introduction to their Electronic Structure and Properties, Clarendon Press2010.

[12] Y. Xia, C.T. Campbell, B. Roldan Cuenya, M. Mavrikakis, Introduction: Advanced Materials and Methods for Catalysis and Electrocatalysis by Transition Metals, Chem. Rev. 121(2) (2021) 563-566. https://doi.org/10.1021/acs.chemrev.0c01269

[13] E. Khare, N. Holten-Andersen, M.J. Buehler, Transition-metal coordinate bonds for bioinspired macromolecules with tunable mechanical properties, Nature Reviews Materials 6(5) (2021) 421-436. https://doi.org/10.1038/s41578-020-00270-z

[14] A.T. Howe, P.J. Fensham, Electronic properties of binary compounds of the first-row transition metals, Quarterly Reviews, Chemical Society 21(4) (1967) 507-524. https://doi.org/10.1039/qr9672100507

[15] D. Wang, D. Astruc, The recent development of efficient Earth-abundant transition-metal nanocatalysts, Chem. Soc. Rev. 46(3) (2017) 816-854. https://doi.org/10.1039/C6CS00629A

[16] P. Du, R. Eisenberg, Catalysts made of earth-abundant elements (Co, Ni, Fe) for water splitting: recent progress and future challenges, Energy & Environmental Science 5(3) (2012) 6012-6021. https://doi.org/10.1039/c2ee03250c

[17] I. Roger, M.A. Shipman, M.D. Symes, Earth-abundant catalysts for electrochemical and photoelectrochemical water splitting, Nature Reviews Chemistry 1(1) (2017) 1-13. https://doi.org/10.1038/s41570-016-0003

Materials Research Forum LLC
https://doi.org/10.21741/9781644903711-2

[18] F. Ma, Z.-M. Luo, J.-W. Wang, B.M. Aramburu-Trošelj, G. Ouyang, Earth-abundant-metal complexes as photosensitizers in molecular systems for light-driven CO2 reduction, Coord. Chem. Rev. 500 (2024) 215529. https://doi.org/10.1016/j.ccr.2023.215529

[19] L. Rotundo, R. Gobetto, C. Nervi, Electrochemical CO2 reduction with earth-abundant metal catalysts, Current Opinion in Green and Sustainable Chemistry 31 (2021) 100509. https://doi.org/10.1016/j.cogsc.2021.100509

[20] A. Budhwar, A. Gautam, P.V. More, C.S. Pant, S. Banerjee, P.K. Khanna, Modified iron oxide nanoparticles as burn rate enhancer in composite solid propellants, Vacuum 156 (2018) 483-491. https://doi.org/10.1016/j.vacuum.2018.08.013

[21] R. Arun Chandru, R.P. Patel, C. Oommen, B. Raghunandan, Initial studies on development of high-performance nano-structured Fe2O3 catalysts for solid rocket propellants, J. Mater. Eng. Perform. 28 (2019) 810-816. https://doi.org/10.1007/s11665-018-3812-x

[22] S.S. Joshi, P.R. Patil, V. Krishnamurthy, Thermal decomposition of ammonium perchlorate in the presence of nanosized ferric oxide, Defence Science Journal 58(6) (2008) 721-727. https://doi.org/10.14429/dsj.58.1699

[23] M. Kohga, S. Togo, Catalytic effect of added Fe2O3 amount on thermal decomposition behaviors and burning characteristics of ammonium nitrate/ammonium perchlorate propellants, Combust. Sci. Technol. 192(9) (2020) 1668-1681. https://doi.org/10.1080/00102202.2019.1620736

[24] M. Kohga, S. Togo, Influence of iron oxide on thermal decomposition behavior and burning characteristics of ammonium nitrate/ammonium perchlorate-based composite propellants, Combust. Flame 192 (2018) 10-24. https://doi.org/10.1016/j.combustflame.2018.01.040

[25] J. Pei, H. Zhao, F. Yang, D. Yan, Graphene Oxide/Fe2O3 Nanocomposite as an Efficient Catalyst for Thermal Decomposition of Ammonium Perchlorate via the Vacuum-Freeze-Drying Method, Langmuir 37(20) (2021) 6132-6138. https://doi.org/10.1021/acs.langmuir.1c00108

[26] G. Mani, P.R. Nair, S. Mathew, Polymeric Carbon Nitride/Iron Oxide Composites: A Novel Class of Catalysts with Reduced Metal Content for Ammonium Perchlorate Thermal Decomposition, ACS omega 7(43) (2022) 38512-38524. https://doi.org/10.1021/acsomega.2c03761

[27] L. Yang, Z. Mi, H. Fang, R. Xu, B. Lv, J. Li, G. Zhang, Fabrication of highly catalytic active α-Fe2O3-carbon nanotube composites for thermal decomposition of ammonium perchlorate by light and temperature control strategy, Surfaces and Interfaces 44 (2024) 103642. https://doi.org/10.1016/j.surfin.2023.103642

[28] S. Paulose, R. Raghavan, B.K. George, Graphite oxide-iron oxide nanocomposites as a new class of catalyst for the thermal decomposition of ammonium perchlorate, RSC Advances 6(51) (2016) 45977-45985. https://doi.org/10.1039/C6RA06860J

[29] P. Majumder, R. Gangopadhyay, Evolution of graphene oxide (GO)-based nanohybrid materials with diverse compositions: an overview, RSC advances 12(9) (2022) 5686-5719. https://doi.org/10.1039/D1RA06731A

[30] A. Dey, J. Athar, P. Varma, H. Prasant, A.K. Sikder, S. Chattopadhyay, Graphene-iron oxide nanocomposite (GINC): an efficient catalyst for ammonium perchlorate (AP) decomposition and burn rate enhancer for AP based composite propellant, RSC Advances 5(3) (2015) 1950-1960. https://doi.org/10.1039/C4RA10812D

[31] M. Zhang, F. Zhao, Y. Yang, J. Zhang, N. Li, H. Gao, Effect of rGO-Fe2O3 nanocomposites fabricated in different solvents on the thermal decomposition properties of ammonium perchlorate, CrystEngComm 20(43) (2018) 7010-7019. https://doi.org/10.1039/C8CE01434E

[32] N.M. Juibari, A. Eslami, Green synthesis of ZnCo2O4 nanoparticles by Aloe albiflora extract and its application as catalyst on the thermal decomposition of ammonium perchlorate, J. Therm. Anal. Calorim. 130 (2017) 1327-1333. https://doi.org/10.1007/s10973-017-6613-9

[33] M. Zhang, F. Zhao, Y. Yang, T. An, W. Qu, H. Li, J. Zhang, N. Li, Catalytic activity of ferrates (NiFe2O4, ZnFe2O4 and CoFe2O4) on the thermal decomposition of ammonium perchlorate, Propellants, Explosives, Pyrotechnics 45(3) (2020) 463-471. https://doi.org/10.1002/prep.201900211

[34] W. Wang, S. Guo, D. Zhang, Z. Yang, One-pot hydrothermal synthesis of reduced graphene oxide/zinc ferrite nanohybrids and its catalytic activity on the thermal decomposition of ammonium perchlorate, Journal of Saudi Chemical Society 23(2) (2019) 133-140. https://doi.org/10.1016/j.jscs.2018.05.001

[35] Y. Zu, Y. Zhao, K. Xu, Y. Tong, F. Zhao, Preparation and comparison of catalytic performance for nano MgFe2O4, GO-loaded MgFe2O4 and GO-coated MgFe2O4 nanocomposites, Ceram. Int. 42(16) (2016) 18844-18850. https://doi.org/10.1016/j.ceramint.2016.09.030

[36] S. Elbasuney, M. Yehia, Thermal decomposition of ammonium perchlorate catalyzed with CuO nanoparticles, Def. Technol. 15(6) (2019) 868-874. https://doi.org/10.1016/j.dt.2019.03.004

[37] M. Nourine, M.K. Boulkadid, S. Touidjine, H. Akbi, S. Belkhiri, Exploring the potential of nitrocellulose for improving the catalytic efficiency of nano copper oxide in the thermal degradation of ammonium perchlorate, Reaction Kinetics, Mechanisms and Catalysis 136(5) (2023) 2769-2783. https://doi.org/10.1007/s11144-023-02469-x

[38] D.L. Sivadas, D. Thomas, M.S. Haseena, T. Jayalatha, G. Rekha Krishnan, S. Jacob, R. Rajeev, Insight into the catalytic thermal decomposition mechanism of ammonium perchlorate, J. Therm. Anal. Calorim. 138(1) (2019) 1-10. https://doi.org/10.1007/s10973-019-08209-5

[39] L. Tan, J. Xu, S. Li, D. Li, Y. Dai, B. Kou, Y. Chen, Direct Growth of CuO Nanorods on Graphitic Carbon Nitride with Synergistic Effect on Thermal Decomposition of Ammonium Perchlorate, Materials 10(5) (2017) 484. https://doi.org/10.3390/ma10050484

[40] J. Xu, S. Li, L. Tan, B. Kou, Enhanced catalytic activity of mesoporous graphitic carbon nitride on thermal decomposition of ammonium perchlorate via copper oxide modification, Mater. Res. Bull. 93 (2017) 352-360. https://doi.org/10.1016/j.materresbull.2017.05.038

[41] S. Saha, A. Chowdhury, N. Kumbhakarna, Effect of copper chromite on ammonium perchlorate decomposition-A TGA-FTIR-MS and FE-SEM study, J. Therm. Anal. Calorim. 149(17) (2024) 9401-9412. https://doi.org/10.1007/s10973-024-13395-y

[42] A.P. Sanoop, R. Rajeev, B.K. George, Synthesis and characterization of a novel copper chromite catalyst for the thermal decomposition of ammonium perchlorate, Thermochimica Acta 606 (2015) 34-40. https://doi.org/10.1016/j.tca.2015.03.006

[43] D. Zhang, Q. Li, R. Li, H. Li, H. Gao, F. Zhao, L. Xiao, G. Zhang, G. Hao, W. Jiang, Significantly Enhanced Thermal Decomposition of Mechanically Activated Ammonium Perchlorate Coupling with Nano Copper Chromite, ACS Omega 6(24) (2021) 16110-16118. https://doi.org/10.1021/acsomega.1c02002

[44] Y. Zhao, X. Zhang, X. Xu, Y. Zhao, H. Zhou, J. Li, H. Jin, Synthesis of NiO nanostructures and their catalytic activity in the thermal decomposition of ammonium perchlorate, CrystEngComm 18(25) (2016) 4836-4843. https://doi.org/10.1039/C6CE00627B

[45] C. Bai, D. Yang, C. Liu, F. Zhu, C. Tu, G. Li, Y. Luo, In situ synthesis NiO@ TiO$_2$/MXene as a promoter for ammonium perchlorate based solid propellants, Appl. Surf. Sci. 652 (2024) 159228. https://doi.org/10.1016/j.apsusc.2023.159228

[46] Y. Lan, B. Jin, J. Deng, Y. Luo, Graphene/nickel aerogel: an effective catalyst for the thermal decomposition of ammonium perchlorate, RSC advances 6(85) (2016) 82112-82117. https://doi.org/10.1039/C6RA15661D

[47] Y. Zhang, Z. Li, M. Ma, Synthesis of Ni nanopowders with cauliflower-like shapes and their catalytic property on the thermal decomposition of ammonium perchlorate, Reaction Kinetics, Mechanisms and Catalysis 136(4) (2023) 2327-2341. https://doi.org/10.1007/s11144-023-02473-1

[48] S. Bao, T. Li, C. Guo, Y. Zhao, H. Zhang, R. Wu, H. Shi, One-Pot Preparation of HTPB/nNi and Its Catalyst for AP, Nanomaterials 12(15) (2022) 2669. https://doi.org/10.3390/nano12152669

[49] L. Bircumshaw, B. Newman, The thermal decomposition of ammonium perchlorate, II. The kinetics of the decomposition, the effect of particle size, and discussion of results, Proceedings of the Royal Society of London. Series A. Mathematical and Physical Sciences 227(1169) (1955) 228-241. https://doi.org/10.1098/rspa.1955.0006

[50] V. Boldyrev, Thermal decomposition of ammonium perchlorate, Thermochimica Acta 443(1) (2006) 1-36. https://doi.org/10.1016/j.tca.2005.11.038

Applications for Earth-Abundant Transition Metals
Materials Research Foundations 179 (2025) 37-78

Materials Research Forum LLC
https://doi.org/10.21741/9781644903711-3

# Chapter 3

# Earth-Abundant Metals in Pharmaceutical Industries

Jaison Jeevanandam[1], Michael K. Danquah[2]*

[1]Department of Experimental Neurobiology, Preclinical research program, National Institute of Mental Health, 250 67 Klecany, Czechia.

[2]Chemical Engineering Department, University of Tennessee, Knoxville, TN 37403, USA

*mdanquah@utk.edu

## Abstract

Several earth-abundant metals, such as iron, zinc, calcium, titanium, magnesium and manganese, are widely utilized in pharmaceutical industry to reduce the production cost and enhance sustainability in recent times. These metals are used in pharmaceutical applications, such as drug delivery systems, antimicrobial, anticancer, antidiabetic, bioimaging agents and biosensors. Some of these metals serve as a standalone medicine for the treatment of common and novel diseases, deficiencies and as nutraceuticals. Thus, this chapter is an overview of earth-abundant metals in numerous pharmaceutical applications. In addition, the pharmaceutical applications of nanosized earth-abundant metal particles were also discussed.

## Keywords

Earth-Abundant Metals; Metal Nanoparticles; Pharmaceutics; Antimicrobial Agents; Drug Delivery Systems

## Contents

## 1.    Introduction

Earth's crust contains various minerals as natural chemical compounds [1], aggregates of minerals in the form of rocks [2], dissolved minerals in the form of soil and water, that contains dissolved minerals [3]. The chemical compounds in the earth's crust mostly consists of metals, such as silicon, calcium, magnesium, sodium, aluminum, potassium and iron [4]. In general, these metals are classified into ferrous, nonferrous and alloys, depending on the existence and nonexistence of iron with/without magnetic properties as well as multiple metals along with other elements (alloys) [5-7]. Among these metals, earth-abundant metals, especially transition metals, are widely used in numerous applications [8-10]. It is noteworthy that metals, such as silicon, aluminum and iron along with oxygen account for about 88.1% of minerals in earth's crust, and hence, they are known as earth abundant metals [11]. In particular, first-row earth-abundant transition metals are widely utilized as homogenous catalysis applications [12, 13]. Further, the remaining 11.9% of the earth's crust was made-up of 90 other elements, which includes metals, such as calcium, magnesium, sodium and potassium [14, 15]. Thus, these metals are also included under the category of earth-abundant metals [16, 17]. These earth-abundant metals are extracted from the mineral ores in the earth's crust via mining [18], smelting [19], crushing [20] and electrolysis [21] along with pyrometallurgy [19], hydrometallurgy [22] and reactivity approaches [23]. The major advantages of metal extraction from their respective ores are economic benefits as they are profitable in mining industry which causes less damage the environment, compared to oil drilling process [24]. Furthermore, bioleaching is an alternative cost-effective approach for the extraction of metals, especially from low grade ores with various residues and less demand of energy, that can improve appearance of extracted metal from ore [25]. Even though methods to extract metals have advantages, there are few limitations, such as damage to land, causation of water, soil and air as well as hazardous to human health [26]. Thus, recent methods have been introduced in recent times for the extraction of metals, especially earth-abundant metals to be beneficial for various applications.

Recently, several earth-abundant metals are widely utilized in pharmaceutical industry to reduce the production cost and enhance sustainability [9, 27]. In particular, iron, zinc, calcium, titanium, magnesium and manganese have been widely used in various biomedical applications [28, 29]. All these metals are trace elements that are required as nutrients to humans and hence are extensively used in nutraceuticals [30]. Further, these earth-abundant metals are highly beneficial as drug delivery system [31], antimicrobial [32], anticancer [33], antidiabetic [34], bioimaging agents [35] and fabrication of biosensors [36] as well as standalone medicine for the treatment of neurodegenerative and rare diseases [17, 37]. Thus, this chapter is an overview of earth-abundant metals in numerous pharmaceutical applications. In addition, the pharmaceutical applications of nanosized earth-abundant metal particles were also discussed.

## 2. Overview of earth-abundant metals

### 2.1 Iron (Fe)

Iron is the most abundant metal (atomic number 26) in the form of magnetite and hematite around 5% in the earth's crust [38]. Further, it has been identified that the inner and outer core of the earth contains about 80% of iron [39]. Likewise, earth contains 35% of iron by mass [40], followed by oxygen (30%) [41], silicon (15%) [42] and magnesium (13%) [43]. Iron is considered as a crucial nutrient in the human body [44], that is extensively utilized to produce steel, which is an alloy of carbon and iron [45]. This metal is highly available in earth's crust due to its nuclear stability in stellar nuclear fusion reactions [46]. Furthermore, iron exists as complex along with other chemical compounds via oxidation-reduction reactions through chemical mobility [47]. It is noteworthy that mononuclear and dinuclear iron complexes, [4Fe-4S] and [2Fe-2S] clusters, iron protophorphyrin IX and [Fe-Ni-S] clusters are the significant iron complexes, that are widely used in protein biochemistry [48].

### 2.2 Zinc (Zn)

Zinc (atomic number 30) is another abundant metal present in earth's crust [49], which makes up to 0.02% by its weight (70 parts per million [ppm] in the earth's crust) [50]. In general, zinc originated in the soil or water and is available in the environment via weathering and erosion of minerals and rocks [51]. The chemical composition of the parent rock decides the concentration of zinc in the soil [52]. It can be noted that sphalerite and zinc sulfide are the most common zinc ore [53]. There are five naturally occurring stable zinc isotopes, namely $^{64}Zn$, $^{66}Zn$, $^{67}Zn$, $^{68}Zn$ and $^{70}Zn$, which makes up 48.6%, 27.9%, 4.1%, 18.8% and 0.6% of natural zinc, respectively [54]. This metal is widely utilized in rubber vulcanization [55] and is used as potential medicine [56]. Likewise, zinc has been identified as an essential trace metal [57], that is required for the growth and development of animals, plants and humans [58].

### 2.3 Calcium (Ca)

Calcium (atomic number 20) is another alkaline earth abundant metal [59], next to zinc, which makes up about 4.1% of its composition in earth's crust by mass [60]. It is a trimorphic metal, which is harder than sodium and is softer than aluminum [61]. It is noteworthy that calcium never occurs as a standalone pure elemental metal state in the environment [62] and is extracted from compounds, such as gypsum (calcium sulfate) [63], limestone (calcium carbonate) [64], apatite (calcium chloro- or fluoro-phosphate) [65] and fluorite (calcium fluoride) [66]. Moreover, dissolved calcium bicarbonate exists in hard water [67]. These calcium particles were highly beneficial as supplements in food and pharmaceuticals [67]. Additionally, this metal is utilized as bleaches in paper industry [68] and are used as components in cement [69] and electrical insulators [70]. Further, calcium is also used in steelmaking [71], refining [72], alloying [73] and soap manufacturing industries [74].

## 2.4 Titanium (Ti)

Titanium (atomic number 22) is also an earth abundant element in the crust of the earth, which makes up to 0.6% [75]. It has widely existed in all sediments and rocks [76], and is primarily identified to be present in minerals, such as ilmenite, rutile and sphene [77]. It can be noted that titanium has been recognized as a strong affinity towards oxygen and is known as lithophile element [78]. This metal has a 3034°F (Fahrenheit) of high melting point [79]. The commercial isolation of titanium wasn't started until 1948, which was initially isolated in the year 1910 [80]. The Kroll approach is identified to be highly beneficial for the isolation of titanium, where titanium tetrachloride was reduced by using magnesium [81]. Titanium is extensively used in numerous industries, such as maritime, aerospace, architecture, chemical and consumer product production [82]. In maritime industries, this metal is used in offshore oil rigs due to their high resistant towards corrosion and salt water [83], whereas they possess inert ceramic layer to be ideal for corrosive environments and are utilized in the valves, pumps and heat exchangers of chemical plants [84]. In the aerospace industry, titanium is used in the manufacturing of landing gear, bolts, aircraft engines, bolts, airframes and leading edges, due to their low density, high strength and light weight [85]. Additionally, the inert nature of titanium makes it nonreactive with skin and is used in the jewelry industry [86]. Various literatures reported that there is no essential biological role for titanium in animals [87]. However, certain organisms sequester titanium and possess few health benefits among plants [88]. It has been reported that titanium is non-toxic and biocompatible with the human body which makes it highly beneficial to produce dentals implants, surgical instruments and prosthetics [89].

## 2.5 Magnesium (Mg)

Magnesium (atomic number 12) is also another abundant element in the earth, which makes up to 2% of the total element quantity in earth's crust [90]. This metal has been identified as the third most abundant metal in seawater, bitterns, lake and well brines [91]. Magnesium possesses various exclusive properties, such as excellent machinability, thermal conductivity, strength-to-weight ratio, dimensional stability, electromagnetic shielding and are easily recyclable [92]. This metal has been identified in over 60 minerals, which includes magnesite, olivine, dolomite, brucite and carnallite [93]. Magnesium is widely used in automobiles and machinery as a structural component and is included along with aluminum as alloys to make beverage cans [94]. Further, this metal is highly beneficial to eliminate sulfur from steel as well as iron and is used to replace steel and aluminum in certain structural applications due to their light weight [95]. Furthermore, magnesium has been identified as nontoxic and possess several significance in biological systems as a vital nutrient [96]. This metal is a cofactor for about 300 enzyme to regulate protein synthesis, control of blood glucose, nerve and muscle function, as well as blood pressure regulation [97]. Moreover, magnesium has been involved in the oxidative phosphorylation, energy production, glycolysis, stabilization of adenosine triphosphate (ATP)-generating reactions, counter ion for ATP and nuclear acids [98].

## 2.6 Manganese (Mn)

Manganese (atomic number 25) is also considered as an earth-abundant element, which makes up about 0.1% of the earth's crust, including soil, air and water [99]. Historically, this metal ore is used by Romans and Egyptians for the removal of color from glasses and provide of tint of black, pink or purple color [100]. This metal has been identified to exist in over 100 minerals, which

Applications for Earth-Abundant Transition Metals                    Materials Research Forum LLC
Materials Research Foundations 179 (2025) 37-78          https://doi.org/10.21741/9781644903711-3

includes rhodochrosite (manganese carbonate) [101] and pyrolusite (manganese dioxide) [102]. This metal is extremely beneficial in the biogeochemical cycle of iron, carbon, sulfur and several trace elements [103]. Likewise, manganese is widely utilized as a desulfurizing and deoxidizing additive in the making of steel [104], welding consumables [105] as well as an alloy constituent [106]. This metal also has several biological roles in the body, which includes formation of connective tissues and bones [107], helps in clotting of blood with vitamin K [108], normal function of brain and nerves [109], regulation of blood sugar [110], carbohydrate and fat metabolism [111], normal function of immune system [112] and reproductive hormone [113], activation of numerous enzymes [114] and scavenge reactive oxygen species (ROS) in mitochondrial oxidative stress [111].

## 2.7    Other earth-abundant metals

Silicon (Si; atomic number 14) is the second most abundant metal, which is about 28% by mass in the earth's crust [115]. This metal possesses a strong affinity towards oxygen for the formation of silicon dioxide (silica) and other silicates, that are chemically inert [116]. Silicon is extensively utilized to produce transistors, computer chips, integrated circuits and other electronic devices as they are semiconductors [117]. Further, silicon is used to make alloys for the manufacturing of cylinder heads, engine blocks, machine tools, ceramics, bricks and Portland cement [118]. Furthermore, this metal is beneficial in silicone formation, that are used in the production of hair conditioners, lubricants, waterproof sealants, surgical procedures and cosmetic products [119], whereas silicon carbides are used to make lasers and abrasives as well as mold release agents [120]. Likewise, aluminum (Al; atomic number 13) is also an abundant metal, which makes up about 8% of its mass in the earth's crust [121]. This element is widely used in packaging food materials as foil to protect and insulate food as they are impermeable and corrosion resistant [122]. Besides, they are utilized in the manufacturing of commercial and military vehicles, freight rail cars, powered flight and aerospace [123]. Additionally, they are beneficial in the production of construction materials, such as windows, doors, roofing, cladding and foil insulation [124], domestic wiring [125], high-voltage power lines [126] and solar stove mirrors [127]. Moreover, sodium (Na; atomic number 11) and potassium (K; atomic number 19), which covers about 2.8% and 2.4% of the earth's crust, respectively, were also considered as earth-abundant metals [14, 128].

## 3.    Pharmaceutical applications of earth-abundant metals

In recent times, earth-abundant metals are widely utilized in biomedical applications, such as antimicrobial, antidiabetic, anticancer agents, to fabricate drug delivery systems and other applications, including fabrication of biosensors and standalone medicines.

## 3.1    Drug delivery systems

Iron has been widely employed in developing effective drug delivery systems. Recently, elemental iron has been included in metal-organic frameworks to delivery potential drugs. Materials of Institute Lavoisier-101 (MIL-101) has been prepared with iron (Fe) as shown in Figure 1 to act as a significant carrier of drugs [129]. It has been identified that Fe-MOF can stabilize and solubilize hydrophilic cidofovir, benzophenone-4, zidovudine triphosphate drugs, hydrophobic ibuprofen, doxorubicin (DOX) and benzophenone-3 drugs, as well as amphiphilic caffeine and busulfan drugs [130]. Further, MIL-101 with Fe has been used to encapsulate DOX via highly stable coordination

Applications for Earth-Abundant Transition Metals                      Materials Research Forum LLC
Materials Research Foundations 179 (2025) 37-78          https://doi.org/10.21741/9781644903711-3

bonds [131], Fe-MIL-101-amine for loading isoniazid [132], ibuprofen loaded in MOF-74 with oxidized $Fe^{3+}$ [133] and dihydroartemisinin encapsulated in MIL-101(Fe) coated with iron oxide-carbon composite [134]. The incorporation of drugs in MOF with Fe improved their clinical efficiency, half-life, stability and bioavailability with less cytotoxicity [135-137]. The daily average dietary iron intake in humans is $10 - 15$ mg, however only $1 - 2$ mg is absorbed in the intestine, failing which leads to iron deficiency [138]. Hence, iron has been chelated and encapsulated via liposomes, hydrolyzed glucomannan, microstructured emulsion particles, maltodextrin microparticles and double emulsion to improve their bioavailability as a potential drug delivery system [139]. Furthermore, Figueiredo et al. (2018) developed a novel iron-based layered double hydroxide (LDH) implants as a potential drug delivery carriers. In this study, the double hydroxides were prepared with iron-aluminum combination and chloride anions as charge-compensating ions via coprecipitation approach. The results revealed that LDH implants with iron has ability to promote tissue bio-integration and preserve blood microcirculation [140].

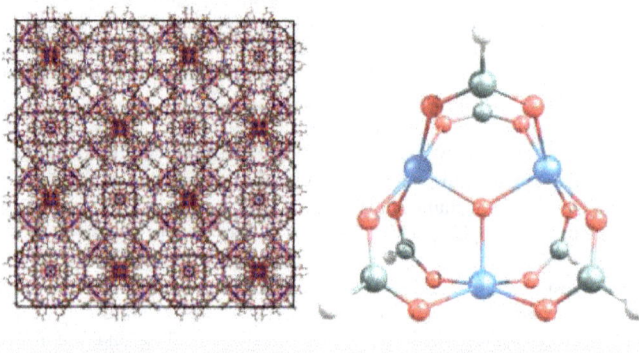

*Figure 1. Unit cell of Metal organic framework (MOF) (Left) and molecular model of MIL-100(Fe) [141]. Reproduced with permission from Keshavarz et al. (2021), ©MDPI, 2021 (Open access)*

In recent times, Jakubowski et al. (2022) prepared a novel drug carriers using X and Y-type zinc zeolites for the formulation of 6-Mercaptopurine, which is a chemotherapeutic drug. In this study, coordination interactions are utilized for the entrapment via nitrogen and sulfur atoms in the drug with zinc cations. The results revealed that 30% of the drug was released from the carrier in the initial 10 h, where the rest of the drug was released in 20 h with enhanced cell viability and biocompatibility [142]. Similarly, the same research group has utilized hydroxyapatite doped with zinc for the encapsulation of anticancer 6-mercaptopurine drug. In this study, biocompatible zinc-doped hydroxyapatite was used to formulate 6-mercaptopurine which is about 130 nm of particle size for potential intravenous delivery. The results showed that the drug was evenly distributed for sustained release in neutral environment and immediate release in acidic environment [143]. Likewise, Meshkini and Oveisi (2017) developed a mesoporous zinc hydroxyapatite, that is conjugated with methotrexate-F127, for enhancing chemotherapeutic drug delivery. In this study, zinc-substituted hydroxyapatite with mesoporous structure was bound with methotrexate-F127 via stable amide linkage to act as a potential drug carrier with high biocompatibility and enhanced drug loading capacity towards osteosarcoma cells [144]. Further, Ye et al. (2023) demonstrated the synthesis of zinc-based MOFs for effective drug delivery applications. In this study, doxorubicin drug was formulated in benzoic acid-mediated zinc-based MOFs via solvothermal approach with 33.74% of drug loading capacity. The results revealed that the drug delivery system is pH-sensitive with elevated antitumor activity and low toxicity towards noncancerous normal cells [145]. Furthermore, Taha et al. (2005) fabricated a novel zinc-crosslinked thiolated alginic acid beads as a significant enteric delivery system with folic acid as a model drug [146].

It has been reported in recent times that calcium silicate has been highly beneficial as a potential drug delivery agent due to their enhanced drug-loading capacity, biodegradability, biocompatibility, bioactivity, highly hollow/porous structure, surface specific area, drug release property and pH-responsive drug release behavior [147]. Furthermore, calcium phosphate cements are utilized as drug delivery materials in synthetic bone grafts due to their significant injectability, osteoconductivity, intrinsic porosity and low-temperature setting reaction. Ions, high molecular weight biomolecules and low molecular weight drugs are the classification of active principles or drugs, that can be formulated in calcium phosphate cements as an efficient drug carrier [148]. Moreover, calcium phosphate ceramic drug delivery systems for bone tissue engineering applications [149], calcium phytate as a drug carrier for prolonged risedronate release [150] and calcium silicate-based microspheres to formulated repaglinide as a gastroretentive floating drug delivery system [151] were also reported. Park et al. (2014) developed a novel titanium modified bone implant with numerous micro-holes and hollow structure to act as a drug delivery system. In this study, dexamethasone was formulated in the modified implant to be released in the target rabbit tibia. The results revealed that the titanium modified implant is highly beneficial for the sustained release of drugs with improved bioavailability [152]. Further, Khandan et al. (2012) demonstrated that fenestrated, in-plane microneedles, that are based on titanium element can be used for passive ocular drug delivery applications. In this study, diffusive transport via fast-dissolving coating of drugs in Ti-based microneedles was identified to effectively delivery the drug load in rabbit cornea [153].

Likewise, Oi et al. (2015) prepared magnesium whitlockite and amorphous calcium magnesium phosphate with porous structure with the help of creatine phosphate via rapid microwave-assisted synthesis for effective drug delivery. In this study, creatine phosphate biomolecules has been formulated in the Mg-based microspheres with 753 mg/g of enhanced drug loading capacity and 94% of high drug encapsulation efficiency [154]. Similarly, Song et al. (2023) developed a novel drug co-delivery platform via Mg-based micromotors to enhance hepatoma carcinoma cell therapy. The micromotor was fabricated with poly (lactic-co-glycolic acid) (PLGA) and chitosan to deliver hydrophobic doxorubicin and hydrophilic curcumin drug. The results showed that the micromotor with Mg possesses 2.9 and 1.5 times higher absorption of hydrophobic and hydrophilic drug towards carcinoma cells, respectively [155]. Recently, Zhang et al. (2019) prepared a novel manganese arsenite delivery system for the formulation of pH-low insertion peptide. The study identified that the drug delivery system possess enhanced controlled release capacity and excellent targeting ability towards cancer cells [156]. Zhong et al. (2022) also demonstrated the synthesis of mitochondrial targeted drug delivery system that are combined with manganese catalyzed Fenton reaction. The system was used as a carrier for doxorubicin modified with triphenylphosphine to release in target mitochondrial site, elevate intracellular reactive oxygen species generation and catalyze highly toxic hydroxyl oxidative radical production [157]. Moreover, silicon [158],

Materials Research Forum LLC
https://doi.org/10.21741/9781644903711-3

aluminum [159], sodium [160] and potassium [161] are the other earth-abundant metals that are utilized to develop drug delivery system.

## 3.2    Antimicrobial agent

Sheta et al. (2022) prepared a novel iron (III)-MOFs via reflux approach to exhibit antimicrobial property. The study revealed that MOFs with iron possess enhanced antibacterial activity against gram negative *Escherichia coli*, gram positive *Staphylococcus aureus*, *Candida* fungal species and *Aspergillus niger* yeast species as shown in Figure 2 [162]. Further, Patamia et al. (2023) utilized chelated iron along with bio-based material to enhance their antimicrobial effect. In this study, halloysite nanotubes with enhanced iron-chelating capability towards kojic acid are utilized as a carrier for curcumin and resveratrol. The results showed that the iron chelated nanotubes possess antibacterial activity against *Enterococcus faecalis* ATCC 29212, *E. coli* ATCC 25922, *S. aureus* ATCC 29213 and *Klebsiella pneumoniae* ATCC 700603 [163]. Furthermore, zinc and copper have been proven to possess the ability in developing resistance and co-resistance towards antimicrobial against bacteria in animal feed. It has been reported that the zinc resistance is linked with the resistance to methicillin in *Staphylococci* species, whereas supplementation of zinc via animal field will lead to an increment in the multi-resistant proportion of *E. coli* in the gut [164]. Moreover, Pasquet et al. (2014) investigated the dissolution of zinc ions from zinc oxide suspensions to exhibit antimicrobial activity. The study emphasized that the zinc ions possess enhanced antibacterial efficacy against *Pseudomonas aeruginosa*, *E. coli* and *S. aureus*, as well as antifungal activity against *Aspergillus brasilensis* and *Candida albicans* [165].

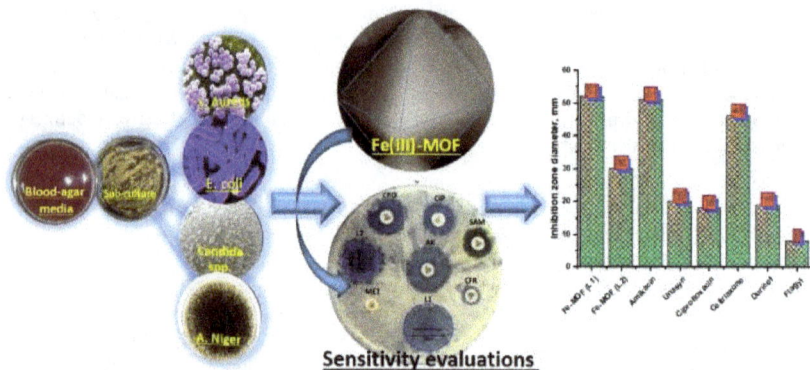

*Figure 2. Antimicrobial activity of iron-included MOFs [162]. Reproduced with permission from Sheta et al. (2022), ©Springer Nature, 2022.*

Recently, Kamphof et al. (2023) summarized the antimicrobial activity of calcium phosphates, that are substituted with ions, such as copper, silver, selenite, samarium, rubidium, zinc and gadolinium [166]. Likewise, Mokabber et al. (2020) developed a calcium phosphate containing silver via electrodeposited coatings for potential antimicrobial properties. The study revealed that calcium phosphate possesses enhanced antibacterial efficacy against gram-positive *S. aureus* with bacterial reduction of 83.7 ± 4.5% via contact inhibition and 76.1 ± 8.3% via silver-ion leaching process

[167]. Similarly, various studies revealed that implants based on titanium produced via additive manufacturing approach possess enhanced metal facilitated antimicrobial agents. It has been identified that laser powder bed fusion-based additive manufacturing technology can be used to form titanium-aluminum-vanadium-copper alloy with elevated antimicrobial property facilitated by their intricate shapes [168]. Later, Kotsakis et al. (2016) reported that chemotherapeutic and antimicrobial agents, that are used in peri-implantitis treatment, can alter the cytocompatibility and physicochemistry of titanium in the surfaces of implants or biofilms. The study emphasized that the Ti-surfaces with the residues of chemotherapeutic agent potentially affect the osteoblastic response with antimicrobial activity against human oral bacterial microbiome from dental plaque [169].

Magnesium has been identified in various studies to be a potential candidate in the formation of biomaterials for biodegradable implant synthesis. These Mg-based implants were highly beneficial in exhibiting antibacterial activity against a broad range of both gram-negative and gram-positive bacterial strains [170]. Further, Lin et al. (2021) prepared Mg-based alloys with metal particles, such as zinc, silver and copper has significantly improved their corrosion resistance and antibacterial activities [171]. Furthermore, Sultana et al. (2003) synthesized cephradine metal complexes with chromium, magnesium, calcium and manganese for potential antibacterial efficacy against 10 gram-positive and gram-negative bacterial strains [172]. Moreover, Revanasiddappa et al. (2012) developed a novel mononuclear manganese (II) complexes with biologically significant drugs, such as Diphenylpyraline hydrochloride, Clomiphene citrate, Embramine hydrochloride, Imipramine hydrochloride and Dothiepin hydrochloride. The results showed that the Mn-based complexes possess enhanced antimicrobial activity against two gram-positive, gram-negative and fungal strains [173]. Also, Khonina et al. (2020) prepared a novel zinc, silicon and boron metal-based glycerohydrogels of biogenic elements via sol-gel approach. The study claimed that silicon-boron-zinc-glycerol hydrogel can be a significant alternative for topical conventional antimicrobial agents for skin and mucous membrane disease treatment [174]. Additionally, Lehmann et al. (2015) deposited silicon oxide films via plasma approach for the controlled permeation of copper as potential antimicrobial agent. The study demonstrated that the silica films helps in the sustained release of copper to exhibit antimicrobial agent against pathogenic dental microbes [175]. Aluminum [176], sodium [177] and potassium [178] were also reported in various studies to possess significant antimicrobial activities.

### 3.3    Anticancer agent

In recent times, compounds, complexes and chelators of iron is proposed to be widely beneficial as a significant anticancer agent [179-181]. There are several iron-based chemotherapeutic agents and iron facilitator LS081 with essential anticancer abilities [182, 183]. Su et al. (2020) summarized that ferroptosis with iron response elements can be a novel pharmacological entity to inhibit cancer cells for effective cancer treatment [184]. Also, Ye et al. (2019) developed a unique iron (II) phenanthroline complex to exhibit anticancer ability against transferrin R (TFR1) protein-overexpressing esophageal squamous cell carcinoma cells via accumulation of reactive oxygen species and damage of deoxyribonucleic acid (DNA) [185]. Recently, Porchia et al. (2020) revealed that the zinc complexes with nitrogen donor ligands possess enhanced anticancer abilities. The study emphasized that the zinc complexes included in quinoline, diamine, 2, 2'-bipyridine, 1, 10-phenanthroline, terpyridine, pyridine, imidazoles, analogous imidazole and Schiff base systems possess crucial anticancer capabilities [186]. Skrajnowska and Bobrowska-Korczak (2019) demonstrated the role of zinc in immune system and anticancer defense mechanisms. The study summarized that zinc exhibits anticancer activity due to their enhanced antioxidant properties, which influences immune system as shown in **Figure 3**, synthesis and repair of DNA and RNA, transcription factors, cell signaling regulation, inhibition/activation of enzymes, as well as cell membrane and structural stabilization [187].

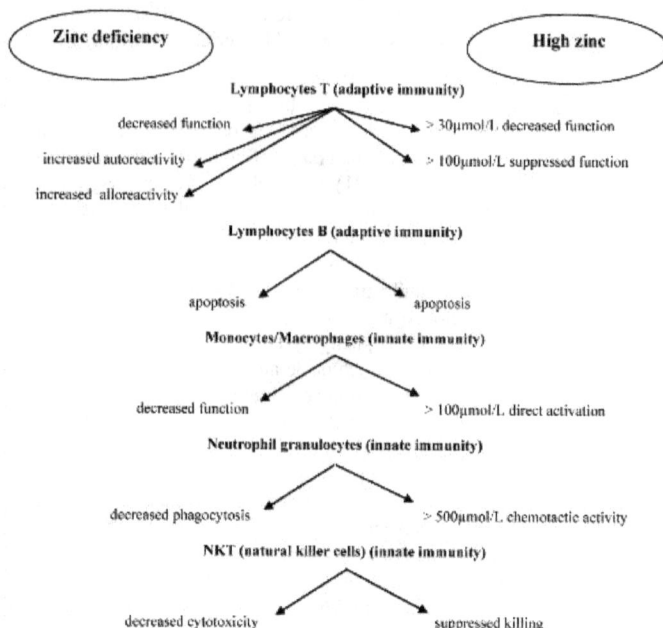

*Figure 3. Anticancer and immune response mechanism of elemental zinc [187]. Reproduced with permission from Skrajnowska and Bobrowska-Korczak (2019), ©MDPI, 2019 (Open access).*

Varghese et al. (2019) revealed the connection between the calcium and anticancer agents in the death of cancer cells and proliferation of noncancerous normal cells. The study explained that the regulation of calcium ion levels through the axis of endoplasmic reticulum-mitochondria and their respective action towards pumps and channels with the plasma membrane to exhibit cancer cell death and survival [188]. Further, Tepedelen et al. (2017) evaluated the anticarcinogenic effects of calcium fructoborate against metastatic MDA-MB-231 breast cancer cell line. The study showed that the complex calcium possesses excellent anticarcinogenic effects with increased p-p53 levels up to 2.4-fold, autophosphorylated ataxia telangiectasisa mutated (pATM) gene levels of 12.5-fold and caspase-9 of 10.7-fold with 2.5-fold of decreased poly (adenosine diphosphate-ribose) polymerase (PARP) levels [189]. Numerous studies also emphasized that titanium complexes possess significant anticancer agents [190, 191]. Ellahioui et al. (2017) summarized the effective anticancer efficiency of titanium, gallium and tin-based metallodrugs. The authors experimentally reported that titanocene (IV) complexes, organotin (IV) derivatives and organogallium (III) compounds possess enhanced cytotoxic activity against cancer cells [192]. Furthermore, metallodrug with titanium tackles the endoplasmic reticulum to be a potential anticancer agent, which was evaluated via genome study for the first time. The results showed that the Ti-based metallodrug induces cell cycle arrest and apoptosis at the G2/M phase in breast cancer cells, which was identified via ribonucleic acid (RNA) sequencing-based technology [193]. Moreover, Li et al. (2021) evaluated the anticancer effect of magnesium with biodegradability towards hepatobiliary carcinoma cells via *in vivo* and *in vitro* study. The study confirmed that magnesium can inhibit cancer cells by elevating apoptosis, carbonic anhydrase 9 expression decrement, and reduce the expression of upstream protein hypoxia-inducible factor 1-alpha [194]. Additionally, Mirmalek et al. (2016) compared the apoptogenic activity and in vitro cytotoxicity of cisplatin and magnesium chloride in MCF-7 breast cancer cell lines. The study revealed that the Mg chloride exhibited dose-dependent cytotoxic effect with 44% of apoptosis at 24 h. Hence, the study proposed that Mg supplement can be a potential chemotherapeutic drug to treat breast cancer, compared to conventional cisplatin-based treatment [195]. Also, manganese Schiff base complexes [196], beta-diiminato manganese III complex [197], manganese-based metal organic framework [198], manganese complex with arginine dithiocarbamate ligand [199] and apotransferrin [200] were identified to have enhanced anticancer effect. Besides, sodium [201], potassium [202], silicon [203] and aluminum [204] were also recognized to contain abilities to inhibit cancer cells.

### 3.4 Antidiabetic agent

Recently, several earth-abundant metals are widely employed and explored as potential antidiabetic agents. Brabha and Anitha (2020) reported that chelating tris n-methylethylenediamine iron complex possesses significant antidiabetic property. The study emphasized that the iron complex has alpha-amylase and glucosidase inhibition assay, compared to standard antidiabetic acarbose drug [205]. Further, Kongot et al. (2019) experimentally proved that iron (III) and oxidovanadium (IV) complexes with $O_2N_2$ donor linkage have significant antidiabetic effects. The study showed that iron complex can increase glucose uptake by inducing N-(7-nitrobenz-2-oxa-1,3-diazol-4-yl) amino-D-glucose (NBDG) uptake up to 95.4%, which is higher compared to conventional antidiabetic metformin drug [206]. Furthermore, Chukwuma et al. (2020) summarized that zinc (II) complexes possess potential antidiabetic activity. The review clearly mentioned that zinc complex combined with organic ligands can exhibit up to 72% of antidiabetic activity [207]. Moreover, Fujimoto et al. (2013) demonstrated the antidiabetic property of zinc complex, that are combined with organoselenium ligand. In this study, selenoferous zinc complex of di(2-selenopyridine-N-oxidato)zinc(II) complex has been identified to possess enhanced tissue penetration and gastrointestinal absorption with enhanced antidiabetic activity in mouse model at lowest dosage [208]. Likewise, Sadeghian et al. (2019) proved the oral supplementation of magnesium have ability to improve lipid profile by marginally decreasing microalbuminaria in diabetic nephropathy patients without affecting blood urine nitrogen, inflammation and serum creatinine levels, however increased resistance towards insulin [209]. Similarly, Razzaghi et al. (2017) reported that supplementation of magnesium can eventually improve healing of wounds by improving glucose metabolism in diabetic foot ulcer patients [210]. Recently, El-Megharbel et al. (2022) included divalent transition metals, such as cobalt and manganese along with antidiabetic sitagliptin drug and revealed that the formulation possess enhanced antidiabetic, antimicrobial, hepatoprotective and antioxidant effect [211]. However, it

Materials Research Forum LLC
https://doi.org/10.21741/9781644903711-3

can be noted that other trace elements, such as calcium, titanium, silicon, aluminum, sodium and potassium are utilized as antidiabetic agents, only when they are in the form of nanoparticles [212, 213].

## 3.5    Other applications

Earth-abundant metals, such as iron [214], zinc complexes [215], calcium [216], magnesium [217], manganese [218] and silicon [219] are extensively utilized as bioimaging applications for the diagnosis of diseases, such as cancer, diabetes, neurodegenerative and rare ailments. Recently, Carrasco (2018) summarized and provided a chart that shows numerous metal organic frameworks are utilized for biosensor development [220] as shown in **Figure 4**. Further, Jain et al. (2020) reported that distinct plasmonic metals coated over the optical fiber can be used to fabricate potential biosensors. The study showed that metals, such as silver, gold, copper and aluminum, possess enhanced surface plasmon resonance (SPR) to detect biomolecules as a significant biosensor [221]. Furthermore, Wang et al. (2017) summarized that electrochemical biosensors can be fabricated using transition-metal dichalcogenides. These biosensors are widely used in the detection of organic and inorganic analytes, such as DNA, proteins, heavy metals and glucose [222]. Moreover, Gelen and Tiekink (2005) published a book on metallotherapeutic drugs that can be used as a potential standalone medicine candidate for the treatment of diseases [223]. Recently, various publications are available which listed several earth-abundant metals to be utilized as medicines for disease treatments [224-226].

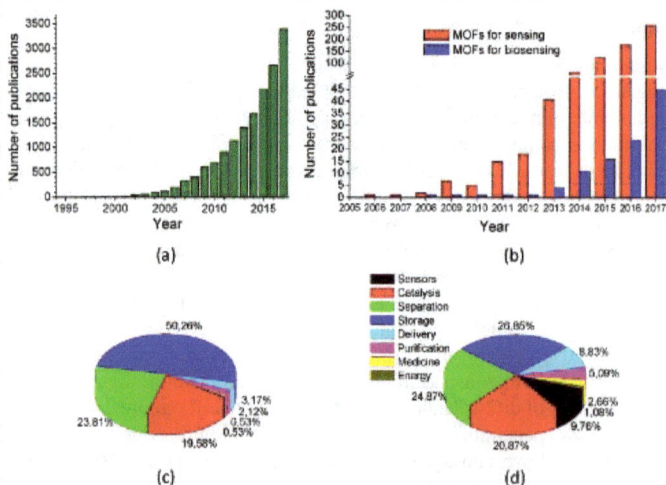

*Figure 4. Bibliographic analysis of MOFs for biosensing applications [220]. Reproduced with permission from Carrasco (2018), ©MDPI, 2018 (Open access).*

## 4. Pharmaceutical applications of earth-abundant metal nanoparticles

In the last decade, the emergence of nanotechnology has led to several nanoparticles, based on earth-abundant metals, that are widely utilized in pharmaceutical applications. **Table 1** is the summary of various earth-abundant metal nanoparticles, that are either used or extensively under research to be utilized in specific pharmaceutical applications.

*Table 1. Pharmaceutical applications of earth-abundant metal nanoparticles*

| Earth-abundant metal nanoparticles | Synthesis method/approach | Size/morphology | Pharmaceutical application | Refs. |
|---|---|---|---|---|
| **Drug delivery systems** | | | | |
| Iron oxide – polymer | Co-precipitation method | 10 - 49 nm | Quercetin carrier for cancer treatment | [227] |
| Iron oxide – chitosan | Co-precipitation method | ~8 – 15 nm | Phytic acid delivery in colon cancer cells | [228] |
| Iron oxide – poly vinyl alcohol | Co-precipitation method | 30 nm | Sorafenib delivery in liver cancer cells | [229] |
| Iron oxide | Commercial source | ~12 nm | Targeted peptide delivery in prostate cancer cells | [230] |
| Iron oxide | Precipitation method | Spherical shape, 8-10 nm in size | Amphotericin B drug delivery in visceral leishmaniasis | [231] |
| Zinc | Metal organic framework | Globular, 100 nm in size | Arsenic trioxide (drug) delivery | [232] |
| Zinc oxide | Precipitation method | Hexagon shape, 21-39 nm in size | Quercetin delivery in breast cancer cells | [233] |
| Zinc oxide | Ethanolic extract of *Camellia sinensis* L. | Spherical, below 100 nm | Paclitaxel delivery in breast cancer cells | [234] |
| Zinc oxide | Precipitation method | Hexagon, ~97 nm | Kappa-Carrageenan carrier to target infectious bacteria | [235] |
| Calcium | Cross linkage of polypeptide via mineralization | Spherical, ~80 – 103 nm | Doxorubicin delivery in osteosarcoma chemotherapy | [236] |
| Calcium carbonate | Layer-by-layer approach | Monodispersed, irregular shapes | Targeted oral curcumin delivery in colon cells | [237] |
| Calcium carbonate | Gas diffusion reaction | ~100 nm | Targeted Kaempferol-3-O-rutinoside delivery in cancer cells | [238] |
| Titanium (IV) folate | Precipitate method | Polydispersed agglomerated particles | Human serum albumin and glutathione delivery in breast cancer cells | [239] |

Materials Research Forum LLC
https://doi.org/10.21741/9781644903711-3

| Titanium dioxide | Hydrolysis, sonication, peptization method | Spherical and rod shape, 27 – 30 nm | Targeted norfloxacin delivery towards infectious bacteria strains and tumour cells | [240] |
|---|---|---|---|---|
| Titanium dioxide | Spontaneous amide coupling and sequential mixing method | Agglomerated, ~ 200 nm size | Formulation of doxorubicin to target cancer cells as sonodynamic chemotherapeutic agent | [241] |
| Magnesium ions stabilized with DNA | Precipitation method | ~90 nm in size | Targeted doxorubicin delivery | [242] |
| Magnesium orotate | Freeze-drying technique | 24 – 458 nm in size | Targeted delivery towards breast and liver cancer cells | [243] |
| Silica doped with magnesium | Sol-gel synthesis | Elliptical shape, ~90 nm | Targeted delivery of antibacterial moxifloxacin drug | [244] |
| Manganese oxide | Solution mixture approach | Distinct shapes of size ~23 nm | Targeted doxorubicin delivery towards murine colon cancer cells | [245] |
| Manganese oxide | Physical interaction approach | Nanosheets, 20 – 60 nm | *In vivo* delivery of cisplatin in mouse models | [246] |
| Alumina | Functionalization approach | Pore size of 2 – 7 nm | Ibuprofen release for anti-inflammatory drug | [247] |
| **Antimicrobial agents** | | | | |
| Iron oxide | *Zea mays* L. (Corn ear) leaves-based biosynthesis | Spherical shape | Antibacterial activity against food borne pathogens (*B. cereus, S. aureus* and *S. typhimurium*) | [248] |
| Iron oxide – gallic acid | Precipitation method | 5 – 11 nm in size, spherical shape | Antimicrobial activity against *E. coli, B. subtilis* and *S. aureus* (bacteria), *C. albican, A. niger, S. cerevisiae* (Fungi) | [249] |
| Iron oxide | Biosynthesis using *Platanus orientalis* leaf extract | Spherical shape, ~38 nm in size | Antifungal activity against *Mucor piriformis* and *A. niger* | [250] |
| Zinc oxide | Biosynthesis using leaves and fruit extract of *Caltropis procera* | 88 – 100 nm in size | Antibacterial activity against *Vibrio cholerae* and enterotoxic *E. coli* | [251] |

Materials Research Forum LLC
https://doi.org/10.21741/9781644903711-3

| Zinc oxide | Commercial source | ~30 nm | Antibacterial activity against *Campylobacter jejuni* | [252] |
|---|---|---|---|---|
| Zinc oxide | Commercial source | ~70 nm | Antifungal activity against *Botrytis cinerea* and *Penicillium expansum* | [253] |
| Calcium oxide | *Mentha pipertia* extract | Agglomerated disc shape, below 100 nm | Antibacterial activity against *E. coli* | [254] |
| Calcium-cadmium ferrite | Thermal treatment method | Spherical, ~37 nm | Antibacterial property against gram-positive and negative bacteria | [255] |
| Calcium oxide | *Trigona* species honey | Spherical, below 100 nm in size | Antifungal activity against *Colletotrichum brevisporum* | [256] |
| Titanium dioxide | Laser ablation method | ~34 nm | Antibacterial activity against *E. coli* | [257] |
| Titanium dioxide | *Azadirachta indica* leaf extract | Spherical, 25 – 87 nm in size | Antibacterial activity against *E. coli*, *B. subtilis*, *S. typhi*, and *K. pneumoniae* | [258] |
| Titanium dioxide | Sol-gel and plant extract synthesis approach | 10 - 13 nm (sol-gel) and 6 – 8 nm (plant) | Antifungal activity against *Ustilago tritici* | [259] |
| Magnesium oxide | Commercial source | 20 nm | Antibacterial activity against *E. coli* and *Salmonella* Stanley | [260] |
| Magnesium oxide | *Amaranthus tricolor*, *A. blitum* and *Andrographis paniculata* leaf extract | Spherical, ~ 78 nm | Antibacterial activity against *E. coli* | [261] |
| Magnesium oxide | Commercial source | - | Antifungal activity against *C. albicans* | [262] |
| Manganese oxide | *Ziziphora clinopodioides* Lam leaf extract | Spherical, ~47 – 71 nm | Antibacterial and antifungal property | [263] |
| Silica | One-pot direct template approach | Pore size of 10.7 nm | Antifungal activity against *Alternaria solani* | [264] |
| Silica | *Punica granatum* leaf extract | Spherical, ~12 nm | Antibacterial property against *E. coli* and *Salmonella* species | [265] |
| **Anticancer agents** | | | | |
| Iron oxide | *Punica granatum* fruit peel extract | Below 11 nm in size | Activity against colon, lung, breast and cervical cancer cells | [266] |

| Zinc | Fenugreek powder and gamma irradiation | 46 nm | Activity against Ehrlich Ascites Carcinoma and human colon adenocarcinoma | [267] |
|---|---|---|---|---|
| Calcium carbonate | Chemical approach functionalized with hyaluronan | 100 nm | Colorectal chemotherapy | [268] |
| Titanium dioxide | Laser ablation method | Spherical, ~19 – 26 nm | Activity against prostate cancer cells | [269] |
| Magnesium oxide | Aqueous *Sargassum wightii* extract | Flower shape, ~68.06 nm | Activity against human lung cancer cells | [270] |
| Manganese oxide | Precipitation method | 35 – 40 nm | Activity against breast and lung cancer cells | [271] |
| Silica | Functionalization method | Spherical mesoporous morphology | Activity against human colon carcinoma | [272] |
| Alumina | Pulsed laser ablation | Round edges, 20 – 60 nm in size | Chronic myelogenous leukaemia | [273] |
| **Antidiabetic agents** | | | | |
| Iron oxide | *Securidaca longipedunculata* stem extract | Crystallite size of 4.07 nm | Reduce blood glucose level in diabetic rats | [274] |
| Zinc oxide | Commercial source | Below 10 nm in size | Improve glucose tolerance in type 1 and 2 diabetic rats | [275] |
| Calcium carbonate | *Ailanthus altissima* extracts | Spherical, globular, rod shape with 100 nm in size | Alpha-amylase, alpha-glucosidase and dipeptidyl peptidase IV enzyme inhibition | [276] |
| Titanium dioxide | *Coleus aromaticus* extract | Hexagon shape, 12 nm in size | Alpha-amylase and alpha-glucosidase enzyme inhibition | [277] |
| Magnesium hydroxide | *Monodora myristica* seed extract | Spherical shape, 5 – 100 nm in size | Alpha-amylase and alpha-glucosidase enzyme inhibition | [278] |
| Manganese | *Syzygium aromaticum* clove extract | Spherical shape, 100 – 500 nm in size | Alpha-amylase and alpha-glucosidase enzyme inhibition | [279] |
| Silica | Stober method with silane polymerization | Mesoporous spherical shape, 50 – 130 nm in size | Alpha-amylase and alpha-glucosidase enzyme inhibition | [280] |
| **Other applications** | | | | |
| Metal nanoparticles | - | - | Fluorescent enhancement and surface enhanced | [281] |

| | | | Raman scattering in biosensors | |
|---|---|---|---|---|
| Metal nanoparticles | - | - | Electrochemical glucose biosensor | [282] |
| Metal nanoparticles | - | - | Disposable biosensors | [283] |
| Metal nanoparticles | *Curcuma longa* tuber powder | - | Bioimaging | [284] |
| Metal nanoparticles | Dendrimer capping agent | - | Bioimaging | [285] |
| Metallic nanoparticles | Algae mediated synthesis | - | Translational medicine | [286] |
| Metallic nanoparticles | Encapsulation method | - | Cancer medicine | [287] |

## 5. Future perspective

It can be noted from section 3 and 4 that standalone earth-abundant metals, their nanosized particles and metals incorporated in nanoformulations are widely used in various pharmaceutical applications. However, toxicity is a major issue, while utilizing them in biological applications [288, 289]. Recently, most of the standalone earth-abundant metals are formulated in phytochemicals or biopolymers to reduce their toxicity [290]. Chemical synthesized metal nanoparticles were also formulated in less/nontoxic biomolecules to reduce their toxicity [291]. Several biosynthesis approaches have been introduced to form nontoxic metal nanoparticles for reducing their toxicity to be utilized for potential pharmaceutical applications [292]. Further, computational approaches [293], such as artificial intelligence [294], machine learning [295], computer simulations [296] and Internet of Things (IoT) [297] were also incorporated with earth-abundant metals along with nanotechnology for their effective pharmaceutical applications. Thus, numerous earth-abundant metals will be utilized for the above-mentioned biomedical as well as novel applications to be beneficial for pharmaceutical industries in the future.

## Conclusion

Earth-abundant metals are widely used in various pharmaceutical applications due to their copious existence in the earth. The introduction of nanotechnology to form metal nanoparticles has led to highly valuable pharmaceutical products, due to their high surface to volume ratio and surface reactivity. Even though standalone earth-abundant metals and their nanoparticles are used or extensively under research to be beneficial in pharmaceutical industries, the toxicity of these metals is a major concern. Each of these earth-abundant metals are trace elements in humans, animals and plants, and increased concentration in the biological environment will eventually lead to enhanced cytotoxicity. Hence, sustained, targeted and controlled delivery of metal-based pharmaceutical entities are highly essential in capturing the wide pharmaceutical market in the future.

## References

[1] C.D. Pérez-Aguilar, M. Cuéllar-Cruz, The formation of crystalline minerals and their role in the origin of life on Earth, Progress in Crystal Growth and Characterization of Materials 68 (2022) 100558. https://doi.org/10.1016/j.pcrysgrow.2022.100558

[2] L.N. Warr, Earth's clay mineral inventory and its climate interaction: A quantitative assessment, Earth-Science Reviews 234 (2022) 104198. https://doi.org/10.1016/j.earscirev.2022.104198

[3] Z.-q. GUO, J.-r. ZHOU, K.-f. ZHOU, J.-f. JIN, X.-j. WANG, Z. Kui, Soil-water characteristics of weathered crust elution-deposited rare earth ores, Transactions of Nonferrous Metals Society of China 31 (2021) 1452-1464. https://doi.org/10.1016/S1003-6326(21)65589-9

[4] S.C. Bhatla, M.A. Lal, Essential and Functional Mineral Elements, Plant Physiology, Development and Metabolism, Springer2023, pp. 25-49. https://doi.org/10.1007/978-981-99-5736-1_2

[5] J.E. Johnson, T.M. Present, J.S. Valentine, Iron: Life's primeval transition metal, Proceedings of the National Academy of Sciences 121 (2024) e2318692121. https://doi.org/10.1073/pnas.2318692121

[6] W. Jintang, X. Kuangdi, Nonferrous Metal, The ECPH Encyclopedia of Mining and Metallurgy, Springer2023, pp. 1-2. https://doi.org/10.1007/978-981-19-0740-1_1358-1

[7] J.L. Cann, A. De Luca, D.C. Dunand, D. Dye, D.B. Miracle, H.S. Oh, E.A. Olivetti, T.M. Pollock, W.J. Poole, R. Yang, Sustainability through alloy design: Challenges and opportunities, Progress in Materials Science 117 (2021) 100722. https://doi.org/10.1016/j.pmatsci.2020.100722

[8] J. Wang, X. Yue, Y. Yang, S. Sirisomboonchai, P. Wang, X. Ma, A. Abudula, G. Guan, Earth-abundant transition-metal-based bifunctional catalysts for overall electrochemical water splitting: A review, Journal of Alloys and Compounds 819 (2020) 153346. https://doi.org/10.1016/j.jallcom.2019.153346

[9] D. Wang, D. Astruc, The recent development of efficient Earth-abundant transition-metal nanocatalysts, Chemical Society Reviews 46 (2017) 816-854. https://doi.org/10.1039/C6CS00629A

[10] O.S. Wenger, Photoactive complexes with earth-abundant metals, Journal of the American Chemical Society 140 (2018) 13522-13533. https://doi.org/10.1021/jacs.8b08822

[11] M.A. Benvenuto, Metals and alloys: industrial applications, Walter de Gruyter GmbH & Co KG2016. https://doi.org/10.1515/9783110441857

[12] L. Ilies, S.P. Thomas, I.A. Tonks, Earth-Abundant Metals in Catalysis, Asian Journal of Organic Chemistry 9 (2020). https://doi.org/10.1002/ajoc.202000111

[13] P. Chirik, R. Morris, Getting down to earth: the renaissance of catalysis with abundant metals, ACS Publications, 2015, pp. 2495-2495. https://doi.org/10.1021/acs.accounts.5b00385

[14] H. Hashizume, Natural Mineral Materials, Springer2022. https://doi.org/10.1007/978-4-431-56924-4

[15] D. Malik, N. Narayanasamy, V. Pratyusha, J. Thakur, N. Sinha, Inorganic Nutrients: Macrominerals, Textbook of Nutritional Biochemistry, Springer2023, pp. 391-446. https://doi.org/10.1007/978-981-19-4150-4_11

[16] Y. Feng, S. Long, X. Tang, Y. Sun, R. Luque, X. Zeng, L. Lin, Earth-abundant 3d-transition-metal catalysts for lignocellulosic biomass conversion, Chemical Society Reviews 50 (2021) 6042-6093. https://doi.org/10.1039/D0CS01601B

[17] W. Xia, A. Mahmood, Z. Liang, R. Zou, S. Guo, Earth-abundant nanomaterials for oxygen reduction, Angewandte Chemie International Edition 55 (2016) 2650-2676. https://doi.org/10.1002/anie.201504830

[18] A. Le Donne, V. Trifiletti, S. Binetti, New earth-abundant thin film solar cells based on chalcogenides, Frontiers in chemistry 7 (2019) 297. https://doi.org/10.3389/fchem.2019.00297

[19] L. Grandell, M. Höök, Assessing rare metal availability challenges for solar energy technologies, Sustainability 7 (2015) 11818-11837. https://doi.org/10.3390/su70911818

[20] S. Shenoy, M.M. Farahat, C. Chuaicham, K. Sekar, B. Ramasamy, K. Sasaki, Mixed-Phase Fe2O3 Derived from Natural Hematite Ores/C3N4 Z-Scheme Photocatalyst for Ofloxacin Removal, Catalysts 13 (2023) 792. https://doi.org/10.3390/catal13050792

[21] A. Li, Y. Sun, T. Yao, H. Han, Earth-abundant transition-metal-based electrocatalysts for water electrolysis to produce renewable hydrogen, Chemistry-A European Journal 24 (2018) 18334-18355. https://doi.org/10.1002/chem.201803749

[22] A.G.R. Toledo, D. Bevilaqua, S. Panda, A. Akcil, Hydrometallurgical Processing of Sulfide Minerals from the Perspective of Semiconductor Electrochemistry: A Review, Minerals Engineering 204 (2023) 108409. https://doi.org/10.1016/j.mineng.2023.108409

[23] B. Bozic-Weber, E.C. Constable, C.E. Housecroft, Light harvesting with Earth abundant d-block metals: Development of sensitizers in dye-sensitized solar cells (DSCs), Coordination Chemistry Reviews 257 (2013) 3089-3106. https://doi.org/10.1016/j.ccr.2013.05.019

[24] T. Igogo, K. Awuah-Offei, A. Newman, T. Lowder, J. Engel-Cox, Integrating renewable energy into mining operations: Opportunities, challenges, and enabling approaches, Applied Energy 300 (2021) 117375. https://doi.org/10.1016/j.apenergy.2021.117375

[25] P. Yaashikaa, B. Priyanka, P.S. Kumar, S. Karishma, S. Jeevanantham, S. Indraganti, A review on recent advancements in recovery of valuable and toxic metals from e-waste using bioleaching approach, Chemosphere 287 (2022) 132230. https://doi.org/10.1016/j.chemosphere.2021.132230

[26] M.S. Sankhla, M. Kumari, M. Nandan, R. Kumar, P. Agrawal, Heavy metals contamination in water and their hazardous effect on human health-a review, Int. J. Curr. Microbiol. App. Sci (2016) 5 (2016) 759-766. https://doi.org/10.20546/ijcmas.2016.510.082

[27] M. Kaushik, A. Moores, New trends in sustainable nanocatalysis: Emerging use of earth abundant metals, Current Opinion in Green and Sustainable Chemistry 7 (2017) 39-45. https://doi.org/10.1016/j.cogsc.2017.07.002

[28] M. Awais, A. Aizaz, A. Nazneen, Q.u.A. Bhatti, M. Akhtar, A. Wadood, M. Atiq Ur Rehman, A review on the recent advancements on Therapeutic effects of ions in the physiological environments, Prosthesis 4 (2022) 263-316. https://doi.org/10.3390/prosthesis4020026

[29] A. Mahapatro, Metals for biomedical applications and devices, Journal of Biomaterials and Tissue Engineering 2 (2012) 259-268. https://doi.org/10.1166/jbt.2012.1059

[30] M.R. Islam, S. Akash, M.H. Jony, M.N. Alam, F.T. Nowrin, M.M. Rahman, A. Rauf, M. Thiruvengadam, Exploring the potential function of trace elements in human health: a therapeutic perspective, Molecular and Cellular Biochemistry 478 (2023) 2141-2171. https://doi.org/10.1007/s11010-022-04638-3

[31] S. Zaib, I. Khan, Recent advances in the sustainable synthesis of quinazolines using earth-abundant first row transition metals, Current Organic Chemistry 24 (2020) 1775-1792. https://doi.org/10.2174/1385272824999200726230848

[32] D. Xia, Z. Shen, G. Huang, W. Wang, J.C. Yu, P.K. Wong, Red phosphorus: an earth-abundant elemental photocatalyst for "green" bacterial inactivation under visible light, Environmental science & technology 49 (2015) 6264-6273. https://doi.org/10.1021/acs.est.5b00531

[33] M. Alavi, R.S. Varma, Phytosynthesis and modification of metal and metal oxide nanoparticles/nanocomposites for antibacterial and anticancer activities: Recent advances, Sustainable Chemistry and Pharmacy 21 (2021) 100412. https://doi.org/10.1016/j.scp.2021.100412

[34] Y. Liu, S. Zeng, W. Ji, H. Yao, L. Lin, H. Cui, H.A. Santos, G. Pan, Emerging theranostic nanomaterials in diabetes and its complications, Advanced Science 9 (2022) 2102466. https://doi.org/10.1002/advs.202102466

[35] S. Shuvaev, P. Caravan, Metal Ions in Bio-Imaging Techniques: A Short Overview, Metal Ions in Bio-Imaging Techniques (2021) 1. https://doi.org/10.1515/9783110685701-007

[36] G. Maduraiveeran, M. Sasidharan, W. Jin, Earth-abundant transition metal and metal oxide nanomaterials: Synthesis and electrochemical applications, Progress in Materials Science 106 (2019) 100574. https://doi.org/10.1016/j.pmatsci.2019.100574

[37] X. Du, J. Zhao, J. Mi, Y. Ding, P. Zhou, B. Ma, J. Zhao, J. Song, Efficient photocatalytic H2 evolution catalyzed by an unprecedented robust molecular semiconductor {Fe11} nanocluster without cocatalysts at neutral conditions, Nano Energy 16 (2015) 247-255. https://doi.org/10.1016/j.nanoen.2015.06.025

[38] E.C. Ferré, I. Kupenko, F. Martín-Hernández, D. Ravat, C. Sanchez-Valle, Magnetic sources in the Earth's mantle, Nature Reviews Earth & Environment 2 (2021) 59-69. https://doi.org/10.1038/s43017-020-00107-x

[39] K. Hirose, B. Wood, L. Vočadlo, Light elements in the Earth's core, Nature Reviews Earth & Environment 2 (2021) 645-658. https://doi.org/10.1038/s43017-021-00203-6

[40] A.N. Pilchin, Iron and its unique role in Earth evolution, UNAM2006.

[41] N. Lane, Oxygen: the molecule that made the world, Oxford University Press, USA2002.

[42] R.B. Georg, A.N. Halliday, E.A. Schauble, B.C. Reynolds, Silicon in the Earth's core, Nature 447 (2007) 1102-1106. https://doi.org/10.1038/nature05927

[43] F.-Z. Teng, Magnesium isotope geochemistry, Reviews in Mineralogy and Geochemistry 82 (2017) 219-287. https://doi.org/10.2138/rmg.2017.82.7

[44] M. Wessling-Resnick, Iron: Basic nutritional aspects, Molecular, genetic, and nutritional Aspects of Major and trace minerals, Elsevier2017, pp. 161-173. https://doi.org/10.1016/B978-0-12-802168-2.00014-2

[45] Z. Fan, S.J. Friedmann, Low-carbon production of iron and steel: Technology options, economic assessment, and policy, Joule 5 (2021) 829-862. https://doi.org/10.1016/j.joule.2021.02.018

[46] R.V. Petrescu, R. Aversa, S. Kozaitis, A. Apicella, F.I. Petrescu, Some basic reactions in nuclear fusion, American Journal of Engineering and Applied Sciences 10 (2017). https://doi.org/10.3844/ajeassp.2017.709.716

[47] A.G. Caporale, A. Violante, Chemical processes affecting the mobility of heavy metals and metalloids in soil environments, Current Pollution Reports 2 (2016) 15-27. https://doi.org/10.1007/s40726-015-0024-y

[48] P.A. Frey, G.H. Reed, The ubiquity of iron, ACS Publications, 2012. https://doi.org/10.1021/cb300323q

[49] K. Kaur, R. Gupta, S.A. Saraf, S.K. Saraf, Zinc: the metal of life, Comprehensive Reviews in Food Science and Food Safety 13 (2014) 358-376. https://doi.org/10.1111/1541-4337.12067

[50] D.I. Kosik-Bogacka, N. Łanocha-Arendarczyk, Zinc, Zn, Mammals and Birds as Bioindicators of Trace Element Contaminations in Terrestrial Environments: An Ecotoxicological Assessment of the Northern Hemisphere (2019) 363-411. https://doi.org/10.1007/978-3-030-00121-6_11

[51] C. Noulas, M. Tziouvalekas, T. Karyotis, Zinc in soils, water and food crops, Journal of Trace Elements in Medicine and Biology 49 (2018) 252-260. https://doi.org/10.1016/j.jtemb.2018.02.009

[52] B.J. Alloway, Sources of heavy metals and metalloids in soils, Heavy metals in soils: trace metals and metalloids in soils and their bioavailability (2013) 11-50. https://doi.org/10.1007/978-94-007-4470-7_2

[53] H. Wang, S. Wen, G. Han, L. Xu, Q. Feng, Activation mechanism of lead ions in the flotation of sphalerite depressed with zinc sulfate, Minerals Engineering 146 (2020) 106132. https://doi.org/10.1016/j.mineng.2019.106132

[54] D.F. Araújo, G.R. Boaventura, W. Machado, J. Viers, D. Weiss, S.R. Patchineelam, I. Ruiz, A.P.C. Rodrigues, M. Babinski, E. Dantas, Tracing of anthropogenic zinc sources in coastal environments using stable isotope composition, Chemical Geology 449 (2017) 226-235. https://doi.org/10.1016/j.chemgeo.2016.12.004

[55] S. Mostoni, P. Milana, B. Di Credico, M. D'Arienzo, R. Scotti, Zinc-based curing activators: new trends for reducing zinc content in rubber vulcanization process, Catalysts 9 (2019) 664. https://doi.org/10.3390/catal9080664

[56] M.T. Ansari, F. Sami, F.A. Khairudiin, M.Z. Atan, T.A. bin Tengku Mohamad, S. Majeed, S. Ali, Applications of Zinc nanoparticles in medical and healthcare fields, Current Nanomedicine (Formerly: Recent Patents on Nanomedicine) 8 (2018) 225-233. https://doi.org/10.2174/2405461503666180709100110

[57] M. Stefanidou, C. Maravelias, A. Dona, C. Spiliopoulou, Zinc: a multipurpose trace element, Archives of toxicology 80 (2006) 1-9. https://doi.org/10.1007/s00204-005-0009-5

[58] A.H. Gondal, A. Zafar, D. Zainab, M. Toor, S. Sohail, S. Ameen, A. Ijaz, B. Ch, I. Hussain, S. Haider, A detailed review study of zinc involvement in animal, plant and human nutrition, Indian Journal of Pure & Applied Biosciences 9 (2021) 262-271. https://doi.org/10.18782/2582-2845.8652

[59] R.C. Ropp, Encyclopedia of the alkaline earth compounds, Newnes2012. https://doi.org/10.1016/B978-0-444-59550-8.00001-6

[60] W. Wu, Y.-G. Xu, Z.-F. Zhang, X. Li, Calcium isotopic composition of the lunar crust, mantle, and bulk silicate Moon: A preliminary study, Geochimica et Cosmochimica Acta 270 (2020) 313-324. https://doi.org/10.1016/j.gca.2019.12.001

[61] P. Naumov, D.P. Karothu, E. Ahmed, L. Catalano, P. Commins, J. Mahmoud Halabi, M.B. Al-Handawi, L. Li, The rise of the dynamic crystals, Journal of the American Chemical Society 142 (2020) 13256-13272. https://doi.org/10.1021/jacs.0c05440

[62] S.V. Dorozhkin, Calcium orthophosphates, Journal of materials science 42 (2007) 1061-1095. https://doi.org/10.1007/s10853-006-1467-8

[63] Q. Guan, Y. Sui, F. Zhang, W. Yu, Y. Bo, P. Wang, W. Peng, J. Jin, Preparation of α-calcium sulfate hemihydrate from industrial by-product gypsum: A review, Physicochemical Problems of Mineral Processing 57 (2021) 168-181. https://doi.org/10.37190/ppmp/130795

[64] M. Al Omari, I. Rashid, N. Qinna, A. Jaber, A. Badwan, Calcium carbonate, Profiles of drug substances, excipients and related methodology 41 (2016) 31-132. https://doi.org/10.1016/bs.podrm.2015.11.003

[65] E. Rosson, A. Rincón Romero, D. Badocco, F. Zorzi, P. Sgarbossa, R. Bertani, P. Pastore, E. Bernardo, Production of porous ceramic materials from spent fluorescent lamps, Applied Sciences 11 (2021) 6056. https://doi.org/10.3390/app11136056

[66] S. You, S. Cao, C. Mo, Y. Zhang, J. Lu, Synthesis of high purity calcium fluoride from fluoride-containing wastewater, Chemical Engineering Journal 453 (2023) 139733. https://doi.org/10.1016/j.cej.2022.139733

[67] B. Müller, J.S. Meyer, R. Gächter, Alkalinity regulation in calcium carbonate-buffered lakes, Limnology and Oceanography 61 (2016) 341-352. https://doi.org/10.1002/lno.10213

[68] K. Li, X.-F. Wang, D.-Y. Li, Y.-C. Chen, L.-J. Zhao, X.-G. Liu, Y.-F. Guo, J. Shen, X. Lin, J. Deng, The good, the bad, and the ugly of calcium supplementation: a review of calcium

intake on human health, Clinical interventions in aging (2018) 2443-2452. https://doi.org/10.2147/CIA.S157523

[69] M.A. Saghiri, J. Orangi, A. Asatourian, J.L. Gutmann, F. Garcia-Godoy, M. Lotfi, N. Sheibani, Calcium silicate-based cements and functional impacts of various constituents, Dental materials journal 36 (2017) 8-18. https://doi.org/10.4012/dmj.2015-425

[70] A.A. Abd, Studying the mechanical and electrical properties of epoxy with PVC and calcium carbonate filler, International Journal of Engineering & Technology 3 (2014) 545-553. https://doi.org/10.14419/ijet.v3i4.3425

[71] Q. Zhao, J. Li, K. You, C. Liu, Recovery of calcium and magnesium bearing phases from iron-and steelmaking slag for CO2 sequestration, Process Safety and Environmental Protection 135 (2020) 81-90. https://doi.org/10.1016/j.psep.2019.12.012

[72] L. Shiqi, Liquation Refining, The ECPH Encyclopedia of Mining and Metallurgy, Springer2024, pp. 1044-1045. https://doi.org/10.1007/978-981-99-2086-0_1378

[73] M. Wang, C. Jiang, S. Zhang, X. Song, Y. Tang, H.-M. Cheng, Reversible calcium alloying enables a practical room-temperature rechargeable calcium-ion battery with a high discharge voltage, Nature chemistry 10 (2018) 667-672. https://doi.org/10.1038/s41557-018-0045-4

[74] M. Saleem, M. Ali, A. Saeed, Preparation of Soap and Detergents with Potential Use of Biochemical Methods, Recent Advances in Industrial Biochemistry, Springer2024, pp. 433-446. https://doi.org/10.1007/978-3-031-50989-6_17

[75] Y.M. Ahmed, K.S.M. Sahari, M. Ishak, B.A. Khidhir, Titanium and its alloy, International Journal of Science and Research 3 (2014) 1351-1361.

[76] X. He, J. Ma, G. Wei, Z. Wang, L. Zhang, T. Zeng, Z. Zhang, Mass-dependent fractionation of titanium stable isotopes during intensive weathering of basalts, Earth and Planetary Science Letters 579 (2022) 117347. https://doi.org/10.1016/j.epsl.2021.117347

[77] H.C.S. Subasinghe, A.S. Ratnayake, General review of titanium ores in exploitation: present status and forecast, Comunicações Geológicas 109 (2022) 21-31.

[78] L. Hoare, M. Klaver, N.S. Saji, J. Gillies, I.J. Parkinson, C.J. Lissenberg, M.-A. Millet, Melt chemistry and redox conditions control titanium isotope fractionation during magmatic differentiation, Geochimica et Cosmochimica Acta 282 (2020) 38-54. https://doi.org/10.1016/j.gca.2020.05.015

[79] F. Froes, Titanium: physical metallurgy, processing, and applications, ASM international2015. https://doi.org/10.31399/asm.tb.tpmpa.9781627083188

[80] C. Leyens, M. Peters, Titanium and titanium alloys: fundamentals and applications, Wiley Online Library2006.

[81] M. El Khalloufi, O. Drevelle, G. Soucy, Titanium: An overview of resources and production methods, Minerals 11 (2021) 1425. https://doi.org/10.3390/min11121425

[82] S.A. Salihu, Y. Suleiman, A. Eyinavi, A. Usman, Classification, Properties and Applications of titanium and its alloys used in aerospace, automotive, biomedical and marine industry-A Review, Int. J. Precious Eng. Res. Appl. 4 (2019) 23-36.

[83] A. Shokri, M.S. Fard, Corrosion in seawater desalination industry: A critical analysis of impacts and mitigation strategies, Chemosphere 307 (2022) 135640. https://doi.org/10.1016/j.chemosphere.2022.135640

[84] A.M. Khorasani, M. Goldberg, E.H. Doeven, G. Littlefair, Titanium in biomedical applications-properties and fabrication: a review, Journal of biomaterials and tissue engineering 5 (2015) 593-619. https://doi.org/10.1166/jbt.2015.1361

[85] P. Singh, H. Pungotra, N.S. Kalsi, On the characteristics of titanium alloys for the aircraft applications, Materials today: proceedings 4 (2017) 8971-8982. https://doi.org/10.1016/j.matpr.2017.07.249

[86] M. Lyapina, M. Cekova, M. Deliverska, J. Galabov, A. Kisselova, Immunotoxicological aspects of biocompatibility of titanium, Journal of IMAB-Annual Proceeding Scientific Papers 23 (2017) 1550-1559. https://doi.org/10.5272/jimab.2017232.1550

[87] K.T. Kim, M.Y. Eo, T.T.H. Nguyen, S.M. Kim, General review of titanium toxicity, International journal of implant dentistry 5 (2019) 1-12. https://doi.org/10.1186/s40729-019-0162-x

[88] D. Kumar, O.P. Dhankher, R.D. Tripathi, C.S. Seth, Titanium dioxide nanoparticles potentially regulate the mechanism (s) for photosynthetic attributes, genotoxicity, antioxidants defense machinery, and phytochelatins synthesis in relation to hexavalent chromium toxicity in Helianthus annuus L, Journal of Hazardous Materials 454 (2023) 131418. https://doi.org/10.1016/j.jhazmat.2023.131418

[89] A. Markowska-Szczupak, M. Endo-Kimura, O. Paszkiewicz, E. Kowalska, Are titania photocatalysts and titanium implants safe? Review on the toxicity of titanium compounds, Nanomaterials 10 (2020) 2065. https://doi.org/10.3390/nano10102065

[90] S.S. Prasad, S. Prasad, K. Verma, R.K. Mishra, V. Kumar, S. Singh, The role and significance of Magnesium in modern day research-A review, Journal of Magnesium and alloys 10 (2022) 1-61. https://doi.org/10.1016/j.jma.2021.05.012

[91] F. Vicari, S. Randazzo, J. López, M.F. de Labastida, V. Vallès, G. Micale, A. Tamburini, G.A. Staiti, J.L. Cortina, A. Cipollina, Mining minerals and critical raw materials from bittern: Understanding metal ions fate in saltwork ponds, Science of the Total Environment 847 (2022) 157544. https://doi.org/10.1016/j.scitotenv.2022.157544

[92] M. Gupta, S.N.M. Ling, Magnesium, magnesium alloys, and magnesium composites, John Wiley & Sons2011. https://doi.org/10.1002/9780470905098

[93] Y. González, A. Navarra, R.I. Jeldres, N. Toro, Hydrometallurgical processing of magnesium minerals-a review, Hydrometallurgy 201 (2021) 105573. https://doi.org/10.1016/j.hydromet.2021.105573

[94] M.K. Kulekci, Magnesium and its alloys applications in automotive industry, The International Journal of Advanced Manufacturing Technology 39 (2008) 851-865. https://doi.org/10.1007/s00170-007-1279-2

[95] N. Khademian, Y. Peimaei, Lightweight materials (LWM) in transportation especially application of aluminum in light weight automobiles (LWA), International conference on interdisciplinary studies in nanotechnology, 2020, pp. 1-22.

Materials Research Forum LLC
https://doi.org/10.21741/9781644903711-3

[96] G.K. Schwalfenberg, S.J. Genuis, Vitamin D, essential minerals, and toxic elements: exploring interactions between nutrients and toxicants in clinical medicine, The Scientific World Journal 2015 (2015) 318595. https://doi.org/10.1155/2015/318595

[97] J. Jeevanandam, M. K Danquah, S. Debnath, V. S Meka, Y. S Chan, Opportunities for nano-formulations in type 2 diabetes mellitus treatments, Current pharmaceutical biotechnology 16 (2015) 853-870. https://doi.org/10.2174/1389201016666150727120618

[98] A.C.R. Souza, A.R. Vasconcelos, D.D. Dias, G. Komoni, J.J. Name, The integral role of magnesium in muscle integrity and aging: A comprehensive review, Nutrients 15 (2023) 5127. https://doi.org/10.3390/nu15245127

[99] M.M. Najafpour, G. Renger, M. Hołynska, A.N. Moghaddam, E.-M. Aro, R. Carpentier, H. Nishihara, J.J. Eaton-Rye, J.-R. Shen, S.I. Allakhverdiev, Manganese compounds as water-oxidizing catalysts: from the natural water-oxidizing complex to nanosized manganese oxide structures, Chemical reviews 116 (2016) 2886-2936. https://doi.org/10.1021/acs.chemrev.5b00340

[100] W.D. Bancroft, R. Nugent, The Manganese Equilibrium in Glasses, The Journal of Physical Chemistry 33 (2002) 481-497. https://doi.org/10.1021/j150298a001

[101] Y. Wu, B. Shi, H. Liang, W. Ge, C.J. Yan, X. Yang, Magnetic properties of low grade manganese carbonate ore, Applied mechanics and materials 664 (2014) 38-42. https://doi.org/10.4028/www.scientific.net/AMM.664.38

[102] F. Teng, S.-h. Luo, X. Kang, Y.-g. Liu, H.-t. Shen, J. Ye, L.-j. Chang, Y.-c. Zhai, Y.-n. Dai, Preparation of manganese dioxide from low-grade pyrolusite and its electrochemical performance for supercapacitors, Ceramics International 45 (2019) 21457-21466. https://doi.org/10.1016/j.ceramint.2019.07.136

[103] R.M. Maier, Biogeochemical cycling, Environmental microbiology, Elsevier2015, pp. 339-373. https://doi.org/10.1016/B978-0-12-394626-3.00016-8

[104] B. Buľko, M. Molnár, P. Demeter, M. Červenka, G. Tréfa, M. Mochnacká, D. Baricová, S. Hubatka, L. Fogaraš, V. Šabík, Deep Steel Desulfurization Practice, Transactions of the Indian Institute of Metals 75 (2022) 2807-2816. https://doi.org/10.1007/s12666-022-02643-0

[105] F. Taube, Manganese in occupational arc welding fumes-aspects on physiochemical properties, with focus on solubility, Annals of occupational hygiene 57 (2013) 6-25.

[106] H. Danninger, C. Gierl-Mayer, M. Prokofyev, M.-C. Huemer, R. de Oro Calderon, R. Hellein, A. Müller, G. Stetina, Manganese-a promising element also in high alloy sintered steels, Powder Metallurgy 64 (2021) 115-125. https://doi.org/10.1080/00325899.2021.1886717

[107] E. Kalisińska, H. Budis, Manganese, Mn, Mammals and birds as bioindicators of trace element contaminations in terrestrial environments: An ecotoxicological assessment of the Northern hemisphere (2019) 213-246. https://doi.org/10.1007/978-3-030-00121-6_7

[108] P. Maseko, M. van Rooy, H. Taute, C. Venter, J.C. Serem, H.M. Oberholzer, Whole blood ultrastructural alterations by mercury, nickel and manganese alone and in combination: An ex vivo investigation, Toxicology and Industrial Health 37 (2021) 98-111. https://doi.org/10.1177/0748233720983114

[109] K.J. Horning, S.W. Caito, K.G. Tipps, A.B. Bowman, M. Aschner, Manganese is essential for neuronal health, Annual review of nutrition 35 (2015) 71-108. https://doi.org/10.1146/annurev-nutr-071714-034419

[110] E.S. Koh, S.J. Kim, H.E. Yoon, J.H. Chung, S. Chung, C.W. Park, Y.S. Chang, S.J. Shin, Association of blood manganese level with diabetes and renal dysfunction: a cross-sectional study of the Korean general population, BMC endocrine disorders 14 (2014) 1-8. https://doi.org/10.1186/1472-6823-14-24

[111] L. Li, X. Yang, The essential element manganese, oxidative stress, and metabolic diseases: links and interactions, Oxidative medicine and cellular longevity 2018 (2018) 7580707. https://doi.org/10.1155/2018/7580707

[112] Q. Wu, Q. Mu, Z. Xia, J. Min, F. Wang, Manganese homeostasis at the host-pathogen interface and in the host immune system, Seminars in cell & developmental biology, Elsevier, 2021, pp. 45-53. https://doi.org/10.1016/j.semcdb.2020.12.006

[113] J.M. Studer, W.P. Schweer, N.K. Gabler, J.W. Ross, Functions of manganese in reproduction, Animal Reproduction Science 238 (2022) 106924. https://doi.org/10.1016/j.anireprosci.2022.106924

[114] M. Chino, L. Leone, G. Zambrano, F. Pirro, D. D'Alonzo, V. Firpo, D. Aref, L. Lista, O. Maglio, F. Nastri, Oxidation catalysis by iron and manganese porphyrins within enzyme-like cages, Biopolymers 109 (2018) e23107. https://doi.org/10.1002/bip.23107

[115] D. Kumar, M. Johari, Characteristics of silicon crystal, its covalent bonding and their structure, electrical properties, uses, AIP Conference Proceedings, AIP Publishing, 2020. https://doi.org/10.1063/5.0003505

[116] J.L. Bañuelos, E. Borguet, G.E. Brown Jr, R.T. Cygan, J.J. DeYoreo, P.M. Dove, M.-P. Gaigeot, F.M. Geiger, J.M. Gibbs, V.H. Grassian, Oxide-and silicate-water interfaces and their roles in technology and the environment, Chemical Reviews 123 (2023) 6413-6544. https://doi.org/10.1021/acs.chemrev.2c00130

[117] S. Wang, X. Liu, P. Zhou, The road for 2D semiconductors in the silicon age, Advanced Materials 34 (2022) 2106886. https://doi.org/10.1002/adma.202106886

[118] L. O'Bannon, Dictionary of ceramic science and engineering, Springer Science & Business Media2012.

[119] V.B. Sardar, N. Rajhans, A. Pathak, T. Prabhu, Developments in silicone material for biomedical applications-A review, 14th international conference on humanizing work and work environment. Punjab, India, 2016.

[120] Y. Cho, J. Hwang, M.-S. Park, B.H. Kim, Fabrication methods for microscale 3D structures on silicon carbide, International Journal of Precision Engineering and Manufacturing 23 (2022) 1477-1502. https://doi.org/10.1007/s12541-022-00717-z

[121] R.A. Yokel, B. Sjögren, Aluminum, Handbook on the Toxicology of Metals, Elsevier2022, pp. 1-22. https://doi.org/10.1016/B978-0-12-822946-0.00001-5

Materials Research Forum LLC
https://doi.org/10.21741/9781644903711-3

[122] F. Debeaufort, K. Galic, M. Kurek, N. Benbettaieb, M. Scetar, Metal packaging, Packaging Materials and Processing for Food, Pharmaceuticals and Cosmetics 13 (2021) 75-104. https://doi.org/10.1002/9781119825081.ch4

[123] M.G. Norton, Aluminum-The Material of Flight, Ten Materials That Shaped Our World, Springer2021, pp. 161-175. https://doi.org/10.1007/978-3-030-75213-2_10

[124] R.K. Gupta, S. Agarwal, P. Mukhopadhyay, Plastics in Buildings and Construction, Applied Plastics Engineering Handbook, Elsevier2024, pp. 683-703. https://doi.org/10.1016/B978-0-323-88667-3.00010-2

[125] A.T. Tabereaux, R.D. Peterson, Aluminum production, Treatise on process metallurgy, Elsevier2024, pp. 625-676. https://doi.org/10.1016/B978-0-323-85373-6.00004-1

[126] G. Chen, X. Wang, J. Wang, J. Liu, T. Zhang, W. Tang, Damage investigation of the aged aluminium cable steel reinforced (ACSR) conductors in a high-voltage transmission line, Engineering Failure Analysis 19 (2012) 13-21. https://doi.org/10.1016/j.engfailanal.2011.09.002

[127] M. Graham, Solar Energy: Let the Sun Shine In, AuthorHouse2023.

[128] J. Park, J. Lee, M.H. Alfaruqi, W.-J. Kwak, J. Kim, J.-Y. Hwang, Initial investigation and evaluation of potassium metal as an anode for rechargeable potassium batteries, Journal of Materials Chemistry A 8 (2020) 16718-16737. https://doi.org/10.1039/D0TA03562A

[129] X. Liu, T. Liang, R. Zhang, Q. Ding, S. Wu, C. Li, Y. Lin, Y. Ye, Z. Zhong, M. Zhou, Iron-based metal-organic frameworks in drug delivery and biomedicine, ACS Applied Materials & Interfaces 13 (2021) 9643-9655. https://doi.org/10.1021/acsami.0c21486

[130] P. Horcajada, T. Chalati, C. Serre, B. Gillet, C. Sebrie, T. Baati, J.F. Eubank, D. Heurtaux, P. Clayette, C. Kreuz, Porous metal-organic-framework nanoscale carriers as a potential platform for drug delivery and imaging, Nature materials 9 (2010) 172-178. https://doi.org/10.1038/nmat2608

[131] R. Anand, F. Borghi, F. Manoli, I. Manet, V. Agostoni, P. Reschiglian, R. Gref, S. Monti, Host-guest interactions in Fe (III)-trimesate MOF nanoparticles loaded with doxorubicin, The Journal of Physical Chemistry B 118 (2014) 8532-8539. https://doi.org/10.1021/jp503809w

[132] G. Wyszogrodzka, P. Dorożyński, B. Gil, W.J. Roth, M. Strzempek, B. Marszałek, W.P. Węglarz, E. Menaszek, W. Strzempek, P. Kulinowski, Iron-based metal-organic frameworks as a theranostic carrier for local tuberculosis therapy, Pharmaceutical research 35 (2018) 1-11. https://doi.org/10.1007/s11095-018-2425-2

[133] Q. Hu, J. Yu, M. Liu, A. Liu, Z. Dou, Y. Yang, A low cytotoxic cationic metal-organic framework carrier for controllable drug release, Journal of medicinal chemistry 57 (2014) 5679-5685. https://doi.org/10.1021/jm5004107

[134] D. Wang, J. Zhou, R. Chen, R. Shi, G. Xia, S. Zhou, Z. Liu, N. Zhang, H. Wang, Z. Guo, Magnetically guided delivery of DHA and Fe ions for enhanced cancer therapy based on pH-responsive degradation of DHA-loaded Fe3O4@ C@ MIL-100 (Fe) nanoparticles, Biomaterials 107 (2016) 88-101. https://doi.org/10.1016/j.biomaterials.2016.08.039

Materials Research Forum LLC
https://doi.org/10.21741/9781644903711-3

[135] V. Agostoni, T. Chalati, P. Horcajada, H. Willaime, R. Anand, N. Semiramoth, T. Baati, S. Hall, G. Maurin, H. Chacun, Towards an Improved anti-HIV Activity of NRTI via Metal-Organic frameworks nanoparticles, Advanced healthcare materials 2 (2013) 1630-1637. https://doi.org/10.1002/adhm.201200454

[136] V. Agostoni, R. Anand, S. Monti, S. Hall, G. Maurin, P. Horcajada, C. Serre, K. Bouchemal, R. Gref, Impact of phosphorylation on the encapsulation of nucleoside analogues within porous iron (iii) metal-organic framework MIL-100 (Fe) nanoparticles, Journal of Materials Chemistry B 1 (2013) 4231-4242. https://doi.org/10.1039/c3tb20653j

[137] M. Marcos-Almaraz, R. Gref, V. Agostoni, C. Kreuz, P. Clayette, C. Serre, P. Couvreur, P. Horcajada, Towards improved HIV-microbicide activity through the co-encapsulation of NRTI drugs in biocompatible metal organic framework nanocarriers, Journal of Materials Chemistry B 5 (2017) 8563-8569. https://doi.org/10.1039/C7TB01933E

[138] S.-R. Pasricha, J. Tye-Din, M.U. Muckenthaler, D.W. Swinkels, Iron deficiency, The Lancet 397 (2021) 233-248. https://doi.org/10.1016/S0140-6736(20)32594-0

[139] E. Piskin, D. Cianciosi, S. Gulec, M. Tomas, E. Capanoglu, Iron absorption: factors, limitations, and improvement methods, ACS omega 7 (2022) 20441-20456. https://doi.org/10.1021/acsomega.2c01833

[140] M.P. Figueiredo, V.R. Cunha, F. Leroux, C. Taviot-Gueho, M.N. Nakamae, Y.R. Kang, R.B. Souza, A.M.C. Martins, I.H.J. Koh, V.R. Constantino, Iron-based layered double hydroxide implants: Potential drug delivery carriers with tissue biointegration promotion and blood microcirculation preservation, Acs Omega 3 (2018) 18263-18274. https://doi.org/10.1021/acsomega.8b02532

[141] F. Keshavarz, M. Kadek, B. Barbiellini, A. Bansil, Electrochemical potential of the metal organic framework MIL-101 (Fe) as cathode material in Li-ion batteries, Condensed Matter 6 (2021) 22. https://doi.org/10.3390/condmat6020022

[142] M. Jakubowski, M. Kucinska, M. Ratajczak, M. Pokora, M. Murias, A. Voelkel, M. Sandomierski, Zinc forms of faujasite zeolites as a drug delivery system for 6-mercaptopurine, Microporous and Mesoporous Materials 343 (2022) 112194. https://doi.org/10.1016/j.micromeso.2022.112194

[143] M. Jakubowski, Ł. Majchrzycki, A. Zarkov, A. Voelkel, M. Sandomierski, Zinc-doped hydroxyapatite as a pH responsive drug delivery system for anticancer drug 6-mercaptopurine, Journal of Biomedical Materials Research Part B: Applied Biomaterials 112 (2024) e35395. https://doi.org/10.1002/jbm.b.35395

[144] A. Meshkini, H. Oveisi, Methotrexate-F127 conjugated mesoporous zinc hydroxyapatite as an efficient drug delivery system for overcoming chemotherapy resistance in osteosarcoma cells, Colloids and Surfaces B: Biointerfaces 158 (2017) 319-330. https://doi.org/10.1016/j.colsurfb.2017.07.006

[145] X. Ye, M. Xiong, K. Yuan, W. Liu, X. Cai, Y. Yuan, Y. Yuan, Y. Qin, D. Wu, Synthesis and characterization of a novel zinc-based metal-organic framework containing benzoic acid: a low-toxicity carrier for drug delivery, Iranian Journal of Pharmaceutical Research: IJPR 22 (2023). https://doi.org/10.5812/ijpr-136238

[146] M. Taha, K. Aiedeh, Y. Al-Hiari, H. Al-Khatib, Synthesis of zinc-crosslinked thiolated alginic acid beads and their in vitro evaluation as potential enteric delivery system with folic acid as model drug, Die Pharmazie-An International Journal of Pharmaceutical Sciences 60 (2005) 736-742.

[147] Y.-J. Zhu, X.-X. Guo, T.-K. Sham, Calcium silicate-based drug delivery systems, Expert opinion on drug delivery 14 (2017) 215-228. https://doi.org/10.1080/17425247.2016.1214566

[148] M.-P. Ginebra, C. Canal, M. Espanol, D. Pastorino, E.B. Montufar, Calcium phosphate cements as drug delivery materials, Advanced drug delivery reviews 64 (2012) 1090-1110. https://doi.org/10.1016/j.addr.2012.01.008

[149] S. Bose, S. Tarafder, Calcium phosphate ceramic systems in growth factor and drug delivery for bone tissue engineering: a review, Acta biomaterialia 8 (2012) 1401-1421. https://doi.org/10.1016/j.actbio.2011.11.017

[150] M. Sandomierski, M. Jakubowski, M. Ratajczak, T. Buchwald, R.E. Przekop, Ł. Majchrzycki, A. Voelkel, Calcium and strontium phytate particles as a potential drug delivery system for prolonged release of risedronate, Journal of Drug Delivery Science and Technology 80 (2023) 104176. https://doi.org/10.1016/j.jddst.2023.104176

[151] S.K. Jain, A. Awasthi, N. Jain, G. Agrawal, Calcium silicate based microspheres of repaglinide for gastroretentive floating drug delivery: Preparation and in vitro characterization, Journal of controlled release 107 (2005) 300-309. https://doi.org/10.1016/j.jconrel.2005.06.007

[152] Y.-S. Park, J.-Y. Cho, S.-J. Lee, C.I. Hwang, Modified titanium implant as a gateway to the human body: the implant mediated drug delivery system, BioMed Research International 2014 (2014) 801358. https://doi.org/10.1155/2014/801358

[153] O. Khandan, A. Famili, M.Y. Kahook, M.P. Rao, Titanium-based, fenestrated, in-plane microneedles for passive ocular drug delivery, 2012 Annual International Conference of the IEEE Engineering in Medicine and Biology Society, IEEE, 2012, pp. 6572-6575. https://doi.org/10.1109/EMBC.2012.6347500

[154] C. Qi, Y.-J. Zhu, F. Chen, J. Wu, Porous microspheres of magnesium whitlockite and amorphous calcium magnesium phosphate: microwave-assisted rapid synthesis using creatine phosphate, and application in drug delivery, Journal of Materials Chemistry B 3 (2015) 7775-7786. https://doi.org/10.1039/C5TB01106J

[155] Q. Song, Y. Liu, X. Ding, M. Feng, J. Li, W. Liu, B. Wang, Z. Gu, A drug co-delivery platform made of magnesium-based micromotors enhances combination therapy for hepatoma carcinoma cells, Nanoscale 15 (2023) 15573-15582. https://doi.org/10.1039/D3NR01548C

[156] K. Zhang, H. Lin, J. Mao, X. Luo, R. Wei, Z. Su, B. Zhou, D. Li, J. Gao, H. Shan, An extracellular pH-driven targeted multifunctional manganese arsenite delivery system for tumor imaging and therapy, Biomaterials science 7 (2019) 2480-2490. https://doi.org/10.1039/C9BM00216B

[157] X. Zhong, X. Bao, H. Zhong, Y. Zhou, Z. Zhang, Y. Lu, Q. Dai, Q. Yang, P. Ke, Y. Xia, Mitochondrial targeted drug delivery combined with manganese catalyzed Fenton reaction for

the treatment of breast cancer, International Journal of Pharmaceutics 622 (2022) 121810. https://doi.org/10.1016/j.ijpharm.2022.121810

[158] E.J. Anglin, L. Cheng, W.R. Freeman, M.J. Sailor, Porous silicon in drug delivery devices and materials, Advanced drug delivery reviews 60 (2008) 1266-1277. https://doi.org/10.1016/j.addr.2008.03.017

[159] N. Abbariki, M. Bagherzadeh, S. Aghili, H. Daneshgar, Green Composite Synthesis Based on Aluminum Fumarate Metal-Organic Framework (MOF) for Doxorubicin Delivery: An in Vitro Study, Journal of Drug Delivery Science and Technology (2024) 106203. https://doi.org/10.1016/j.jddst.2024.106203

[160] D. Nikolova, M. Simeonov, C. Tzachev, A. Apostolov, L. Christov, E. Vassileva, Polyelectrolyte complexes of chitosan and sodium alginate as a drug delivery system for diclofenac sodium, Polymer International 71 (2022) 668-678. https://doi.org/10.1002/pi.6273

[161] P. Biernat, W. Musiał, D. Gosławska, J. Pluta, The impact of selected preparations of trace elements-magnesium, potassium, calcium, and zinc on the release of diclofenac sodium from enteric coated tablets and from sustained release capsules, Advances in Clinical and Experimental Medicine 23 (2014) 205-213. https://doi.org/10.17219/acem/37059

[162] S.M. Sheta, S.R. Salem, S.M. El-Sheikh, A novel Iron (III)-based MOF: Synthesis, characterization, biological, and antimicrobial activity study, Journal of Materials Research 37 (2022) 2356-2367. https://doi.org/10.1557/s43578-022-00644-9

[163] V. Patamia, C. Zagni, R. Fiorenza, V. Fuochi, S. Dattilo, P.M. Riccobene, P.M. Furneri, G. Floresta, A. Rescifina, Total Bio-Based Material for Drug Delivery and Iron Chelation to Fight Cancer through Antimicrobial Activity, Nanomaterials 13 (2023) 2036. https://doi.org/10.3390/nano13142036

[164] S. Yazdankhah, K. Rudi, A. Bernhoft, Zinc and copper in animal feed-development of resistance and co-resistance to antimicrobial agents in bacteria of animal origin, Microbial ecology in health and disease 25 (2014) 25862. https://doi.org/10.3402/mehd.v25.25862

[165] J. Pasquet, Y. Chevalier, J. Pelletier, E. Couval, D. Bouvier, M.-A. Bolzinger, The contribution of zinc ions to the antimicrobial activity of zinc oxide, Colloids and Surfaces A: Physicochemical and Engineering Aspects 457 (2014) 263-274. https://doi.org/10.1016/j.colsurfa.2014.05.057

[166] R. Kamphof, R.N. Lima, J.W. Schoones, J.J. Arts, R.G. Nelissen, G. Cama, B.G. Pijls, Antimicrobial activity of ion-substituted calcium phosphates: A systematic review, Heliyon 9 (2023). https://doi.org/10.1016/j.heliyon.2023.e16568

[167] T. Mokabber, H. Cao, N. Norouzi, P. Van Rijn, Y. Pei, Antimicrobial electrodeposited silver-containing calcium phosphate coatings, ACS applied materials & interfaces 12 (2020) 5531-5541. https://doi.org/10.1021/acsami.9b20158

[168] T.C. Dzogbewu, W.B. du Preez, Additive manufacturing of titanium-based implants with metal-based antimicrobial agents, Metals 11 (2021) 453. https://doi.org/10.3390/met11030453

[169] G.A. Kotsakis, C. Lan, J. Barbosa, K. Lill, R. Chen, J. Rudney, C. Aparicio, Antimicrobial agents used in the treatment of peri-implantitis alter the physicochemistry and

cytocompatibility of titanium surfaces, Journal of periodontology 87 (2016) 809-819. https://doi.org/10.1902/jop.2016.150684

[170] V. Luque-Agudo, M.C. Fernández-Calderón, M.A. Pacha-Olivenza, C. Perez-Giraldo, A.M. Gallardo-Moreno, M.L. González-Martín, The role of magnesium in biomaterials related infections, Colloids and surfaces B: Biointerfaces 191 (2020) 110996. https://doi.org/10.1016/j.colsurfb.2020.110996

[171] Z. Lin, X. Sun, H. Yang, The role of antibacterial metallic elements in simultaneously improving the corrosion resistance and antibacterial activity of magnesium alloys, Materials & Design 198 (2021) 109350. https://doi.org/10.1016/j.matdes.2020.109350

[172] N. Sultana, M.S. Arayne, M. Afzal, Synthesis and antibacterial activity of cephradine metal complexes: part I complexes with magnesium, calcium, chromium and manganese, Pakistan journal of pharmaceutical sciences 16 (2003) 59-72.

[173] H. Revanasiddappa, L. Shivakumar, K. Prasad, B. Vijay, B. Jayalakshmi, Synthesis, structural characterization and antimicrobial activity evaluation of manganese (II) complexes with biologically important drugs, Chemical Sciences Journal (2012). https://doi.org/10.1155/2013/760754

[174] T.y.G. Khonina, N.V. Kungurov, N.y.V. Zilberberg, N.y.P. Evstigneeva, M.M. Kokhan, A.I. Polishchuk, E.V. Shadrina, E.Y. Nikitina, V.V. Permikin, O.N. Chupakhin, Structural features and antimicrobial activity of hydrogels obtained by the sol-gel method from silicon, zinc, and boron glycerolates, Journal of Sol-Gel Science and Technology 95 (2020) 682-692. https://doi.org/10.1007/s10971-020-05328-6

[175] A. Lehmann, S. Rupf, A. Schubert, I.-M. Zylla, H.J. Seifert, A. Schindler, T. Arnold, Plasma deposited silicon oxide films for controlled permeation of copper as antimicrobial agent, Clinical Plasma Medicine 3 (2015) 3-9. https://doi.org/10.1016/j.cpme.2015.01.001

[176] S.C. Londono, H.E. Hartnett, L.B. Williams, Antibacterial activity of aluminum in clay from the Colombian Amazon, Environmental Science & Technology 51 (2017) 2401-2408. https://doi.org/10.1021/acs.est.6b04670

[177] J. Slots, Selection of antimicrobial agents in periodontal therapy, Journal of periodontal research 37 (2002) 389-398. https://doi.org/10.1034/j.1600-0765.2002.00004.x

[178] T. Baygar, N. Saraç, Ö. Ceylan, A. Uğur, R. Boran, U. Balcı, In vitro biological activities of potassium metaborate; antioxidative, antimicrobial and antibiofilm properties, Journal of Boron 7 (2022) 475-481.

[179] A. Vessieres, Iron compounds as anticancer agents, (2019). https://doi.org/10.1039/9781788016452-00062

[180] W.A. Wani, U. Baig, S. Shreaz, R.A. Shiekh, P.F. Iqbal, E. Jameel, A. Ahmad, S.H. Mohd-Setapar, M. Mushtaque, L.T. Hun, Recent advances in iron complexes as potential anticancer agents, New Journal of Chemistry 40 (2016) 1063-1090. https://doi.org/10.1039/C5NJ01449B

[181] D. Richardson, Iron chelators as therapeutic agents for the treatment of cancer, Critical reviews in oncology/hematology 42 (2002) 267-281. https://doi.org/10.1016/S1040-8428(01)00218-9

Materials Research Forum LLC
https://doi.org/10.21741/9781644903711-3

[182] Z. Li, H. Tanaka, F. Galiano, J. Glass, Anticancer activity of the iron facilitator LS081, Journal of Experimental & Clinical Cancer Research 30 (2011) 1-10. https://doi.org/10.1186/1756-9966-30-1

[183] U. Basu, M. Roy, A.R. Chakravarty, Recent advances in the chemistry of iron-based chemotherapeutic agents, Coordination Chemistry Reviews 417 (2020) 213339. https://doi.org/10.1016/j.ccr.2020.213339

[184] Y. Su, B. Zhao, L. Zhou, Z. Zhang, Y. Shen, H. Lv, L.H.H. AlQudsy, P. Shang, Ferroptosis, a novel pharmacological mechanism of anti-cancer drugs, Cancer letters 483 (2020) 127-136. https://doi.org/10.1016/j.canlet.2020.02.015

[185] J. Ye, J. Ma, C. Liu, J. Huang, L. Wang, X. Zhong, A novel iron (II) phenanthroline complex exhibits anticancer activity against TFR1-overexpressing esophageal squamous cell carcinoma cells through ROS accumulation and DNA damage, Biochemical Pharmacology 166 (2019) 93-107. https://doi.org/10.1016/j.bcp.2019.05.013

[186] M. Porchia, M. Pellei, F. Del Bello, C. Santini, Zinc complexes with nitrogen donor ligands as anticancer agents, Molecules 25 (2020) 5814. https://doi.org/10.3390/molecules25245814

[187] D. Skrajnowska, B. Bobrowska-Korczak, Role of zinc in immune system and anti-cancer defense mechanisms, Nutrients 11 (2019) 2273. https://doi.org/10.3390/nu11102273

[188] E. Varghese, S.M. Samuel, Z. Sadiq, P. Kubatka, A. Liskova, J. Benacka, P. Pazinka, P. Kruzliak, D. Büsselberg, Anti-cancer agents in proliferation and cell death: the calcium connection, International journal of molecular sciences 20 (2019) 3017. https://doi.org/10.3390/ijms20123017

[189] B.E. Tepedelen, M. Korkmaz, E. Tatlisumak, E.T. Uluer, E. Ölmez, İ. Değerli, E. Soya, S. Inan, A study on the anticarcinogenic effects of calcium fructoborate, Biological Trace Element Research 178 (2017) 210-217. https://doi.org/10.1007/s12011-016-0918-6

[190] I. Kostova, Titanium and vanadium complexes as anticancer agents, Anti-Cancer Agents in Medicinal Chemistry (Formerly Current Medicinal Chemistry-Anti-Cancer Agents) 9 (2009) 827-842. https://doi.org/10.2174/187152009789124646

[191] M. Cini, T.D. Bradshaw, S. Woodward, Using titanium complexes to defeat cancer: the view from the shoulders of titans, Chemical Society Reviews 46 (2017) 1040-1051. https://doi.org/10.1039/C6CS00860G

[192] Y. Ellahioui, S. Prashar, S. Gomez-Ruiz, Anticancer applications and recent investigations of metallodrugs based on gallium, tin and titanium, Inorganics 5 (2017) 4. https://doi.org/10.3390/inorganics5010004

[193] M. Miller, A. Mellul, M. Braun, D. Sherill-Rofe, E. Cohen, Z. Shpilt, I. Unterman, O. Braitbard, J. Hochman, E.Y. Tshuva, Titanium tackles the endoplasmic reticulum: a first genomic study on a titanium anticancer metallodrug, Iscience 23 (2020). https://doi.org/10.1016/j.isci.2020.101262

[194] T. Li, W. Xu, C. Liu, J. He, Q. Wang, D. Zhang, K. Sui, Z. Zhang, H. Sun, K. Yang, Anticancer effect of biodegradable magnesium on hepatobiliary carcinoma: an in vitro and in

vivo study, ACS Biomaterials Science & Engineering 7 (2021) 2774-2782.
https://doi.org/10.1021/acsbiomaterials.1c00288

[195] S.A. Mirmalek, E. Jangholi, M. Jafari, S. Yadollah-Damavandi, M.A. Javidi, Y. Parsa, T. Parsa, S.A. Salimi-Tabatabaee, H.G. Kolagar, S.K. Jalil, Comparison of in vitro cytotoxicity and apoptogenic activity of magnesium chloride and cisplatin as conventional chemotherapeutic agents in the MCF-7 cell line, Asian Pacific Journal of Cancer Prevention 17 (2016) 131-134. https://doi.org/10.7314/APJCP.2016.17.S3.131

[196] A.T. Odularu, Manganese Schiff base complexes, crystallographic studies, anticancer activities, and molecular docking, Journal of Chemistry 2022 (2022) 7062912. https://doi.org/10.1155/2022/7062912

[197] R. Farghadani, J. Rajarajeswaran, N.B.M. Hashim, M.A. Abdulla, S. Muniandy, A novel β-diiminato manganese III complex as the promising anticancer agent induces G 0/G 1 cell cycle arrest and triggers apoptosis via mitochondrial-dependent pathways in MCF-7 and MDA-MB-231 human breast cancer cells, RSC Advances 7 (2017) 24387-24398. https://doi.org/10.1039/C7RA02478A

[198] R. Zheng, J. Guo, X. Cai, L. Bin, C. Lu, A. Singh, M. Trivedi, A. Kumar, J. Liu, Manganese complexes and manganese-based metal-organic frameworks as contrast agents in MRI and chemotherapeutics agents: Applications and prospects, Colloids and Surfaces B: Biointerfaces 213 (2022) 112432. https://doi.org/10.1016/j.colsurfb.2022.112432

[199] R. Irfandi, I. Raya, Potential anticancer activity of Mn (II) complexes containing arginine dithiocarbamate ligand on MCF-7 breast cancer cell lines, Annals of Medicine and Surgery 60 (2020) 396-402. https://doi.org/10.1016/j.amsu.2020.11.018

[200] L. Yao, Q.-Y. Chen, X.-L. Xu, Z. Li, X.-M. Wang, Interaction of manganese (II) complex with apotransferrin and the apotransferrin enhanced anticancer activities, Spectrochimica Acta Part A: Molecular and Biomolecular Spectroscopy 105 (2013) 207-212. https://doi.org/10.1016/j.saa.2012.12.029

[201] B. Lipinski, Sodium selenite as an anticancer agent, Anti-Cancer Agents in Medicinal Chemistry (Formerly Current Medicinal Chemistry-Anti-Cancer Agents) 17 (2017) 658-661. https://doi.org/10.2174/1871520616666160607011024

[202] Y. Meng, J. Sun, J. Yu, C. Wang, J. Su, Dietary intakes of calcium, iron, magnesium, and potassium elements and the risk of colorectal cancer: a meta-analysis, Biological trace element research 189 (2019) 325-335. https://doi.org/10.1007/s12011-018-1474-z

[203] N.F. Lazareva, V.P. Baryshok, I.M. Lazarev, Silicon-containing analogs of camptothecin as anticancer agents, Archiv der Pharmazie 351 (2018) 1700297. https://doi.org/10.1002/ardp.201700297

[204] C. Gan, P. Liu, Adsorption behavior of anticancer drug on the aluminum nitride surface: Density functional theory evolution, Phosphorus, Sulfur, and Silicon and the Related Elements 196 (2021) 1061-1070. https://doi.org/10.1080/10426507.2021.1966428

[205] M.J. Brabha, M.A. Malbi, Antidiabetic activity of chelating tris n-methylethylenediamine iron complex, Drug Invention Today 14 (2020).

[206] M. Kongot, D.S. Reddy, V. Singh, R. Patel, N.K. Singhal, A. Kumar, Oxidovanadium (IV) and iron (III) complexes with O2N2 donor linkage as plausible antidiabetic candidates: Synthesis, structural characterizations, glucose uptake and model biological media studies, Applied Organometallic Chemistry 34 (2020) e5327. https://doi.org/10.1002/aoc.5327

[207] C.I. Chukwuma, S.S. Mashele, K.C. Eze, G.R. Matowane, S.M. Islam, S.L. Bonnet, A.E. Noreljaleel, L.M. Ramorobi, A comprehensive review on zinc (II) complexes as anti-diabetic agents: The advances, scientific gaps and prospects, Pharmacological research 155 (2020) 104744. https://doi.org/10.1016/j.phrs.2020.104744

[208] S. Fujimoto, H. Yasui, Y. Yoshikawa, Development of a novel antidiabetic zinc complex with an organoselenium ligand at the lowest dosage in KK-Ay mice, Journal of Inorganic Biochemistry 121 (2013) 10-15. https://doi.org/10.1016/j.jinorgbio.2012.12.008

[209] M. Sadeghian, L. Azadbakht, N. Khalili, M. Mortazavi, A. Esmaillzadeh, Oral magnesium supplementation improved lipid profile but increased insulin resistance in patients with diabetic nephropathy: a double-blind randomized controlled clinical trial, Biological Trace Element Research 193 (2020) 23-35. https://doi.org/10.1007/s12011-019-01687-6

[210] R. Razzaghi, F. Pidar, M. Momen-Heravi, F. Bahmani, H. Akbari, Z. Asemi, Magnesium supplementation and the effects on wound healing and metabolic status in patients with diabetic foot ulcer: a randomized, double-blind, placebo-controlled trial, Biological trace element research 181 (2018) 207-215. https://doi.org/10.1007/s12011-017-1056-5

[211] S.M. El-Megharbel, N.M. Al-Baqami, E.H. Al-Thubaiti, S.H. Qahl, B. Albogami, R.Z. Hamza, Antidiabetic drug sitagliptin with divalent transition metals manganese and cobalt: synthesis, structure, characterization antibacterial and antioxidative effects in liver tissues, Current Issues in Molecular Biology 44 (2022). https://doi.org/10.3390/cimb44050124

[212] W. Arika, P. Ogola, D. Nyamai, A. Mawia, F. Wambua, N.G. Kiboi, J. Wambani, S. Njagi, H. Rachuonyo, K. Emmah, Mineral elements content of selected Kenyan antidiabetic medicinal plants, (2016).

[213] F. Al-Awadi, J. Anim, T. Srikumar, M. Al-Rustom, Possible role of trace elements in the hypoglycemic effect of plants extract in diabetic rats, The Journal of Trace Elements in Experimental Medicine: The Official Publication of the International Society for Trace Element Research in Humans 17 (2004) 31-44. https://doi.org/10.1002/jtra.10048

[214] Z. Liu, S. Wang, W. Li, Y. Tian, Bioimaging and biosensing of ferrous ion in neurons and HepG2 cells upon oxidative stress, Analytical chemistry 90 (2018) 2816-2825. https://doi.org/10.1021/acs.analchem.7b04934

[215] X. Tian, S. Hussain, C. de Pace, L. Ruiz-Pérez, G. Battaglia, ZnII complexes for bioimaging and correlated applications, Chemistry-An Asian Journal 14 (2019) 509-526. https://doi.org/10.1002/asia.201801437

[216] C. Qi, J. Lin, L.-H. Fu, P. Huang, Calcium-based biomaterials for diagnosis, treatment, and theranostics, Chemical Society Reviews 47 (2018) 357-403. https://doi.org/10.1039/C6CS00746E

[217] H. Helmholz, O. Will, T. Penate-Medina, J. Humbert, T. Damm, B. Luthringer-Feyerabend, R. Willumeit-Römer, C.C. Glüer, O. Penate-Medina, Tissue responses after

implantation of biodegradable Mg alloys evaluated by multimodality 3D micro-bioimaging in vivo, Journal of Biomedical Materials Research Part A 109 (2021) 1521-1529. https://doi.org/10.1002/jbm.a.37148

[218] S. Roy, J. Gu, W. Xia, C. Mi, B. Guo, Advancements in manganese complex-based MRI agents: Innovations, design strategies, and future directions, Drug Discovery Today (2024) 104101. https://doi.org/10.1016/j.drudis.2024.104101

[219] D.Ş. Karaman, M.P. Sarparanta, J.M. Rosenholm, A.J. Airaksinen, Multimodality imaging of silica and silicon materials in vivo, Advanced materials 30 (2018) 1703651. https://doi.org/10.1002/adma.201703651

[220] S. Carrasco, Metal-organic frameworks for the development of biosensors: a current overview, Biosensors 8 (2018) 92. https://doi.org/10.3390/bios8040092

[221] S. Jain, A. Paliwal, V. Gupta, M. Tomar, SPR studies on optical fiber coated with different plasmonic metals for fabrication of efficient biosensors, Materials Today: Proceedings 33 (2020) 2180-2186. https://doi.org/10.1016/j.matpr.2020.03.710

[222] Y.-H. Wang, K.-J. Huang, X. Wu, Recent advances in transition-metal dichalcogenides based electrochemical biosensors: A review, Biosensors and Bioelectronics 97 (2017) 305-316. https://doi.org/10.1016/j.bios.2017.06.011

[223] M. Gielen, E.R. Tiekink, Metallotherapeutic drugs and metal-based diagnostic agents: the use of metals in medicine, John Wiley & Sons2005. https://doi.org/10.1002/0470864052

[224] B. Meunier, A. Robert, G. Crisponi, V. Nurchi, J.I. Lachowicz, M. Nairz, G. Weiss, A.B. Pai, R.J. Ward, R.R. Crichton, Essential metals in medicine: Therapeutic use and toxicity of metal ions in the clinic, Walter de Gruyter GmbH & Co KG2019.

[225] A. Basu, Metals in medicine: an overview, Sci. Revs. Chem. Commun 5 (2015) 77-87.

[226] M. Rai, A.P. Ingle, S. Medici, Biomedical applications of metals, Springer2018. https://doi.org/10.1007/978-3-319-74814-6

[227] A.M. Malekzadeh, A. Ramazani, S.J.T. Rezaei, H. Niknejad, Design and construction of multifunctional hyperbranched polymers coated magnetite nanoparticles for both targeting magnetic resonance imaging and cancer therapy, Journal of colloid and interface science 490 (2017) 64-73. https://doi.org/10.1016/j.jcis.2016.11.014

[228] F. Barahuie, D. Dorniani, B. Saifullah, S. Gothai, M.Z. Hussein, A.K. Pandurangan, P. Arulselvan, M.E. Norhaizan, Sustained release of anticancer agent phytic acid from its chitosan-coated magnetic nanoparticles for drug-delivery system, International journal of nanomedicine (2017) 2361-2372. https://doi.org/10.2147/IJN.S126245

[229] M. Ebadi, K. Buskaran, S. Bullo, M.Z. Hussein, S. Fakurazi, G. Pastorin, Drug delivery system based on magnetic iron oxide nanoparticles coated with (polyvinyl alcohol-zinc/aluminium-layered double hydroxide-sorafenib), Alexandria Engineering Journal 60 (2021) 733-747. https://doi.org/10.1016/j.aej.2020.09.061

[230] M.S.U. Ahmed, A.B. Salam, C. Yates, K. Willian, J. Jaynes, T. Turner, M.O. Abdalla, Double-receptor-targeting multifunctional iron oxide nanoparticles drug delivery system for

the treatment and imaging of prostate cancer, International journal of nanomedicine (2017) 6973-6984. https://doi.org/10.2147/IJN.S139011

[231] R. Kumar, K. Pandey, G.C. Sahoo, S. Das, V. Das, R. Topno, P. Das, Development of high efficacy peptide coated iron oxide nanoparticles encapsulated amphotericin B drug delivery system against visceral leishmaniasis, Materials Science and Engineering: C 75 (2017) 1465-1471. https://doi.org/10.1016/j.msec.2017.02.145

[232] J. Schnabel, R. Ettlinger, H. Bunzen, Zn-MOF-74 as pH-responsive drug-delivery system of arsenic trioxide, ChemNanoMat 6 (2020) 1229-1236. https://doi.org/10.1002/cnma.202000221

[233] P. Sathishkumar, Z. Li, R. Govindan, R. Jayakumar, C. Wang, F.L. Gu, Zinc oxide-quercetin nanocomposite as a smart nano-drug delivery system: Molecular-level interaction studies, Applied Surface Science 536 (2021) 147741. https://doi.org/10.1016/j.apsusc.2020.147741

[234] M. Akbarian, S. Mahjoub, S.M. Elahi, E. Zabihi, H. Tashakkorian, Green synthesis, formulation and biological evaluation of a novel ZnO nanocarrier loaded with paclitaxel as drug delivery system on MCF-7 cell line, Colloids and Surfaces B: Biointerfaces 186 (2020) 110686. https://doi.org/10.1016/j.colsurfb.2019.110686

[235] S. Vijayakumar, K. Saravanakumar, B. Malaikozhundan, M. Divya, B. Vaseeharan, E.F. Durán-Lara, M.-H. Wang, Biopolymer K-carrageenan wrapped ZnO nanoparticles as drug delivery vehicles for anti MRSA therapy, International journal of biological macromolecules 144 (2020) 9-18. https://doi.org/10.1016/j.ijbiomac.2019.12.030

[236] K. Li, D. Li, L. Zhao, Y. Chang, Y. Zhang, Y. Cui, Z. Zhang, Calcium-mineralized polypeptide nanoparticle for intracellular drug delivery in osteosarcoma chemotherapy, Bioactive Materials 5 (2020) 721-731. https://doi.org/10.1016/j.bioactmat.2020.04.010

[237] N.M. Elbaz, A. Owen, S. Rannard, T.O. McDonald, Controlled synthesis of calcium carbonate nanoparticles and stimuli-responsive multi-layered nanocapsules for oral drug delivery, International journal of pharmaceutics 574 (2020) 118866. https://doi.org/10.1016/j.ijpharm.2019.118866

[238] Y. Li, S. Zhou, H. Song, T. Yu, X. Zheng, Q. Chu, CaCO3 nanoparticles incorporated with KAE to enable amplified calcium overload cancer therapy, Biomaterials 277 (2021) 121080. https://doi.org/10.1016/j.biomaterials.2021.121080

[239] A. Alsaed, F.I. Elshami, M.M. Ibrahim, H. Shereef, H. Mohany, R. van Eldik, S.Y. Shaban, Folate titanium (IV)-chitosan nanocomposites as drug delivery system for active-targeted cancer therapy: Design, HSA/GSH binding, mechanistic, and biological investigations, Journal of Drug Delivery Science and Technology 97 (2024) 105826. https://doi.org/10.1016/j.jddst.2024.105826

[240] N. Salahuddin, M. Abdelwahab, M. Gaber, S. Elneanaey, Synthesis and Design of Norfloxacin drug delivery system based on PLA/TiO2 nanocomposites: Antibacterial and antitumor activities, Materials Science and Engineering: C 108 (2020) 110337. https://doi.org/10.1016/j.msec.2019.110337

[241] S. Kim, S. Im, E.-Y. Park, J. Lee, C. Kim, T.-i. Kim, W.J. Kim, Drug-loaded titanium dioxide nanoparticle coated with tumor targeting polymer as a sonodynamic chemotherapeutic agent for anti-cancer therapy, Nanomedicine: Nanotechnology, Biology and Medicine 24 (2020) 102110. https://doi.org/10.1016/j.nano.2019.102110

[242] H. Zhao, X. Yuan, J. Yu, Y. Huang, C. Shao, F. Xiao, L. Lin, Y. Li, L. Tian, Magnesium-stabilized multifunctional DNA nanoparticles for tumor-targeted and pH-responsive drug delivery, ACS applied materials & interfaces 10 (2018) 15418-15427. https://doi.org/10.1021/acsami.8b01932

[243] A. Hassani, S.A. Hussain, N. Abdullah, S. Kamaruddin, R. Rosli, Characterization of magnesium orotate-loaded chitosan polymer nanoparticles for a drug delivery system, Chemical Engineering & Technology 42 (2019) 1816-1824. https://doi.org/10.1002/ceat.201800478

[244] G.K. Pouroutzidou, L. Liverani, A. Theocharidou, I. Tsamesidis, M. Lazaridou, E. Christodoulou, A. Beketova, C. Pappa, K.S. Triantafyllidis, A.D. Anastasiou, Synthesis and characterization of mesoporous mg-and sr-doped nanoparticles for moxifloxacin drug delivery in promising tissue engineering applications, International Journal of Molecular Sciences 22 (2021) 577. https://doi.org/10.3390/ijms22020577

[245] Y. Miao, X. Zhou, J. Bai, W. Zhao, X. Zhao, Hollow polydopamine spheres with removable manganese oxide nanoparticle caps for tumor microenvironment-responsive drug delivery, Chemical Engineering Journal 430 (2022) 133089. https://doi.org/10.1016/j.cej.2021.133089

[246] Y. Hao, L. Wang, B. Zhang, D. Li, D. Meng, J. Shi, H. Zhang, Z. Zhang, Y. Zhang, Manganese dioxide nanosheets-based redox/pH-responsive drug delivery system for cancer theranostic application, International journal of nanomedicine (2016) 1759-1778. https://doi.org/10.2147/IJN.S98832

[247] S. Kapoor, R. Hegde, A.J. Bhattacharyya, Influence of surface chemistry of mesoporous alumina with wide pore distribution on controlled drug release, Journal of controlled release 140 (2009) 34-39. https://doi.org/10.1016/j.jconrel.2009.07.015

[248] J.K. Patra, M.S. Ali, I.-G. Oh, K.-H. Baek, Proteasome inhibitory, antioxidant, and synergistic antibacterial and anticandidal activity of green biosynthesized magnetic Fe3O4 nanoparticles using the aqueous extract of corn (Zea mays L.) ear leaves, Artificial cells, nanomedicine, and biotechnology 45 (2017) 349-356. https://doi.org/10.3109/21691401.2016.1153484

[249] S.T. Shah, W.A. Yehye, O. Saad, K. Simarani, Z.Z. Chowdhury, A.A. Alhadi, L.A. Al-Ani, Surface functionalization of iron oxide nanoparticles with gallic acid as potential antioxidant and antimicrobial agents, Nanomaterials 7 (2017) 306. https://doi.org/10.3390/nano7100306

[250] H.S. Devi, M.A. Boda, M.A. Shah, S. Parveen, A.H. Wani, Green synthesis of iron oxide nanoparticles using Platanus orientalis leaf extract for antifungal activity, Green Processing and Synthesis 8 (2019) 38-45. https://doi.org/10.1515/gps-2017-0145

[251] W. Salem, D.R. Leitner, F.G. Zingl, G. Schratter, R. Prassl, W. Goessler, J. Reidl, S. Schild, Antibacterial activity of silver and zinc nanoparticles against Vibrio cholerae and

Materials Research Forum LLC
https://doi.org/10.21741/9781644903711-3

enterotoxic Escherichia coli, International journal of medical microbiology 305 (2015) 85-95. https://doi.org/10.1016/j.ijmm.2014.11.005

[252] Y. Xie, Y. He, P.L. Irwin, T. Jin, X. Shi, Antibacterial activity and mechanism of action of zinc oxide nanoparticles against Campylobacter jejuni, Applied and environmental microbiology 77 (2011) 2325-2331. https://doi.org/10.1128/AEM.02149-10

[253] L. He, Y. Liu, A. Mustapha, M. Lin, Antifungal activity of zinc oxide nanoparticles against Botrytis cinerea and Penicillium expansum, Microbiological research 166 (2011) 207-215. https://doi.org/10.1016/j.micres.2010.03.003

[254] U. Ijaz, I.A. Bhatti, S. Mirza, A. Ashar, Characterization and evaluation of antibacterial activity of plant mediated calcium oxide (CaO) nanoparticles by employing Mentha pipertia extract, Materials Research Express 4 (2017) 105402. https://doi.org/10.1088/2053-1591/aa8603

[255] A. Hashemi, M. Naseri, S. Ghiyasvand, E. Naderi, S. Vafai, Evaluation of physical properties, cytotoxicity, and antibacterial activities of calcium-cadmium ferrite nanoparticles, Applied Physics A 128 (2022) 236. https://doi.org/10.1007/s00339-022-05294-6

[256] B. Maringgal, N. Hashim, I.S.M.A. Tawakkal, M.H. Hamzah, M.T.M. Mohamed, Biosynthesis of CaO nanoparticles using Trigona sp. Honey: Physicochemical characterization, antifungal activity, and cytotoxicity properties, Journal of Materials Research and Technology 9 (2020) 11756-11768. https://doi.org/10.1016/j.jmrt.2020.08.054

[257] M. Zimbone, M. Buccheri, G. Cacciato, R. Sanz, G. Rappazzo, S. Boninelli, R. Reitano, L. Romano, V. Privitera, M. Grimaldi, Photocatalytical and antibacterial activity of TiO2 nanoparticles obtained by laser ablation in water, Applied Catalysis B: Environmental 165 (2015) 487-494. https://doi.org/10.1016/j.apcatb.2014.10.031

[258] B. Thakur, A. Kumar, D. Kumar, Green synthesis of titanium dioxide nanoparticles using Azadirachta indica leaf extract and evaluation of their antibacterial activity, South African Journal of Botany 124 (2019) 223-227. https://doi.org/10.1016/j.sajb.2019.05.024

[259] M.A. Irshad, R. Nawaz, M.Z. ur Rehman, M. Imran, J. Ahmad, S. Ahmad, A. Inam, A. Razzaq, M. Rizwan, S. Ali, Synthesis and characterization of titanium dioxide nanoparticles by chemical and green methods and their antifungal activities against wheat rust, Chemosphere 258 (2020) 127352. https://doi.org/10.1016/j.chemosphere.2020.127352

[260] T. Jin, Y. He, Antibacterial activities of magnesium oxide (MgO) nanoparticles against foodborne pathogens, Journal of Nanoparticle Research 13 (2011) 6877-6885. https://doi.org/10.1007/s11051-011-0595-5

[261] J. Jeevanandam, Y. San Chan, M.K. Danquah, Evaluating the antibacterial activity of MgO nanoparticles synthesized from aqueous leaf extract, Med One 4 (2019).

[262] F. Kong, J. Wang, R. Han, S. Ji, J. Yue, Y. Wang, L. Ma, Antifungal activity of magnesium oxide nanoparticles: effect on the growth and key virulence factors of Candida albicans, Mycopathologia 185 (2020) 485-494. https://doi.org/10.1007/s11046-020-00446-9

[263] B. Mahdavi, S. Paydarfard, M.M. Zangeneh, S. Goorani, N. Seydi, A. Zangeneh, Assessment of antioxidant, cytotoxicity, antibacterial, antifungal, and cutaneous wound healing activities of green synthesized manganese nanoparticles using Ziziphora

Materials Research Forum LLC
https://doi.org/10.21741/9781644903711-3

clinopodioides Lam leaves under in vitro and in vivo condition, Applied organometallic chemistry 34 (2020) e5248. https://doi.org/10.1002/aoc.5248

[264] A. Derbalah, M. Shenashen, A. Hamza, A. Mohamed, S. El Safty, Antifungal activity of fabricated mesoporous silica nanoparticles against early blight of tomato, Egyptian journal of basic and applied sciences 5 (2018) 145-150. https://doi.org/10.1016/j.ejbas.2018.05.002

[265] R. Periakaruppan, R. P, J. Danaraj, Biosynthesis of silica nanoparticles using the leaf extract of Punica granatum and assessment of its antibacterial activities against human pathogens, Applied Biochemistry and Biotechnology 194 (2022) 5594-5605. https://doi.org/10.1007/s12010-022-03994-6

[266] M. Yusefi, K. Shameli, R.R. Ali, S.-W. Pang, S.-Y. Teow, Evaluating anticancer activity of plant-mediated synthesized iron oxide nanoparticles using Punica granatum fruit peel extract, Journal of Molecular Structure 1204 (2020) 127539. https://doi.org/10.1016/j.molstruc.2019.127539

[267] A.I. El-Batal, F.M. Mosalam, M. Ghorab, A. Hanora, A.M. Elbarbary, Antimicrobial, antioxidant and anticancer activities of zinc nanoparticles prepared by natural polysaccharides and gamma radiation, International journal of biological macromolecules 107 (2018) 2298-2311. https://doi.org/10.1016/j.ijbiomac.2017.10.121

[268] J. Bai, J. Xu, J. Zhao, R. Zhang, Hyaluronan and calcium carbonate hybrid nanoparticles for colorectal cancer chemotherapy, Materials research express 4 (2017) 095401. https://doi.org/10.1088/2053-1591/aa822d

[269] A.J. Hadi, U.M. Nayef, M.S. Jabir, F.A. Mutlak, Titanium dioxide nanoparticles prepared via laser ablation: evaluation of their antibacterial and anticancer activity, Surface Review and Letters 30 (2023) 2350066. https://doi.org/10.1142/S0218625X2350066X

[270] A. Pugazhendhi, R. Prabhu, K. Muruganantham, R. Shanmuganathan, S. Natarajan, Anticancer, antimicrobial and photocatalytic activities of green synthesized magnesium oxide nanoparticles (MgONPs) using aqueous extract of Sargassum wightii, Journal of Photochemistry and Photobiology B: Biology 190 (2019) 86-97. https://doi.org/10.1016/j.jphotobiol.2018.11.014

[271] M.R. Shaik, R. Syed, S.F. Adil, M. Kuniyil, M. Khan, M.S. Alqahtani, J.P. Shaik, M.R.H. Siddiqui, A. Al-Warthan, M.A. Sharaf, Mn3O4 nanoparticles: Synthesis, characterization and their antimicrobial and anticancer activity against A549 and MCF-7 cell lines, Saudi Journal of Biological Sciences 28 (2021) 1196-1202. https://doi.org/10.1016/j.sjbs.2020.11.087

[272] P. Jänicke, C. Lennicke, A. Meister, B. Seliger, L.A. Wessjohann, G.N. Kaluđerović, Fluorescent spherical mesoporous silica nanoparticles loaded with emodin: synthesis, cellular uptake and anticancer activity, Materials Science and Engineering: C 119 (2021) 111619. https://doi.org/10.1016/j.msec.2020.111619

[273] A.J. Talaei, N. Zarei, A. Hasan, S.H. Bloukh, Z. Edis, N.A. Gamasaee, M. Heidarzadeh, M.M.N. Babadaei, K. Shahpasand, M. Sharifi, Fabrication of inorganic alumina particles at nanoscale by a pulsed laser ablation technique in liquid and exploring their protein binding, anticancer and antipathogenic activities, Arabian Journal of Chemistry 14 (2021) 102923. https://doi.org/10.1016/j.arabjc.2020.102923

[274] A.I. Daniel, M.B. Umar, O.J. Tijani, R. Muhammad, Antidiabetic potentials of green-synthesized alpha iron oxide nanoparticles using stem extract of Securidaca longipedunculata, International Nano Letters 12 (2022) 281-293. https://doi.org/10.1007/s40089-022-00377-x

[275] R.D. Umrani, K.M. Paknikar, Zinc oxide nanoparticles show antidiabetic activity in streptozotocin-induced Type 1 and 2 diabetic rats, Nanomedicine 9 (2014) 89-104. https://doi.org/10.2217/nnm.12.205

[276] N. Samad, U. Ejaz, S. Kousar, A.A. Al-Mutairi, A. Khalid, Z.S. Amin, S. Bashir, S.A. Al-Hussain, A. Irfan, M.E. Zaki, A novel approach to assessing the antioxidant and anti-diabetic potential of synthesized calcium carbonate nanoparticles using various extracts of Ailanthus altissima, Frontiers in Chemistry 12 (2024) 1345950. https://doi.org/10.3389/fchem.2024.1345950

[277] W. Anupong, R. On-Uma, K. Jutamas, S.H. Salmen, S.A. Alharbi, D. Joshi, G. Jhanani, Antibacterial, antifungal, antidiabetic, and antioxidant activities potential of Coleus aromaticus synthesized titanium dioxide nanoparticles, Environmental Research 216 (2023) 114714. https://doi.org/10.1016/j.envres.2022.114714

[278] O.B. Afolabi, O.I. Oloyede, B.T. Aluko, J.A. Johnson, Biosynthesis of magnesium hydroxide nanomaterials using Monodora myristica, antioxidative activities and effect on disrupted glucose metabolism in streptozotocin-induced diabetic rat, Food Bioscience 41 (2021) 101023. https://doi.org/10.1016/j.fbio.2021.101023

[279] A. Ahmad, S. Noor, S.A. Muhammad, S.B. Hussain, Eco-Friendly synthesis of manganese nanoparticles from Syzygium aromaticum and study of their biological activities, Inorganic and Nano-Metal Chemistry (2024) 1-14. https://doi.org/10.1080/24701556.2024.2313229

[280] R.G. Kerry, K.R. Singh, S. Mahari, A.B. Jena, B. Panigrahi, K.C. Pradhan, S. Pal, B. Kisan, J. Dandapat, J. Singh, Bioactive potential of morin loaded mesoporous silica nanoparticles: A nobel and efficient antioxidant, antidiabetic and biocompatible abilities in in-silico, in-vitro, and in-vivo models, OpenNano 10 (2023) 100126. https://doi.org/10.1016/j.onano.2023.100126

[281] M.T. Yaraki, Y.N. Tan, Metal nanoparticles-enhanced biosensors: synthesis, design and applications in fluorescence enhancement and surface-enhanced Raman scattering, Chemistry-An Asian Journal 15 (2020) 3180-3208. https://doi.org/10.1002/asia.202000847

[282] A.A. Saei, J.E.N. Dolatabadi, P. Najafi-Marandi, A. Abhari, M. de la Guardia, Electrochemical biosensors for glucose based on metal nanoparticles, TrAC Trends in Analytical Chemistry 42 (2013) 216-227. https://doi.org/10.1016/j.trac.2012.09.011

[283] S. Malathi, I. Pakrudheen, S.N. Kalkura, T. Webster, S. Balasubramanian, Disposable biosensors based on metal nanoparticles, Sensors International 3 (2022) 100169. https://doi.org/10.1016/j.sintl.2022.100169

[284] R. Sankar, P.K. Rahman, K. Varunkumar, C. Anusha, A. Kalaiarasi, K.S. Shivashangari, V. Ravikumar, Facile synthesis of Curcuma longa tuber powder engineered metal nanoparticles for bioimaging applications, Journal of Molecular Structure 1129 (2017) 8-16. https://doi.org/10.1016/j.molstruc.2016.09.054

[285] S.R. Barman, A. Nain, S. Jain, N. Punjabi, S. Mukherji, J. Satija, Dendrimer as a multifunctional capping agent for metal nanoparticles for use in bioimaging, drug delivery and sensor applications, Journal of Materials Chemistry B 6 (2018) 2368-2384. https://doi.org/10.1039/C7TB03344C

[286] B. Uzair, A. Liaqat, H. Iqbal, B. Menaa, A. Razzaq, G. Thiripuranathar, N. Fatima Rana, F. Menaa, Green and cost-effective synthesis of metallic nanoparticles by algae: Safe methods for translational medicine, Bioengineering 7 (2020) 129. https://doi.org/10.3390/bioengineering7040129

[287] B. Skóra, K.A. Szychowski, J. Gmiński, A concise review of metallic nanoparticles encapsulation methods and their potential use in anticancer therapy and medicine, European Journal of Pharmaceutics and Biopharmaceutics 154 (2020) 153-165. https://doi.org/10.1016/j.ejpb.2020.07.002

[288] F.O. Ohiagu, P. Chikezie, C. Ahaneku, C. Chikezie, Human exposure to heavy metals: toxicity mechanisms and health implications, Material Sci Eng 6 (2022) 78-87. https://doi.org/10.15406/mseij.2022.06.00183

[289] S. Medici, M. Peana, A. Pelucelli, M.A. Zoroddu, An updated overview on metal nanoparticles toxicity, Seminars in cancer biology, Elsevier, 2021, pp. 17-26. https://doi.org/10.1016/j.semcancer.2021.06.020

[290] W. Khan, S. Subhan, D.F. Shams, S.G. Afridi, R. Ullah, A.A. Shahat, A.S. Alqahtani, Antioxidant potential, phytochemicals composition, and metal contents of Datura alba, BioMed research international 2019 (2019) 2403718. https://doi.org/10.1155/2019/2403718

[291] Y. Zhang, R. Huang, X. Zhu, L. Wang, C. Wu, Synthesis, properties, and optical applications of noble metal nanoparticle-biomolecule conjugates, Chinese Science Bulletin 57 (2012) 238-246. https://doi.org/10.1007/s11434-011-4747-x

[292] J. Jeevanandam, J. Rodrigues, Sustainable synthesis of bionanomaterials using non-native plant extracts for maintaining ecological balance: A computational bibliography analysis, Journal of Environmental Management 358 (2024) 120892. https://doi.org/10.1016/j.jenvman.2024.120892

[293] A. Manuja, B. Kumar, R. Kumar, D. Chhabra, M. Ghosh, M. Manuja, B. Brar, Y. Pal, B. Tripathi, M. Prasad, Metal/metal oxide nanoparticles: Toxicity concerns associated with their physical state and remediation for biomedical applications, Toxicology Reports 8 (2021) 1970-1978. https://doi.org/10.1016/j.toxrep.2021.11.020

[294] K. Kasture, P. Shende, Amalgamation of artificial intelligence with nanoscience for biomedical applications, Archives of Computational Methods in Engineering 30 (2023) 4667-4685. https://doi.org/10.1007/s11831-023-09948-3

[295] E.A. Bamidele, A.O. Ijaola, M. Bodunrin, O. Ajiteru, A.M. Oyibo, E. Makhatha, E. Asmatulu, Discovery and prediction capabilities in metal-based nanomaterials: An overview of the application of machine learning techniques and some recent advances, Advanced Engineering Informatics 52 (2022) 101593. https://doi.org/10.1016/j.aei.2022.101593

[296] J. Li, C. Wang, L. Yue, F. Chen, X. Cao, Z. Wang, Nano-QSAR modeling for predicting the cytotoxicity of metallic and metal oxide nanoparticles: A review, Ecotoxicology and Environmental Safety 243 (2022) 113955. https://doi.org/10.1016/j.ecoenv.2022.113955

[297] P.K.D. Pramanik, A. Solanki, A. Debnath, A. Nayyar, S. El-Sappagh, K.-S. Kwak, Advancing modern healthcare with nanotechnology, nanobiosensors, and internet of nano things: Taxonomies, applications, architecture, and challenges, IEEE Access 8 (2020) 65230-65266. https://doi.org/10.1109/ACCESS.2020.2984269

Applications for Earth-Abundant Transition Metals
Materials Research Foundations 179 (2025) 79-96

Materials Research Forum LLC
https://doi.org/10.21741/9781644903711-4

# Chapter 4

# Earth-abundant metals-Based Therapeutics: Transport, Design, and Applications

Aisha Yasin, Urwa Muaaz, Irtaza Ashraf, Atifa Batool, Muhammad Ramzan Saeed Ashraf Janjua, Syed Ali Raza Naqvi*

Department of Chemistry, Government College University Faisalabad, Faisalabad-38040, Pakistan

* draliraza@gcuf.edu.pk

## Abstract

At present, a total of 76 elements have been systematically incorporated into compounds that have either obtained clinical approval or are employed for diagnostic and therapeutic purposes. The advanced analytical techniques and instruments in cellular and molecular biology, spanning proteomics and genomics, have the potential to significantly influence the development and application of essential elements for therapeutic and diagnostic objectives. This includes both stable and radioactive compounds. The use of abundant earth metals in the pharmaceutical industry is attracting considerable attention due to their potential in creating cost-effective, sustainable, and innovative therapeutic solutions. Furthermore, abundant earth metals often demonstrate unique and desirable chemical properties that can be leveraged in drug design and synthesis. However, challenges exist in identifying elemental speciation in biological ecosystems. The integration of abundant earth metals into drug development and manufacturing processes presents an opportunity to create more affordable, effective, and sustainable therapeutic solutions. This ultimately contributes to improved healthcare outcomes globally.

## Keywords

Medicinal Inorganic Chemistry, Metal Based Drugs, Transition Metals, Metallopharmaceuticals, Earth Abundant Metal Complexes, Drug Development

## Contents

## 1.     Introduction

Organic and inorganic compounds have played a significant role in the field of medicine throughout human history [1-3]. Dating back approximately 5000 years before Christ (BC), ancient Indian cultures utilized Ayurvedic medical practices that incorporated various metals and minerals, including mercury, gold, silver, lead, zinc, and copper [4]. In China, mineral-based "stone" medicines have been used for almost two millennia. These remedies include processed minerals such as mirabilite and calomel, as well as naturally occurring minerals like pyrite (Pyritum), cinnabar (Cinnabaris), and realgar ($As_4S_4$). Stone remedies also involve fossils of fauna and dragon bones. Many stone medicines contain trace elements such as mica and often incorporate supplementary elements like lithium and zirconium, which are added to muscovite derived from pegmatitic sources.

At present, not more than 76 elements which are being used in compounds approved for the use in clinical settings or are undergoing clinical trials for diagnostic or therapeutic purposes [5-7].

Traditional empirical applications of all the 76 elements which are the part of thousands of medicines, have been identifying with promising biological activity through phenotypic screening, are now being improved by more advanced screening methods and techniques aimed at understanding activity and target sites. Modern advancements in physical techniques such as photoactivation, ultrasound, X-ray Absorption Spectroscopy/X-ray Fluorescence (XAS/XRF), Positron Emission Tomography (PET), Single Photon Emission Computed Tomography (SPECT), and Magnetic Resonance Imaging (MRI), are contributing to the progression of contemporary approaches. Additionally, developments in synthesis and characterization methods, including combinatorial chemistry, further support the advancement of these techniques. The use of medicinal or diagnostic agents, whether approved pharmaceuticals or under clinical evaluation, is expanding within the medical field [8-10]. Compounds containing beryllium and nickel are known as major allergens. T-cells that recognize beryllium and nickel complexes of self-peptides may cause hypersensitivity reactions in humans [11]. While rhodium-based antibacterial and anticancer compounds show promise, they have not yet been used in clinical practice [12].

The primary aim of this chapter is to uncover the potential metal complexes for therapeutic and diagnostic purposes that can effectively modulate their reactivity within biological systems. This involves considering the varying oxidation states exhibited by transition metal ions and the presence of redox-active sites in their ligands. Furthermore, the role of metal complexes as catalysts is emphasized, with the metal serving as the catalytic nucleus and the ligands contributing to the overall catalytic activity. It is noteworthy that even in minimal concentrations, catalytic metal complexes can exert a more substantial influence on biological processes compared to their native enzymatic counterparts [13-15]. The kinetic lability and thermodynamic stability of metal complexes are influenced by a variety of factors, including the characteristics of the ligands such as their coordination geometry, and outer-sphere interactions. The properties of metal ions and their oxidation states largely determine their characteristics.

## 2.    Transport of iron and metallodrugs by transferrin

Human serum contains a vital protein known as transferrin, which not only binds to a variety of metallic ions but also has a strong affinity for iron. Transferrin plays a crucial role in the administration and transportation of specific medications and diagnostic tools. It is essential for the precise delivery and activation of iron in metal complexes with antibacterial and anticancer properties, ensuring that the iron is effectively transported to target cells and released or activated upon cellular penetration. Additionally, iron has been fundamental for the emergence of life, and pyrite ($FeS_2$), the most common metal sulfide on Earth, may have served as the initial energy source for life. In some microorganisms, a protein called ferric-ion-binding protein (Fbp) serves as the bacterial equivalent of transferrin [16]. Within the cellular environment, iron is released from vesicles when the pH is reduced to approximately 5.6, at which point apo-transferrin is recycled back into the serum. While transferrin maintains a strong binding affinity for iron(III) (Fe(III)), it is also capable of binding and transporting a variety of other metals, including medicinal Bi(III) antiulcer agents, radio diagnostic gallium(III) (Ga(III)) and indium(III) (In(III)) ions, as well as toxic metal ions such as aluminium(III) (Al(III)) and plutonium(IV) (Pu(IV)). Some researchers posit that substantial binding to hydroxides may indicate "covalency" or a degree of softness in metal-ligand interactions.

## 2.1    Titanium based transferrin

Role of titanium in biological systems is still a fascinating mystery waiting to be unraveled as it is the tenth abundant element in Earth's crust. Lately, there has been growing interest in Ti(IV) complexes within the scientific community due to their potential anticancer properties. However, certain Ti(IV) complexes like budotitane and titanocene dichloride have been withdrawn from clinical trials due to their high reactivity in water, posing challenges in formulation. Future research focusing on alternative chelated Ti(IV) complexes with improved hydrolytic stability holds promise for various applications. It's worth noting that Ti(IV) has a stronger binding affinity to serum transferrin compared to Fe(III). The overexpression of transferrin receptors in neoplastic cells suggests that Ti(IV) may be taken up at the cellular level through a transferrin-mediated mechanism. Additionally, Ti(IV) demonstrates a strong affinity for bacterial transferrin, indicating the potential for the development of metallo-antibiotics based on Ti(IV) [17].

## 2.2    Bismuth based transferrin

The use of bismuth compounds for treating gastrointestinal issues has been well established. However, their recent attention has been due to their antibacterial properties. It has been found that medications containing bismuth can effectively eliminate Helicobacter pylori, which is known to hinder the healing of ulcers. When it comes to the binding of metals, the transferrin binding constants at the N- and C-lobe sites are comparable, although there are usually slight differences. Fe(III) exhibits a preference for binding to the C-lobe [18].

## 2.3    Ruthenium based transferrin

Ruthenium (Ru) is a key element in drug delivery, especially in the field of anticancer treatments. This document explores the development of ruthenium complexes as metal-based anticancer drugs, which are currently undergoing extensive preclinical research. These complexes have unique photophysical properties that can be used in light-activated anticancer drugs and photodynamic therapy [19]. Moreover, certain Ru complexes have displayed potential for cancer immunotherapy due to their immunogenic properties. The report also delves into the debated role of transferrin (Tf) in the transportation and cellular uptake of ruthenium anticancer drugs, indicating that the cellular uptake of specific Ru complexes, such as KP1019, occurs mainly through passive diffusion rather than Tf-mediated pathways [20].

Furthermore, Ru complexes may be suitable for intratumoral injections, which hold significant implications for treating incurable malignancies. The report also discusses the interactions of Ru complexes with proteins like transferrin and albumin, which can influence the drugs bioavailability and therapeutic effectiveness. Additionally, it addresses how the poor stability of Ru complexes in biological media could be advantageous for direct tumor treatment, given their high cytotoxicity, rapid cellular uptake, and production of fewer hazardous breakdown products [18].

## 2.4    Cobalt based transferrin

The term "nanoparticles" (NPs) refers to very tiny particles that have a diameter of 1 to 1000 nm. Metal nanoparticles (MNPs) have been a significant focus in biomedical applications, with cobalt nanoparticles (Co NPs) gaining particular attention due to their innovative qualities. Co NPs are utilized in both the medical and engineering fields, and research has indicated their potential for applications such as phototherapy, thermotherapy, and chemotherapy, while also noting their capability to cause cell death [21].

## 3.    Metal complexes as probes for imaging

Various imaging techniques are commonly used in clinical settings, such as computed tomography, optical imaging using fluorescence/phosphorescence, and magnetic resonance imaging (MRI). These techniques often involve metal complexes [10, 11, 22-24]. This section predominantly delves into MRI and radionuclide imaging, which serve as the primary clinical applications of imaging probes comprised of metal complexes [25].

### 3.1    Employing metal complexes as contrasting agents for MRI

Metal complexes play a crucial role in Magnetic Resonance Imaging (MRI) by helping us understand the behavior of water molecules in different parts of the body. Compounds with a paramagnetic metal core can speed up the relaxation rate of nearby water molecules, which enhances tissue contrast in MRI scans. Gd(III), a lanthanide ion with strong magnetic properties, is commonly used in MRI contrast agents due to its seven unpaired 4f electrons. However, when creating Gd(III) MRI contrast agents, it's important to find the right balance between the complex's stability and the number of coordinated water molecules [26].

To reduce potential toxicity, Gd(III) is tightly bound in contrast agents, as it can replace Ca(II) in biological systems. Gadolinium (III)-based MRI contrast agents that are clinically approved typically have a single water molecule directly linked to the Gd(III) –DTPA complex generally termed as Gadovist (**compound 1, Fig. 1**) or Gd(III)-DOTA complex which is generally termed as Dotarem (**compound 2, Fig. 1**) . These compounds are mainly used for visualizing anatomy and extracellular imaging. Gadolinium (III) and other lanthanide-based pH-responsive contrast agents are designed to enhance their effectiveness in specific low pH environments, particularly in oncological settings. They contain ligands that, under normal pH conditions, bind to the metal core, preventing water from entering and enhancing contrast. However, at low pH, the ligand is released. Alternative substances can be used to monitor changes in the concentration of specific endogenous metal ions associated with various pathological states. These agents include a metal-specific chelating ligand and can respond to the presence of those ions by activating the contrast [27].

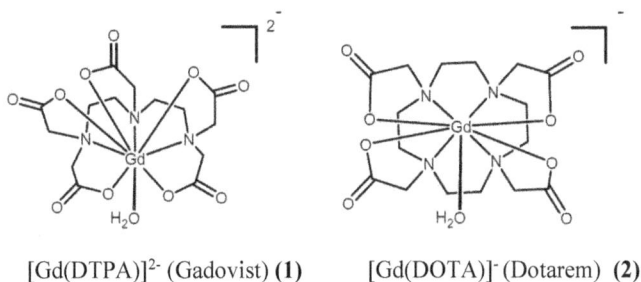

[Gd(DTPA)]$^{2-}$ (Gadovist) **(1)**        [Gd(DOTA)]$^-$ (Dotarem)  **(2)**

*Fig 1    Gd (III) contrast agents*

## 4. Design of metallodrug

Finding the right locations to target for therapy is a crucial part of the drug discovery process. This helps to focus on tumor cells instead of healthy ones, reducing unintended side effects. Metal complexes often have the disadvantage of targeting multiple locations, but this can also be advantageous in combating resistance. While organic drugs are usually designed to target a single location in cells, they may not always be single-targeted. It is essential to remember that hydrogen-bond acceptors play a major role in the creation of organic drugs. These sites that bind protons, such as amine nitrogens, can also bind metals, but there has been limited research into this aspect of organic drug action [5].

### 4.1 Gold complexes

The Au(I) cyanide complex, also known as $Au(CN)_2$, was widely used to treat tuberculosis approximately a century ago. Safer Au(I) thiolate complexes, such as aurothioglucose (Solganal) and aurothiomalate (Myocrisin), were initially introduced as treatments for rheumatoid arthritis in the 1930s. Compound 3 an oral medication introduced in 1985, is currently considered the gold standard for treating osteoarthritis. In 2012, a study involving over 13,000 bioactive compounds, including approved medications, revealed the anti-parasitic effects of compound 3 (**Fig.2**). This discovery led to a phase I clinical investigation to confirm the drug's safety for potential use as a broad-spectrum anti-parasitic treatment, reigniting interest in this medication. Additionally, ongoing research is exploring the anti-cancer properties of auranofin, as initially reported by Lorber. Although strong oxidants are typically found in gold(III) complexes, the appropriate chelating ligands can effectively stabilize square-planar Au(III), which shares the same electronic structure as Pt(II). Several Au(III) complexes have exhibited promising pharmacological activity, utilizing ligands such as dithiocarbamate and porphyrins [28].

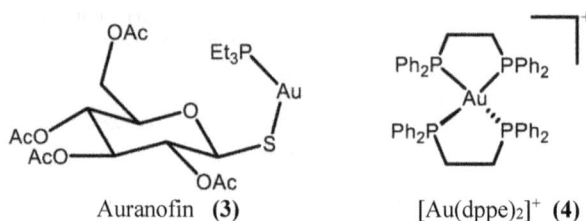

Auranofin (**3**)          $[Au(dppe)_2]^+$ (**4**)

*Fig.2 Structures of the linear Au(I) oral anti-arthritic drug auranofin and the tetrahedral Au(I) anticancer complex $[Au(dppe)_2]^+$*

### 4.2 Ruthenium complexes

Ru(III) complexes, known as compound **5** and compound **6** (along with its more soluble form, compound 7/IT139), have been the subject of recent clinical trials as potential candidates for cancer treatment (**Fig. 3**). These complexes have shown potential for cancer treatment, but the specific target sites within the body are not yet fully understood. It is believed that the activation mechanism of these complexes in the body involves their reduction to Ru(II). Compound 5 is particularly notable for its anti-metastatic properties, while compound **6** has demonstrated cytotoxic effects. However, clinical studies of compound **5** have not yet provided conclusive evidence of its therapeutic efficacy in cancer treatment. Researchers have also been exploring ways to enhance the stability of Ru(II) complexes, such as by incorporating a $\pi$-bound arene, although the effectiveness of this enhancement depends on the specific additional ligands involved. The optimal level of stability required for effective anticancer activity is contingent upon the intended mechanism of action. The complex compound **7** (**Fig. 3**) undergoes rapid hydrolysis in aqueous conditions, leading to the loss of the Cl ligand. Subsequently, it interacts with DNA, forming mono-functional adducts, with a preference for binding to guanine at its $N_7$ position. This study provides valuable new insights into the potential mechanisms of action of these Ru(III) and Ru(II) complexes for cancer treatment [13].

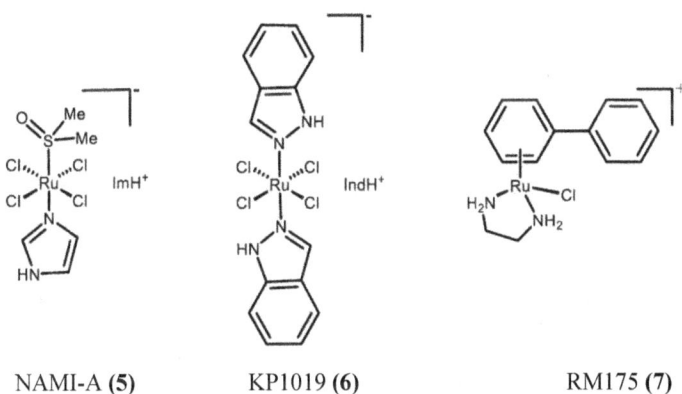

NAMI-A (**5**)          KP1019 (**6**)          RM175 (**7**)

*Fig.3  Ru(II) anticancer complexes KP1019, NAMI-A and RM175*

### 4.3    Platinum Complexes

After the discovery of cisplatin's effectiveness in treating cancer in the 1960s, platinum-based medications became widely used in cancer treatment. Platinum-based drugs, such as cisplatin and its derivatives, have been pivotal in the treatment of various cancers including ovarian, colon, lung, testicular, cervical, and others. These drugs share structural features such as neutrality, planar-square shape, and cis isomerism, and typically include two ligands. While these drugs have shown promise in cancer treatment, they also come with drawbacks such as nephrotoxicity, neurotoxicity, ototoxicity, and the potential for cancer cells to develop resistance. To address these limitations, novel cisplatin derivatives are being considered. Heptaplatin, oxaliplatin, and carboplatin are used globally as anticancer medications, while lobaplatin and nedaplatin are used as local anticancer medications [29].

### 4.4    Copper complexes

The discovery of cisplatin, a potent anticancer drug, has opened up new possibilities for developing safer and more effective metal-based chemotherapy treatments. Utilizing endogenous metals instead of gold, silver, or platinum is proposed to be less harmful to normal cells. Copper-based compounds are being considered as a potential alternative for chemotherapy due to the lower oxygen levels in tumor cells, which trigger their invasion, metastasis, and the anaerobic process known as the Warburg effect. Taking advantage of tumor hypoxia, new prodrugs that activate specifically in low-oxygen environments of cancer cells can be developed [29]. Copper ions, which can exist in two distinct oxidation states, are highly desirable for use within cells. In hypoxic cancer cells, copper(II) ions are converted to Cu(I), enabling copper compounds to target malignancies at the cellular level. Once created, Cu(I) ions stimulate the generation of reactive oxygen species (ROS) and reactive nitrogen species (RNS), leading to cell death.

The main ways that Cu(II) complexes work against cancer are:

• Intercalation: the van der Waals link that allows a copper complex to be inserted between adjacent DNA bases

• Interaction with DNA at the "minor groove," or "small cavity," level of the DNA molecule

• Oxidative mechanism: prompt production of reactive oxygen species (ROS) in the vicinity of the DNA molecule. Included in this is the production of hydroxyl radicals when Cu(II) is reduced to Cu(I) in the presence of reducing chemicals such as $H_2O_2$.

$$LCu\,(II) + H_2O_2 \rightarrow LCu\,(I) + {}^{\bullet}OOH + H^{+}$$

$$LCu\,(I) + H_2O_2 \rightarrow LCu\,(II) + {}^{\bullet}OH + OH$$

where an organic ligand is denoted by L.

The hydrolytic mechanism involves the breaking of phosphodiester bonds and subsequent destruction of nucleic acid molecules as a result of interactions between Cu(II) ions and the phosphate anion from the nucleotides. Proteasome inhibition, on the other hand, refers to the selective degradation and recycling of intracellular proteins by multiprotein complexes found in both the cytoplasm and nucleus. Topoisomerase I or II is involved in DNA replication and transcription [29].

## 5.    Metallo-antibiotics

Today, there is a growing concern about microbial resistance to prescribed medications. To tackle this issue, we have explored two innovative approaches. Firstly, we have used photoactivation to introduce a modern anti-tuberculosis medication into bacteria. This process has been incredibly successful, with the medication becoming 5.5 times more effective against Mycobacterium smegmatis when exposed to blue light. The medication release occurs in less than 500 nanoseconds, and it specifically targets the bacteria while posing minimal risk to healthy human cells. Secondly, we have employed organometallic fragments to modify organic antimicrobial medications. One exciting development is the organometallation of the antimalarial medication sulfadoxine, which has resulted in new compounds that combat drug resistance and show promise in targeting novel locations. Moreover, our research has revealed strong antibacterial activity in several complexes, including activity against significant pathogens such as Candida albicans, Candida neoformans, and methicillin-resistant Staphylococcus aureus (MRSA) [30-32].

Overall, our findings suggest that the conjugation of organic pharmaceuticals with organometallic fragments such as compounds 8 & 9 (Fig. 4) holds great potential for repurposing existing medications. Clinical studies on chloroquine conjugate ferroquine for treating resistant malaria parasites are already underway, and the exploration of ferrocene derivatives such as ferrocifens, a potent combination of tamoxifen and ferrocene, shows promise for future clinical testing. This research highlights the exciting possibilities for combating microbial resistance through innovative approaches [33].

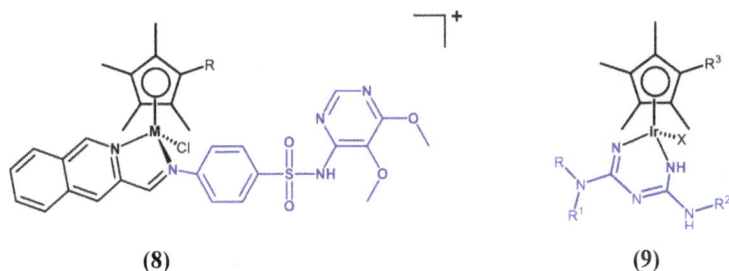

*Fig.4    Organomettalic complexes: (8) Organometallic conjugates of the drug sulfadoxine (blue), M=Rh(III), Ir(III), with antimicrobial activity (9) Organo-iridium(III) antimicrobial biguanide complexes*

## 6.    Metal based radiopharmaceuticals

A significant proportion of radiopharmaceuticals sanctioned by Plate Number 1 are metal complexes, with numerous additional candidates currently in the clinical trial phase. These agents can be utilized for either radiotherapy (beta, alpha, electron capture) or diagnostic purposes (positron or gamma-ray emission), depending on the nature of the radioactive emissions involved. $^{99m}$Tc generators, which are widely available and possess suitable decay properties, have become the preferred radioisotope for clinical SPECT applications. This radioisotope has a half-life of six hours and emits a single γ ray with an energy of 140 keV. Complexes based on Tc(I) and Tc(V) have been extensively utilized in the fabrication of radiopharmaceuticals, providing various structural advantages [34, 35]. Alongside $^{99m}$Tc, other SPECT radio-metals like $^{67}$Ga and $^{111}$In were also previously employed. Additionally, chelators based on iron-sequestering hydroxypyridinone units and other compounds have been extensively studied. Radio-metals like $^{89}$Zr and $^{64}$Cu have been investigated in PET chemistry, each offering specific benefits such as lengthy half-life and distinctive applications [36].

### 6.1    Metals in radiotherapy

In the field of nuclear medicine, theranostic pairings that target SSTR2 receptors, such as $^{68}$Ga/$^{177}$Lu-DOTATATE, have become very popular. $^{177}$Lu-dotatate, or lucathera, is approved to treat gastric enteropancreatic neuroendocrine tumors. $^{177}$Lu-DOTA-PSMA variants are being used in 188 active clinical studies. These same targeted chemicals have been used in clinical investigations of the beta emitter $^{89}$Y, and radio-immunotherapy has found use for it [37]. Recent studies have shown that the use of $^{225}$Ac PSMA-617, a DOTA derivative, may be beneficial in treating individuals with metastatic prostate cancer. Some patients have experienced total remission, sparking increased interest in using alpha emission in medical settings. Alpha emission has a limited range and cannot penetrate biological tissues, even at high energy, but it can deliver focused, potent anti-cancer effects. While the traditional management of bone metastases has involved the use of $^{223}$Ra for palliative care, advancements in chelators for alpha-emitting isotopes, such as $^{225}$Ac and its daughter radioisotope $^{213}$Bi, are paving the way for more targeted radiotherapy using alpha emitters [38].

## 6.2    Gallium based radiopharmaceutical

Gallium-67 ($^{67}$Ga) isotope is produced by bombarding $^{68}$Zn with protons at a centralized cyclotron facility, resulting in γ radiation with a half-life of 78 hours. It is then delivered to the user's location for γ-scintigraphy and SPECT imaging, particularly for imaging infections, inflammation, lymphomas, and other cancers. $^{67}$Ga formulations, especially citrate-based ones, are commonly used for this purpose. However, the use of $^{67}$Ga for lymphoma imaging has largely been replaced by $^{18}$F-fluorodeoxyglucose (FDG) imaging with positron emission tomography (PET). Despite FDG imaging's lower specificity, it is still utilized for infection imaging [38].

Gallium has various applications in nuclear medicine. Gallium-3+ ($Ga^{3+}$) has a strong affinity for fluoride ions, widely used in nuclear medicine as fluorine-18, a radioactive isotope that emits positrons. Chemistry is being explored to incorporate gallium radionuclides into biomolecules for PET imaging. Gallium may also be involved in anti-cancer metallodrugs such as tris(8-hydroxyquinolinate) and gallium nitrate (Ganite). Research is ongoing regarding the use of radionuclide imaging with $^{67}$Ga and $^{68}$Ga to study the mechanisms and pharmacokinetics of these medications [39].

## 6.3    Indium based radiopharmaceutical

Another isotope, $^{111}$In, with a physical half-life of 67 hours, is created in a cyclotron and decays to yield two gamma photons at 173 and 247 KeV, allowing for late imaging. However, there are drawbacks, including low-quality images and an 18–30 hour imaging delay after injection due to incorrect photon energy. $^{111}$In is also used in the radiolabeling of various materials. Furthermore, the synthetic long-acting cyclic octapeptide octreotide (Sandostatin®) exhibits pharmacological properties with the hormone somatostatin. It binds to somatostatin receptors and is a more potent inhibitor of insulin, glucagon, and growth hormone than somatostatin. SPECT imaging using $^{111}$In-octreotide is employed in somatostatin receptor scintigraphy for detecting neuroendocrine tumors. The FDA-approved radiopharmaceutical Octreoscan$^{TM}$ contains phenetreotide, a complex compound with specific chemical properties [40].

## 6.4    Yttrium-90 based radiopharmaceutical

Yttrium-90 ($^{90}$Y) radioembolization, also known as selective internal radiation treatment (SIRT) or transarterial radioembolization (TARE), involves injecting millions of $^{90}$Y-labeled microspheres directly into the liver to induce radiation-induced necrosis. This process can lead to ischemia and embolism. The $^{90}$Y in the microspheres decays to stable $^{90}$Zr through β-decay, which does not offer any therapeutic benefits. During the decay of $^{90}$Y, high-energy β-particles with an average energy of 0.9267 MeV (maximum of 2.28 MeV) are released, with a half-life of 64.04 hours (2.67 days). Approximately 94% of the radiation is discharged within 11 days. The β-particles aim to cause radiation damage to nearby tissues by penetrating up to a depth of 2.5 mm, with a maximum penetration of 11 mm [41].

## 6.5 Copper based radiopharmaceutical

There are various copper isotopes used for cancer imaging and treatment. Among these, $^{64}$Cu, produced by a cyclotron, is considered one of the most versatile radionuclides with high potential for therapeutic applications $^{64}$Cu is suitable for tracers with fast pharmacokinetics such as small compounds and peptides, as well as slow pharmacokinetic agents like mAbs and stem cell monitoring, due to its intermediate half-life of 12.7 hours. Moreover, the relatively low dose of $^{64}$Cu emitting positron ($\beta^+$) reduces patient burden and concerns about dosimetry. Its 12.7-hour half-life allows for both therapeutic and high-resolution PET imaging due to the simultaneous emission of auger electrons and beta particles ($\beta^-$) [42].

Recent research suggests that $^{64}$CuCl$_2$, the simplest form of $^{64}$Cu, has advantages over other $^{64}$Cu-labeled compounds. Preclinical research in cellular and animal models has shown $^{64}$CuCl$_2$'s potential as a theranostic drug in glioblastoma, PCa, and melanoma, among other human malignancies. It enables the investigation of in vivo copper transport and its variations in different disease conditions using PET, shedding light on the biological handling of this crucial component and potentially leading to valuable $^{64}$Cu imaging applications. However, not all of its potential has been explored. Some research on animal tumor models, Alzheimer's disease, Wilson's disease, and Menkes disease have shown documented alterations in copper distribution thought to be related to the activity of the Cu transporter Ctr1. Most of these studies involved intravenous injections of acetate or $^{64}$Cu-dichloride. However, this behavior deviates from the typical biological pathway for copper transit. It is believed that understanding biological copper transit fully requires different delivery methods, particularly oral delivery, as well as exploring additional biological forms of copper, such as copper bonded to amino acids and albumin. Studying ceruloplasmin is one of the promising outcomes to consider [43].

## 7. Application of metal nanoparticles in pharmacology

Nanoparticles synthesized differently have been extensively used in various in vitro diagnostic applications. Gold and silver nanoparticles have been found to be extensively used as antibacterial agents against a wide range of pathogens in humans and animals. Silver nanoparticles have a wide range of applications as an antibacterial agent in commonly used medicinal and consumer products. Another application for nanoparticles is as a biosensor. Gold nanoparticles have applications in disease management, enamel discoloration, and in the production of spectacles. A novel drug delivery method involves the use of metal nanoparticles, which provide a large surface area for delivering medications and genes at the nanoscale, resulting in adjustable stability and minimal toxicity. Citrate-stabilized gold nanoparticles have been widely used to specifically target human oral squamous cell carcinoma (HSC$_3$) cancer cells. These gold nanoparticles were coated with EGFR (anti-epidermal growth factor) to specifically target cancer cells (receptor). The photothermal action enhanced the effectiveness of the gold nanoparticles, leading to a twenty-fold expansion as a result of local heating caused by light radiation [44]. Additional medication applications are covered in the section below.

Applications for Earth-Abundant Transition Metals    Materials Research Forum LLC
Materials Research Foundations 179 (2025) 79-96    https://doi.org/10.21741/9781644903711-4

## 7.1    Antimicrobial activity of metal nanoparticles

Silver nanoparticles are commonly used as antimicrobial agents due to their proven antiviral, antifungal, and anti-inflammatory properties. These nanoparticles interact with proteins, DNA, or microbial enzymes to render microorganisms inactive and halt their growth. Studies have shown that silver nanoparticles exhibit strong antibacterial properties against various types of bacteria. The effectiveness of silver nanoparticles is dependent on their form, and research has revealed their ability to effectively prevent HIV infections. Additionally, research has demonstrated that silver and gold nanoparticles, ranging in size from 1 to 10 nm, can interact with the HIV-1 glycoprotein, impeding the virus's ability to bind to host cells [45]. Moreover, biosynthesized metal nanoparticles, especially silver nanoparticles derived from plant extracts, have shown potential fungicidal effects greater than those of commonly used antibiotics such as fluconazole and amphotericin. These nanoparticles have been applied to combat Candida sp., showing damage to the fungal membrane and intracellular components, ultimately leading to the death of the fungal cell. Compared to current antifungal medications, these nanoparticles could offer a more effective and safer alternative, as current medications may have limited efficacy and significant adverse effects, including nausea, renal failure, liver damage, elevated body temperature, and diarrhea. Overall, nanoparticles are a promising and improved approach for combating microbial infections [46].

### 7.1.1    Mechanism of Antimicrobial activity

The mechanism of action of metal nanoparticles as antibacterial agents is not yet fully understood, but several publications have outlined the potential process. The antibacterial properties of metal nanoparticles depend on various factors, including size, shape, and surface charge. These nanoparticles can penetrate bacterial cell walls, particularly in Gram-positive bacteria, and subsequently inhibit protein production, deactivate DNA and enzymes, and ultimately lead to bacterial cell death [47-49]. The large surface area of nanoparticles also enables effective interaction with microbes. The shape of nanoparticles also influences their antibacterial activity. For instance, research has shown that silver nanoparticles with different shapes but the same surface area can exhibit varying levels of antibacterial activity. Triangle-shaped nanoparticles, due to their greater surface area to volume ratios and crystallographic frameworks, demonstrate the highest antibacterial action. Nanoparticles can also generate free radicals, which have the potential to harm microbial cells by damaging their respiratory systems and cell walls, ultimately driving the microorganisms out. Furthermore, nanoparticles have been found to control bacterial signal transduction processes and exhibit larvicidal action against vectors of filariasis and malaria. Additionally, they show activity against some cancer cells and plasmodial infections [50].

## 7.2    Anti-inflammatory activity

The anti-inflammatory action plays a crucial role in the wound healing process. The biological mechanism of anti-inflammation generates several compounds, including cytokines, interleukins, and inflammatory mediators such as enzymes and antibodies released by B cells, T cells, and macrophages. Main organs of immunity release anti-inflammatory substances like IL-1 and IL-2 [51, 52]. These inflammatory mediators take part in metabolic processes and regulate the growth of illnesses. Research has shown that nanoparticles of gold and platinum produced using plant extracts have a strong beneficial effect on wound healing and tissue regeneration through their anti-inflammatory functions. This suggests that platinum and gold nanoparticles are being used to naturally prevent inflammation [53].

## 7.3 Anticancer activity

The study used human lung cancer cells to investigate the toxicity of silver nanoparticles encapsulated in plant latex. The results demonstrated that the silver nanoparticles are toxic to AS49 cells in a dose-dependent manner. Additionally, plant latex can act as a stabilizing agent for the water-soluble silver nanoparticles and can facilitate their delivery to the target cells. Gold nanoparticles with a size of 2 nm, have been incorporated into the chemotherapy drug Paclitaxel. According to the TGA evaluation data, it has been observed that around 70 paclitaxel molecules can bind to a single gold nanoparticle [54, 55]. This particular property of the gold nanoparticles makes them suitable to serve as a scaffold for the transportation of heavy biomolecules, including DNA, RNA, proteins, and peptides [56].

## 7.4 Antiviral activity

The discovery of metallic nanoparticles with multiple binding sites that interact with the viral membrane's gp120 proteins has revealed a promising mechanism for combatting HIV. These bio-based nanoparticles have been found to effectively inhibit both cell-associated and cell-free viruses. Recent studies have also identified gold and silver nanoparticles as effective disruptors of the HIV-1 life cycle, particularly in the post-entry stage, making them potential antiviral agents against retroviruses. Additionally, research has shown that plant-derived nanoparticles have the potential to serve as alternative treatments for inhibiting the spread of viral infections. Viruses can rapidly infect hosts and expand their populations. It has been found that silver nanoparticles derived from plant extracts can act as potent antiviral medications for a variety of viral diseases. Metallic nanoparticles have also been reported to possess strong antiviral properties by preventing viruses from entering the host's system [46].

## 7.5 Anti-diabetic activity

Metallic nanoparticles are being investigated for their potential in managing diabetes mellitus (DM), a metabolic disorder characterized by elevated blood sugar levels. Traditionally, controlling DM involves a balanced diet, artificial insulin, or medications, presenting a significant challenge. However, recent research by Daisy and Saipriya has shown promising therapeutic effects of gold nanoparticles in diabetic models. Their study, using mice as animal models, demonstrated a significant reduction in blood creatinine, uric acid, and alkaline phosphatase levels following treatment with gold nanoparticles. This research suggests a potential alternative approach to managing diabetes [57].

## 8. Future Aspects

The use of Earth-abundant metals (EAMs) holds significant promise in the pharmaceutical industry. There is a growing demand for cost-effective, sustainable, and environmental friendly alternatives to rare and precious metals. Basic metals such as iron, copper, and zinc are anticipated to assume a more prominent role in green chemistry, particularly as catalysts in drug synthesis, thereby reducing the environmental footprint of pharmaceutical manufacturing. Safer, metal-based medications with reduced toxicity could potentially revolutionize new drug categories and cancer treatments, spurred by advancements in biocompatibility research. Furthermore, progress in materials science and nanotechnology may facilitate the integration of EAMs into targeted drug delivery systems, ultimately enhancing therapeutic precision. Regulatory bodies are likely to advocate for the use of EAMs as the industry transitions towards eco-friendlier practices, which will significantly shape the future landscape of medication development and production.

**Acknowledgment**

We are very thankful to Government College University Faisalabad, Faisalabad, Pakistan to support in completing this book chapter.

**References**

[1] Sajid, Z., et al., Novel Armed Pyrazolobenzothiazine Derivatives: Synthesis, X-Ray Crystal Structure and POM analyses of Biological Activity Against Drug Resistant Clinical Isolate of Staphylococcus aureus. Pharmaceutical Chemistry Journal, 2016. 50(3): p. 172-180. https://doi.org/10.1007/s11094-016-1417-y

[2] Shagufta and I. Ahmad, An insight into the therapeutic potential of quinazoline derivatives as anticancer agents. Medchemcomm, 2017. 8(5): p. 871-885. https://doi.org/10.1039/C7MD00097A

[3] Mohsin, N.U.A., et al., Cyclooxygenase-2 (COX-2) as a Target of Anticancer Agents: A Review of Novel Synthesized Scaffolds Having Anticancer and COX-2 Inhibitory Potentialities. Pharmaceuticals (Basel), 2022. 15(12). https://doi.org/10.3390/ph15121471

[4] Ono, H., et al., Hydrogen gas inhalation treatment in acute cerebral infarction: a randomized controlled clinical study on safety and neuroprotection. Journal of Stroke and Cerebrovascular Diseases, 2017. 26(11): p. 2587-2594. https://doi.org/10.1016/j.jstrokecerebrovasdis.2017.06.012

[5] Imberti, C. and P.J. Sadler, 150 years of the periodic table: New medicines and diagnostic agents, in Advances in inorganic chemistry. 2020, Elsevier. p. 3-56. https://doi.org/10.1016/bs.adioch.2019.11.001

[6] Bukhari, M.H., et al., Synthesis and anti-bacterial activities of some novel pyrazolobenzothiazine-based chalcones and their pyrimidine derivatives. Medicinal Chemistry Research, 2012. 21(10): p. 2885-2895. https://doi.org/10.1007/s00044-011-9820-0

[7] Khalid, Z., et al., Anti-HIV activity of new pyrazolobenzothiazine 5,5-dioxide-based acetohydrazides. Med Chem Res, 2015. 24(10): p. 3671-3680. https://doi.org/10.1007/s00044-015-1411-z

[8] Naqvi, S.A.R., 99mTc-labeled antibiotics for infection diagnosis: Mechanism, action, and progress. Chemical Biology & Drug Design. n/a(n/a). https://doi.org/10.1111/cbdd.13923

[9] Rasheed, R., et al., Development of 99mTc-SDP-choline SPECT radiopharmaceutical for imaging of cerebrovascular diseases. Pak J Pharm Sci, 2020. 33(1(Supplementary)): p. 241-244. https://doi.org/10.36721/PJPS.2020.33.1.SUP.241-244.1

[10] Rasheed, R., et al., (99m) Tc-tazobactam, a novel infection imaging agent: Radiosynthesis, quality control, biodistribution, and infection imaging studies. J Labelled Comp Radiopharm, 2017. 60(5): p. 242-249. https://doi.org/10.1002/jlcr.3494

[11] Roohi, S., S.K. Rizvi, and S.A.R. Naqvi, (177)Lu-DOTATATE Peptide Receptor Radionuclide Therapy: Indigenously Developed Freeze Dried Cold Kit and Biological Response in In-Vitro and In-Vivo Models. Dose Response, 2021. 19(1): p. 1559325821990147. https://doi.org/10.1177/1559325821990147

[12] Bhagi-Damodaran, A. and Y. Lu, The periodic table's impact on bioinorganic chemistry and biology's selective use of metal ions. The periodic table II: catalytic, materials, biological and medical applications, 2019: p. 153-173. https://doi.org/10.1007/430_2019_45

[13] Sigel, H. and A. Sigel, The bio-relevant metals of the periodic table of the elements. Zeitschrift für Naturforschung B, 2019. 74(6): p. 461-471. https://doi.org/10.1515/znb-2019-0056

[14] Javed, K., et al., Fabrication of a ZnFe2O4@Co/Ni-MOF nanocomposite and photocatalytic degradation study of azo dyes. RSC Advances, 2024. 14(42): p. 30957-30970. https://doi.org/10.1039/D4RA05283H

[15] Zaib, S., et al., Monoamine Oxidase Inhibition and Molecular Modeling Studies of Piperidyl-thienyl and 2-Pyrazoline Derivatives of Chalcones. Med Chem, 2015. 11(5): p. 497-505. https://doi.org/10.2174/1573406410666141229101130

[16] Li, H., R. Wang, and H. Sun, Systems approaches for unveiling the mechanism of action of bismuth drugs: new medicinal applications beyond Helicobacter pylori infection. Accounts of chemical research, 2018. 52(1): p. 216-227. https://doi.org/10.1021/acs.accounts.8b00439

[17] Ogun, A.S. and A. Adeyinka, Biochemistry, transferrin, in StatPearls [Internet]. 2022, StatPearls Publishing.

[18] Rosário, J.d.S., et al., Biological activities of bismuth compounds: an overview of the new findings and the old challenges not yet overcome. Molecules, 2023. 28(15): p. 5921. https://doi.org/10.3390/molecules28155921

[19] Saeed, M., et al., Green and eco-friendly synthesis of Co3O4 and Ag-Co3O4: Characterization and photo-catalytic activity. Green Processing and Synthesis, 2019. 8: p. 382-390. https://doi.org/10.1515/gps-2019-0005

[20] Ong, Y.C., L. Kedzierski, and P.C. Andrews, Do bismuth complexes hold promise as antileishmanial drugs? Future Medicinal Chemistry, 2018. 10(14): p. 1721-1733. https://doi.org/10.4155/fmc-2017-0287

[21] Ma, Y., et al., Advances of cobalt nanomaterials as anti-infection agents, drug carriers, and immunomodulators for potential infectious disease treatment. Pharmaceutics, 2022. 14(11): p. 2351. https://doi.org/10.3390/pharmaceutics14112351

[22] Saleem, S.M., et al., Radiosynthesis and Preclinical Evaluation of [99mTc]Tc-Tigecycline Radiopharmaceutical to Diagnose Bacterial Infections. 2024. 17(10): p. 1283. https://doi.org/10.3390/ph17101283

[23] Rasheed, R., et al., (177) Lu-5-Fluorouracil a potential theranostic radiopharmaceutical: radiosynthesis, quality control, biodistribution, and scintigraphy. J Labelled Comp Radiopharm, 2016. 59(10): p. 398-403. https://doi.org/10.1002/jlcr.3423

[24] Usmani, S., et al., 225Ac Prostate-Specific Membrane Antigen Posttherapy α Imaging: Comparing 2 and 3 Photopeaks. Clin Nucl Med, 2019. 44(5): p. 401-403. https://doi.org/10.1097/RLU.0000000000002525

[25] Wahsner, J., et al., Chemistry of MRI contrast agents: current challenges and new frontiers. Chemical reviews, 2018. 119(2): p. 957-1057. https://doi.org/10.1021/acs.chemrev.8b00363

[26] Clough, T.J., et al., Synthesis and in vivo behaviour of an exendin-4-based MRI probe capable of β-cell-dependent contrast enhancement in the pancreas. Dalton Transactions, 2020. 49(15): p. 4732-4740. https://doi.org/10.1039/D0DT00332H

[27] Jodal, A., et al., A comparison of three 67/68 Ga-labelled exendin-4 derivatives for β-cell imaging on the GLP-1 receptor: the influence of the conjugation site of NODAGA as chelator. EJNMMI research, 2014. 4: p. 1-10. https://doi.org/10.1186/s13550-014-0031-9

[28] Needham, R.J. and P.J. Sadler, A Periodic Table for Life and Medicines. The Periodic Table II: Catalytic, Materials, Biological and Medical Applications, 2019: p. 175-201. https://doi.org/10.1007/430_2019_51

[29] Lucaciu, R.L., et al., Metallo-drugs in cancer therapy: Past, present and future. Molecules, 2022. 27(19): p. 6485. https://doi.org/10.3390/molecules27196485

[30] Ude, Z., et al., A new class of prophylactic metallo-antibiotic possessing potent anti-cancer and anti-microbial properties. Dalton Transactions, 2019. 48(24): p. 8578-8593. https://doi.org/10.1039/C9DT00250B

[31] Rizvi, S.U., et al., Antimicrobial and antileishmanial studies of novel (2E)-3-(2-chloro-6-methyl/methoxyquinolin-3-yl)-1-(aryl)prop-2-en-1-ones. Chem Pharm Bull (Tokyo), 2010. 58(3): p. 301-6. https://doi.org/10.1248/cpb.58.301

[32] Shahzad, A., et al., Therapeutic potential of quinoa seed extract as regenerative and hepatoprotective agent in induced liver injury wistar rat model Pak. J. Pharm. Sci., 2021. 34(6(Suppl)): p. 2309-2315.

[33] Čongrádyová, A., et al., Antimicrobial activity of copper (II) complexes. 2014.

[34] Geraldes, C.F. and M.H. Delville, Iron Oxide Nanoparticles for Bio-Imaging. Metal Ions in Bio-Imaging Techniques, 2021: p. 271. https://doi.org/10.1515/9783110685701-015

[35] Naqvi, S.A.R., et al., Susceptibility of (99m)Tc-Ciprofloxacin for Common Infection Causing Bacterial Strains Isolated from Clinical Samples: an In Vitro and In Vivo Study. Appl Biochem Biotechnol, 2019. 188(2): p. 424-435. https://doi.org/10.1007/s12010-018-2915-z

[36] Fernández Barahona, I., Nanomaterials for multimodal molecular imaging. 2023.

[37] Imberti, C., et al., Manipulating the in vivo behaviour of 68Ga with tris (hydroxypyridinone) chelators: pretargeting and blood clearance. International Journal of Molecular Sciences, 2020. 21(4): p. 1496. https://doi.org/10.3390/ijms21041496

[38] Blower, P.J., et al., Gallium: New developments and applications in radiopharmaceutics. Advances in Inorganic Chemistry, 2021. 78: p. 1-35. https://doi.org/10.1016/bs.adioch.2021.04.002

[39] Blower, J.E., et al., The radiopharmaceutical chemistry of the radionuclides of gallium and indium. Radiopharmaceutical Chemistry, 2019: p. 255-271. https://doi.org/10.1007/978-3-319-98947-1_14

[40] Saatçi, M., M. Ekinci, and E. Gündoğdu, Radiopharmaceuticals Used in Molecular Imaging. Düzce Üniversitesi Sağlık Bilimleri Enstitüsü Dergisi, 2021. 11(1): p. 115-122.

[41] Villalobos, A., et al. Yttrium-90 radioembolization dosimetry: what trainees need to know. in Seminars in Interventional Radiology. 2020. Thieme Medical Publishers, Inc. https://doi.org/10.1055/s-0041-1722877

[42] Radioisotopes, I. and R.S. No, 7. Copper-64 Radiopharmaceuticals: Production, Quality Control and Clinical Applications. 2022, IAEA: Austria.

[43] Zhou, Y., et al., 64Cu-based radiopharmaceuticals in molecular imaging. Technology in cancer research & treatment, 2019. 18: p. 1533033819830758. https://doi.org/10.1177/1533033819830758

[44] Tehri, N., et al., Biosynthesis, antimicrobial spectra and applications of silver nanoparticles: Current progress and future prospects. Inorganic and Nano-Metal Chemistry, 2022. 52(1): p. 1-19. https://doi.org/10.1080/24701556.2020.1862212

[45] Gold, K., et al., Antimicrobial activity of metal and metal-oxide based nanoparticles. Advanced Therapeutics, 2018. 1(3): p. 1700033. https://doi.org/10.1002/adtp.201700033

[46] Khandel, P., et al., Biogenesis of metal nanoparticles and their pharmacological applications: present status and application prospects. Journal of Nanostructure in Chemistry, 2018. 8: p. 217-254. https://doi.org/10.1007/s40097-018-0267-4

[47] Rafique, A., et al., Chia seed-mediated fabrication of ZnO/Ag/Ag2O nanocomposites: structural, antioxidant, anticancer, and wound healing studies. 2024. 12. https://doi.org/10.3389/fchem.2024.1405385

[48] Naqvi, S.A.R., et al., Antioxidant, Antibacterial, and Anticancer Activities of Bitter Gourd Fruit Extracts at Three Different Cultivation Stages. Journal of Chemistry, 2020. 2020: p. 7394751. https://doi.org/10.1155/2020/7394751

[49] Naqvi, S.A.R., et al., Antioxidant and antibacterial evaluation of honey bee hive extracts using in vitro models. Mediterranean Journal of Nutrition and Metabolism, 2013. 6: p. 247-253. https://doi.org/10.3233/s12349-013-0139-x

[50] Nisar, P., et al., Antimicrobial activities of biologically synthesized metal nanoparticles: an insight into the mechanism of action. JBIC Journal of Biological Inorganic Chemistry, 2019. 24: p. 929-941. https://doi.org/10.1007/s00775-019-01717-7

[51] Khan, N.U., et al., Technetium-99m labeled Ibuprofen: Development and biological evaluation using sterile inflammation induced animal models. Mol Biol Rep, 2019. 46(3): p. 3093-3100. https://doi.org/10.1007/s11033-019-04762-2

[52] Khan, Z., et al., Synthesis and Antimicrobial Activity of 2-Aryl-4H-3,1-benzoxazin-4-ones. Asian Journal of Chemistry, 2012. 25: p. 152-156. https://doi.org/10.14233/ajchem.2013.12846

[53] Jain, A., R. Anitha, and S. Rajeshkumar, Anti inflammatory activity of Silver nanoparticles synthesised using Cumin oil. Research Journal of Pharmacy and Technology, 2019. 12(6): p. 2790-2793. https://doi.org/10.5958/0974-360X.2019.00469.4

[54] Khan, N.U.H., et al., Technetium-99m radiolabeling and biological study of epirubicin for in vivo imaging of multi-drug-resistant Staphylococcus aureus infections via single photon emission computed tomography. Chem Biol Drug Des, 2019. 93(2): p. 154-162. https://doi.org/10.1111/cbdd.13393

[55] Rasheed, R., et al., Alpha Therapy with (225)Actinium Labeled Prostate Specific Membrane Antigen: Reporting New Photopeak of 78 Kilo-electron Volts for Better Image Statistics. Indian journal of nuclear medicine : IJNM : the official journal of the Society of Nuclear Medicine, India, 2019. 34(1): p. 76-77. https://doi.org/10.4103/ijnm.IJNM_115_18

[56] Andleeb, A., et al., A systematic review of biosynthesized metallic nanoparticles as a promising anti-cancer-strategy. Cancers, 2021. 13(11): p. 2818. https://doi.org/10.3390/cancers13112818

[57] Patra, N., et al., Antibacterial, anticancer, anti-diabetic and catalytic activity of bio-conjugated metal nanoparticles. Advances in Natural Sciences: Nanoscience and Nanotechnology, 2018. 9(3): https://doi.org/10.1088/2043-6254/aad12d

Applications for Earth-Abundant Transition Metals
Materials Research Foundations 179 (2025) 97-117

Materials Research Forum LLC
https://doi.org/10.21741/9781644903711-5

# Chapter 5

# Transition Metal Chalcogenides for Dye-Sensitized Solar Cells

M.M. Iqbal[1], Abdul Hafeez[1], S.S. Ali[1,*]

[1]School of Physical Sciences, University of the Punjab, Lahore 54590, Pakistan

*shahbaz.sps@pu.edu.pk

## Abstract

Dye-sensitized solar cells (DSSCs), categorized under $3^{rd}$ generation technology in the photovoltaic, have attracted a lot of interest since the attainment of efficiency ranging just above 7 % in 1991. More recently, transition metal chalcogenides including metal tellurides, metal selenides and metal sulfides have been studied due to their low expense, excellent conductivity, tunable band gap, durability and good stability. Yet, more work must be done regarding the mechanism and utilization of metal chalcogenides in DSSCs particularly as counter electrodes (CEs). The development of metal chalcogenide materials being mixed within highly electrocatalytic materials is crucial to increase their efficiency for DSSCs. A layout of DSSCs is included that is designed with metal chalcogenides as the component of CEs and the properties of CEs for better photovoltaic conversion efficiencies (PCEs).

## Keywords

Transition Metal Chalcogenides, Dye-Sensitized Solar Cells, Counter Electrode, Third Generation Technology, Power Conversion Efficiency

## Contents

## 1.    Introduction

### 1.1    Solar Cells (SCs)

In the last few decades, human beings have fulfilled their demands using non-renewable sources like petrol, gas, diesel and coal. The need for energy increases with the increase in population. The negative impact of non-renewable energy resources on the environment and the limited supply of energy needs leads the researchers toward renewable energy resources. Renewable energy sources play an essential role in fulfilling the energy needs. Among the sources of renewable energy, solar cells stand out as the most beneficial products due to their cost-effective and easy manufacturing features. Solar cells can be categorized into three distinct generations: (i) wafer-based, (ii) thin film-based and (iii) organic/inorganic [1, 2].

Among these three generations, solar cells whose wafer is based on silicon (SWSCs) are commercially available, but others are under development and research. Silicon solar cells have many problems like high cost and the need for highly purified single crystals of silicon during manufacturing. The silicon is collected from the sand of the beach and maximum reported efficiency of silicon solar cells is around 25 %. Solar cells entered their 2nd generation to reduce the production hurdles of silicon single-crystalline solar cells. But solar cells of second-generation were also not able to overcome these limitations due to some drawbacks including being toxic to the environment, low heat stability, less efficiency and high-cost rare earth metals. The competitive efficiency in laboratory-based productions of the 3rd generation solar cells with silicon solar cells is capturing the attention of scientists. Moreover, their cost-effectiveness, humidity countering, environmental friendliness, time-saving, easy processing and indoor lighting also increase the interest of the researchers [2, 3].

## 1.2    Dye-sensitized Solar Cells (DSSCs)

In 1991 marking a significant breakthrough O Regan and Gratzel first introduced the DSSCs, third and most advanced generation of solar cells. DSSCs are composed of inorganic and organic materials. DSSCs can be categorized into organic and inorganic based on different light-absorbing layers [1]. The inorganic or synthetic metal-based complex dye includes porphyrin, rhodium and ruthenium dye, on the other hand, organic solar cells are fabricated from various parts of plants including fruits, flowers, seeds, leaves, barks, peels, pulps, roots and petals. The most prevalent and widely occurring organic dyes are carotene, anthocyanin, flavonoids, chlorophyll and betalein. In terms of efficiency and stability, inorganic dyes are more convenient as compared to organic dyes, but environmental aspects and accessibility of organic dyes are comparable to those inorganic dyes. In DSSCs, the ruthenium dye is highly efficient due to high stability, low fabrication cost, ease of synthesis, extraordinary performance of photoelectron conversion, pollution-free and electron-hole generation but requires high cost and large synthesis time. In the last 20-25 years, the synthesis method, stability, and efficiency of DSSCs have increased rapidly. *Gratzel et al.* synthesized the monolayer light harvesting charge transfer dye and gained power conversion efficiency (PCE) up to 7.12 % when irradiated by stimulated solar light upon coating the colloidal $TiO_2$ films in 1991 [4]. In 2011, a group enlarged the efficiency of DSSCs up to 12.3 % through the substitution of traditional synthesized redox electrolyte and custom donor-p-bridge-acceptor, $[Co^{(II/III)}$ tris(bipyridyl)] and sensitizer (zinc porphyrin) respectively [5]. Afterward, the highest efficiency of DSSCs was reported in 2017, by obtaining a PCE of 28.9 % using indoor conditions at 1000 lux [6].

## 1.3    Parts and Working of DSSCs

Typically, a DSSC consists of five parts including a substrate, nanocrystalline semiconductor oxide (used as the photoanode), an electrolyte to facilitate the flow of ions, counter electrode (CE) which serves as dye molecule and a redox electrode. Dye molecules work as a sensitizer to absorb the solar radiations and inject photogenerated electrons into the semiconductor oxide conduction band by leaving dye in an oxidized state. Injected electrons pass through the semiconductor to the conductive fluorine-doped tin oxide (FTO) glass and then continue their path to external circuit, which requires a suitable band structure of the semiconductor oxide. External circuit electrons are collected by counter electrode material which then convey electrons to electrolytes for catalytic reduction of the redox category ($Co^{+3}$ or $I_3^-$). The electrolyte manages the sending of electrons and the dye is brought back to its ground state by reducing its oxidation state. These processes are very important for the proper usage of photoexcited electrons for reduction process (catalytic) of molecules in the electrolyte, especially influencing the PCEs of DSSC by enhancing the fill factor and short-circuit current [7, 8]. The parts, working, and the mechanism involved in the DSSCs are illustrated in Fig. 1 (a-b).

The basic working principle of the DSSCs can be written as:

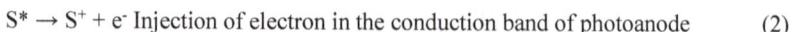

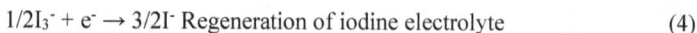

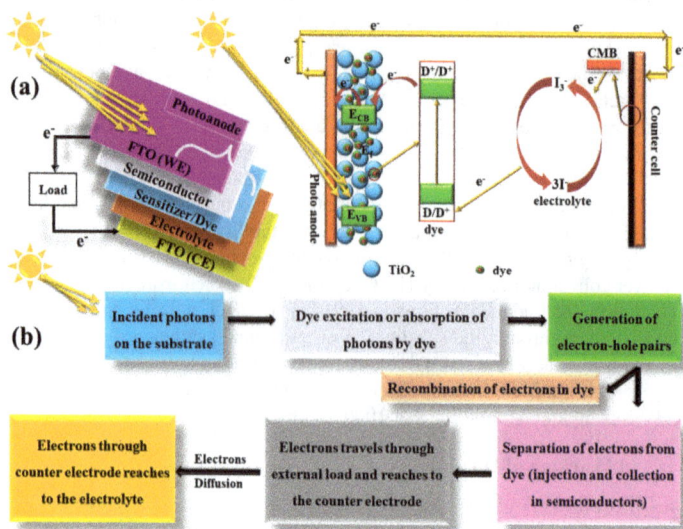

$$S + h\upsilon \rightarrow S^* \text{ Sunlight absorption} \tag{1}$$

$$S^* \rightarrow S^+ + e^- \text{ Injection of electron in the conduction band of photoanode} \tag{2}$$

$$S + 3/2I^- \rightarrow S + 1/2I_3^- \text{ Regeneration of dye} \tag{3}$$

$$1/2I_3^- + e^- \rightarrow 3/2I^- \text{ Regeneration of iodine electrolyte} \tag{4}$$

Figure 1. (a) Parts, working and (b) mechanism of dye-sensitized solar cells.

## 1.4    Properties of CEs for Better Efficiency

The best properties of the CE materials for greater photovoltaic efficiency include; (i) The excellent diffusion ability and high conductivity of electrons. CE materials permit quick charge transfer through electrons and also give diffusion channels for ions because the diffusion of electron transfer and electrolyte molecules that occur in CE materials are the key processes of electrocatalytic reactions. (ii) Reduction of electrolyte molecules requires greater catalytic activity as it improves the reaction rate, inhibits adverse reactions and polarization and ultimately increases the current density of DSSC devices. (iii) A band with suitable structure to enhance the exchange of electrons at the interface of electrolyte and CE materials, where energy levels of redox electrolyte and CE materials overlap. The optimal condition required for this overlapping is equally contribution of the CE materials valance and conduction band to redox the species. (iv) Excellent stability of electrochemicals to avoid the formation of harmful and inactive byproducts within the specific range of DSSCs [7, 8].

Applications for Earth-Abundant Transition Metals                                    Materials Research Forum LLC
Materials Research Foundations 179 (2025) 97-117                          https://doi.org/10.21741/9781644903711-5

## 2. Chalcogenide

The name "chalcogenide" is derived from the Greek expressions "χαλκ'ος" and "γενν'ω" manifesting "copper birth". The term chalcogenide was originated by the two expert German scientists Wilhelm Biltz and Werner Fische, and was officially used in 1938, after the suggestion of Heinrich Remy who was a member of IUPAC. After that, compounds like oxygen, sulfur, selenium, polonium and tellurium were included in the chalcogenides. Polonium is metal, tellurium and selenium are semiconductors but sulfur and oxygen are categorized as non-metals [9]. Chalcogens also called the "oxygen family", come from the periodic table in the chemical elements of group IV A or 16. The word "chalcogens" also originates from the Greek word meaning "ore formers" and hence, the word "chalcogenides" is derived from the compounds. Chalcogenides have attracted great interest because of their low phonon energy, infrared transparency, exceptional photo-sensitivity and a high value of refractive index [10].

### 2.1 Transition Metal Chalcogenides (TMCs)

TMCs have a general formula "MX" and "$MX_2$", where "M" stands for transition metals from groups 4-10,16 (such as Mo, Nb, V, Ti, Fe, W, Ta and Hf) and "X" stands for the chalcogens (Se, S, and Te). Each molecule of transition metal (TM) has six chalcogenide atoms surrounding it through a covalent bond, but individual layers are attached by Van der Walls force [9, 11]. TMCs are found in many structures and stoichiometries like the simplest and most important ones are 1:2 and 1:1. The highly investigated chalcogenides are $MoS_2$, Zinc (ZnTe, ZnS, and ZnSe), mercury (HgCdTe) and cadmium (CdSe, CdTe and CdS). They mostly show nonionic and highly covalent behavior indicated by their wurtzite structure and zinc blend [10].

### 2.2 Synthesis Methods of TMCs and their Composites

There are several methods for the synthesis of TMCs and their composites, including the one-pot-heat-up method, electrospinning, hot plate method, hydrothermal method or solvothermal and some other methods as shown in Fig. 2.

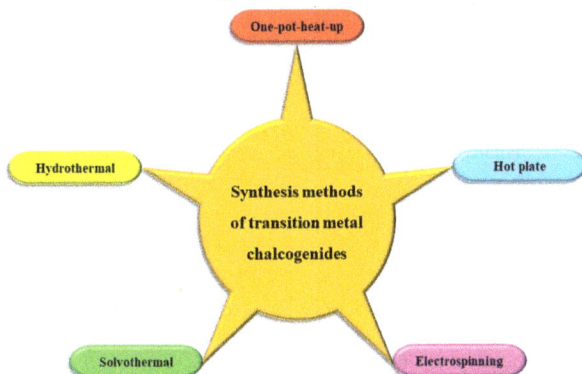

*Figure 2. Synthesis methods of TMCs and their nanocomposites.*

Materials Research Forum LLC
https://doi.org/10.21741/9781644903711-5

## 2.3    One-pot-heat-up Method

For the controllable and highly amenable large-scale synthesis of chalcogenides the one- pot-heat-up approach is used. Fig. 3 displays the synthesis mechanism of transition metal chalcogenides nanomaterials using a one-pot-heat-up method which means that the reaction took place in one single pot and the dopant precursor material was filled in the syringe. This method is operated at a very low temperature of 120 °C, a single pot is used for the chemical reactions and the production of nanosheets.

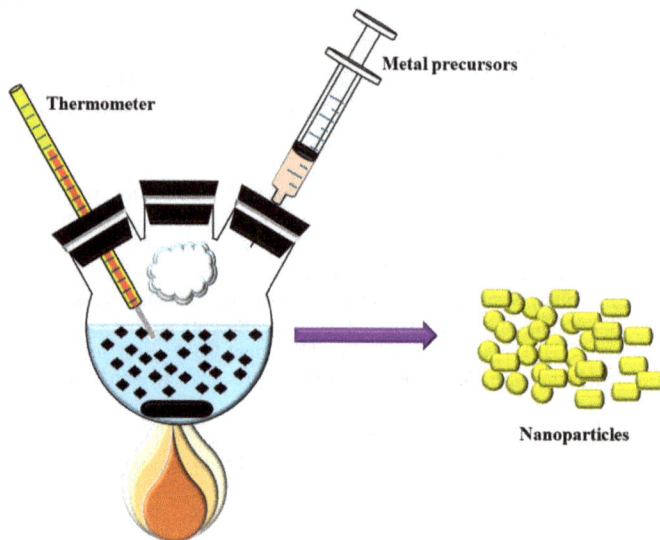

*Figure 3. Synthesis mechanism for the production of TMCs using a one-pot-heat-up method.*

## 2.4    Hot plate Method

This method is quite basic and convenient in which initial precursors are exposed to a certain degree of heat to oxidize and cause the creation of metal oxides. The temperature used for growth points was kept low, and there is no specific count for the vapors to tally as given by Zhu and Sow for the general growth mechanism of the transition metal dichalcogenides as illustrated in Fig. 4. The necessary parameters for getting the controlled layered nanostructures of TMCs are time, molar ratio, temperature and capping agents with a perfect choice of the metal precursor. It is one of the most beneficial chemical preparation techniques for the production of controlled size and shape of the TMCs nanostructures [9, 12].

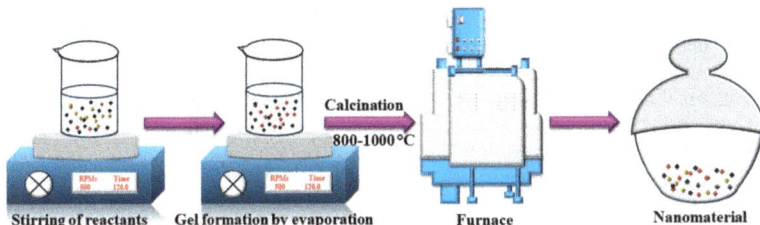

*Figure 4. Synthesis of transition metal chalcogenides by a hot plate method.*

## 2.5 Electrospinning

A versatile and resourceful synthesis method used for the generation of TMCs and their composites is called electrospinning. In this method, the syringe is loaded with a desired viscous precursor and the syringe tip is provided with a required high-voltage current. Fig. 5 displays the design for the synthesis of nanofiber by the electrospinning method. The electro-spun fibers of TMCs not only deliver the ions and electron transfer but also can help the expansion of large-volume, which is used in energy storage devices and batteries [9].

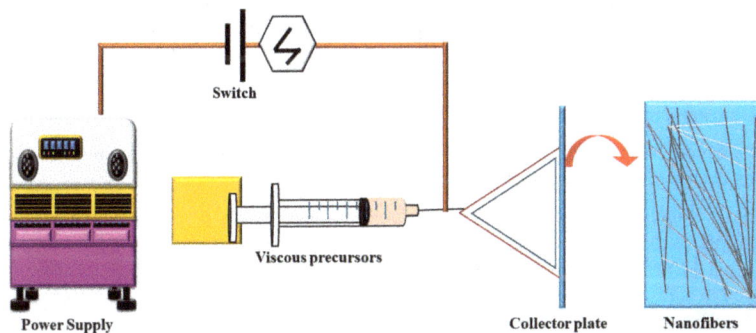

*Figure 5. Synthesis of transition metal chalcogenides by electrospinning method.*

## 2.6 Solvothermal/Hydrothermal Method

The most employed method for the synthesis of nanoparticles and TMCs is called hydrothermal. This method usually takes place in a chemically sealed autoclave, which promotes the growth and formation of crystals at higher temperatures and pressure. The precursors are filled into the autoclave, kept at a higher temperature to get the nanomaterials and calcined but for some high crystalline materials, the temperature should be low. The hydrothermal method can be controlled by adjusting the time, pressure, pH, organic materials, concentrations and temperature of the reactive reagents.

Applications for Earth-Abundant Transition Metals                Materials Research Forum LLC
Materials Research Foundations 179 (2025) 97-117          https://doi.org/10.21741/9781644903711-5

Another method for the synthesis of nanomaterials and TMCs is solvothermal, which occurs when pressure is above 1 bar, and temperature exceeds the boiling point. The solvent material can be organic, inorganic, water or alcohol. However, now a days most of the fabrication is done by hydrothermal method as compared to solvothermal. Hydrothermally synthesized nanohybrid materials are useful in a number of ways such as CEs for DSSCs, photocurrent, photocatalytic, and photoelectrochemical processes [9, 13].

## 2.7    Other Synthesis Methods

The microwave-assisted method and template-assisted method are used for the control of size and shape, indorsing the nucleation and showing synthesis for reducing the time, as compared to the conventional methods such as sol-gel, solvent mixing and cation exchange method [9, 14].

## 3.    Properties and Applications of TMCs

TMCs have gained much attention for DSSCs due to their tunable optical absorbance, optical band gap, excellent electrical conductivity (1-2.4 eV), high carrier mobility, good stability, corrosion resistance, durability, flexible nature and excellent catalytic activity which increases the redox coupling, ultimately leading to better efficiency of DSSCs. Transition metal chalcogenides also have several applications in optoelectronics (photodetectors), energy storage devices (supercapacitors and solar cells), catalysis (photocatalysis and electrocatalysis), sensors (chemical and biosensors), biomedical (drug delivery and bioimaging) and water purification [1, 3, 4, 6, 8, 15].

## 4.    Characterization Techniques for DSSCs

To get a deeper insight into the properties and activities of TMCs, scientists employ a range of complex characterizations techniques.

To identify widely the size, shape and orientation of the material at the nano-level, scanning electron microscopy (SEM) technique is available. However, transmission electron microscopy (TEM) and FESEM are also used to gain further insights. X-ray diffraction technique is then employed to study the phase formation and size of crystallites of the material. XPS stands for X-ray photoelectron spectroscopy, is used to identify the elemental composition and oxidation state of the material. Energy-dispersive X-ray spectroscopy (EDX) gives qualitative elemental information such as existence and semi-quantitative information regarding concentration. UV-visible spectrophotometer is used to study optical properties and calculate optical band gap energy and/or absorbance of the material. The biochemical composition and structure of chemical molecular bonding in the substance is determined through Raman spectroscopy.

Current density (J) and voltage (V) analysis is a procedure applied to determine the photovoltaic behaviour of DSSCs. To boost the efficiency and performance of solar cells it is essential to uncover the hidden barriers to current flow in solar cells and electrochemical impedance spectroscopic (EIS) analysis is the most promising and advanced method for unveiling these barriers in DSSCs, perovskites, and organic solar cells. Unlocking better and more efficient solar cells demands will require a good understanding of the basic factors that restrict the current through the cell. To explain the C-V characteristics of a solar device it is necessary to measure the charge transport resistance. Literature studies point to the fact that C-V measurements are

Materials Research Forum LLC
https://doi.org/10.21741/9781644903711-5

usually performed at the scanning of bias voltage from -1 V to 1 V and frequency range 2 KHz to 10 MHz [1, 9, 16].

## 5. Application of TMCs in DSSCs

### 5.1 Quantum Dot (QD) TMCs Based DSSCs

In QD-DSSCs, the metal chalcogenides of binary, ternary, as well as quaternary compositions have been employed for light absorption applications. First, cadmium sulfide (CdS) and cadmium selenide (CdSe) QDs were the most popular metal chalcogenide sensitizers. Later, more complicated QDs, for example, alloyed $CdSe_xTe_{1-x}$ quantum dots and core/shell structures like CdSe/CdS, and NBG binary QDs i.e, PbS and PbSe, were used as light sensitizers. Also, QDs have been incorporated in several research investigations as co-sensitizers to dyes of solar cells. Namely, CdS and CdSe QDs are employed as co-sensitizers with organic dyes in DSSCs. In recent works, *J. W. AlGhamdi et al.* synthesized active CdSe QDs by using pulse laser ablation in liquid (PLAL) and used it as a co-sensitizer with N719 organic dye in $TiO_2$-based DSSCs. This approach led to a rise in PCE, which moved from 5.17 % to 7.09 %. This is due to a relatively 40 % rise in the short-circuit current density ($J_{sc}$) as shown in Fig. 6 [17, 18].

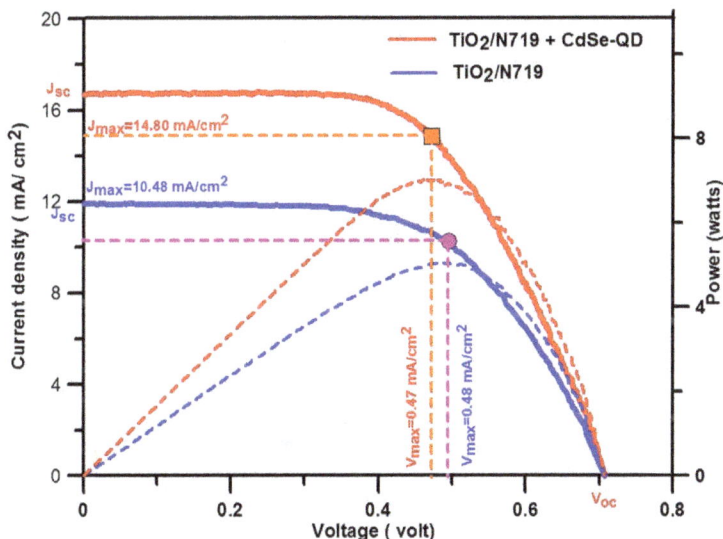

*Figure 6. Current-Voltage measurement of $TiO_2/N719$ and $TiO_2/N719/CdSe$ quantum dots [18].*

Materials Research Forum LLC
https://doi.org/10.21741/9781644903711-5

## 5.2 TMCs Progress Toward Flexible DSSCs

It is further observed that the trend is slowly but steadily moving towards the preparation of flexible and fibrous energy devices. Despite this, different challenges affect the efficiency and working performance of fiber-shaped energy devices (FSEDs) like flexible dye-sensitized solar cells (FDSSCs) and supercapacitors which are inferior to planar ones. *Ali et al.* also synthesized multi-walled carbon nanotubes (MWCNTs) fibers with cobalt selenium (CoSe) nanoparticles by hydrothermal technique. These CoSe nanoparticle-decorated fibers with better electrocatalytic performance were then wrapped around $TiO_2$/N719-coated Ti wire or photoanode as the CE for FDSSCs. The FDSSCs CEs made from CoSe/MWCNT composite had a PCE of 6.42 % [17, 19].

## 5.3 Sulfides Based TMCs for DSSCs

The advantages offered by sulfides based TMCs have raised a lot of interest in the DSSC field towards the use of metal sulfides (MSs). Among these materials the greatest stability of bandgaps, better electrical conductivity and efficient light absorption are the properties that are exhibited. Some of these MSs including cobalt sulfides (CoS), zinc sulfides (ZnS), copper sulfides ($Cu_2S$), and nickel sulfides (NiS) have some distinct properties that are beneficial for DSSCs enhancement.

MSs are now the leading candidates for DSSCs. Using a straightforward chemical process, laminar tungsten sulfides ($WS_2$) and molybdenum sulfides ($MoS_2$) were created as CEs for DSSCs. The study found that $WS_2$ and $MoS_2$ were efficient at lowering triiodide. $MoS_2$ PCE value was 7.59 %, while for $WS_2$ its value was 7.73 % for DSSCs. These numbers corresponded to the results of Pt based CE for DSSCs (7.64 %). However, the FFs of $WS_2$ and $MoS_2$ CEs were higher, at 0.70 and 0.73, respectively. Compared to the FF of 0.66 generated by Pt CE, these FFs showed the related high catalytic activity for triiodide reduction. For $MoS_2$-DSSC and $WS_2$-DSSC, the correspondingly high values of open voltage ($V_{oc}$) and $J_{sc}$ were 0.76 V and 13.84 $mAcm^{-2}$ and 0.78 V and 14.13 $mAcm^{-2}$, respectively [20-22].

## 5.4 Vanadium Based TMCs for DSSCs

Vanadium chalcogenides (VCs) are a class of TMCs that include vanadium oxide ($V_xO_y$), vanadium selenide ($VSe_2$) and vanadium sulfides ($VS_2$). These compounds have exceptional optical and electrical properties, including low charge-transfer resistance, high electrical conductivity and excellent catalytic activity. These properties make these compounds useful for a range of energy applications, like electrocatalysts and energy conversion. With a production cost one-tenth that of Ag, *Xia et al.* created a $V_2O_5$/Al counter electrode for a solid DSSC that is economically favorable. In that investigation, a vapor deposition process was used to produce $V_2O_5$ on the Spiro-Omitted layer and Al was used as a back contact. Better photovoltaic characteristics were shown by a 10 nm thick $V_2O_5$ CE on Al, which had an efficiency of around 2 % and FF 0.34. These characteristics were comparable to those of the innovative Ag CE; as a result, the $V_2O_5$/Al counter electrode is helpful for solid DSSC. *Elbohy et al.* used $V_2O_5$ as the charge recombination and blocking layer in DSSCs for the first time to increase their efficiency. They found that the incident photon-to-electron conversion efficiency (IPCE) of a $V_2O_5$ layered DSSC photoanode was higher than that of a $TiO_2$ layered DSSC photoanode, ranging from 88 % to 89 %. The $V_2O_5$-based interface charge recombination resistance ($R_{ct}$) is lower than the $TiO_2$/electrolyte interface according to an EIS analysis of both DSSCs. Therefore, the $V_2O_5$

layer-modified DSSCs performed better than standard cell technology because the charge transfer in back was prevented by altering the reactivity of the $TiO_2$ surface [23, 24].

## 5.5 Manganese Based TMCs for DSSCs

Hot injection synthesis was used to create manganese-substituted earth-abandoned copper (Cu) and tin (Sn) based quaternary metal chalcogenides (QMCs) QDs, such as $Cu_2MnSnS_4$, $Cu_2MnSnS_2Se_2$ and $Cu_2MnSnSe_4$. These three QD's structural and optoelectrical characteristics were investigated and compared. Owing to their improved optical and electrical characteristics, these QDs were used as QDSSC sensitizers. Compared to the other two QDSSCs, it was found that the $Cu_2MnSnS_2Se_2/TiO_2$-based QDSSCs had better photovoltaic performance in terms of photocurrent, open circuit and efficiency. This is because the combination of selenium and sulfides in $Cu_2MnSnS_2Se_2$ QDs creates a synergistic effect. Fig. 7 illustrates the current density and voltage measurements of the $CMTS/TiO_2$, $CMTSSe/TiO_2$ and $CMTSe/TiO_2$ electrode-based QSSDs [25].

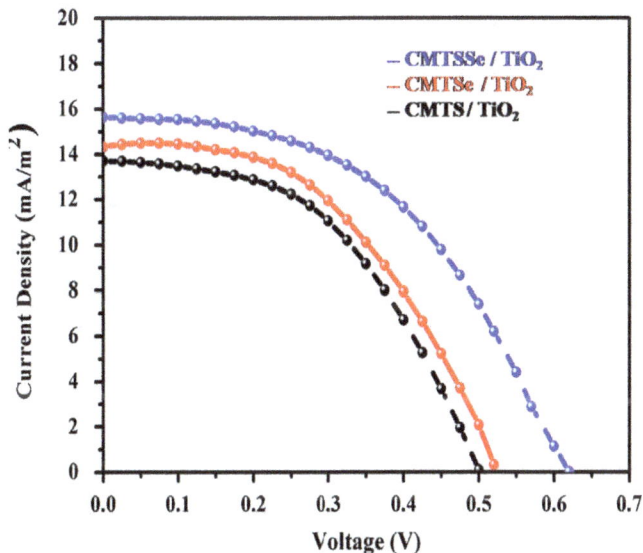

*Figure 7. Current density-voltage measurements of CMTSe/TiO₂, CMTS/TiO₂ and CMTSSe/TiO₂ electrode-based QSSDs [25].*

## 5.6 Iron Based TMCs for DSSCs

As of right now, several CE materials show potential as platinum substitutes, including conductive polymers, carbon materials and metal alloys and compounds. TMCs are thought to be the most visually appealing of them because of how closely their electronic structures resemble platinum. Furthermore, TMCs with exceptional thermal and electrical conductivities and good durability at first demonstrate remarkable electrochemical activity as electrocatalysts. Alloys and compounds of bimetal transition metal (BTM) have been effectively employed including $MIn_2S_4$ (M = Fe, Co, Ni) and $NiCo_2S_4$ as counter electrodes for DSSCs. Additionally, ternary sulfides such as $CoFeS_2$ and $NiS_2$ have demonstrated remarkable performances. FeSe shows 5.05 % efficiency [8, 26].

## 5.7    Cobalt-Nickel Based TMCs for DSSCs

Creating inexpensive CEs with strong electrocatalytic performance is a significant difficulty for DSSCs. Ternary selenides based on cobalt and nickel synthesized via solvothermal technique reported in the literature, with varying Co and Ni ratios. Simple spray deposition was used to create ternary selenide films based on cobalt-nickel, which can be used as high-performing and reasonably priced CEs in DSSCs. Due to the synergistic impact between cobalt and nickel ions the cobalt-nickel ternary selenide CEs display great charge-transfer ability and high electrocatalytic activity for the reduction of $I_3^-$. Specifically, $Co_{0.42}Ni_{0.58}Se$ CE has more electrocatalytic activity than platinum (Pt) CE and other ternary selenide CEs which based on cobalt-nickel. Additionally, DSSC with $Co_{0.42}Ni_{0.58}Se$ CE photoelectrical conversion efficiency (PCE) is 6.15 %, which is higher than the DSSC which based on Pt (5.53 %) as displayed in Fig. 8. The outcomes confirmed once more that a novel class of ternary chalcogenides which based on cobalt-nickel, can be considered as a possibly quite effective catalytic material in the DSSCs [27].

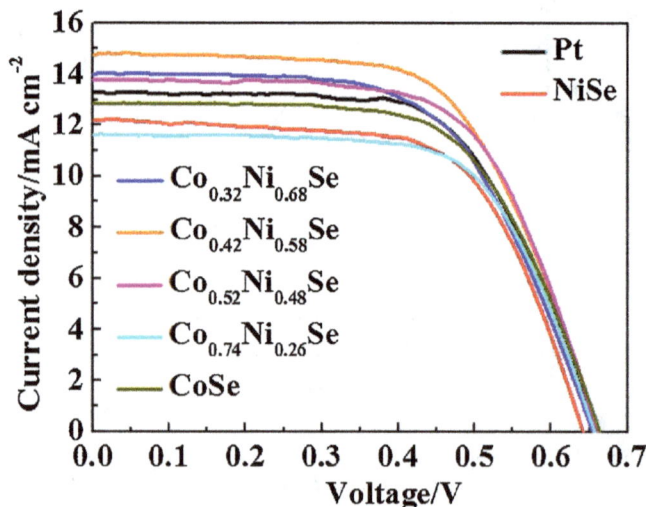

*Figure 8. Photocurrent density-voltage measurements of $Co_xNi_{1-x}Se$ for DSSCs [27].*

## 5.8 Copper Based TMCs for DSSCs

Copper can act as a redox mediator in addition to be used as dye in the complex chemicals found in DSSCs. Even though there aren't many solar cells in electronic devices, copper mediators acting as redox couples have found use in them. This is because they can support and provide a big cell output voltage of up to 1 V, owing to their great ability to regenerate the dye in the cell with just 100 mV of driving force. In contrast to the redox mediators $I^-/I_3^-$ that were previously discussed, copper compounds are not corrosive. Furthermore, in systems with comparable features, their longer diffusion duration can result in greater fill factors and photocurrents. In contrast to complex compounds based on iodine/iodide/cobalt, those containing copper offer less [28].

The development of cost-effective, stable and high-performing CE is essential in advancing quantum dot-sensitized solar cells from the laboratory to the real world. For preparing ternary chalcogenide based on copper ($Cu_{2-x}S_ySe_{1-y}$) alloyed semiconductors on fluorine-dopped tin oxide (FTO) substrate, a two-step deposition process was used. To acquire $Cu_{2-x}S$ CE, the first stage is to deposit binary copper chalcogenides CuS nanostructures developed by the microwave-irradiation course of action through screen-printing on FTO substrate and subsequent annealing in a nitrogen atmosphere. The second step is the application of the composition engineering method for the preparation of $Cu_{2-x}S_ySe_{1-y}$ alloyed electrocatalyst. This is done by employing the drop-casting method to deposit elemental Se on the as-synthesized $Cu_{2-x}S$ nanostructures. A tunable crystal structure, composition and morphology of as prepared $Cu_{2-x}S_ySe_{1-y}$ CE has been established in contrast to $Cu_{2-x}S$ CE.

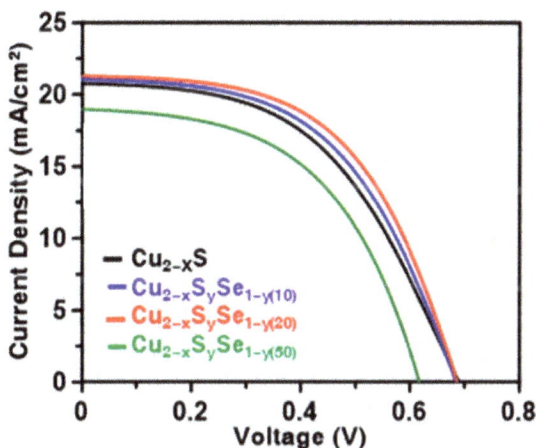

*Figure 9. Current density and voltage measurements of copper based chalcogenides for DSSCs [29].*

The enhanced $Cu_{2-x}S_ySe_{1-y}$ CE showed decent Rct and superior reduction ability of $Sn_2$ of polysulfide electrolyte from the electrochemical analysis. $Cu_{2-x}S_ySe_{1-y}$ CE-assembled QDSSCs have thereby generated conversion efficiencies of 8.02 %, higher than virgin $Cu_{2-x}S$ CE's (7.24 %) as shown in Fig. 9. Among the samples labeled as $Cu_{2-x}S_ySe_{1-y}$ CE has proved to have outstanding stability of electrochemical for the polysulfide redox pair, demonstrating minor fluctuations in current density, as well as in the profile of the curve, after 200 consecutive CV cycles. Furthermore, after 120 hours of illumination, the finest cell devices made with $Cu_{2-x}S_ySe_{1-y}$ CE demonstrated exceptional stability in an open environment, maintaining less than 60 % of their initial performance. Their alloyed semiconductors film has outstanding electrocatalytic qualities and could be an interesting CE material for applications in photovoltaics. due to easy synthesis, time efficiency and low cost [29].

## 5.9    Zinc Based TMCs for DSSCs

Due to their optoelectronic qualities and tunable band gaps, TMCs have drawn a lot of interest lately as viable options for energy conversion applications. For use in DSSC applications, undoped, indium ($In^{3+}$) and zinc ($Zn^{2+}$) doped cadmium sulfide/telluride (CST) nanocomposites (NCs) synthesized using a hydrothermal technique from chalcogenide materials. CST-NCs polycrystalline nature with hexagonal phases was validated by the XRD data. The nanocomposites' optical band gap exhibited variation ranging from 1.99 eV to 2.30 eV, contingent upon the kind and quantity of doping materials utilized. Higher power conversion efficiency was demonstrated by current-voltage (I-V) characteristics for Zn (3 %) and Ln (1 %) doped CST-NCs as displayed in Fig. 10. This demonstrates that doping is essential for enhancing the performance of the DSSC. Because of its large surface area and the creation of defect levels, that offer a direct carrier transport channel to the electrons, zinc and indium doped CST DSSC performs well. As a result, the recombination rate was lowered, increasing the $V_{oc}$ and ultimately, the solar cell device's efficiency [30].

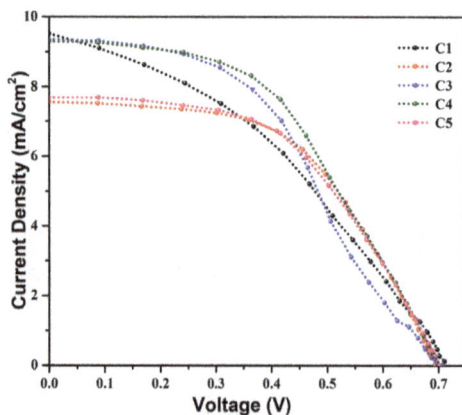

*Figure 10. Current density-voltage measurements of undoped, $In^{3+}$, $Zn^{2+}$ doped CST NCs based DSSC [30].*

### 5.10 Palladium Based TMCs for DSSCs

The binary alloy of iridium-palladium (Ir-Pd) employed as counter electrode for DSSC. Liquid phase deposition (LPD) method was used to fabricate the CE. The effects of hydrogen hexachloroiridate (IV) hydrate ($H_2Cl_6Ir \cdot H_2O$) concentration on the device's characteristics and functionality were studied. The iridium came from $H_2Cl_6Ir \cdot H_2O$. The presence of Ir-Pd dominant phase in the sample was verified by XRD analysis. Until an ideal concentration of 0.7 mM was attained, the grain size of Ir-Pd rose as the concentration of $H_2Cl_6Ir \cdot H_2O$ increased. The maximum PCE of 5.84 % was demonstrated by the device composed of Ir–Pd CE at 0.7 mM $H_2Cl_6Ir \cdot H_2O$, surpassing the PCE of 5.04 % for the device using Pt CE. This is because the apparatus has the longest carrier lifetime ($\tau$), highest recombination resistance ($R_{cr}$), lowest $R_{ct}$ and highest reduction current ($J_{pc}$) and IPCE. Ir concentration in the binary alloy of Ir-Pd notably impacted the PCE. Further the result of PCE indicated that Ir-Pd CE would be a good replacement of platinum in the gadget as counter electrode [31].

### 5.11 Cadmium Based TMCs for DSSCs

Composites, namely: cadmium sulfides or carbon dots co-doped with sulfur and nitrogen (CdS/S, N-CDs) were synthesized through hydrothermal method. Citric acid and thiourea were used as a source of carbon and sulfur while cadmium acetate dihydrate was the source of cadmium. Thiourea being a sulphur precursor helps the synthesis of cadmium sulfides and helps in doping of carbon dots. The shape of cadmium sulfides is significantly influenced by temperature, according to the results. At 200 °C, homogeneous flower-like shape of CdS was transformed into non-homogenous worm-like particles. Furthermore, because the inclusion of citric acid alters the acidity of the solution, the morphology of CdS which was flower-like transformed into a particle morphology in the synthesis of CdS/S, N-CDs. DSSCs were modified with CdS/S, N-CDs and carbon dots co-doped with nitrogen and sulfur. Short circuit current density value obtained for carbon dots co-doped with nitrogen was 14.07 mA/cm$^2$ and for CdS/S, N-CDs it was 18.38 mA/cm$^2$. The $V_{oc}$ for CDs was equivalent to 0.61 V according to the results, but it dropped to 0.59 V in the case of CdS/S, N-CDs. While $V_{oc}$ declined, PCE improved in the case of CdS/S, and N-CDs rose from 3.44 % to 5.7 % as shown in Fig. 11 [32]. Fig. 12 illustrates the maximum PCE and the total number of publications per year related to TMCs taken from Google Scholar for the last 10 years. The J, $V_{oc}$, FF and PCE of different TMCs and their composites are listed in Table 1.

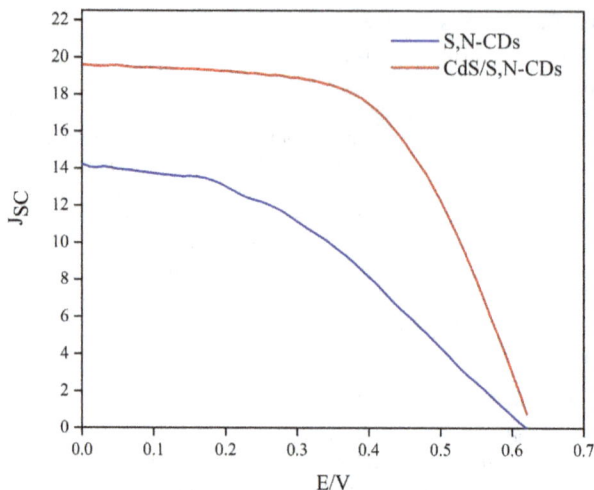

*Figure 11. Current density-voltage measurements of S, N-CDs and CdS/S, N-CDs for DSSCs [32].*

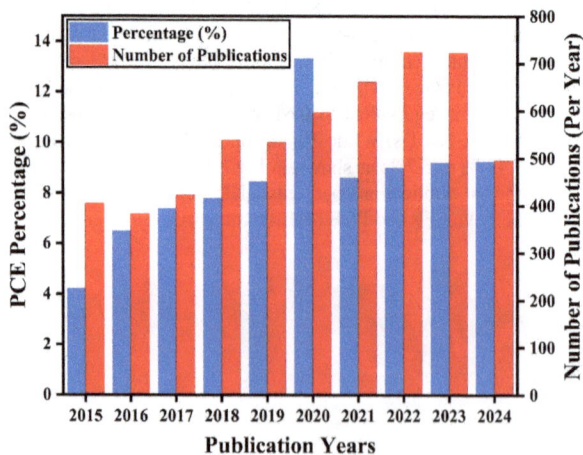

*Figure 12. Power conversion efficiency and number of publications for the last ten years of transition metal-based material for DSSCs.*

Applications for Earth-Abundant Transition Metals          Materials Research Forum LLC
Materials Research Foundations 179 (2025) 97-117          https://doi.org/10.21741/9781644903711-5

*Table 1. The CE, $J_{sc}$, $V_{oc}$, FF and PCE for different TMCs.*

| Sr. No. | Anode or counter electrode | $J_{sc}$ [mAcm$^{-2}$] | $V_{oc}$ [V] | FF | PCE [%] | References |
|---------|---------------------------|---------|------|-----|---------|------------|
| 01 | CoSe/MWCNT fibre | 13.78 | 0.72 | 0.65 | 6.42±0.3 | [19] |
| 02 | TiO$_2$-CdSe Ds-N719 | 16.7±0.4 | 0.71±0.02 | 0.59±0.02 | 7.1±0.3 | [18] |
| 03 | MoS$_2$ | 13.84 | 0.76 | 0.73 | 7.59 | [22] |
| 04 | WS$_2$ | 14.13 | 0.78 | 0.70 | 7.73 | [22] |
| 05 | V$_2$O$_5$/Al | - | - | 0.34 | 2 | [24] |
| 06 | CMTS/TiO$_2$ | 13.92 | 0.51 | 0.52 | 3.60 | [25] |
| 07 | CMTSe/TiO$_2$ | 14.21 | 0.53 | 0.52 | 3.91 | [25] |
| 08 | CMTSSe/TiO$_2$ | 15.93 | 0.62 | 0.55 | 5.39 | [25] |
| 09 | CoS | 11.91 | 0.75 | 0.73 | 6.5 | [33] |
| 10 | NiS | 18.71 | 0.74 | 0.63 | 8.62 | [34] |
| 11 | CoSe | 12.81 | 0.66 | 0.63 | 5.33 | [27] |
| 12 | Co$_{0.74}$Ni$_{0.26}$Se | 11.62 | 0.66 | 0.65 | 4.98 | [27] |
| 13 | Co$_{0.52}$Ni$_{0.48}$Se | 13.77 | 0.66 | 0.64 | 5.82 | [27] |
| 14 | Co$_{0.42}$Ni$_{0.58}$Se | 14.79 | 0.66 | 0.63 | 6.15 | [27] |
| 15 | Co$_{0.32}$Ni$_{0.68}$Se | 14.02 | 0.65 | 0.61 | 5.56 | [27] |
| 16 | NiSe | 12.19 | 0.64 | 0.64 | 5.00 | [27] |
| 17 | Cu$_{2-x}$S | 21.0 | 0.696 | 0.496 | 7.24 | [29] |
| 18 | Cu$_{2-x}$S$_y$Se$_{1-y(10)}$ | 21.2 | 0.69 | 0.519 | 7.60 | [29] |
| 19 | Cu$_{2-x}$S$_y$Se$_{1-y(20)}$ | 21.3 | 0.69 | 54.5 | 8.02 | [29] |
| 20 | Cu$_{2-x}$S$_y$Se$_{1-y(50)}$ | 19.1 | 0.622 | 0.517 | 6.13 | [29] |
| 21 | CST | 9.24 | 0.711 | 0.391 | 2.619 | [30] |
| 22 | CST-1% Zn$^{2+}$ | 7.54 | 0.69 | 0.54 | 2.82 | [30] |
| 23 | CST-3% Zn$^{2+}$ | 9.34 | 0.707 | 0.442 | 2.92 | [30] |
| 24 | CST-1% In$^{3+}$ | 9.29 | 0.703 | 0.485 | 3.17 | [30] |
| 25 | CST-3% In$^{3+}$ | 7.68 | 0.702 | 0.512 | 2.76 | [30] |
| 26 | S, N-CDs | 14.07 | 0.61 | 0.4 | 3.44 | [32] |
| 27 | Cds/S, N-CDs | 18.38 | 0.59 | 0.52 | 5.7 | [32] |

**Challenges and Future Perspectives**

The challenges that lead to the optoelectronic degradation of the various DSSCs components include poor light trapping efficiency, formation of thin films that crack, poor layer adhesion and tendency of thin films to degrade when exposed to the dye solution. The type of materials selected also comes into play in several aspects of DSSC where the suitable material enhances the right optical and electrical properties. Moreover, if corrosive or unstable materials are used in the electrolyte it affects the cell and electrodes and lowers the efficiency of the whole device. More work must be done on such materials and more investigation of the processes occurring on structures of this type to enhance properties of the developed structures and to have a deeper understanding of molecular-level interaction of all developing materials to make the best of developed structures. An area where literature review reveals a need for more research is in understanding how factors such as surface roughness, crystal structure and other impurities affect the electrocatalytic characterizing transition metal sulfides. Therefore, further research regarding the process of scale-up of transition metal sulfides for use in solar devices, which may be suitable for use in the commercial market is the next logical progression [1, 21]. To develop the synthesis of chalcogenides further, one must investigate several deposition techniques like chemical bath and electrodeposition methods. Further methods including microwave synthesis and successive ionic layer adsorption and reaction processes can also be explored. These techniques have remained underutilized, particularly in the sector of solar cells. Furthermore, there is a need to extend the stability of solar cells by improving various properties through testing [23].

## References

[1] P.S. Saud, A. Bist, A.A. Kim, A. Yousef, A. Abutaleb, M. Park, S.-J. Park, B. Pant, Dye-sensitized solar cells: Fundamentals, recent progress, and Optoelectrical properties improvement strategies, Optical Materials 150 (2024) 115242. https://doi.org/10.1016/j.optmat.2024.115242

[2] M. Farji, Development of photovoltaic cells: A materials prospect and next-generation futuristic overview, Brazilian Journal of Physics 51(6) (2021) 1916-1928. https://doi.org/10.1007/s13538-021-00981-w

[3] A. Bist, S. Chatterjee, Review on Efficiency Enhancement Using Natural Extract Mediated Dye-Sensitized Solar Cell for Sustainable Photovoltaics, Energy Technology 9(8) (2021) 2001058. https://doi.org/10.1002/ente.202001058

[4] B. O'regan, M. Grätzel, A low-cost, high-efficiency solar cell based on dye-sensitized colloidal TiO2 films, nature 353(6346) (1991) 737-740. https://doi.org/10.1038/353737a0

[5] A. Yella, H.-W. Lee, H.N. Tsao, C. Yi, A.K. Chandiran, M.K. Nazeeruddin, E.W.-G. Diau, C.-Y. Yeh, S.M. Zakeeruddin, M. Grätzel, Porphyrin-sensitized solar cells with cobalt (II/III)-based redox electrolyte exceed 12 percent efficiency, science 334(6056) (2011) 629-634. https://doi.org/10.1126/science.1209688

[6] M. Freitag, J. Teuscher, Y. Saygili, X. Zhang, F. Giordano, P. Liska, J. Hua, S.M. Zakeeruddin, J.-E. Moser, M. Grätzel, Dye-sensitized solar cells for efficient power generation under ambient lighting, Nature Photonics 11(6) (2017) 372-378. https://doi.org/10.1038/nphoton.2017.60

[7] X. Zhang, Y. Yang, S. Guo, F. Hu, L. Liu, Mesoporous Ni0. 85Se nanospheres grown in situ on graphene with high performance in dye-sensitized solar cells, ACS Applied Materials & Interfaces 7(16) (2015) 8457-8464. https://doi.org/10.1021/acsami.5b00464

[8] P. Wei, X. Chen, G. Wu, J. Li, Y. Yang, Z. Hao, X. Zhang, L. Liu, Recent advances in cobalt-, nickel-, and iron-based chalcogen compounds as counter electrodes in dye-sensitized solar cells, Chinese Journal of Catalysis 40(9) (2019) 1282-1297. https://doi.org/10.1016/S1872-2067(19)63361-9

[9] S. Yadav, S.R. Yashas, H.P. Shivaraju, Transitional metal chalcogenide nanostructures for remediation and energy: a review, Environmental Chemistry Letters 19(5) (2021) 3683-3700. https://doi.org/10.1007/s10311-021-01269-w

[10] M.M. Khan, Introduction and fundamentals of chalcogenides and chalcogenides-based nanomaterials, 2021, pp. 1-6. https://doi.org/10.1016/B978-0-12-820498-6.00001-9. https://doi.org/10.1016/B978-0-12-820498-6.00001-9

[11] Y. Jung, Y. Zhou, J.J. Cha, Intercalation in two-dimensional transition metal chalcogenides, Inorganic Chemistry Frontiers 3(4) (2016) 452-463. https://doi.org/10.1039/C5QI00242G

[12] Y. ZHU, C.H. SOW, HOTPLATE TECHNIQUE FOR NANOMATERIALS, COSMOS 04(02) (2008) 235-255. https://doi.org/10.1142/s0219607708000354. https://doi.org/10.1142/S0219607708000354

[13] Y. Liu, Y. Li, J. Zhao, R. Zhang, M. Ji, Z. You, Y. An, Solvothermal syntheses, characterizations and semiconducting properties of four quaternary thioargentates Ba2AgInS4, Ba3Ag2Sn2S8, BaAg2MS4 (M = Sn, Ge), Journal of Alloys and Compounds 815 (2020) 152413. https://doi.org/https://doi.org/10.1016/j.jallcom.2019.152413. https://doi.org/10.1016/j.jallcom.2019.152413

[14] M. Alhaddad, A. Shawky, CuS assembled rGO heterojunctions for superior photooxidation of atrazine under visible light, Journal of Molecular Liquids 318 (2020) 114377. https://doi.org/https://doi.org/10.1016/j.molliq.2020.114377. https://doi.org/10.1016/j.molliq.2020.114377

[15] J. Gong, S. Krishnan, Mathematical modeling of dye-sensitized solar cells, Dye-Sensitized Solar Cells, Elsevier2019, pp. 51-81. https://doi.org/10.1016/B978-0-12-814541-8.00002-1

[16] G. Hariharan, A. Dharani, N.S.V. Moorthi, K. Saravanan, Design and fabrication of mixed phase transition metal chalcogenide hybrid composite (M= Ni & Co) for Pt-free efficient dye-sensitized solar cell applications, Journal of Materials Science: Materials in Electronics 34(3) (2023) 188. https://doi.org/10.1007/s10854-022-09665-w

[17] A.R. Tapa, W. Xiang, X. Zhao, Metal chalcogenides (M x E y; E= S, Se, and Te) as counter electrodes for dye-sensitized solar cells: an overview and guidelines, Advanced Energy and Sustainability Research 2(10) (2021) 2100056. https://doi.org/10.1002/aesr.202100056

[18] J.M. AlGhamdi, S. AlOmar, M.A. Gondal, R. Moqbel, M.A. Dastageer, Enhanced efficiency of dye co-sensitized solar cells based on pulsed-laser-synthesized cadmium-selenide quantum dots, Solar Energy 209 (2020) 108-117. https://doi.org/10.1016/j.solener.2020.08.091

[19] A. Ali, K. Shehzad, F. Ur-Rahman, S.M. Shah, M. Khurram, M. Mumtaz, R.U.R. Sagar, Flexible, low cost, and platinum-free counter electrode for efficient dye-sensitized solar cells, ACS Applied Materials & Interfaces 8(38) (2016) 25353-25360. https://doi.org/10.1021/acsami.6b08826

[20] R. Sasikumar, S. Thirumalaisamy, B. Kim, B. Hwang, Dye-sensitized solar cells: Insights and research divergence towards alternatives, Renewable and Sustainable Energy Reviews 199 (2024) 114549. https://doi.org/10.1016/j.rser.2024.114549

[21] L.H. Kharboot, N.A. Fadil, T.A.A. Bakar, A.S.M. Najib, N.H. Nordin, H. Ghazali, A review of transition metal sulfides as counter electrodes for dye-sensitized and quantum dot-sensitized solar cells, Materials 16(7) (2023) 2881. https://doi.org/10.3390/ma16072881

[22] M. Wu, Y. Wang, X. Lin, N. Yu, L. Wang, L. Wang, A. Hagfeldt, T. Ma, Economical and effective sulfide catalysts for dye-sensitized solar cells as counter electrodes, Physical chemistry chemical physics 13(43) (2011) 19298-19301. https://doi.org/10.1039/c1cp22819f

[23] P.P. Sanap, S.P. Gupta, S.S. Kahandal, J.L. Gunjkar, C.D. Lokhande, B.R. Sankapal, Z. Said, R.N. Bulakhe, J.M. Kim, A.B. Bhalerao, Exploring vanadium-chalcogenides toward solar cell application: A review, Journal of Industrial and Engineering Chemistry (2023). https://doi.org/10.1016/j.jiec.2023.09.004

[24] J. Xia, C. Yuan, S. Yanagida, Novel counter electrode V2O5/Al for solid dye-sensitized solar cells, ACS Applied Materials & Interfaces 2(7) (2010) 2136-2139. https://doi.org/10.1021/am100380w

[25] R. Kottayi, S. Ilangovan, V. Ilangovan, R. Sittaramane, Manganese incorporated earth-abundant copper and tin-based metal chalcogenides QDs on solar absorption of photovoltaic cells, Chemical Physics Letters 833 (2023) 140913. https://doi.org/https://doi.org/10.1016/j.cplett.2023.140913. https://doi.org/10.1016/j.cplett.2023.140913

[26] Y. Duan, Q. Tang, J. Liu, B. He, L. Yu, Transparent metal selenide alloy counter electrodes for high-efficiency bifacial dye-sensitized solar cells, Angewandte chemie international edition 53(52) (2014) 14569-14574. https://doi.org/10.1002/anie.201409422

[27] Q. Jiang, K. Pan, C.-S. Lee, G. Hu, Y. Zhou, Cobalt-nickel based ternary selenides as high-efficiency counter electrode materials for dye-sensitized solar cells, Electrochimica Acta 235 (2017) 672-679. https://doi.org/10.1016/j.electacta.2017.03.100

[28] K. Pawlus, T. Jarosz, Transition metal coordination compounds as novel materials for dye-sensitized solar cells, Applied Sciences 12(7) (2022) 3442. https://doi.org/10.3390/app12073442

[29] A.S. Rasal, T.-W. Chang, C. Korupalli, J.-Y. Chang, Composition engineered ternary copper chalcogenide alloyed counter electrodes for high-performance and stable quantum dot-sensitized solar cells, Composites Part B: Engineering 232 (2022) 109610. https://doi.org/10.1016/j.compositesb.2021.109610

[30] M. Prabhu, M. Marikkannan, M.S. Pandian, P. Ramasamy, K. Ramachandran, Effect of zinc and indium doping in chalcogenide (CdS/Te) nanocomposites towards dye-sensitized solar cell applications, Journal of Physics and Chemistry of Solids 168 (2022) 110802. https://doi.org/10.1016/j.jpcs.2022.110802

[31] N. Aziz, M. Rahman, A. Umar, E. Mawarnis, Iridium-palladium binary alloy as a counter electrode in dye-sensitized solar cells, Dalton Transactions 52(48) (2023) 18354-18361. https://doi.org/10.1039/D3DT03375A

[32] H. Safardoust-Hojaghan, O. Amiri, M. Salavati-Niasari, M. Hassanpour, H. Khojasteh, L.K. Foong, Performance improvement of dye sensitized solar cells based on cadmium sulfide/S, N co doped carbon dots nanocomposites, Journal of Molecular Liquids 301 (2020) 112413. https://doi.org/10.1016/j.molliq.2019.112413

[33] M. Wang, A.M. Anghel, B. Marsan, N.-L. Cevey Ha, N. Pootrakulchote, S.M. Zakeeruddin, M. Grätzel, CoS supersedes Pt as efficient electrocatalyst for triiodide reduction in dye-

Applications for Earth-Abundant Transition Metals          Materials Research Forum LLC
Materials Research Foundations 179 (2025) 97-117          https://doi.org/10.21741/9781644903711-5

sensitized solar cells, Journal of the American Chemical Society 131(44) (2009) 15976-15977. https://doi.org/10.1021/ja905970y

[34] Y. Li, H. Wang, H. Zhang, P. Liu, Y. Wang, W. Fang, H. Yang, Y. Li, H. Zhao, A {0001} faceted single crystal NiS nanosheet electrocatalyst for dye-sensitised solar cells: sulfur-vacancy induced electrocatalytic activity, Chemical communications 50(42) (2014) 5569-5571. https://doi.org/10.1039/c4cc01691b

Applications for Earth-Abundant Transition Metals
Materials Research Foundations 179 (2025) 118-144

Materials Research Forum LLC
https://doi.org/10.21741/9781644903711-6

# Chapter 6

# Transition Metal Composite for Supercapacitors

Mehdi Zarei[1,2], Aziz Babapoor[1] *, Mehdi Eskandarzade[1]

[1]Department of Chemical Engineering, University of Mohaghegh Ardabili, Ardabil, Iran

[2]Department of Mechanical Engineering, University of Mohaghegh Ardabili, Ardabil, Iran

* babapoor@uma.ac.ir

**Abstract**

The unique electrochemical properties, elevated energy density, and excellent cycle durability of transition metal composites have made them attractive supercapacitor electrode choices. Research on the effectiveness of transition metal composites in supercapacitors has focused on a number of different systems, including those based on ruthenium, manganese, cobalt, nickel, and others. This chapter explores the role of transition metal composites in high-performance supercapacitors advancement, emphasizing their pseudocapacitive characteristics and potential to improve energy storage systems. Also, the new advancements in the investigation of the supercapacitor performance of the material are evaluated.

**Keywords**

Transition Metals, Supercapacitors, Composites, EDLCs, Pseudocapacitors

## Contents

## 1.   Introduction

Energy storage systems (ENS) are essential amidst contemporary society. The primary characteristics evaluated while selecting a storage device are particular energy, specific power, longevity, dependability, and protection. Most of our energy requirements are actualized by fossil fuels, which are significantly adverse for the ecosystem. Conversely, wind and solar energies are pristine and abundant energy sources. The unpredictable and erratic nature of wind and sunlight results in fluctuation and inconsistency in energy production from wind turbines and solar devices. Consequently, the advancement of energy storage techniques is imperative, necessitating the creation of ENSs. Numerous nations accumulate substantial quantities of water to harness hydroelectric power. Nevertheless, evidence indicates that the construction of major dams adversely affects the Earth [1,2].

Lead acid batteries are the most prevalent energy storage systems, characterized by poor energy density, restricted charge-discharge cycles, and a brief lifespan. Furthermore, they comprise hazardous lead and detrimental acid. Li-ion batteries are utilized in portable gadgets, including smartphones and tablets. Batteries have the capacity to store substantial energy. They exhibit limitations, including self-discharge, limited lifespan, and sensitivity to temperature [1]. High-efficiency novel technologies necessitate substantial energy storage, achievable just through alternative storage solutions like supercapacitors. Supercapacitors have garnered considerable interest owing to their capacity to store substantial energy and their extended lifespan, presenting enormous potential for the advancement of hybrid energy storage systems across diverse applications [3].

While supercapacitors may deliver power hundreds or thousands of times greater than batteries within a comparable volume, they are unable to retain equivalent amounts of energy. Consequently, supercapacitors are employed in applications that necessitate significant energy bursts without the requirement for substantial energy storage. Supercapacitors may also be utilized in battery-based energy storage systems [3]. Additionally, supercapacitors possess several uses in renewable energy power facilities, memory backup systems, industrial equipment, and hybrid electric vehicles. Supercapacitors demonstrate rapid charge and discharge rates and extended operational lifespans in comparison to batteries [4]. Supercapacitors are attractive because they connect electrolytic capacitors and batteries [3].

A Ragone plot graphically illustrates energy density (ED) vs. power density (PD). Fig. 1 presents a Ragone plot that compares supercapacitors with various electrical energy storage systems. Nonetheless, the Ragone plot fails to consider additional factors such as cost, safety, and longevity. It is important to acknowledge that while supercapacitors discharge rapidly, they may also be recharged in brief intervals [3,5].

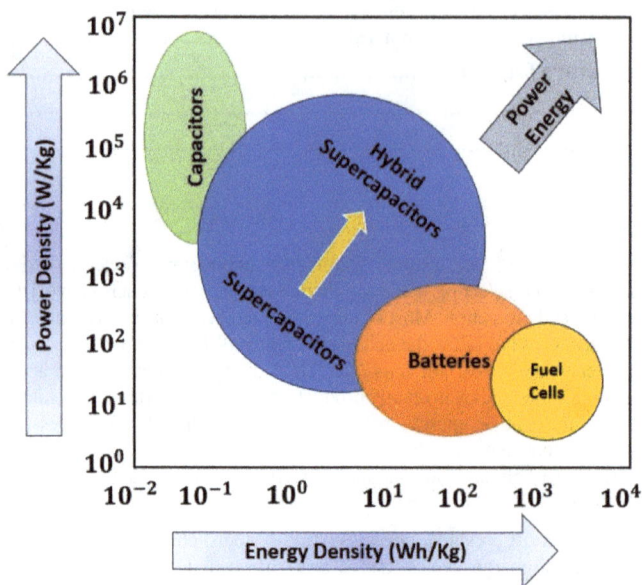

*Fig. 1: Ragone plot that compares supercapacitors with various electrical energy storage systems [6]*

Another benefit of supercapacitors is their longevity in cycles. Owing to their efficient mechanism, these devices can function for millions of cycles, whereas batteries typically endure only a few thousand cycles. Furthermore, the disparity in operating temperatures between batteries and supercapacitors must be considered. Supercapacitors can operate efficiently at temperatures as low as -40 °C, a capability not attainable by batteries. In addition to these benefits, supercapacitors are often safer owing to their rapid charging and discharging speeds [7]. Fig. 2 illustrates different types of supercapacitors.

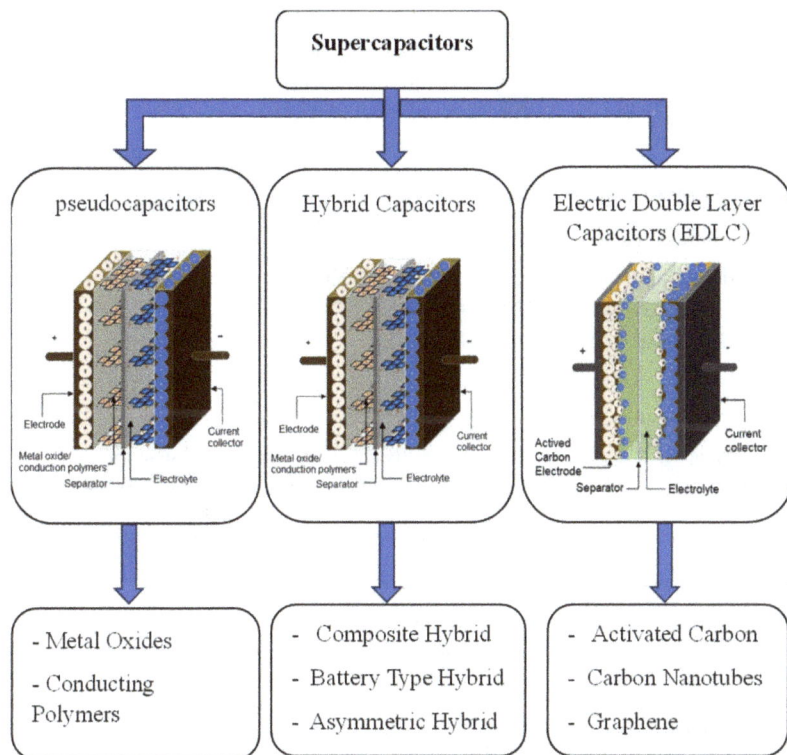

*Insert Fig. 2*

*Fig. 2: Types of supercapacitors and their respective operational methods [8,9]*

In the past few years, numerous nanostructured metal oxide composites have been employed as supercapacitors. The microstructure of the electrodes, in addition to the inherent features of the materials, substantially affects energy storage characteristics. The anodization process for generating an oxide layer on supercapacitors, by deriving the oxide layer from the substrate, can markedly diminish the interfacial resistance between the oxide coating and the substrate relative to alternative methods, thus improving its electrochemical performance[10,11]. The nanopores with an exposed surface enhance electrolyte infiltration and improve accessibility to the oxide surface. Moreover, highly porous oxide offers multiple unoccupied sites to facilitate expansion and contraction during its operation as a battery and pseudo-capacitor [12,13].

The specific surface area is crucial as the pseudo-capacitive process transpires on the electrode surfaces. Furthermore, an increased effective surface area grants additional active regions. This rising is contingent upon their microstructure. Homogeneous pore diameters and enhanced surface area can augment the capacity of the metal oxide electrode. An additional significant factor to consider is their electrical conductivity. Most metal oxides generally possess greater band gaps than metals, demonstrating semiconductor characteristics. The utilization of these materials as electrodes is constrained by inadequate electrical conductivity, which restricts their capacity [14,15].

## 2.     Fundamentals of supercapacitors

### 2.1     Energy storage mechanisms

Energy storage in supercapacitors relies on charge accumulation or reversible reduction processes. Supercapacitors are divided into three classifications [16,17]:

- Electrochemical Double-Layer Capacitors (EDLCs)

- Pseudocapacitors

- Hybrid supercapacitors, integrating the features of the former two.

Hybrid supercapacitor devices are distinguished by enhanced capacity and superior energy storage characteristics. These supercapacitors present a favorable perspective by integrating the properties of both pseudo-capacitors and EDLCs. In pseudo-capacitors, energy storage relies on recurrent oxidation processes occurring in the active interface of the electrolyte and electrode [16,18,19].

#### 2.1.1     Electric double-layer capacitors (EDLCs)

In these electrodes, charge storage is established via a dual structure created at the interface of electrolyte-electrode via ions. Throughout the charging procedure, the surfaces of the two electrodes acquire positive or negative charges due to an external electric current, thereby attracting opposite charged ions. Upon removing the electrical power, the charges reciprocal attraction facilitates the generation of a stable electrochemical double layer, thereby enabling energy storage by this mechanism. During discharge, electrons migrate negative-to-positive electrode, producing a current and releasing energy. Porous carbon compounds serve as specialized electrodes for this process [18,20].

#### 2.1.2     Pseudocapacitors

This technique relies on adsorption and desorption or reduction processes to produce capacitance. During the adsorption process, ions aggregate at the electrolyte-electrode interface under an applied electrical power, proceeded by reversible reduction reactions. Energy is concurrently stored. During the discharge process, ions are emitted, and the retained energy is transmitted to the external circuit. Consequently, since ions in pseudo-capacitors can infiltrate the electrodes and engage in the reduction reaction, the capacity of these types of supercapacitors is 10-100 times exceeding that of EDLC capacitors [18,20,21].

### 2.2     Electrolyte

The electrolyte is essential for supercapacitors, alongside the electrodes. Electrolyte requirements encompass a broad potential range, excellent electrochemical stability, little resistance, and low toxicity [22]. The potential decomposition of electrolytes generally governs the voltage of the breakdown of supercapacitors, hence constraining power density (PD) and energy density (ED).

An elevated operating voltage can result in enhanced PD and ED. The electrolyte ionic conductivity is a crucial component influencing power density. The electrolytes are categorized into ionic liquid, organic, and aqueous, each with distinct advantages and disadvantages [14].

Aqueous electrolytes have advantages like enhanced ionic conductivity, affordability, and non-flammability. Conversely, organic electrolytes demonstrate reduced conductivity, are costly, frequently combustible, and exhibit severe toxicity. Nevertheless, the possible range in aqueous solutions (about 1.2V) is significantly inferior to that of organic electrolytes. The inclusion of oxidative active elements in aqueous electrolytes significantly enhances capacity by facilitating more extensive redox interactions between the electrode and electrolyte [22]. Aqueous electrolytes are classified into three categories: neutral, alkaline, and acidic solutions (e.g., $Na_2SO_4$, KOH, and $H_2SO_4$, respectively). Aqueous solutions yield elevated ionic concentration and reduced resistance, which is advantageous for capacity and performance metrics. Where cost supersedes energy density, the utilization of aqueous electrolytes is warranted. Conversely, organic electrolytes present a broader electrochemical window, yielding enhanced energy and power density, albeit at increased costs. Certain organic electrolytes encounter constraints owing to their elevated flammability. Ionic liquid electrolytes, or molten salts at ambient temperature, consist entirely of ions and exist in a liquid state at room temperature without the need for solvents. They constitute the third group of electrolytes, in addition to aqueous and organic electrolytes. These electrolytes exhibit exceptional thermal and chemical stability, minimal flammability, and an extensive electrochemical window. Although exhibiting considerable promise, their elevated viscosity leads to diminished ionic conductivity [14,17].

## 2.3    Electrode materials

The electrode materials choice and production are vital for improving supercapacitor performance. Electrodes for supercapacitors must possess thermal stability, excellent surface wettability, chemical stability, appropriate electrical conductivity, and corrosion resistance. Furthermore, the capacity to enable faradaic charge transfer is crucial for enhancing capacitive performance. The capacitance factor is determined by surface area and surface morphology [19,23].

Two critical criteria for the design of supercapacitor devices are:

1. Enhancing electrochemically active areas by the selection of proper electrode materials.

2. Optimizing pore dimensions and morphology to facilitate ion movement inside the electrolyte.

Electrodes with reduced pore size yield more capacity and, therefore, enhanced energy density. However, diminished pore size also results in diminishing power density. Consequently, the selection of the electrode is contingent upon the application. Furthermore, a suitable pore size distribution enhances capacity. Optimal distribution of micro and mesopores throughout the electrode enhances mass and ion transfer rates while improving electrolyte accessibility. This part of supercapacitors is categorized into three classifications [23,24]:

- **Carbon-Based electrodes:** Carbon-based materials were initially employed as EDLCs in 1975. Their application as EDLCs is owing to their simplicity of synthesis, eco-friendliness, affordability, and extensive surface area. The characteristics of these materials are contingent upon the dimensions of their pores, often approximately 1 nm. Moreover, their electrochemical, chemical, and thermal stability in diverse electrolytes, ranging from concentrated acids to concentrated bases, coupled with high electrical conductivity, symmetrical charge-discharge profiles, and rectangular cyclic potential curves, render them viable candidates for supercapacitors. Diverse forms of carbonaceous materials are employed in supercapacitors. Carbon nanotube electrodes, characterized by their accessible exterior structures with extensive surface area, distinctive internal mesoporous networks, and exceptional thermal and chemical durability, are highly advanced for supercapacitors. Carbon nanotubes exhibit reduced ESR and, hence, enhanced power compared to activated carbons since the mesoporous architecture of the carbon nanotube network facilitates more efficient and rapid ion penetration in the electrolyte. Extensive research on diverse carbon compounds is currently being conducted [15,22,23,25]

- **Metal Oxides:** This group are prevalent materials employed in the construction of supercapacitors that utilize redox processes [23]. Metal oxides, via the pseudocapacitive energy storage mechanism, offer superior specific capacitance (SC) relative to traditional carbon materials and enhanced cycle stability compared to polymeric materials [15,22,25].

- **Conductive Polymers:** Conductive polymers can demonstrate capacitive behavior via faradaic processes. Conductive polymers frequently utilized in supercapacitors encompass polypyrrole (PPy), polythiophene (PTh), polyaniline (PANI), and their byproducts, selected for their attributes of high voltage ranges and affordability. Nonetheless, elevated resistance and diminished stability restrict their utilization. Swelling and contraction may specifically occur during the using process, causing mechanical electrode breakdown and diminishing the performance of the supercapacitor [15,22,25].

## 2.4 Pseudocapacitive mechanism in transition metal oxides

Carbon-based electrodes present numerous advantages as charge storage materials in supercapacitors. Nevertheless, these electrodes have nearly attained their specific capacitance (SC) threshold. Their dependence on double-layer capacitive processes constrains the charge storage capacity of carbon electrodes. In contrast, metal oxides generally exhibit significantly greater capacitance. While double-layer capacitance is present in these materials, many are synthesized with high surface area morphologies, and their main benefit lies in charge storage facilitated by rapid and reversible redox processes. The charge storage in these electrodes is not solely due to capacitive processes but is mostly facilitated by electron transport mechanisms [6,26].

Recent investigations into metal oxide-based supercapacitors have concentrated on the design and advancement of electrodes while ensuring stable charge retention. The intrinsic characteristics of metal oxides dictate storage operational efficiency. Conversely, the external engineering of materials, including the incorporation of reinforcements and the creation of composites with other substances, is anticipated to markedly improve electrochemical performance [27].

Cations of metallic elements that transfer a greater number of electrons in oxidation/reduction reactions within aqueous electrolytes are anticipated to yield enhanced charge storage capacity. The spectrum of oxidation/reduction potentials and the potential for water decomposition establish the operational potential window. Cations with diminished electronegativity can enhance electron transfer and augment power density, whereas cations with elevated electronegativity can elevate the charge storage capacity [6,27].

It is essential to recognize that these metal oxides exhibit changing oxidation states, hence augmenting the capacity of supercapacitors. Despite increasing interest in these oxides, they exhibit deficiencies, including inadequate cycle stability and insufficient power density, which

Materials Research Forum LLC
https://doi.org/10.21741/9781644903711-6

continue to pose significant barriers to their current utilization. The diminished PD of metal oxides is due to inadequate electrical conductivity, which restricts the swift transmission of electrons. Instability in electrode materials during cycling is frequently attributed to swelling and contraction phenomena occurring during charge/discharge cycles [28].

Theoretical specific capacitance, presuming uniform charge storage over the full potential range, can be determined by the quantity of electrons stored or released during oxidation and reduction reactions:

$$C = [\frac{nF}{m}] \cdot \frac{X}{E} \tag{1}$$

n: electrons in oxidation/reduction reactions,

F: Faraday constant,

X: surface fraction

m: the metal oxide molar mass, and

E: the potential window

The morphologies can be regulated through diverse preparation methods that yield distinct crystal growth and orientations, resulting in varied crystal faces for the procedure. The surface energy of various crystal planes differs, leading to distinct oxidation/reduction activities and, therefore, modified charge storage ability. Morphologies must demonstrate micro/nano-porosity which improves the electrolyte ions diffusion ions into the electrodes [6,27].

## 3. Transition metal composite for supercapacitors

Transition metals display elevated density, melting points, and boiling points. The features are attributed to the metallic bonding facilitated by delocalized d electrons. This results in cohesiveness, which then intensifies with an increased quantity of shared electrons [29]. Numerous transition metal oxides (TMOs) have been thoroughly investigated. The suboptimal rate stability and efficiency of these materials stem from the low electrical conductivity, a critical deficiency that must be remedied for effective employment in supercapacitors [29,30].

Metal oxides provide superior specific capacitance relative to traditional carbonaceous materials due to reversible redox processes (Faradaic reactions) [31]. Transition metal oxides are categorized into base ($Fe_3O_4$, NiO, $MnO_2$) and noble ($IrO_2$ and $RuO_2$) metal oxides [8,32,33].

Nonetheless, these materials exhibit favorable energy density. Nonetheless, their low conductivity, unpredictable volumetric changes, and sluggish ion transport in the bulk phase hinder their practical and industrial uses [34,35].

### 3.1 Ruthenium-based composites

In the last decade, Ruthenium oxide has been the predominant Ru-based electrode material in supercapacitors. Nonetheless, $RuO_2$ is somewhat costly. Consequently, numerous researchers are rigorously investigating $RuO_2$ electrode materials to identify alternative methods for reducing the cost of supercapacitors. A crucial method is to regulate the film's thickness [36]. The film's structure, morphology, and homogeneity can be regulated through applied voltage, bath temperature, and the concentration of precursors and additional agents [37]. Crystalline and amorphous forms of RuO2 exhibit electroactivity, hence justifying their classification as effective electrode materials[38]. The cyclic voltammetry arises from the Faradaic processes in RuO2, triggered by the $H^+$ coupled $e^-$ mechanism, as delineated in the following equation [39]:

$$RuO_2 + mH^+ + me^- = RuO_{2-m}(OH)_m \qquad (2)$$

RGO-RuO$_2$ composite as a flexible electrode was investigated[40]. The electrochemical analysis of the RuO$_2$/rGO electrode revealed a SC of 1371 F/g. The flexible solid-state asymmetric supercapacitor rGO-RuO$_2$-PVA/H$_2$SO$_4$-WO3 demonstrated an SC of 114F/g and maintained 88% capacitance conservation (5000 GCD) cycles, which is attributable to the cooperative effects of rGO and RuO$_2$.

An improvement in the capacitive operation of hybrid ruthenium oxide materials developed by Guo et al. [41]. RuO2 encapsulated by sorbitol+ Tween, and also by glucose+ Tween. The capacitances of these compounds were significantly above the pristine RuO2. Also, capacitance conservation for new compounds was 30.2% more than pristine RuO2. The enhancement is due to the diminutive size and uniform distribution of nanoparticles. Yan et al. [42] devised a straightforward and effective method for decorating multiwalled carbon nanotubes (MWCNTs) with RuO$_2$ nanoparticles for application in supercapacitors. The process involved the direct synthesis of RuO$_2$ nanoparticles, which were subsequently affixed to MWCNTs using a RuCl$_3$ solution through microwave-assisted irradiation. The cyclic voltammetry results indicated that RuO$_2$/MWCNTs exhibited markedly higher capacity compared to pristine MWCNTs in the identical medium. RuO$_2$/TiO$_2$ hybrid composites demonstrated exceptional specific capacitances(1200 F/g) compare to commercial RuO2-based electrode (~600 F/g) [43].

RuO2 nanoparticles can be affixed to carbon nanofibers(CNF) by a hydrothermal deposition technique[44]. The nanoparticles, averaging 2 nm in diameter, are uniformly distributed over the CNF surfaces. This leads to the fabrication of supercapacitors employing RuO2-grafted carbon nanofiber nanocomposites as electrodes. The incorporation of CNFs diminishes contact resistance and enhances quick electron transfer, hence augmenting electrochemical performance. The RuO2-grafted CNF presents stability and a short relaxation period, rendering it appropriate for quick charging and discharging.

Ruthenium oxide/graphene nanocomposites have been created for supercapacitor applications[45]. The ruthenium oxide nanoparticles are uniformly distributed over the graphene sheets, improving conductivity and preventing restacking and agglomeration. The composites exhibit notable electrochemical activation capabilities, with graphene substantially influencing performance. The composites demonstrate a SC of 441.1F/g at 0.1A/g, exceptional cycle stability, 1.6V for voltage window, and improved ED.

Hsu et al. [46] constructed a supercapacitor with flexible transparent properties utilizing RuO2-coated CsxWO3(nanorods) as the anode for acidic electrolytes. The electrodes demonstrated remarkable pseudocapacitive operation, with appropriate RuO2 treatment substantially improving pseudocapacitance. The enhanced capacitance stemmed from limited diffusion and surface-limited mechanisms, rendering the composite particularly advantageous for transparent photothermal supercapacitors.

A study [47] examined a nanocomposite of $RuO2$ nanoparticles and $MoS2$ nanowires, evaluating its electrochemical performance. The composite electrode demonstrated SC values of 972 F/g, along with 100% cycling durability (10,000 cycles), rendering it appropriate for energy storage applications. Korkmaz et al. [48] synthesized an aerogel utilizing double-walled carbon nanotubes and ruthenium chloride then employing it to construct symmetrical supercapacitor devices. The cyclic voltammetry analysis showed SC of 423 F/g, while the charge-discharge profiles indicated 420.3 F/g. The aerogel maintained 96.38% retention after 5000 cycles. Also, other researchers [49] examined cobalt ruthenium sulfides as a supercapacitor electrode, demonstrating enhanced SC, and outstanding cycling stability.

### 3.2 Manganese-based composites

Mn-base composites have been extensively examined as a prospective electrode material [50]. $MnO_2$ possesses a broader operating potential range compared to other transition metal oxides, leading to enhanced ED [51]. In $MnO_2$ supercapacitors, significant attributes were observed [52]. Manganese-based supercapacitors can be manufactured using several fabrication methods, incorporating other metals, nonmetals, polymers, and crystals [53].

$MnO2$ electrodes accumulate charge by the surface adsorption of cations from the electrolyte, including $Li^+$, $Na^+$, $K^+$, and $H_3O^+$, alongside their integration into the electrode [54]:

$$(MnO_2)_{surface} + C^+ + e^- \leftrightarrow (MnOOC)_{surface} \qquad (3)$$

The inadequate electrical conductivity and sluggish cation and anion mobility of the solid manganese dioxide confine the reaction of pseudocapacitive to a superficial layer, hence diminishing rate capacity and specific capacitance. Initiatives are underway to rectify these inadequacies through the amalgamation with other materials [55]. One of the most appealing characteristics is its elevated theoretical SC of up to 1370 F/g, determined by the single-electron redox reaction between $Mn^{3+}$ and $Mn^{4+}$. Nonetheless, it is observed that $MnO_2$ frequently exhibits very low real specific capacitances (200-300 F/g), which can be ascribed to its significant phase transition and inadequate electrical conductivity [56,57].

As mentioned, $Mn_2O_3$ is a potential catalyst and energy storage material. Nevertheless, its inadequate conductivity restricts its application as a supercapacitors. A multi-tiered designed architecture prevents nanowire stacking, ensuring high ED and cycle stability in asymmetric supercapacitors (ASC). The $\alpha$-$MnO_2$/$Mn_3O_4$@CC configuration has a SC of 181F/g at 2A/g, whereas the asymmetric supercapacitor with activated carbon demonstrates an ED of 118.3 Wh/kg [58].

A novel composite material, $MnO2$@$SnO2$ hollow spherical, has been studied[59]. It comprises $MnO_2$ hollow spheres and scattered nano-$SnO2$. The manganese dioxide structure augments the electrochemical effective surface area, while the tin oxide nanoparticles collaborate with $MnO2$ frameworks, influencing charge and mass transfer. The composite has remarkable cyclability, delivering 541.6F/g at 1A/g and sustaining 498F/g after 1500 cycles. It can be utilized to fabricate an electrochemical capacitor (EC) with elevated ED. The materials Indicated efficient redox cycling based on the following reactions:

$$MnO_2 + K^+ + e^- \leftrightarrow MnOOK \qquad (4)$$

$$SnO_2 + 2e^- + H_2O \leftrightarrow SnO + 2OH^- \qquad (5)$$

A research [60] Investigates flower-like spheroidal δ-MnO2/Mn2O3 composites characterized by a distinctive nanosheet structure. The composite improves pseudocapacitance and lowers interfacial resistance. The electrode demonstrates a prominent SC of 182F/g and outstanding cycle stability. The composites have a notable ED of 9.1Wh/kg, rendering them suitable for different kinds of energy storage systems.

MnO2/CuO composites were generated as electrode materials, with wheat-shaped composites exhibiting enhanced electrochemical performance[61]. The wheat-shaped composite electrodes demonstrated a capacitance of 261.4mF/cm² and maintained 90% retention (1000 cycles). The symmetric aqueous supercapacitors, utilizing wheat-shaped CuO anode and manganese dioxide cathode, showed an outstanding SC of 152.7mF/cm² and 74.2% cycling stability (500 cycles). A 3D hierarchical MnO2@NiCo2O4 core-shell nanoflower electrode was investigated[62], attaining a SC of 634.37F/g at 0.25A/g. Also, this electrode exhibited 83.1% retention (3000 cycles), indicating remarkable capacitance stability and presenting an excellent supercapacitor.

The application of layered-MnO2/activated carbon composites as efficient and cost-effective electrode materials for supercapacitors was investigated[63]. The stacked α-MnO2 nanoparticles, produced via lignin activation, double the SC of activated carbon to 294F/g at 5 mV/s. The surface chemistry and pore architectures substantially affect the electrochemical properties, and the composites exhibit remarkable cycle stability and rate capability. A study on MnO2-MWCNT nanocomposites as supercapacitor electrodes [64] revealed that elevated CNT concentration decreased the adsorption of crystalline MnO2, causing CNT aggregation. Electrochemical analyses demonstrated that the SC of manganese dioxide climbed from 124F/g to 145F/g with the incorporation of 1mg/ml MWCNTs. Vargheese et al. [65] created a supercapacitor utilizing nitrogen-doped carbon composites generated from a triazine framework decorated with dual metal oxides. The composite exhibited elevated capacitance and cycle stability, with a symmetric system attaining 91.1F/g at 1A/g. The device capacitance preserved about 92.3% after 7000 cycles, due to metal oxide redox reactions and efficient electrolyte transport through porous carbon. Rao et al. [66] created graphene-metal oxide electrodes via a hydrothermal technique, utilizing manganese dioxide and Laser-Induced Graphene (LIG) as electrodes in flexible supercapacitors. The all-flexible Supercapacitor (FSC) demonstrated remarkable areal capacitance, substantial ED, and PD. It exhibited 82% capacitance retention after 2000 cycles and 80% flexibility during bending tests, indicating its potential for diverse ESD applications.

Manganese oxide and molybdenum oxide exhibit elevated theoretical capacitance, as evidenced by Teli et al. [67], which investigated the co-deposition of Mo-Mn. The porous, cubic structure exhibited a notable capacitance of 88.6mF/cm², with an ED of 3.08μWh/cm² at a PD of 125μW/cm². Hu et al. [68] employed an electrodeposition process to alter the morphology and structure of CoMnS nanomaterials, yielding distinctive flower-like architectures. The SC of CoMnS was 3469.5F/g at 1A/g. After 10,000 cycles, it maintained 95.78% at 20A/g. Also, CoMnS-4//AC supercapacitor possesses considerable application potential.

Materials Research Forum LLC
https://doi.org/10.21741/9781644903711-6

## 3.3 Cobalt-based composites

Co3O4 is notably appealing among metal oxides because to its affordability, superior dispositive activity, facile fabrication as a nanostructure, and elevated capacitance computational ratio [69]. Researchers have augmented the ED of battery-type capacitor materials. They achieved this by producing graphene composites, synthesizing nanostructures, and creating composite metal oxides, resulting in beneficial outcomes [70].

The design of electrodes is essential for supercapacitors. Ultrathin Co3O4 nanosheets, produced on RGO, have opportunities for supercapacitor purposes [71]. The Co3O4/RGO electrode has exceptional electrochemical properties and exhibits a SC of 3344.1F/g at 1.25A/g. Also, it demonstrates exceptional capability and maintains a capacity of 87.9% after 6000 cycles. Sheng et al. [72] produced a composite nanomaterial incorporating carbon nanotubes within cobalt oxide, causing a 69.7% increase in specific capacitance. This material has an exceptional ED of 58.46Wh/Kg at ambient temperature and retains 83.2% capacity after 65,000 cycles. The incorporation of CNTs enhances the cohesion of cobalt oxide nanowires, hence improving conductivity, active sites, and stability.

Faradaic charge storage necessitates the redox activity of electrode materials and the mobility of electrons and ions, achievable through a combination of battery materials with diverse dimensions. A heterostructured Co3O4@Ni3S4 composite electrode was fabricated, including Ni3S4 nanosheets on Co3O4 nanowire arrays[73]. This composite shows elevated faradaic activity, effective electron/ion transport, and structural stability. The hybrid supercapacitor demonstrated elevated energy density, minimal self-discharge rate, and superior cycle stability, indicating potential for faradaic energy storage.

Graphene/Co3O4 composites represent encouraging electrode components for high-performance supercapacitors. Nevertheless, their practical application is constrained by production techniques and operational voltage range. A straightforward technique was employed for extensive production, yielding a SC of 570F/g at 1A/g, preserving 93% of the initial capacity. The outstanding performance results from the cooperative interaction between graphene and Co3O4.

Zhu et al. [74] generated hollow Co3O4/NiCo2O4 electrodes exhibiting abundant active sites, elevated specific surface area, and superior conductivity. The electrode demonstrated an impressive SC of 1788F/g at 1A/g, and a constructed hybrid supercapacitor with a potential of 43.9Wh/kg. The formation of Co3O4 nanoparticles on porous carbon yielded a SC of 781F/g at 0.5A/g, exceeding that of the p-Co3O4 and EPC electrodes[75]. The device had a cyclic stability of 96.7% (20,000 cycles). The advancement of nanoporous Co3O4/Ag and Co3O4/Ag/CuO supercapacitor electrodes has demonstrated encouraging outcomes[76]. The Faradaic capacitance, associated with redox processes, produced pseudo-capacitance behavior. The nanoporous CuO network enhanced the SC of the electrode. The electrodes exhibited remarkable ED and PD, together with outstanding cycling and robust rate capability.

A study [77] examined the synthesis of CuO/Co3O4 composites featuring wheat spike spherical and spherical morphologies on a nickel foam substrate. The results indicated that the load cycle substantially affects performance, with optimal electrochemical performance attained after 10 cycles. The W–CuO/Co3O4 electrode surpassed the S–CuO/Co3O4 electrode, exhibiting a SC of 86.0mAh/g and a capacitance preservation rate of 76.9% (5000 cycles). This material exhibits considerable potential for supercapacitors.

Sahoo et al. synthesized a Co3O4/SnO2/rGO nanocomposite in 30 seconds, achieving a high capacity and enhanced cycling stability of 103.2% (15,000 cycles)[78]. The supercapacitor demonstrated elevated energy density and cycling stability. Another research [79] illustrates a symmetric activated carbon AC/Co3O4//AC/Co3O4 (AC= activated carbon) supercapacitor displayed a SC of 125F/g and a significant ED of 55Wh/kg at a PD of 650W/kg. Also, $Co(OH)_2$-CdO nanoplates have been fabricated for supercapacitor applications[80]. The electrodes have a polycrystalline structure and mushroom-like planar morphology, demonstrating specific capacitances of up to 312F/g and 1119F/g. Han et al. [81] have created hierarchical cobalt sulfide-copper nanobelt array topologies. The CuCo2S4 nanobelt arrays exhibit improved accessibility and exceptional conductivity of electrochemically active sites, achieving a high SC. The device exhibited an ED of 40.2Wh/kg at a PD of 799.1W/kg.

## 3.4    Nickel-based composites

Nickel Oxide is a compelling material that possesses small benefits over other transition metal oxides, meeting the requisite requirements for good performance. Nickel oxide shows a significant theoretical capacity of 2584F/g and is cost-effective, rendering it a promising choice for supercapacitor applications. Nevertheless, this supercapacitor electrode is exceptionally stable, with a wide band gap, environmentally safe, and highly electrochemically active. At low crystallinity, NiO exhibits a higher capacitance compared to its highly crystalline counterpart due to efficient diffusion channels that promote the portability of ions [82,83]. Nonetheless, the deficiencies of NiO are similarly complex. NiO, being a quintessential inherent p-type semiconductor with low electronic conductivity, significantly impedes electron migration. Consequently, the experimental capacitance is inferior to the theoretical value. To address these issues, several techniques have been sequentially developed, including hybridization with carbonaceous materials, the introduction of vacancies, and cathodic/anodic doping [84,85]. The efficacy of this material as a supercapacitor electrode is governed by the redox reaction of NiO in alkaline electrolytes by the following equations:

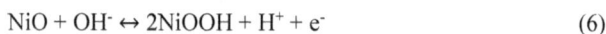

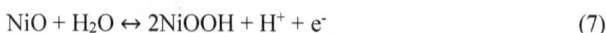

$$NiO + OH^- \leftrightarrow 2NiOOH + H^+ + e^- \tag{6}$$

$$NiO + H_2O \leftrightarrow 2NiOOH + H^+ + e^- \tag{7}$$

Two primary ideas exist about the creation of NiO: one hypothesis proposes that the energy storage mechanism arises between NiOOH and NiO, while another suggests that NiO first transforms within Ni(OH) in alkaline electrolytes, subsequently accompanied by electrochemical reactions between NiOOH and $Ni(OH)_2$. The majority of scientists desire the first one. However, the second theory is also pragmatic [86,87]. A schematic of NiO-based electrode Faradaic reaction is illustrated in Fig. 3.

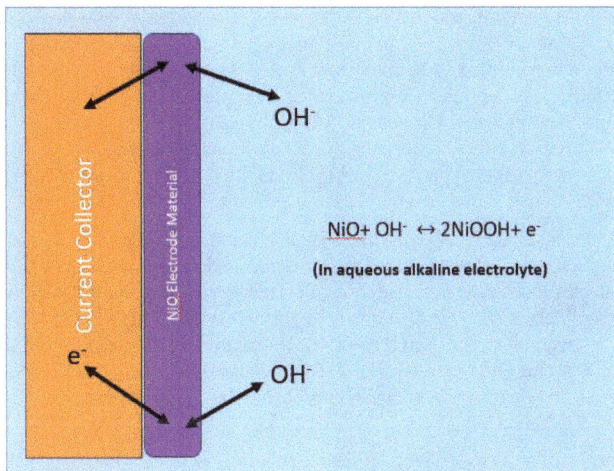

*Fig. 3: A schematic of charge storage processes on nickel oxide-based electrode [88]*

A carbon nanofiber (CNF) electrode material containing dilute NiO particles has been examined as an electrode[88]. The composite electrodes demonstrated exceptional properties, withstanding up to 5000 cycles with a 10% degradation. The rGO@NiO composites were engineered as supercapacitor electrode materials[89], demonstrating a SC of 1093F/g at 1A/g and remarkable cycle durability, preserving 87% of initial capacitance (5000 cycles). The fabrication of binary oxide nanocomposites, including NiO and ZnO, provides a process for generating high-performance supercapacitors [90]. The SC of NiO is 295.5F/g, but ZnO/NiO nanocomposites exhibit a SC of 561.75F/g, surpassing the individual values.

The work by Rafiq et al. [91] on NiO/SnO2 for supercapacitor applications revealed SC of 1035.71F/g and 980.76F/g, respectively, along with a capacitance retention of 64%. The research determined that 1 g of SnO2 represented the ideal concentration for enhanced electrochemical activity. The remarkable functionality of the nanocomposites is ascribed to their enhanced surface area and elevated electrical conductivity, stemming from synergistic interactions between SnO2 and NiO. Xu et al. [92] discovered that the integration of graphene with NiO enhances the reversibility of NiO as a supercapacitor electrode. The unusual graphene structure facilitates an uninterrupted network and enhances redox reactions. The NiO-rGO composite attained a SC of 171.3F/g at 0.5A/g, exhibiting a mere 20.93% decline after 2000 cycles. The CNT-NiO composite has been confirmed for its versatile properties, exhibiting remarkable cyclic stability and a SC of 878.19F/g at 2 mV/s. An asymmetric supercapacitor was produced utilizing the activated carbon and composite material as the negative electrode and the positive electrode, respectively, which yield a capacitance of 197.7F/g, along with ED of 85.7Wh/kg and PD of 11.2kW/kg [93].

The NiO/NiCo2O4 electrode exhibited a SC of 717.8mF/cm² at 2mA/cm², with a retention rate of 74.8%. Aqueous and all-solid symmetric supercapacitors were fabricated with this electrode exhibiting a PD of 880μW/cm² and an ED of 123.0μWh/cm² [94]. The synthesized NMO/NiO nanocomposites exhibited remarkable supercapacitor performance owing to dual Faradaic redox mechanisms[95]. The heterojunction formed between NiO and NMO enhanced the separation of charges, causing higher efficacy. The optimal composite exhibited a SC of 943F/g and exceptional cycling stability, maintaining 83.1% after 4000 cycles. Jiang et al. [96] synthesized Ni@C/NiO composites, exhibiting exceptional electrochemical performance attributed to many active sites, oxygen vacancies, and elevated Ni content. The composites attained a peak SC of 805.3F/g at 0.5A/g and 576.2F/g at 10A/g. The constructed asymmetric supercapacitor (ASC) exhibited elevated specific energy, specific power, and cycle stability, maintaining 69.7% capacitance after 10,000 cycles.

Nickel hydroxide is an advantageous material characterized by specific capacitance, highly diversified morphologies, substantial surface area, and various oxidation states. This compound encounters obstacles such as the augmentation of current density and the stabilization of vanadium following doping. Mane et al. [97] manufactured $Ni(OH)_2$-vanadium microflowers by a chemical bath deposition technique. The $Ni(OH)_2$-0.3% V electrode exhibited an outstanding SC of 1456.8F/g at 3mA/cm2 and demonstrated commendable efficiency with 88.17% capacity maintenance over 2000 cycles, notwithstanding challenges related to increasing current density and stabilizing post-doping vanadium in $Ni(OH)_2$.

Hexagonal nickel sulfide is extensively utilized in supercapacitors due to its elevated theoretical SC and cost-effectiveness. Nonetheless, its electrical conductivity and aggregation impede its practical application. Ouyang et al. [98] synthesized composite materials of NiS-activated carbon nanotubes, yielding a hybrid electrode with a SC of 1266F/g and 81% retention following 2000 cycles at 2.0A/g.

### 3.5    Other transition metal-based composites

Vanadium pentoxide, an essential element in electrochemical energy storage, has been thoroughly investigated owing to its several oxidation states and potential for effective electron transmission [99,100]. The ideal approach to address challenges in the electrochemical charge storage mechanism of supercapacitors is to enhance charge carrier mobility, band gap, structural flexibility, and electrical conductivity via structural modifications and heterostructure nanomaterials [101,102].

The V2O5 nanostructure, with limited graphene integration, displayed a SC of 450.5F/g at 0.5A/g, retaining 73.3% of its initial capacitance. Symmetric supercapacitors utilizing these electrode materials attained a remarkable ED of 33.5Wh/Kg at 425.6W/Kg, exceeding the performance of most existing symmetric supercapacitors [103]. The binary nanocomposite V2O5/TiO2 (VT) has elevated SC and cyclic stability, presenting it as a compelling electrode for commercial energy storage purposes [104]. A study [105] assesses the porous structure of g-C3N4/V2O5, which serves as an encouraging electrode for advanced supercapacitors. If the porous morphology ratio falls below 1:1, the specific capacity diminishes. However, when the ratio surpasses 1:1, the pores get clogged, diminishing the specific capacity. The 3D PCN@V2O5 has an exceptional SC of 457F/g at 0.5A/g. Another study [106] indicates that V2O5/graphene composites can be utilized to fabricate viable supercapacitors with SC of 673.2F/g, exceeding the performance of previous analogous devices. Jia et al. [107] produced V2O5 nanobelt arrays on nickel foam and NiO nanosheet arrays on V2O5 nanobelt arrays. The solitary electrode displayed a SC of 875F/g and a cycle durability of 96.1%. This composite provides enhanced electrochemical performance for supercapacitors.

ZnO is an active material characterized by a high ED of 650A/g, thereby garnering significant attention in recent years [70]. The rGO/ZnO nanocomposite serves as an exceptional electrode, exhibiting a SC of 312F/g and improved cycling stability over 1000 cycles [108]. Another research [109] examines ZnO nanorods within functionalized carbon nanotubes, demonstrating their electric double layer capacitor properties. The hybrid ZnO/CNT nanocomposites showed a SC of 189F/g, a quick charge-discharge performance of 95F/g, and 96%cyclic stability over 1000 cycles. Joshi et al. [110] fabricated ZnO/CoOx electrodes exhibiting an optimal sample of 1.95F/cm² at 5mA/cm² with 94% capacitance retention (12,000 cycles). The ZnO/NiS nanocomposite at a 5:5 ratio exhibited a remarkable capacitance of 1322 F/g, 76.66% retention after 1000 cycles, and enabled calculations of power and energy density, rendering it perfect for supercapacitor performance [111].

Nagarani et al. [112] synthesized rGO/ZnO-CuO nanocomposites comprising spherical and hexagonal nanoparticles on GO sheets. The nanocomposite presented a SC of 260.7F/g at 0.5A/g, substantial ED, and outstanding cycle stability, retaining 98.7% of its initial supercapacitance (10,000 cycles). The integration of ZnO with FeAl2O4 electrode improves cycling stability, ED, and PD, yielding a composite electrode with exceptional capacitance[113]. The composite demonstrated a capacitance of 1149.2F/g in cyclic voltammetry study and 1332.7F/g in galvanostatic charge-discharge analysis within a 2 M KOH electrolytic solution. Gaire et al. [114] developed flexible, nanoporous Fe2O3-rGO electrodes, improving electrochemical performance. The nanoporous architecture accommodates volumetric variations and facilitates electrolyte ion transport. The electrode attained an initial SC of 179F/g and retention after 5000 cycles. Another study [115]examined the electrochemical efficacy of SnO2/ Graphene composites for supercapacitors. The composite attained a peak SC of 818.6F/g, signifying improved performance ascribed to the integration of a graphene matrix. The cycle stability of the composite electrode was also confirmed. Hanif et al. [116] fabricated TiO2-A/GO composites for use as electrodes in electrochemical energy storage. The composite electrode exhibited superior specific capacitances relative to the TiO2-A electrode, with a peak SC of 338.2F/g at 10mV/s, and displaying remarkable cycle stability after 2000 cycles.

**Conclusion and Future Outlooks**

Transition metal composites are known as attractive materials for supercapacitor applications owing to their distinctive amalgamation of electrical, mechanical, and electrochemical properties. Materials, particularly those derived from ruthenium, manganese, cobalt, nickel, and other transition metals, exhibit diverse pseudocapacitive characteristics that augment energy storage capacity while preserving high power density. This chapter illustrates that the collaboration of metal oxides, conductive polymers, and carbon-based materials in composite structures is essential for enhancing electrochemical performance, increasing cycle stability, and expanding the operational range of supercapacitors. These developments render transition metal composites appropriate for diverse applications, spanning handheld electronics to grid energy storage.

Ongoing research into supercapacitors reveals various areas with potential for substantial advancements, including material innovation, sophisticated electrolytes, and morphological regulation. Future endeavors will likely concentrate on identifying more economical alternatives that have equivalent performance. Furthermore, the development of hybrid systems that integrate various transition metals with sophisticated nanomaterials will persist as a prominent area of research. A comprehensive understanding of electrolyte interactions with transition metal composites would facilitate enhanced efficiency and durability in next-generation devices. Moreover, customizing the dimensions, morphology, and dispersion of nanoparticles in composite materials is anticipated to enhance performance at the nanoscale.

In conclusion, transition metal composites for supercapacitors possess significant potential for future energy storage technologies. Ongoing interdisciplinary research emphasizing material innovation and practical applications will be crucial to fully harness their promise in energizing the next generation of electronic gadgets and renewable energy systems.

## Acknowledgments

The financial support from the University of Mohaghegh Ardabili, Iran, is highly appreciated. Also, the authors would like to thank all the people who helped carry out this chapter's research. Acknowledgments are also made to all publications that granted copyright permission to use images.

## References

[1] P. Sharma, V. Kumar, Current Technology of Supercapacitors: A Review, J. Electron. Mater. 49 (2020) 3520–3532. https://doi.org/10.1007/s11664-020-07992-4

[2] P. Kurzweil, HISTORY | Electrochemical Capacitors, in: J.B.T.-E. of E.P.S. Garche (Ed.), Elsevier, Amsterdam, 2009: pp. 596–606. https://doi.org/https://doi.org/10.1016/B978-044452745-5.00006-X

[3] M.A.A. Mohd Abdah, N.H.N. Azman, S. Kulandaivalu, Y. Sulaiman, Review of the use of transition-metal-oxide and conducting polymer-based fibres for high-performance supercapacitors, Mater. Des. 186 (2020) 108199. https://doi.org/https://doi.org/10.1016/j.matdes.2019.108199

[4] M. Zhi, C. Xiang, J. Li, M. Li, N. Wu, Nanostructured carbon–metal oxide composite electrodes for supercapacitors: a review, Nanoscale 5 (2013) 72–88. https://doi.org/10.1039/C2NR32040A

[5] Y. Xu, W. Lu, G. Xu, T.-W. Chou, Structural supercapacitor composites: A review, Compos. Sci. Technol. 204 (2021) 108636. https://doi.org/https://doi.org/10.1016/j.compscitech.2020.108636

[6] M. Wayu, Manganese Oxide Carbon-Based Nanocomposite in Energy Storage Applications, Solids 2 (2021) 232–248. https://doi.org/10.3390/solids2020015

[7] A. González, E. Goikolea, J.A. Barrena, R. Mysyk, Review on supercapacitors: Technologies and materials, Renew. Sustain. Energy Rev. 58 (2016) 1189–1206. https://doi.org/https://doi.org/10.1016/j.rser.2015.12.249

[8] S. Jayakumar, P.C. Santhosh, M.M. Mohideen, A. V Radhamani, A comprehensive review of metal oxides ($RuO_2$, $Co_3O_4$, $MnO_2$ and $NiO$) for supercapacitor applications and global market trends, J. Alloys Compd. 976 (2024) 173170. https://doi.org/https://doi.org/10.1016/j.jallcom.2023.173170

[9] F. Ahmad, A. Shahzad, M. Danish, M. Fatima, M. Adnan, S. Atiq, M. Asim, M.A. Khan, Q.U. Ain, R. Perveen, Recent developments in transition metal oxide-based electrode composites for supercapacitor applications, J. Energy Storage 81 (2024) 110430. https://doi.org/https://doi.org/10.1016/j.est.2024.110430

[10] M. Zarei, S. Nourouzi, R. Jamaati, I.G. Cano, M. Sarret, S. Dosta, S.H. Esmaeili-Faraj, Electrochemical properties of CNT doped nanoporous tin oxide hybrid electrode formed on cold spray tin coating for supercapacitor application, Diam. Relat. Mater. 139 (2023) 110318. https://doi.org/https://doi.org/10.1016/j.diamond.2023.110318

[11] D. V Shinde, D.Y. Lee, S.A. Patil, I. Lim, S.S. Bhande, W. Lee, M.M. Sung, R.S. Mane, N.K. Shrestha, S.-H. Han, Anodically fabricated self-organized nanoporous tin oxide film as a supercapacitor electrode material, RSC Adv. 3 (2013) 9431–9435. https://doi.org/10.1039/C3RA22721A

[12] H. Bian, J. Zhang, M.-F. Yuen, W. Kang, Y. Zhan, D.Y.W. Yu, Z. Xu, Y.Y. Li, Anodic nanoporous SnO2 grown on Cu foils as superior binder-free Na-ion battery anodes, J. Power Sources 307 (2016) 634–640. https://doi.org/https://doi.org/10.1016/j.jpowsour.2015.12.118

[13] M. Zarei, S. Nourouzi, R. Jamaati, S.H. Esmaeili-Faraj, I.G. Cano, S. Dosta, M. Sarret, Electrochemical characterization of nanoporous SnO2 formed by anodization on cold spray tin coating for supercapacitor application, J. Electroanal. Chem. 931 (2023) 117201. https://doi.org/https://doi.org/10.1016/j.jelechem.2023.117201

[14] C. An, Y. Zhang, H. Guo, Y. Wang, Metal oxide-based supercapacitors: progress and prospectives, Nanoscale Adv. 1 (2019) 4644–4658. https://doi.org/10.1039/C9NA00543A

[15] K. Dissanayake, D. Kularatna-Abeywardana, A review of supercapacitors: Materials, technology, challenges, and renewable energy applications, J. Energy Storage 96 (2024) 112563. https://doi.org/https://doi.org/10.1016/j.est.2024.112563

[16] A. Muzaffar, M.B. Ahamed, K. Deshmukh, J. Thirumalai, A review on recent advances in hybrid supercapacitors: Design, fabrication and applications, Renew. Sustain. Energy Rev. 101 (2019) 123–145. https://doi.org/https://doi.org/10.1016/j.rser.2018.10.026

[17] A. Tyagi, S. Banerjee, J. Cherusseri, K.K. Kar, Characteristics of Transition Metal Oxides BT - Handbook of Nanocomposite Supercapacitor Materials I: Characteristics, in: K.K. Kar (Ed.), Springer International Publishing, Cham, 2020: pp. 91–123. https://doi.org/10.1007/978-3-030-43009-2_3

[18] M.G. Ashritha, K. Hareesh, Chapter 9 - Electrode materials for EDLC and pseudocapacitors, in: C.M. Hussain, M.B.B.T.-S.S. Ahamed (Eds.), Elsevier, 2023: pp. 179–198. https://doi.org/https://doi.org/10.1016/B978-0-323-90530-5.00033-2

[19] A. Rajapriya, S. Keerthana, N. Ponpandian, Chapter 4 - Fundamental understanding of charge storage mechanism, in: C.M. Hussain, M.B.B.T.-S.S. Ahamed (Eds.), Elsevier, 2023: pp. 65–82. https://doi.org/https://doi.org/10.1016/B978-0-323-90530-5.00034-4

[20] F. Bu, W. Zhou, Y. Xu, Y. Du, C. Guan, W. Huang, Recent developments of advanced micro-supercapacitors: design, fabrication and applications, Npj Flex. Electron. 4 (2020) 31. https://doi.org/10.1038/s41528-020-00093-6

[21]  A. Joseph, T. Thomas, Pseudocapacitance: An Introduction BT - Pseudocapacitors: Fundamentals to High Performance Energy Storage Devices, in: R.K. Gupta (Ed.), Springer Nature Switzerland, Cham, 2024: pp. 1–17. https://doi.org/10.1007/978-3-031-45430-1_1

[22]  C. Zhao, W. Zheng, A Review for Aqueous Electrochemical Supercapacitors, Front. Energy Res. 3 (2015). https://www.frontiersin.org/journals/energy-research/articles/10.3389/fenrg.2015.00023

[23]  P. Forouzandeh, V. Kumaravel, S.C. Pillai, Electrode Materials for Supercapacitors: A Review of Recent Advances, Catalysts 10 (2020). https://doi.org/10.3390/catal10090969

[24]  H.A. Khan, M. Tawalbeh, B. Aljawrneh, W. Abuwatfa, A. Al-Othman, H. Sadeghifar, A.G. Olabi, A comprehensive review on supercapacitors: Their promise to flexibility, high temperature, materials, design, and challenges, Energy 295 (2024) 131043. https://doi.org/https://doi.org/10.1016/j.energy.2024.131043

[25]  A. Afif, S.M.H. Rahman, A. Tasfiah Azad, J. Zaini, M.A. Islan, A.K. Azad, Advanced materials and technologies for hybrid supercapacitors for energy storage – A review, J. Energy Storage 25 (2019) 100852. https://doi.org/https://doi.org/10.1016/j.est.2019.100852

[26]  T. Brousse, O. Crosnier, D. Bélanger, J.W. Long, 1 - Capacitive and Pseudocapacitive Electrodes for Electrochemical Capacitors and Hybrid Devices, in: D.P. Dubal, P.B.T.-M.O. in S. Gomez-Romero (Eds.), Met. Oxides, Elsevier, 2017: pp. 1–24. https://doi.org/https://doi.org/10.1016/B978-0-12-810464-4.00001-2

[27]  T. Nguyen, M. de F. Montemor, Metal Oxide and Hydroxide–Based Aqueous Supercapacitors: From Charge Storage Mechanisms and Functional Electrode Engineering to Need-Tailored Devices, Adv. Sci. 6 (2019) 1801797. https://doi.org/https://doi.org/10.1002/advs.201801797

[28]  M. Kebede, Electrochemical Devices for Energy Storage Applications, 2020.

[29]  B. De, S. Banerjee, K.D. Verma, T. Pal, P.K. Manna, K.K. Kar, Transition Metal Oxides as Electrode Materials for Supercapacitors BT - Handbook of Nanocomposite Supercapacitor Materials II: Performance, in: K.K. Kar (Ed.), Springer International Publishing, Cham, 2020: pp. 89–111. https://doi.org/10.1007/978-3-030-52359-6_4

[30]  G. Tang, J. Liang, W. Wu, Transition Metal Selenides for Supercapacitors, Adv. Funct. Mater. 34 (2024) 2310399. https://doi.org/https://doi.org/10.1002/adfm.202310399

[31]  G. Wang, L. Zhang, J. Zhang, A review of electrode materials for electrochemical supercapacitors, Chem. Soc. Rev. 41 (2012) 797–828. https://doi.org/10.1039/C1CS15060J

[32]  P. Sinha, S. Banerjee, K.K. Kar, Transition Metal Oxide/Activated Carbon-Based Composites as Electrode Materials for Supercapacitors BT - Handbook of Nanocomposite Supercapacitor Materials II: Performance, in: K.K. Kar (Ed.), Springer International Publishing, Cham, 2020: pp. 145–178. https://doi.org/10.1007/978-3-030-52359-6_6

[33]  J.B. Goodenough, Perspective on Engineering Transition-Metal Oxides, Chem. Mater. 26 (2014) 820–829. https://doi.org/10.1021/cm402063u

Materials Research Forum LLC
https://doi.org/10.21741/9781644903711-6

[34]    S. Raj, P. Kar, P. Roy, Facile synthesis of flower-like morphology Cu0.27Co2.73O4 for a high-performance supercapattery with extraordinary cycling stability, Chem. Commun. 54 (2018) 12400–12403. https://doi.org/10.1039/C8CC04625E

[35]    C. An, Y. Wang, Y. Huang, Y. Xu, L. Jiao, H. Yuan, Porous NiCo2O4 nanostructures for high performance supercapacitors via a microemulsion technique, Nano Energy 10 (2014) 125–134. https://doi.org/https://doi.org/10.1016/j.nanoen.2014.09.015

[36]    Q. Li, S. Zheng, Y. Xu, H. Xue, H. Pang, Ruthenium based materials as electrode materials for supercapacitors, Chem. Eng. J. 333 (2018) 505–518. https://doi.org/https://doi.org/10.1016/j.cej.2017.09.170

[37]    V.D. Patake, S.M. Pawar, V.R. Shinde, T.P. Gujar, C.D. Lokhande, The growth mechanism and supercapacitor study of anodically deposited amorphous ruthenium oxide films, Curr. Appl. Phys. 10 (2010) 99–103. https://doi.org/https://doi.org/10.1016/j.cap.2009.05.003

[38]    D. Majumdar, T. Maiyalagan, Z. Jiang, Recent Progress in Ruthenium Oxide-Based Composites for Supercapacitor Applications, ChemElectroChem 6 (2019) 4343–4372. https://doi.org/https://doi.org/10.1002/celc.201900668

[39]    S. Chalupczok, P. Kurzweil, H. Hartmann, C. Schell, The Redox Chemistry of Ruthenium Dioxide: A Cyclic Voltammetry Study—Review and Revision, Int. J. Electrochem. 2018 (2018) 1273768. https://doi.org/https://doi.org/10.1155/2018/1273768

[40]    A.G. Bagde, D.B. Malavekar, A.C. Lokhande, S.D. Khot, C.D. Lokhande, Flexible solid-state asymmetric supercapacitor based on reduced graphene oxide (rGO)/ruthenium oxide (RuO2) composite electrode, J. Alloys Compd. 980 (2024) 173591. https://doi.org/https://doi.org/10.1016/j.jallcom.2024.173591

[41]    Y. Guo, X. Zou, Y. Wei, L. Shu, A. Li, J. Zhang, R. Wang, Synthesis of organic hybrid ruthenium oxide nanoparticles for high-performance supercapacitors, Electrochim. Acta 443 (2023) 141938. https://doi.org/https://doi.org/10.1016/j.electacta.2023.141938

[42]    S. Yan, H. Wang, P. Qu, Y. Zhang, Z. Xiao, RuO2/carbon nanotubes composites synthesized by microwave-assisted method for electrochemical supercapacitor, Synth. Met. 159 (2009) 158–161. https://doi.org/https://doi.org/10.1016/j.synthmet.2008.07.024

[43]    S. Park, D. Shin, T. Yeo, B. Seo, H. Hwang, J. Lee, W. Choi, Combustion-driven synthesis route for tunable TiO2/RuO2 hybrid composites as high-performance electrode materials for supercapacitors, Chem. Eng. J. 384 (2020) 123269. https://doi.org/https://doi.org/10.1016/j.cej.2019.123269

[44]    C.-M. Chuang, C.-W. Huang, H. Teng, J.-H. Ting, Hydrothermally synthesized RuO2/Carbon nanofibers composites for use in high-rate supercapacitor electrodes, Compos. Sci. Technol. 72 (2012) 1524–1529. https://doi.org/https://doi.org/10.1016/j.compscitech.2012.05.024

[45]    R. Thangappan, M. Arivanandhan, R. Dhinesh Kumar, R. Jayavel, Facile synthesis of RuO2 nanoparticles anchored on graphene nanosheets for high performance composite electrode for supercapacitor applications, J. Phys. Chem. Solids 121 (2018) 339–349. https://doi.org/https://doi.org/10.1016/j.jpcs.2018.05.049

[46]    C.-H. Hsu, K.-H. Tseng, C.-Y. Hsu, D.-H. Chen, RuO2-decorated CsxWO3 composite nanorods as transparent photothermal negative electrode material for enhancing supercapacitor performance in acid electrolyte, Compos. Part B Eng. 252 (2023) 110497. https://doi.org/https://doi.org/10.1016/j.compositesb.2022.110497

[47]    M. Manuraj, J. Chacko, K.N. Narayanan Unni, R.B. Rakhi, Heterostructured MoS2-RuO2 nanocomposite: A promising electrode material for supercapacitors, J. Alloys Compd. 836 (2020) 155420. https://doi.org/https://doi.org/10.1016/j.jallcom.2020.155420

[48]    S. Korkmaz, İ.A. Kariper, C. Karaman, O. Karaman, MWCNT/Ruthenium hydroxide aerogel supercapacitor production and investigation of electrochemical performances, Sci. Rep. 12 (2022) 12862. https://doi.org/10.1038/s41598-022-17286-w

[49]    R. Bolagam, S. Um, Hydrothermal Synthesis of Cobalt Ruthenium Sulfides as Promising Pseudocapacitor Electrode Materials, Coatings 10 (2020). https://doi.org/10.3390/coatings10030200

[50]    B. Liu, Y. Sun, L. Liu, S. Xu, X. Yan, Advances in Manganese-Based Oxides Cathodic Electrocatalysts for Li–Air Batteries, Adv. Funct. Mater. 28 (2018) 1704973. https://doi.org/https://doi.org/10.1002/adfm.201704973

[51]    Z.-H. Huang, Y. Song, D.-Y. Feng, Z. Sun, X. Sun, X.-X. Liu, High Mass Loading MnO2 with Hierarchical Nanostructures for Supercapacitors, ACS Nano 12 (2018) 3557–3567. https://doi.org/10.1021/acsnano.8b00621

[52]    J. Yan, A. Sumboja, X. Wang, C. Fu, V. Kumar, P.S. Lee, Insights on the Fundamental Capacitive Behavior: A Case Study of MnO2, Small 10 (2014) 3568–3578. https://doi.org/https://doi.org/10.1002/smll.201303553

[53]    T. Van Nguyen, L.T. Son, V. Van Thuy, V.D. Thao, M. Hatsukano, K. Higashimine, S. Maenosono, S.-E. Chun, T.V. Thu, Facile synthesis of Mn-doped NiCo2O4 nanoparticles with enhanced electrochemical performance for a battery-type supercapacitor electrode, Dalt. Trans. 49 (2020) 6718–6729. https://doi.org/10.1039/D0DT01177K

[54]    M.N. Sakib, S. Ahmed, S.M.S.M. Rahat, S.B. Shuchi, A review of recent advances in manganese-based supercapacitors, J. Energy Storage 44 (2021) 103322. https://doi.org/https://doi.org/10.1016/j.est.2021.103322

[55]    T.S. Bhat, S.A. Jadhav, S.A. Beknalkar, S.S. Patil, P.S. Patil, MnO2 core-shell type materials for high-performance supercapacitors: A short review, Inorg. Chem. Commun. 141 (2022) 109493. https://doi.org/https://doi.org/10.1016/j.inoche.2022.109493

[56]    M. Huang, F. Li, F. Dong, Y.X. Zhang, L.L. Zhang, MnO2-based nanostructures for high-performance supercapacitors, J. Mater. Chem. A 3 (2015) 21380–21423. https://doi.org/10.1039/C5TA05523G

[57]    X. Tang, S. Zhu, J. Ning, X. Yang, M. Hu, J. Shao, Charge storage mechanisms of manganese dioxide-based supercapacitors: A review, New Carbon Mater. 36 (2021) 702–710. https://doi.org/https://doi.org/10.1016/S1872-5805(21)60082-3

[58]    J. Jia, X. Lian, M. Wu, F. Zheng, Y. Gao, H. Niu, Self-assembly of α-MnO2/Mn3O4 hierarchical structure on carbon cloth for aymmetric supercapacitors, J. Mater. Sci. 56 (2021) 3246–3255. https://doi.org/10.1007/s10853-020-05475-9

[59]   W. Feng, G. Liu, P. Wang, J. Zhou, L. Gu, L. Chen, X. Li, Y. Dan, X. Cheng, Template Synthesis of a Heterostructured MnO2@SnO2 Hollow Sphere Composite for High Asymmetric Supercapacitor Performance, ACS Appl. Energy Mater. 3 (2020) 7284–7293. https://doi.org/10.1021/acsaem.0c00388

[60]   F. Qiao, W. Liu, S. Yu, Q. Zhang, G. Li, H. Chu, Construction of δ-MnO2/Mn2O3 flower sphere composite with two-dimensional hierarchical nanosheets as supercapacitor electrode with enhanced performance, Ceram. Int. 50 (2024) 351–359. https://doi.org/https://doi.org/10.1016/j.ceramint.2023.10.108

[61]   B. Liu, L. Tian, X. Zheng, Z. Xing, MnO2 Films deposited on CuO nanomaterials as electrode materials for supercapacitors, J. Alloys Compd. 911 (2022) 165003. https://doi.org/https://doi.org/10.1016/j.jallcom.2022.165003

[62]   N. Zhang, C. Xu, H. Wang, J. Zhang, Y. Liu, Y. Fang, Assembly of the hierarchical MnO2@NiCo2O4 core–shell nanoflower for supercapacitor electrodes, J. Mater. Sci. Mater. Electron. 32 (2021) 1787–1799. https://doi.org/10.1007/s10854-020-04947-7

[63]   Y. Liu, S. Zuo, B. Shen, Y. Wang, H. Xia, Fabrication of Nanosized Layered-MnO2/Activated Carbon Composites Electrodes for High-performance Supercapacitor, Int. J. Electrochem. Sci. 15 (2020) 7646–7662. https://doi.org/https://doi.org/10.20964/2020.08.87

[64]   R. Singhal, T. Sadowski, M. Chaudhary, R. V Tucci, J. Scanley, R. Patel, P. Kumar Patel, S. Gagnon, A. Koni, K. Singhal, P.K. LeMaire, R. Kumar Sharma, B.P. Singh, C.C. Broadbridge, Optimization of manganese dioxide-multiwall carbon nanotube composite electrodes for supercapacitor applications, Mater. Sci. Energy Technol. 7 (2024) 228–236. https://doi.org/https://doi.org/10.1016/j.mset.2023.12.001

[65]   S. Varghese, R.S. Kumar, R.T.R. Kumar, J.-J. Shim, Y. Haldorai, Binary metal oxide (MnO2/SnO2) nanostructures supported triazine framework-derived nitrogen-doped carbon composite for symmetric supercapacitor, J. Energy Storage 68 (2023) 107671. https://doi.org/https://doi.org/10.1016/j.est.2023.107671

[66]   A. Rao, S. Bhat, S. De, Binder-free laser induced graphene-MnO2 composite electrodes for high areal energy density flexible supercapacitors, Electrochim. Acta 487 (2024) 144152. https://doi.org/https://doi.org/10.1016/j.electacta.2024.144152

[67]   A.M. Teli, S.A. Beknalkar, T.S. Bhat, S.M. Mane, J.C. Shin, Molybdenum–Manganese hydroxide microcubes based electrode via hydrothermal method for asymmetric supercapacitor, Ceram. Int. 48 (2022) 29386–29393. https://doi.org/https://doi.org/10.1016/j.ceramint.2022.06.002

[68]   X. Hu, S. Liu, Y. Chen, J. Jiang, H. Cong, J. Tang, Y. Sun, S. Han, H. Lin, Rational design of flower-like cobalt–manganese-sulfide nanosheets for high performance supercapacitor electrode materials, New J. Chem. 44 (2020) 11786–11795. https://doi.org/10.1039/D0NJ01727B

[69]   S. Aloqayli, C.K. Ranaweera, Z. Wang, K. Siam, P.K. Kahol, P. Tripathi, O.N. Srivastava, B.K. Gupta, S.R. Mishra, F. Perez, X. Shen, R.K. Gupta, Nanostructured cobalt oxide and cobalt sulfide for flexible, high performance and durable supercapacitors, Energy Storage Mater. 8 (2017) 68–76. https://doi.org/https://doi.org/10.1016/j.ensm.2017.05.006

[70]  J. Pan, C. Li, Y. Peng, L. Wang, B. Li, G. Zheng, M. Song, Application of transition metal (Ni, Co and Zn) oxides based electrode materials for ion-batteries and supercapacitors, Int. J. Electrochem. Sci. 18 (2023) 100233. https://doi.org/https://doi.org/10.1016/j.ijoes.2023.100233

[71]  K. Ding, P. Yang, P. Hou, X. Song, T. Wei, Y. Cao, X. Cheng, Ultrathin and Highly Crystalline $Co_3O_4$ Nanosheets In Situ Grown on Graphene toward Enhanced Supercapacitor Performance, Adv. Mater. Interfaces 4 (2017) 1600884. https://doi.org/https://doi.org/10.1002/admi.201600884

[72]  W. Sheng, W. Hanbo, P. Dongyu, W. Ziming, F. Zhitian, Y. Mingrui, L. Kechang, L. Haiyan, $Co_3O_4$ nanowire modified with carbon nanotubes to be used as improved asymmetric supercapacitor electrode, Surfaces and Interfaces 46 (2024) 104049. https://doi.org/https://doi.org/10.1016/j.surfin.2024.104049

[73]  J. Chang, S. Zang, Y. Wang, C. Chen, D. Wu, F. Xu, K. Jiang, Z. Bai, Z. Gao, $Co_3O_4$@$Ni_3S_4$ heterostructure composite constructed by low dimensional components as efficient battery electrode for hybrid supercapacitor, Electrochim. Acta 353 (2020) 136501. https://doi.org/https://doi.org/10.1016/j.electacta.2020.136501

[74]  Y. Zhang, P. Ding, W. Wu, H. Kimura, Y. Shen, D. Wu, X. Xie, C. Hou, X. Sun, X. Yang, W. Du, Facile synthesis of reduced graphene oxide@$Co_3O_4$ composites derived from assisted liquid-phase plasma electrolysis for high-performance hybrid supercapacitors, Appl. Surf. Sci. 609 (2023) 155188. https://doi.org/https://doi.org/10.1016/j.apsusc.2022.155188

[75]  R. Zou, B. Wang, L. Zhu, L. Yan, F. Shi, Y. Sun, B. Shao, S. Zhang, W. Sun, Biomass derived porous carbon supported nano-$Co_3O_4$ composite for high-performance supercapacitors, Diam. Relat. Mater. 126 (2022) 109060. https://doi.org/https://doi.org/10.1016/j.diamond.2022.109060

[76]  G.A. Naikoo, M.A.S. Tabook, B.A.M. Tabook, M. Bano, I.U. Hassan, R.A. Dar, T.A. Saleh, Electrochemical performance of $Co_3O_4$/Ag/CuO electrodes for supercapacitor applications, J. Energy Storage 85 (2024) 111047. https://doi.org/https://doi.org/10.1016/j.est.2024.111047

[77]  K. Xue, L. Tian, X. Zheng, J. Ding, M. Ali, S. Xiao, M. Song, S. Kumar, CuO/$Co_3O_4$ materials grown directly on nickel foam for high-performance supercapacitor electrode, Mater. Chem. Phys. 313 (2024) 128712. https://doi.org/https://doi.org/10.1016/j.matchemphys.2023.128712

[78]  S. Sahoo, A. Milton, A. Sood, R. Kumar, S. Choi, C.K. Maity, S.S. Han, Microwave-assisted synthesis of perovskite hydroxide-derived $Co_3O_4$/$SnO_2$/reduced graphene oxide nanocomposites for advanced hybrid supercapacitor devices, J. Energy Storage 99 (2024) 113321. https://doi.org/https://doi.org/10.1016/j.est.2024.113321

[79]  B.A. Al Jahdaly, A. Abu-Rayyan, M.M. Taher, K. Shoueir, Phytosynthesis of $Co_3O_4$ Nanoparticles as the High Energy Storage Material of an Activated Carbon/$Co_3O_4$ Symmetric Supercapacitor Device with Excellent Cyclic Stability Based on a $Na_2SO_4$ Aqueous Electrolyte, ACS Omega 7 (2022) 23673–23684. https://doi.org/10.1021/acsomega.2c02305

[80]    K.K. Tehare, M.K. Zate, S.T. Navale, S.S. Bhande, S.L. Gaikwad, S.A. Patil, S.K. Gore, M. Naushad, S.M. Alfadul, R.S. Mane, Electrochemical supercapacitors of cobalt hydroxide nanoplates grown on conducting cadmium oxide base-electrodes, Arab. J. Chem. 10 (2017) 515–522. https://doi.org/https://doi.org/10.1016/j.arabjc.2016.01.006

[81]    L. Han, X. Liu, Z. Cui, Y. Hua, C. Wang, X. Zhao, X. Liu, Hierarchical copper cobalt sulfide nanobelt arrays for high performance asymmetric supercapacitors, Inorg. Chem. Front. 8 (2021) 3025–3036. https://doi.org/10.1039/D1QI00352F

[82]    Q. Lu, J.G. Chen, J.Q. Xiao, Nanostructured Electrodes for High-Performance Pseudocapacitors, Angew. Chemie Int. Ed. 52 (2013) 1882–1889. https://doi.org/https://doi.org/10.1002/anie.201203201

[83]    S.H. Sutar, B.M. Babar, K.B. Pisal, A.I. Inamdar, S.H. Mujawar, Feasibility of nickel oxide as a smart electrochromic supercapacitor device: A review, J. Energy Storage 73 (2023) 109035. https://doi.org/https://doi.org/10.1016/j.est.2023.109035

[84]    P. Giannakou, M.G. Masteghin, R.C.T. Slade, S.J. Hinder, M. Shkunov, Energy storage on demand: ultra-high-rate and high-energy-density inkjet-printed NiO micro-supercapacitors, J. Mater. Chem. A 7 (2019) 21496–21506. https://doi.org/10.1039/C9TA07878A

[85]    Y. Li, J. Zhang, Z. Chen, M. Chen, Nickel-based materials: Toward practical application of the aqueous hybrid supercapacitors, Sustain. Mater. Technol. 33 (2022) e00479. https://doi.org/https://doi.org/10.1016/j.susmat.2022.e00479

[86]    Y.Q. Zhang, X.H. Xia, J.P. Tu, Y.J. Mai, S.J. Shi, X.L. Wang, C.D. Gu., Self-assembled synthesis of hierarchically porous NiO film and its application for electrochemical capacitors, J. Power Sources 199 (2012) 413–417. https://doi.org/https://doi.org/10.1016/j.jpowsour.2011.10.065

[87]    X. Xia, J. Tu, Y. Mai, R. Chen, X. Wang, C. Gu, X. Zhao, Graphene Sheet/Porous NiO Hybrid Film for Supercapacitor Applications, Chem. – A Eur. J. 17 (2011) 10898–10905. https://doi.org/https://doi.org/10.1002/chem.201100727

[88]    Y. Yang, F. Yang, H. Hu, S. Lee, Y. Wang, H. Zhao, D. Zeng, B. Zhou, S. Hao, Dilute NiO/carbon nanofiber composites derived from metal organic framework fibers as electrode materials for supercapacitors, Chem. Eng. J. 307 (2017) 583–592. https://doi.org/https://doi.org/10.1016/j.cej.2016.08.132

[89]    Y. Zhang, Y. Shen, X. Xie, W. Du, L. Kang, Y. Wang, X. Sun, Z. Li, B. Wang, One-step synthesis of the reduced graphene oxide@NiO composites for supercapacitor electrodes by electrode-assisted plasma electrolysis, Mater. Des. 196 (2020) 109111. https://doi.org/https://doi.org/10.1016/j.matdes.2020.109111

[90]    J. Angadi V, V. Molahalli, G. Soman, G. Hegde, S. Wang, N. Roy, S.W. Joo, V. Pattar, S.F. Shaikh, C. Prakash, A. Kumar, M. Ubaidullah, M. Zhang, Synthesis of ZnO and NiO nano ceramics composite high-performance supercapacitor and its catalytic capabilities, Ceram. Int. 50 (2024) 39732–39738. https://doi.org/https://doi.org/10.1016/j.ceramint.2024.07.352

[91]    S. Rafiq, A.K. Alanazi, S. Bashir, A.Y. Elnaggar, G.A.M. Mersal, M.M. Ibrahim, S. Yousaf, K. Chaudhary, Optimization studies for nickel oxide/tin oxide (NiO/Xg SnO2, X: 0.5,

1) based heterostructured composites to design high-performance supercapacitor electrode, Phys. B Condens. Matter 638 (2022) 413931. https://doi.org/https://doi.org/10.1016/j.physb.2022.413931

[92]    J. Xu, L. Wu, Y. Liu, J. Zhang, J. Liu, S. Shu, X. Kang, Q. Song, D. Liu, F. Huang, Y. Hu, NiO-rGO composite for supercapacitor electrode, Surfaces and Interfaces 18 (2020) 100420. https://doi.org/https://doi.org/10.1016/j.surfin.2019.100420

[93]    A. Roy, A. Ray, S. Saha, M. Ghosh, T. Das, B. Satpati, M. Nandi, S. Das, NiO-CNT composite for high performance supercapacitor electrode and oxygen evolution reaction, Electrochim. Acta 283 (2018) 327–337. https://doi.org/https://doi.org/10.1016/j.electacta.2018.06.154

[94]    M. Song, X. Jin, L. Tian, L. Liu, H. Feng, J. Ding, M. Ali, Z. Xing, S. Han, Electrochemical deposition of NiO/NiCo2O4 nanostructures for high-performance supercapacitors, Mater. Chem. Phys. 321 (2024) 129514. https://doi.org/https://doi.org/10.1016/j.matchemphys.2024.129514

[95]    K.-C. Lee, J.-H. Huang, Y.-J. Wu, K.-S. Wang, E.-C. Cho, S.-C. Hsu, T.-Y. Liu, Crystal structure-controlled synthesis of NiMoO4/NiO hierarchical microspheres for high-performance supercapacitors and photocatalysts, J. Energy Storage 97 (2024) 112639. https://doi.org/https://doi.org/10.1016/j.est.2024.112639

[96]    X. Jiang, S. Deng, J. Liu, N. Qi, Z. Chen, Metal–organic framework derived NiO/Ni@C composites as electrode materials with excellent electrochemical performance for supercapacitors, J. Energy Storage 37 (2021) 102426. https://doi.org/https://doi.org/10.1016/j.est.2021.102426

[97]    D.B. Mane, O.C. Pore, D.S. Sawant, D. V Rupnavar, R. V Shejwal, S.H. Mujawar, L.D. Kadam, R. V Dhekale, G.M. Lohar, Supercapacitor performance of vanadium-doped nickel hydroxide microflowers synthesized using the chemical route, Appl. Phys. A 129 (2023) 158. https://doi.org/10.1007/s00339-023-06449-9

[98]    Y. Ouyang, Y. Chen, J. Peng, J. Yang, C. Wu, B. Chang, X. Guo, G. Chen, Z. Luo, X. Wang, Nickel sulfide/activated carbon nanotubes nanocomposites as advanced electrode of high-performance aqueous asymmetric supercapacitors, J. Alloys Compd. 885 (2021) 160979. https://doi.org/https://doi.org/10.1016/j.jallcom.2021.160979

[99]    G. Sun, G. Ren, Z. Shi, L. Zhang, Z. Wang, K. Zhan, Y. Yan, J. Yang, B. Zhao, V2O5/vertically-aligned carbon nanotubes as negative electrode for asymmetric supercapacitor in neutral aqueous electrolyte, J. Colloid Interface Sci. 588 (2021) 847–856. https://doi.org/https://doi.org/10.1016/j.jcis.2020.11.126

[100]    H. Lee, V.S. Kumbhar, J. Lee, Y. Choi, K. Lee, Highly reversible crystal transformation of anodized porous V2O5 nanostructures for wide potential window high-performance supercapacitors, Electrochim. Acta 334 (2020) 135618. https://doi.org/https://doi.org/10.1016/j.electacta.2020.135618

[101]    X. Yang, L. Zhang, F. Zhang, T. Zhang, Y. Huang, Y. Chen, A high-performance all-solid-state supercapacitor with graphene-doped carbon material electrodes and a graphene oxide-doped ion gel electrolyte, Carbon N. Y. 72 (2014) 381–386. https://doi.org/https://doi.org/10.1016/j.carbon.2014.02.029

[102]  H. Gamal, A.M. Elshahawy, S.S. Medany, M.A. Hefnawy, M.S. Shalaby, Recent advances of vanadium oxides and their derivatives in supercapacitor applications: A comprehensive review, J. Energy Storage 76 (2024) 109788. https://doi.org/https://doi.org/10.1016/j.est.2023.109788

[103]  H. Liu, W. Zhu, D. Long, J. Zhu, G. Pezzotti, Porous V2O5 nanorods/reduced graphene oxide composites for high performance symmetric supercapacitors, Appl. Surf. Sci. 478 (2019) 383–392. https://doi.org/https://doi.org/10.1016/j.apsusc.2019.01.273

[104]  A. Sherin steena, S. Harini, S.R. Niranjana, V.A.R. M, J. Madhavan, Electrochemical performance of V2O5/TiO2 nanocomposite: As an emerging electrode material for supercapacitor application, Chem. Phys. Impact 9 (2024) 100654. https://doi.org/https://doi.org/10.1016/j.chphi.2024.100654

[105]  Y. Zhou, L. Sun, D. Wu, X. Li, J. Li, P. Huo, H. Wang, Y. Yan, Preparation of 3D porous g-C3N4@V2O5 composite electrode via simple calcination and chemical precipitation for supercapacitors, J. Alloys Compd. 817 (2020) 152707. https://doi.org/https://doi.org/10.1016/j.jallcom.2019.152707

[106]  M. Fu, Q. Zhuang, Z. Zhu, Z. Zhang, W. Chen, Q. Liu, H. Yu, Facile synthesis of V2O5/graphene composites as advanced electrode materials in supercapacitors, J. Alloys Compd. 862 (2021) 158006. https://doi.org/https://doi.org/10.1016/j.jallcom.2020.158006

[107]  D. Jia, F. Zheng, Y. Niu, X. Mao, Y. Yang, Q. Zhen, P. Li, Y. Yu, S. Zhang, Preparation of V2O5 nanobelt arrays/NiO nanosheet arrays composite as supercapacitor electrode material, J. Alloys Compd. 969 (2023) 172283. https://doi.org/https://doi.org/10.1016/j.jallcom.2023.172283

[108]  J. Jayachandiran, J. Yesuraj, M. Arivanandhan, A. Raja, S.A. Suthanthiraraj, R. Jayavel, D. Nedumaran, Synthesis and Electrochemical Studies of rGO/ZnO Nanocomposite for Supercapacitor Application, J. Inorg. Organomet. Polym. Mater. 28 (2018) 2046–2055. https://doi.org/10.1007/s10904-018-0873-0

[109]  R. Ranjithkumar, S.E. Arasi, S. Sudhahar, N. Nallamuthu, P. Devendran, P. Lakshmanan, M.K. Kumar, Enhanced electrochemical studies of ZnO/CNT nanocomposite for supercapacitor devices, Phys. B Condens. Matter 568 (2019) 51–59. https://doi.org/https://doi.org/10.1016/j.physb.2019.05.025

[110]  B. Joshi, E. Samuel, J. Huh, S. Kim, A. Ali, M. Rahaman, H.-S. Lee, S.S. Yoon, Hierarchically decorated ZnO/CoOx nanoparticles on carbon fabric for high energy–density and flexible supercapacitors, Ceram. Int. 50 (2024) 43324–43333. https://doi.org/https://doi.org/10.1016/j.ceramint.2024.08.187

[111]  S.K. Godlaveeti, V.R. G, S. Sangaraju, A.A.A. Mohammed, S.K. Arla, R. Nirlakalla, A.R. Somala, R.R. Nagireddy, High-performance of the ZnO/NiS nanocomposite electrode materials for supercapacitor, Colloids Surfaces A Physicochem. Eng. Asp. 680 (2024) 132749. https://doi.org/https://doi.org/10.1016/j.colsurfa.2023.132749

[112]  S. Nagarani, G. Sasikala, M. Yuvaraj, R.D. Kumar, S. Balachandran, M. Kumar, ZnO-CuO nanoparticles enameled on reduced graphene nanosheets as electrode materials for supercapacitors applications, J. Energy Storage 52 (2022) 104969. https://doi.org/https://doi.org/10.1016/j.est.2022.104969

[113]  M. Imtiaz, B.M. Alotaibi, A. Gassoumi, A.W. Alrowaily, H.A. Alyousef, M.F. Alotiby, A.M.A. Henaish, Facile synthesis of FeAl2O4@ZnO electrode material for supercapacitor application, J. Phys. Chem. Solids 188 (2024) 111941. https://doi.org/https://doi.org/10.1016/j.jpcs.2024.111941

[114]  M. Gaire, N. Khatoon, B. Subedi, D. Chrisey, Flexible iron oxide supercapacitor electrodes by photonic processing, J. Mater. Res. 36 (2021) 4536–4546. https://doi.org/10.1557/s43578-021-00346-8

[115]  V. Velmurugan, U. Srinivasarao, R. Ramachandran, M. Saranya, A.N. Grace, Synthesis of tin oxide/graphene (SnO2/G) nanocomposite and its electrochemical properties for supercapacitor applications, Mater. Res. Bull. 84 (2016) 145–151. https://doi.org/https://doi.org/10.1016/j.materresbull.2016.07.015

[116]  M.A. Hanif, Y.-S. Kim, L.K. Kwac, S. Ameen, A.M. Abdullah, M.S. Akhtar, Titanium oxide aerogel/graphene oxide based electrode for electrochemical supercapacitors, Energy Storage 6 (2024) e617. https://doi.org/https://doi.org/10.1002/est2.617

Applications for Earth-Abundant Transition Metals
Materials Research Foundations 179 (2025) 145-175

Materials Research Forum LLC
https://doi.org/10.21741/9781644903711-7

# Chapter 7

# Transition Metal based Dye-Sensitized Solar Cells

Kayal Kumari, Navneet Kaur, Kailash Devi, Deepika Jamwal*

Department of Chemistry, University Institute of Sciences, Chandigarh University, Gharuan, Punjab, 140413, India

jamwaldeepika@gmail.com, deepika.e15926@cumail.in,

## Abstract

The dye-sensitized solar cell (DSSCs), a next-generation solar technology has received more interest because it achieved about 7% efficiency in 1991. The exceptional electrical, optical, and catalytic capabilities of transition metal chalcogenides (TMCs) like $MoS_2$, $WS_2$, ZnTe, or ZnSe are making them increasingly popular in DSSCs. Dye-sensitized solar cells are considered an ideal indoor photovoltaic (PV) technology because of their outstanding performance despite scattered indoor light, affordability, remarkable adaptability, semi-transparent nature, along the accessibility of a wide range of colors. To minimize the expense of the typically utilized platinum (Pt) counter electrode (CE), many compounds exhibiting strong electrocatalytic performance are being applied to CEs in DSSCs. As a result of their potential for usage across a broad range of applications, such as industrial lubricants as well as electronics, transition metal chalcogenides represent a significant class material that has recently received a great deal of intrigue.

## Keywords

Dye-Sensitized Solar Cell, Transition Metal Chalcogenides, Photovoltaic, Counter Electrode

## Contents

## 1.    Introduction

Metals are among the most important materials; they are well-known for their unique applications in many different domains [1]. Oxygen, sulfur, selenium, tellurium, and polonium are also known as chalcogenides because they all belong to the oxygen family. When transition metals combine with these metals the resulting metals formed are known as transition metal chalcogenides [2,3]. Recent breakthroughs in materials chemistry including nanotechnology have raised expectations for the design along with synthesis for functional materials because of its widespread applicability in a variety of technical disciplines or industries [4]. Transition metal chalcogenides have been the focus of extensive investigation for several year as a result of their potential applications in thermoelectric, photovoltaics, magnetic semiconductors, quantum dots (QDs), sensors, electronics, and superconductors [5,6]. A highly promising technological advancement for "smart windows" is dye-sensitized solar cells. Solar-powered windows harness solar radiation to generate electricity, potentially meeting all of a building's energy needs in the future [7]. Also, the growing world population is driving up the need for energy. The energy needed in many different sectors today comes from non-renewable sources. Many nations including China, India, France, Japan, and Ukraine are dependent on nuclear power plants for the fabrication of electricity generation. However, they also encounter various hazardous problems associated with these facilities, which contribute to environmental pollution [8]. The primary energy sources used worldwide are fossil fuels like coal, natural gas, and oil. Because these sources (except coal) are so easily applied in transportation, they will eventually run out. In addition, the combustion of fossil fuels releases significant amounts of carbon dioxide into the atmosphere, leading to environmental pollution and climate change. The advancement regarding clean green technologies for a healthy environment could result from the growth of renewable energy sources [9]. Consequently, scientists' primary focus is on energy conversion devices based on renewable energy sources. Among the renewable energy sources found on Earth are geothermal, biomass, solar, wind, and hydroelectric power [10]. Because solar energy is available year-round and powers photosynthesis, which is a process necessary for the growth and development of all living things on Earth, it is a widely used renewable energy source. Not only do these sources have restricted accessibility, but they also don't break down naturally [11,12]. These resources are only expected to be accessible until 2024. These factors have led to a rise in interest in clean, renewable energy sources like the sun. DSSCs are a relatively new technology that uses solar power and transforms it into electrical energy [13]. For example, $TiO_2$, ZnO, $SnO_2$, and other oxides with significant band gaps are photoanodes serving as the building blocks of DSSCs. Many methods, including dye sensitization, activation of narrow bandgap inorganic semiconductors, along with transition metal ion doping sensitization, are used to make dye-sensitized solar cells [14]. Because of its advantages over other solar cells, such as low preparation costs, lightweight materials, better processing capabilities, and a large compound structure design space, DSSC has more promise for commercialization [15,16]. PV technology is the most enticing of all sustainable energy technologies because it allows sunlight to be directly converted into efficient electrical energy. The current silicon-based solar cells, however, they're limited by the PV market on land because of their expensive production and environmental cost [17]. Due to the low cost of components and the straightforward manufacturing process, DSSCs are widely considered as a cost-effective PV technology in comparison to expensive conventional silicon photovoltaic cells Figure 1 highlights the key components concluding the methods of preparation, properties, along with the use of dye-sensitized solar cells.

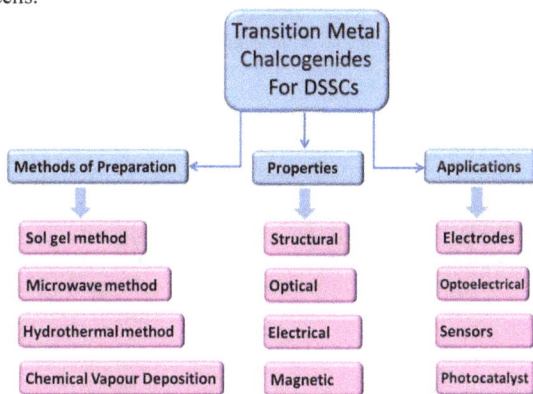

*Figure 1. Fabrication techniques, characteristics, and uses of DSSCs: a comprehensive overview*

Materials Research Forum LLC
https://doi.org/10.21741/9781644903711-7

DSSCs provide a conceptually appealing different to the standard solid-state photovoltaic junction device is clear both technologically and economically. Here, the roles of transporting charge carriers and absorbing light are divided. A light-absorbing broad-band semiconductor with a sensitizer connected to it. Photo-induced electron injection from the dye to the solid conduction band at its interface is the process by which charge separation happens. In addition, the charged carriers are moved toward that charge collector via the semiconductor conduction band. The broad absorption band of the sensitizers allows for substantial sunlight to be harvested [18,19]. DSSCs has several attractive features in the scientific community inexpensive, simple processability, extensive manufacturing as well and effective solar cell conversion into electrical power [20]. DSSCs are regarded as a superior PV technology because they convert light energy into electricity via the photovoltaic effect [21]. Natural dye and light sensitizers are assembled into a process called dye-sensitized solar cells. An effective technique for light gathering would be a DSSCs, which features a large surface area photoanode (around 1200 times that of a flat electrode) and a dye with moderate extinction. The combination of a high extinction dye, a high surface area photoanode, and a flat electrode area gives rise to sufficient absorbance over a large part of the visible spectrum, which includes red wavelengths having the most room for improvement [22].

DSSCs may be created with a considerable degree of freedom in terms of structure, color and scale, in addition appropriateness for different implementation circumstances. For several reasons, distributed solar cells (DSSCs) continue to be a strong third-generation option for solar energy, such as: (i) simple preparation methods that support the conversion of solar power into an environmentally friendly source; (ii) manufacturing not having using hazardous substances; and (iii) Flexibility in design allows DSSCs to be utilized in various setting, ranging from consumer electronics and indoor applications to transparent smart windows. This versatility enables broadly deployed sensors to power the internet-connected devices, heralding the next digital revolution [23,24].

## 1.1 Components of DSSC

The DSSCs is a novel kind of solar cell that employs a photochemical system based on sensitization Using semiconductors with large band gaps to transform visible light into electrical energy. The main constituents of DSSCs are also previously reported by Geon.et.al [25]. The following components are photosensitizer, electrolyte, conductive glass or substrate, photoelectrode, and counter electrode [26]. To increase the total efficiency, it is crucial to optimize each of them.

### 1.1.1 Photosensitizer

The photosensitizer is the DSSC component that ensures the maximal absorption of incident light. The "black dye" has been identified by Naseeruddin et al. as a potential charge transfer sensitizer in DSSC. Wang and researchers achieved an energy conversion efficiency of 7.6% using coumarin dyes based on the donor-pi-conjugated bridge-acceptor concept, with -CN groups functioning as the electron acceptors molecules [27]. Any substance used as a dye should possess electrochemical and photophysical characteristics. The dye should, first and foremost, have an absorption spectrum that encompasses the UV-visible and near IR areas. Second, several -OH groups in the dye should be able to chelate to the Ti (IV) sites present on the $TiO_2$ surface along the dye itself should be luminous. Thirdly, the highest occupied molecular orbital (HOMO) should be positioned far from the $TiO_2$ conductivity band to reduce face-to-face interaction between the electrolyte or anode and improve durability-focused. Additionally, the dye's perimeter should be hydrophobic. If not, water-induced dye distortion from the $TiO_2$ surface may develop, thereby decreasing cell stability. The lowest occupied molecular orbital (LUMO) must lie near the surface of $TiO_2$ and to stay away from the accumulation of dye beyond the $TiO_2$ surface, absorbents such as chenodeoxycholic acid or binding groups like alkoxy-silly and phosphoric acid are inserted between the $TiO_2$ and dye. This results in the limited recombination among the redox electrolyte as well as electrons in the $TiO_2$ nanolayer [28,29].

### 1.1.2 Electrolyte:

The electrolyte is an additional essential part that is necessary to DSSCs. Due to its ability to expedite the charge transfer during this period the photo-electrode as well as counter electrode, the electrolyte is a crucial component of the DSSC [30]. Open circuit voltage ($V_{oc}$) as well as short circuit voltage ($J_{sc}$) is considerably depending upon the electrolyte. The electrolyte mostly applied in the DSSCs is $I^-/I^-$ (redox) ion, $Br^-/Br^-$(solvent) and Co (II)/Co (III) (cations). Electrolyte should possess the property such as dye's absorption spectrum and an electrolyte's absorption spectrum shouldn't overlap. The oxidized dye should be readily regenerated by the redox pair. It must be able to improve conductivity, facilitate quick diffusion of charge carriers, and establish a strong connection between the working and counter electrodes and should not corrode when used with DSSC components. Organic, ionic-based liquid and aqueous are the types of electrolytes that are used in the DSSCs [31,32].

### 1.1.3 Conductive glass or substrate

The transparent conductive oxides (TCO) glass substrate in DSSCs shall be transparent (>80%) also electrically conductive. These features enable the best possible penetration of sunlight and effective, low-energy charge transfer, losses in DSSCs. Fluorine-doped tin oxide ($SnO_2$: F) was discovered by Bavarian et al. to be one of the most popular transparent conductive glass substrates. The most favoured material in DSSCs for broad bandgap semiconducting material oxides functioning as sensitizers was $TiO_2$ depending on nanoparticles [33,34].

### 1.1.4 Photo Electrode

The most common photoanode material is semiconductor oxide material covered with dye molecules adsorbed on a TCO substrate, which serves in the role of the working electrode in DSSCs [35]. In a dye-sensitized solar cells, the photo-electrode is made out of semiconductor nanostructure materials ZnO, $Nb_2O_5$, $TiO_2$, $SnO_2$ (n-type), and NiO (p-type) that are clipped to a transparent conducting substrate. Due to its low price, lack of toxicity, and abundance, $TiO_2$ is the most often used semiconductor material[36].

### 1.1.5 Counter Electrode

In the DSSC, the counter electrode is a crucial element where the mediator is reduced. To decrease the oxidized form of charge mediator, a high level of exchange current density and minimal charge transfer resistance are required in the material used as the CE in deep-sealing solar cells. Several materials have been used as a counter electrode in DSSC like carbon-oriented, platinum, electrode polymers, transition metal complexes as well as metal alloys and compounds. Platinum, on the other hand, is thought to be the greatest substance for catalysing reactions and offering a high exchange current [37]. The counter electrode makes it easier for electrons to return from the

Applications for Earth-Abundant Transition Metals         Materials Research Forum LLC
Materials Research Foundations 179 (2025) 145-175     https://doi.org/10.21741/9781644903711-7

external circuit to the redox electrolyte. Moreover, it facilitates the transfer of photocurrent throughout the whole width of every solar cell. For the redox pair to reduce, the counter electrode must thus be conducting effectively or displaying a low over-voltage. Due to its superior $I^{3-}$ reduction capabilities, platinum has become the preferred material for counter electrodes up to this point [38]. The counter electrode in DSSCs is capable of becoming ready by many methods such as hydrothermal, chemical vapor deposition, thermal decomposition, and in situ polymerization, along with other methods. The processes of preparation have a significant impact on the surface coverage, shape, and catalytic and electrochemical properties of the electrode [39].

## 2. Working principle of dye-sensitized solar cell

The generation of electricity in a DSSCs is driven by the movement of electrons from the photoanode to the CEs, which is a fascinating operating principle. The functioning mechanisms of DSSCs are as follows. To produce excited electrons, dye is used in place of chlorophyll as the light-gathering element in DSSCs. Water is replaced with electrolytes, carbon dioxide replaces $TiO_2$ as the electron acceptor, oxygen functions as both an electron donor and an oxidation product, and the cells are multi-layered to maximize light absorption with electron collection efficiency[8]. In DSSC, light photons are absorbed in the dye molecule layer. The essential primary surface area for the absorption of dye molecules is provided by nanocrystalline $TiO_2$. Dye molecules are stimulated from the HOMO states to the LUMO states after photon absorption [40,41]. Since the dye's electrons are excited from the ground state ($S^+/S$) to the excited state ($S^+/S^*$), most of the dye absorbs light around 700 nm, or about comparable to a photon energy of approximately 1.72 eV. The conduction band of the nanoporous $TiO_2$ electrode, located beneath the dye-excited state, captures a limited amount of UV light from solar photons. Excited electrons, which have a lifespan of nanoseconds, are subsequently injected into this conduction band. Consequently, the dye becomes oxidized [42]. The injected electron travels between the $TiO_2$ nanoparticles before being retrieved and applied to a load, where it is converted into electrical energy. Electric current flows to the counter electrode via the external circuit [43]. As a result, the photosensitizer (oxidized dye molecules) is restored by obtaining electrons based on the I ion redox mediator, which oxidizes them to $I_3$ ions (tri-iodide ions) [38]. The oxidized mediator($I_3^-$) once again diffuses towards the counter electrode, where it is mitigated to $I^-$ ion [44]. The working of DSSCs is shown in Figure 2.

The operation cycle is given below.

$S^+/S + h\nu = S^+/S^*$

$S^+/S^* \rightarrow S^+/S + e^- (TiO_2)$

$S^+/S^* + e^- \rightarrow S^+/S$

$I_3^- + 2e \rightarrow 3I$

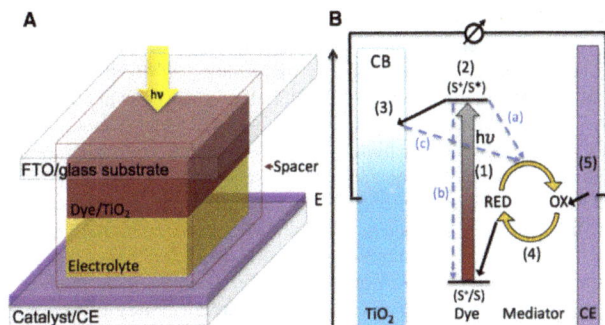

*Figure 2. A) Schematic structure and B) Key processes and operating principles of dye-sensitized solar cells.*

## 3. Classification

Transition metal chalcogenides are a kind of inorganic compound that consists of at least one chalcogen anion and one or more electropositive metals. This category of chalcogen includes elements such as sulfur, selenium, and tellurium. Transition metal chalcogenides are a type of material that has been a considerable focus in the area of DSSCs because to their distinct electrical, optical, and chemical features. These materials interact with transition metals (Mo, W, Ti) with chalcogens (S, Se, and Te). There are several methods to classify the extremely diverse class of metal chalcogenides. As per the count of components, they can be categorized as sulfides, selenites, tellurites, double, triple, quartet, or multiple chalcogenides [45,46].

### 3.1 Sulphur

Naturally occurring in large quantities, sulfur is used by industry to create a wide range of derivative goods. When it comes to morphologies, sulphur might be amorphous with few allotropes or crystalline with a hexagonal structure [47]. Transition metal sulfides are common mineral groupings found in nature, including chalcocite ($Cu_2S$), pyrite ($FeS_2$), and heazlewoodite ($Ni_3S_2$). Transition metal sulfides have high catalytic activity and conductivity, making them appropriate for replacing platinum counter electrodes in DSSCs [48–50]. For use in DSSCs, Qian and her students developed a straightforward two-stage solvothermal method for producing counter electrodes based on multimetallic sulfides particularly quaternary metal sulfide-based yolk-shell nanospheres ($CoNiMoS_x$) [51]. Han et al. design metal sulfide nanoparticles (PbS, $Ag_2S$, CuS, CdS, and ZnS) to be synthesized to take the place of the expensive as well as inexperienced Pt counter electrode in DSSCs and improve energy conversion efficiency [52].

### 3.2 Selenides

Transition Metal selenides are significant metal chalcogenides that have characteristics that are initially used in areas like catalysts. The initial usage of transition metal selenides is for an outstanding performance counter electrode material in DSSCs. Using a straightforward one-spot hydrothermal method, Gong et al. created the initial metal selenide counter electrodes in binary form for DSSCs [53]. Metal selenides are widely utilized as electrocatalysts for the reduction of oxygen and as components for solar cells that collect light. The presentation discusses the application of transparent metal selenide counter electrodes (CEs) in bifacial dye-sensitized solar cells (DSSCs) and highlights the benefits of in situ-grown metal selenide nanosheet fiber CEs for fiber DSSCs [54]. DSSCs with bifacial CEs demonstrate both appropriate PCEs with rear irradiation and increased PCEs when employing front illumination. The study found that bifacial metal selenide-based CEs are more effective in allowing dye-sensitized solar cells to produce energy from front, rear, or both sides. The area of fiber-shaped solar cells, particularly DSSCs designed for integrating into wearable energy textiles for solar energy harvesting, is paying more and more attention to in situ development of metal selenide-based counter electrode [55–58]. Venkateswaran et al. produced polycrystalline CuZnSe2 ternary compounds by metallurgical means or deposited them as thin-film counter electrodes of DSSCs by thermal evaporation. The effectiveness of the $Cu_{1-x}Zn_{1-y}Se_2$-d counter electrode is 1.5%, which is lower than the 4.2% efficiency of the platinum counter electrode. However, this finding has spurred further investigation into other ternary metal selenides to enhance efficiency [59].

*Telluride:* Tellurium (Te) is also known as a chalcogenide element. Tellurium is the heaviest and non-radioactive chalcogenide element, shares similarities with S and Se due to its electron configuration [60]. There has been limited study on CEs based on telluride in DSSCs using metal chalcogenides. Tellurides greatly increased photoelectric conversion efficiencies when they were initially employed as photoanodes to absorb sunlight in solar cells. That being said, Guo et al. created the first telluride-based CEs in 2013. The researchers utilized a composite-hydroxide-mediated approach to create CoTe and NiTe2 nanostructures, which they then used as CEs for DSSCs [61,62]. The extensive development of MoTe2 thin films for Pt-free CEs of DSSCs is described by Hussain et al. using chemical vapor deposition sputtering on conductive glass substrates. The cyclic voltameter results indicate that MoTe2 CEs have strong electrocatalytic activity along with quick reaction kinetics for reducing triiodide to iodide. The MoTe2-based DSSC achieved an optimal power conversion efficiency of 7.25%, similar to the Pt-based DSSC's power conversion efficiency of 8.15%. Also, they incorporated metal tellurides into metal sulfides enhancing their performance as CEs for DSSCs [63,64]. Zhang et al. developed a CE using Co0.85Se nanosheets on PANI-functionalized carbon fibres. The fibre-shaped dye-sensitized solar obtained the greatest record PCE of 10.28% and showed exceptional adaptability [65]. The benefits of metal selenide-based CEs in fiber-shaped DSSCs can be applied to various metal chalcogenides, including metal sulfides and metal tellurides, in fibrous or flexible PV systems [66].

## 4. Synthesis and fabrication of TMC-based DSSCs

The various methods used for integrating TMCs into DSSCs synthesis are the solvothermal, hydrothermal, electrochemical deposition, chemical vapor deposition, spin coating, carbothermal reduction, sol-gel, atmospheric pressure, atomic layer deposition, and sono chemical methods [67–71].

### 4.1 Sol-gel method

In the conventional sol-gel procedure, the precursors, which consist of organic metal compounds like metal alkoxides or inorganic metal salts, hydrolyse as well as polymerize to produce a colloidal suspension [72]. A sol-gel approach for the manufacture of metal sulfides was disclosed by Stanic et al. The production of monoclinic germanium disulfide by the reaction of the sol-gel product with sulfur served as an example of how sulfides are synthesized. In turn, the oxidation of $H_2S$ in the presence of concentrated sulphuric acid produced elemental sulfur. After being transferred into a toluene solution containing germanium ethoxide, the sulfur was discovered to be evenly dispersed throughout the gel. $GeS_2$ in a single phase was produced by heat treating the sol-gel product. A chemical kinetics analysis of this type of sol-gel processing of $GeS_2$, using hydrogen sulfide and germanium ethoxide in toluene, was also published [73].

## 4.2. Hydrothermal and solvothermal method

Hydrothermal synthesis method grows crystals ranging from submicron to nanoscale scale by chemical reactions with a water-based solution during high temperatures as well as pressures [74]. Hydrothermal synthesizing method occurs with steel pressure vessels (autoclaves) at specific temperatures as well as pressures, for reactions in aqueous solutions. In the hydrothermal method, water serves exclusively as a solvent [75–77]. Elevating the temperature above the water's boiling point causes a rise in vapor saturation pressure. This method is frequently used to produce nanoparticles [78]. Nanocrystal synthesis involves adjusting factors like water pressure, temperature, precursor-product system, and reaction time to achieve synchronized nucleation and excellent size distribution. The technique has several advantages over other growing methods, including simple equipment, environmental friendliness, cheap cost, homogeneous output across broad areas, and lower risk [79]. $MoS_2$ NPs were synthesized by Chaudhary N. et al. via a hydrothermal process. By adjusting the hydrothermal temperature and reaction time, $MoS_2$ NPs with excellent crystalline nature and form are produced. After 24 hours and an optimal temperature of 220°C, the hydrothermal process forms $MoS_2$ in the form of a sheet [80]. According to Min Li et al., the production of $MoS_2$ nanospheres at low temperatures (120–150°C) produced amorphous nanospheres devoid of any layered material. Furthermore, the effects of temperature and duration on morphology and crystalline size are considerable [81].

Solvothermal synthesis produces nano-sized powders without using organometallic or hazardous procurers and at low temperatures. The solvothermal process refers to the act of performing chemical reactions under sealed containers containing solvents when simultaneous heating can raise the liquid's temperature to approximately their critical points using autogenous pressures [82]. In solvothermal synthesis, a solvent is used to create nanomaterials at high pressures and temperatures (often between 100 and 1000 atm). The solvothermal process refers to the use of organic solvents instead of water. Solvothermal techniques often employ organic solvents such as toluene, 1,4-butanediol, methanol, and amines [83]. The solvothermal method of TMCs onto graphene sheets is the most extensively used method for creating TMC [84]. Solvothermal synthesis typically involves a chalcogen source such as S, Se, Te, $Na_2SeO_3$, $Na_2TeO_3$, etc. The solvothermal technique, some MC nanocrystals have been synthesized with exquisite control over the size, shape, and crystallinity distributions., including wire-like $FeS_2$, $Cu_2Te$, $Bi_2S_3$, rod-like of MnS, belt-like ZnSe, flowerlike $FeSe_2$, plate-like $Sb_2Te_3$ as well as CuS, sphere-like $In_2S_3$ and $Bi_2S_3$, $FeS_2$ nano webs, CoTe nanotubes [85].

## 4.3. Microwave method

Microwave method assisted in creating an easier, less time-consuming, and atom-efficient way for the synthesis of nanoparticles [86]. The microwave approach uses electromagnetic radiation with a relative wavelength ranging from 1 mm to 1 m. It falls among radio waves and infrared frequencies. The best solvents to utilize in microwave-based syntheses are thought to be water or ionic liquids having a significant dipole moment [87]. Shorter treatment periods are made feasible by rapid heating rates, which not only saves time and energy but also has the potential to inhibit unwanted side reactions and open up new reaction pathways. The basic concept of microwave heating is the idea that heat is produced when a substance absorbs electromagnetic radiation [88]. LT Gularte et al. microwave-assisted chemical bath deposition is used to create CoS counter electrodes (figure 3). First, 0.2 M Co(NO$_3$)$_2$·6H$_2$O and 0.2 M CH$_3$CSNH$_2$ are dissolved in 40 mL of 95% ethanol while being vigorously stirred for 20 minutes at room temperature. There was no usage of complexing or pH-regulating agents. Because thioacetamide reacted quickly with metal salts, it was used as the source of sulphur. FTO glass that had been cleaned beforehand was put in a PTFE vessel, and the combination was heated to 100 °C (CoS-100) over the course of 30 minutes using microwave radiation (800 W, 2.45 GHz) at a rate of 4 °C/min. The item was dried, cleaned, and allowed to cool. For temperatures of 110 °C (CoS-110) and 120 °C (CoS-120), this process was repeated [89].

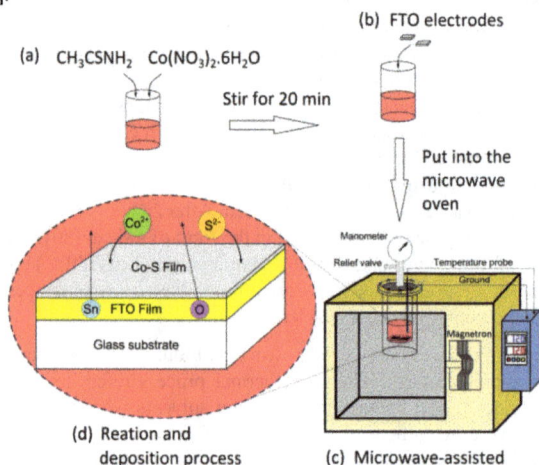

*Figure 3. An illustrated procedure for the in-situ synthesis and deposition of CoS into the FTO substrate using microwave irradiation.*

## 4.4. Electrochemical deposition

This is a process in which metallic ions can solidify and deposit on the cathode surface when a significant quantity of electric current flows through the electrolyte solution. Although electrochemical deposition is a very effective approach for producing metal nanoparticles, it tends to be less applied than the wet-chemical process [90]. Many electrochemical methods, including double-pulse deposition, potential step, and cyclic voltammetry, are used extensively in electrodeposition [91] as shown in figure 4. Electrodeposition has been widely utilized to create several thermoelectric materials based on metal chalcogenides like $Bi_2Te_3$ nanowire arrays, $Bi_2Te_3$ nanotube arrays, $Bi_2Te_3$ films, $Cu_2Te$ films, $Cu_2Te$ nanowires, and SnSe films [92–98].

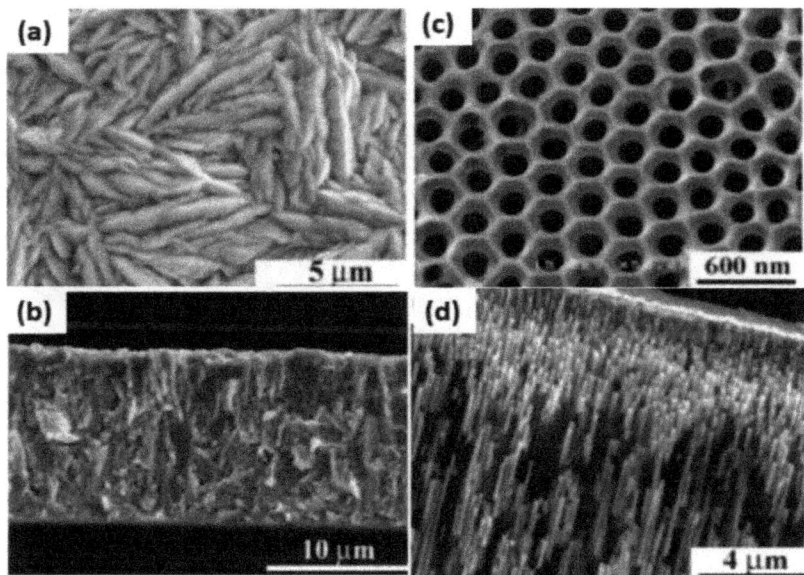

*Figure 4. Scanning Electron Microscopy (SEM) images showing (a) planar and (b) cross-sectional views of the electrodeposited $Bi_2Te_3$ films; (c) highly ordered nanopores in an anodic alumina membrane template; and (d) a side view of the electrodeposited $Bi_2Te_3$ nanowire array partially embedded in the AAM. Reproduced with permission from ref. 91. Copyright: 2019, MDPI.*

Quy et al. described a simple electrodeposition technique to produce nickel sulfide ($Ni_3S_4$) films onto FTO substrates. The substance had a dimensions of 110 nm, and covered an entire FTO substrate, with a special network of highly permeable nanoscale linked nanoparticles that offered a large number of electrochemically active regions for interaction with the electrolyte [99]. Huo et al. utilized electrodeposition (EPD) to put the CoS layer onto the FTO glass substrate. CoS is prepared by using the EPD method by treating the film with $NaBH_4$ with an $H_2SO_4$ aqueous solution. Thus, the DSSCs with CoS CE obtained a power conversion efficiency of 7.72%. The schematic diagram of the preparation of CoS using the EPD method is shown in Figure 5 [100].

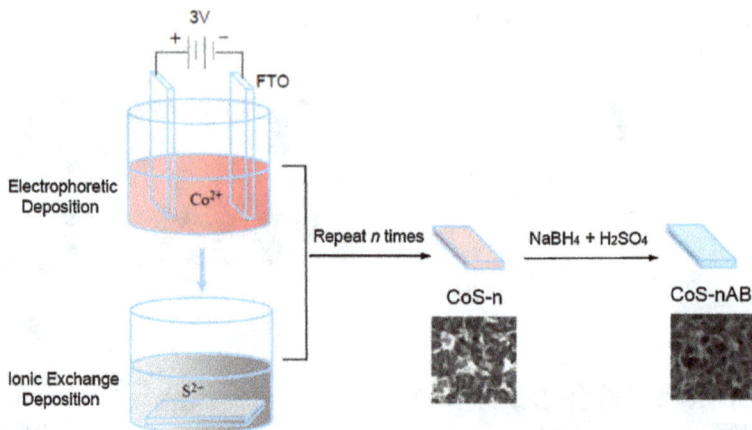

*Figure 5. Schematic representation of the electrophoretic deposition (EPD) and ion exchange deposition processes used for fabricating CoS counter electrode. Reproduced with permission from ref. 100. Copyright: 2015, Elsevier.*

## 4.5. Chemical vapor deposition

The chemical vapor deposition (CVD) technique is a vacuum deposition method that produces high-grade, high-performance rigid materials. CVD is a technique that typically employs high-purity refractory metals. To achieve this, reagents must be able to evaporate without breaking down, form a significant concentration of the metal-containing compound's vapor, along yield only gaseous as well as chemically inert substances when the metal is deposited and the compound decomposes. One newly developed technology that provides a flexible means of obtaining nanostructures for various purposes is the sono-chemical method. This illustrates the application of high-intensity ultrasounds, which provide circumstances for synthesis that are different from those of other traditional techniques such as wet chemical synthesis, spray pyrolysis, and hydrothermal synthesis [101]. S. et al. synthesized $MoS_2$ film directly on the FTO to eliminate the need to transfer synthesized films and establish a one-step process. $MoS_2$ thin films were created by the CVD process. 1gm of sulfur powder was added to the furnace's upstream entrance, and 30 mg of powdered molybdenum chloride ($MoCl_5$) was deposited in the furnace's core. The gap between them is approximately 22.5 cm. In a vacuum atmosphere, experiments were carried out. A reaction time of fifteen minutes and a flow velocity of around fifty sccm are required to achieve the superior grade $MoS_2$ thin films at 600 °C. The present study indicates that $MoS_2$ CEs exhibit superior stability in comparison to Pt CEs. When comparing the obtained efficiency to Pt and $MoS_2$ electrodes synthesized in various ways, including hydrothermal, wet chemical, and electrodeposition procedures, it is much higher [102].

## 4.6    Spin coating

This is a simple solution-based process for creating thin films ranging from nm to µm. Spin coating applies to a variety of organic as well as inorganic materials, including composite materials [103]. The carbothermal reduction method is favourable because it is convenient, cost-effective, and produces consistent results. The carbothermal reduction technique for nanoparticle synthesis is a high-temperature procedure that makes use of carbon sources to reduce metal oxides and make metal nanoparticles [104,105].

## 4.7    Atomic layer deposition

Atomic layer deposition or ALD, has become increasingly significant for thin-film formation in a variety of applications. An atomic layer deposition can create transition metal chalcogenides in thin-film morphology DSSCs. This allows for atomic-level control over film composition and thickness, which makes it easier to synthesize consistent, superior TMC layers that can improve the characteristics of DSSCs. ALD is a gas-phase process made up of a series of self-limiting, surface-controlled processes [106]. TMC is obtained by different methods as mentioned in Table 1.

*Table 1*: TMC obtained by various methods:

| Compound | Solvent | Method | Morphology | Size/nm | References |
|---|---|---|---|---|---|
| SnTe | $H_2O$ | Hydro-solvothermal | Nanorods | 30-50 | [107] |
| $Ag_2Te$ | Water, CTAB, Hydrazine | Hydro-solvothermal | Nanorod, nanowire | ~70 | [108] |
| $FeS_2$ | $H_2O$ | Hydro-solvothermal | nanoparticle | 25-40 | [109] |
| CuTe | Stainless steel, ITO | Electrodeposition | Nano ribbons | 50 | [110] |
| $Bi_2Te_3$ | ITO | Electrodeposition | Nanowires, nanofilms | 20 | [111] |
| CoS | $H_2O$ | Electrodeposition | nanorods | 10-100 | [112] |
| $Bi_2Te_3$ | ethylene glycol | Microwave | nanosheet | 50-150 | [113] |
| PbTe | ethylene glycol | Microwave | Hexahedron nanostructures | 8-15 | [114] |

## 5.    Properties of transition metal chalcogenides nanostructures

### 5.1    Structural properties

Transition metal chalcogenides nanoparticles exhibit unique structural features that render them appealing prospects for DSSCs. These materials consist of transition metals and chalcogens and offer a parallel layered structure similar to that of graphene which guarantees good conductivity properties for the charge [115,116]. The structure of TMCs can be tuned to form quantum dots, nanowires, nanorods, nanoflowers, nanotubes, nanoplates, nanoclusters, nanocages, nanocomposites, nanosheets, core-shell structure, porous structure, Hierarchical Structures, etc. that offer different advantages concerning surface area and catalytic activity [117–119]. $MoS_2$ quantum dots show excellent light absorption properties and modulable wavelength absorptions. On the other hand, small size provides a high surface area for dye loading, which enhances the efficiency of light harvesting in DSSCs [120,121].

*5.2 Optical properties:* Dye-sensitized solar cells are recognized to be a suitable indoor photovoltaic technology due to their outstanding efficiency even in dispersed indoor lighting, affordability, great adaptability, semi-transparent, and availability of a wide range of colors [122–126]. A wide variety of band gaps (1.0-2.0 eV) makes these materials capable of absorbing different parts of the solar spectrum at an effective rate. The high absorption coefficient is a primary distinguishing feature for TMCs that enable appreciable sunlight absorption even in thin layers; this broad spectrum also helps to boost their efficiency when converting solar energy into other forms. They have varying indices of refraction that can be adjusted to enhance light trapping, thus affecting light propagation [127–129]. Chhowalla et al. 2013 commented that these TMCs have varying refractive indices that would affect light propagation, which can be optimized for light trapping to further enhance DSSC efficiency [130]. Eda and Meier studied the photoluminescence (PL) with spectra of absorption of $WS_2/MoS_2/MoSe_2/WSe_2$ for DSSCs (Figure 6) [131]. TMCs are thus quite able to improve DSSC performance much further using such features, making it a very important focus for research and development into renewable energy, as explained by Tang et al. [132].

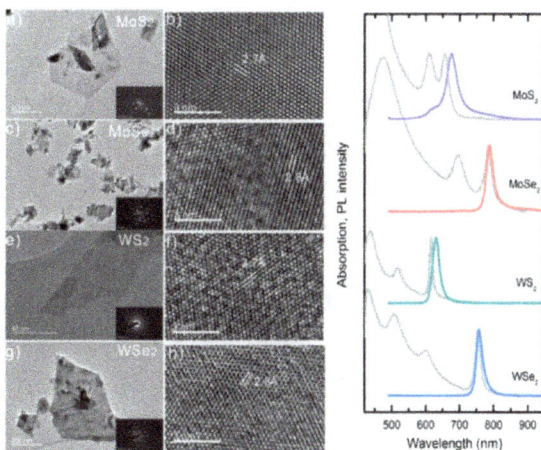

***Figure 6.*** *The spectra of photoluminescence (shown as color lines) and absorption (depicted by gray lines) for mechanically exfoliated single-layer $MoS_2$, $MoSe_2$, $WS_2$, and $WSe_2$ are illustrated. Reproduced with permission from ref. 131. Copyright: 2013, American Chemical Society.*

## 5.3 Electrical properties

DSSCs using transition metal chalcogenides for photoanodes display promising electric properties, increasing the effectiveness of photovoltaic cells. It includes some, among them, but not limited to $MoS_2$, $WS_2$, and $FeS_2$ which possess very high electronic conductivity coupled with layered structures that help in efficient charge separation and transport. The majority of these materials have a direct bandgap meaning that they absorb light better allowing them to respond to sunlight even at very low spectrums like near-infrared rays. Their large surface areas also facilitate easy attachment of common dyes and electrolytes used in DSSCs which enhances overall verbosity charge collection efficiency. The use of transition metal chalcogenides in DSSCs leads to lower rates of charge recombination and higher electron transport rates which yields enhanced short circuit current density with in total efficiency power transformation from the solar cell systems [133–135]. TMDs have been used traditionally as solid lubricants, however in the photonics and electronics sectors, they have demonstrated potential for use in recent times. TMDs are a potential 2D graphene counterpart that may find use in bulk-heterojunction solar cells and DSSCs. Singh and colleagues published a paper on atomically thin $MoS_2$ layers utilized as the interfacial layer, protective layer, electron-transport layer, as well as hole-transport layer in the fabrication of bulk-heterojunction solar cells. This section discusses the uses of $MoS_2$ thin films as counter electrodes for the synthesis of platinum-free DSSCs [102,121]. These findings suggest that materials such as $MoS_2$ have unique electric properties which makes them ideal to optimize DSSC efficiency hence making them viable solar energy families with an emphasis on cheaper sources of sunlight harnessing [136,137]. Wei et al. created DSSCs using metastable 1T metallic phase $MoS_2$. In that study, hydrothermal processes were utilized to build $MoS_2$ films on FTO glass substrates for 24 hours at 180 or 200 °C. MoS$_2$ films developed at 180 °C exhibited structures resembling flowers, but those made at 200 °C had lumps. The formation of both 2H and metastable 1T metallic phases of MoS$_2$ was additionally verified using X-ray photoelectron spectroscopy (XPS), Raman spectroscopy, and images obtained from high-angle annular dark-field scanning transmission electron microscopy. Figure 7 illustrates the J-V curves of 2H-MoS$_2$ and the 1T metallic phase MoS$_2$ with a flower-like structure [138,139].

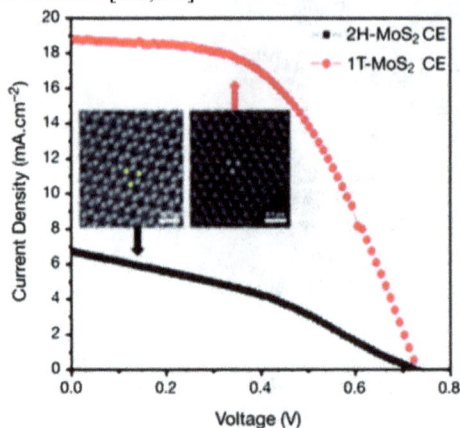

**Figure 7.** *Photocurrent density–voltage (J–V) curves of 2H-type MoS$_2$ and flower-structured 1T metallic phase MoS$_2$ are presented. Reproduced with permission from ref. 138. Copyright: 2017, Royal Society of Chemistry.*

## 5.4 Catalytic properties

Catalytic characteristics of TMC have attracted ever-increasing attention from researchers. For example, compounds of transition metals such as molybdenum or tungsten and their chalcogenide counterparts are among the catalysts that have been improved. These materials manifest unique electronic properties facilitating efficient charge transfer and catalytic activities respectively. When used with other materials in DSSCs, transition metal chalcogenides favourably increase the rate of redox reactions taking place within the counter-electrode region [140–142].

## 5.5 Magnetic properties

Essentially, transition metal chalcogenides are an interesting group of materials that attract attention, especially with DSSCs. These substances are capable of exhibiting distinct magnetic properties. The presence of partially filled d-orbitals in transition metals is responsible for magnetism in TMCs which leads to different forms of magnetism such as ferromagnetism, antiferromagnetism, or ferrimagnetism. In DSSCs, transition metal chalcogenides are commonly used as photoelectrodes or catalysts that help to improve light absorption and charge transfer processes. The way magnetic properties affect electronic structure and optical characteristics can ultimately enhance their performance in terms of power generation by photovoltaics like solar panels – this is achieved mainly through band structure modifications that promote light trapping and electron conduction processes [143–145].

## 6. Applications

Transition metal chalcogenides are also used as a solid lubricant [146]. $MoS_2$, tungsten disulfide ($WS_2$), and other TMCs are often used in DSSCs. Because of their special qualities, which include high surface areas, tunable bandgaps, superior electrical conductivity, and catalytic activity. Therefore, TMCs can have various applications in DSSCs [147] .

### 6.1 As a counter electrode

The CEs of DSSCs function as catalysts. The highest electrolytic action for the tri-iodide/iodide is produced by Pt counter electrodes in DSSCs [148]. Pt exhibits the highest photovoltaic performance among electrocatalysts for DSSCs because of its excellent electrical conductivity, stability in chemical reaction, as well as electrochemical activity, despite being a rare material. The present plan is to investigate new, readily available, reasonably priced, extremely conductive, along with highly electrocatalytic activity compound to substitute the traditional Pt counter electrodes in DSSCs [149]. Transition metal sulfides such as $MoS_2$, CoS, and NiS are the alternative materials that have replaced Pt-based counter electrodes.

$MoS_2$: To create a bulk heterojunction and an electron transport layer in a solar cell, Singh et al. identified the $MoS_2$ layer as a hole-transport, interfacial, and protective layer [121]. In his work he uses the hydrothermal process in which $MoS_2$ films were formed on an FTO glass substrate after 24 hours of reactions at 180 or 200 C. $MoS_2$ films developed at 180°C exhibited structures resembling flowers, but those prepared at 200°C had lumps [138]. Chang et al. utilized a one-stage non-injection process to synthesize $Co_9S_8$ nanocrystals, which were subsequently placed on substrates as the CE of DSSCs. Different characteristics such as electrochemical and photoelectrical were measured. The average efficiency of the DSSC with $Co_9S_8$ CE fabrication was lower at 7.02% than the corresponding DSSC with platinum CE (7.2%) [150]. J. Theerthagiri.et.al synthesized nickel selenide using a one-step hydrothermal reduction reaction to serve as a CEs in DSSCs [151].

### 6.2 As electrocatalytic behaviour

Vanadium chalcogenides have excellent behaviour as electrocatalytic in dye-sensitized solar cells. Unlike other semiconducting TMCs, $VSe_2$ exhibits exceptionally high electrical conductivity and is metallic. Its metallic quality makes it more reflective. This is why more photo-generated carriers are collected by depositing a $VSe_2$ layer placed between the back contact and the absorber layer to lessen light loss due to re-absorption. Additional research on vanadium chalcogenides reveals that the S-V-S-layered structure of $VS_2$, which is held together by weak van der Waals forces, suggests that it can be used as an electrocatalyst in solar energy applications. Given all of these characteristics, vanadium chalcogenides appear to have a lot of potential applications as solar cell materials [152,153].

*NiS as a cathode material in DSSCs:* NiS has been considered one of the most important metal chalcogenides as a cathode material in DSSC because of its enhanced electrochemical activity, an excellent chemical along with environmental stability, as well as availability and affordability of nickel resources. NiS could be synthesized by different processes such as electrodeposition,

CVD, and hydrothermal reaction [154]. Mahuli et.al has synthesized atomic layer deposition of NiS utilizing successive exposures to hydrogen sulfide at 175°C and bis(2,2,6,6-tetramethylheptane-3,5-dionate) nickel (II) [Ni(thd)$_2$]. To fully understand the chemistry of deposition and the characteristics of film growth, complementary combinations of in situ with ex-situ characterization methods are employed. High catalytic activity is demonstrated by the NiS thin film produced in ALD [155].

## Conclusion

DSSCs that use transition metal chalcogenides are a significant development in photovoltaic technology. As PV technology, DSSCs can overcome conventional solar cells. This work's main goal was to present a thorough analysis of innovative chemicals for photoanodes, counter electrodes, electrolytes, along with sensitizers to create DSSCs that are affordable, versatile, safe for the environment, and easy to assemble. For a better understanding of the operation and parts of the DSSCs employing transition metal chalcogenides, this chapter offers a straightforward description. Transition metal chalcogenides have garnered significant interest as potential materials for DSSCs due to their unique electronic, optical, as well as chemical properties. Different scientists have drawn attention towards focusing on creating less expensive materials to replace Pt counter electrodes. In place of the Pt CE, a variety of affordable materials have been studied thus far; metal chalcogenides, such as sulfides, selenides, & tellurides, show great promise. Transition metal chalcogenides have shown promising potential in enhancing the performance of DSSCs. The future scope is to improve photocatalytic properties such as high band gap high carrier mobility along strong photocatalytic activity, making them ideal candidates for improving light absorption, charge transport, and overall efficiency in DSSCs.

## Conflict of Interest

The authors declare no conflicts of interest.

## Acknowledgment

Deepika Jamwal acknowledges financial support from Shastri Indo-Canadian Institute for Shastri Post-Doctoral Fellowship (SRSF-2019) and Chandigarh University for support.

## Reference

[1]D. Wang, Y. Li, Bimetallic Nanocrystals: Liquid-Phase Synthesis and Catalytic Applications, Advanced Materials 23 (2011) 1044–1060. https://doi.org/10.1002/adma.201003695

[2]Chemistry of the Elements, n.d

[3]Z. Ali, T. Zhang, M. Asif, L. Zhao, Y. Yu, Y. Hou, Transition metal chalcogenide anodes for sodium storage, Materials Today 35 (2020) 131–167. https://doi.org/10.1016/j.mattod.2019.11.008

[4]G.R. Patzke, Y. Zhou, R. Kontic, F. Conrad, Oxide Nanomaterials: Synthetic Developments, Mechanistic Studies, and Technological Innovations, Angew Chem Int Ed 50 (2011) 826–859. https://doi.org/10.1002/anie.201000235

[5] J.N. Coleman, M. Lotya, A. O'Neill, S.D. Bergin, P.J. King, U. Khan, K. Young, A. Gaucher, S. De, R.J. Smith, I.V. Shvets, S.K. Arora, G. Stanton, H.-Y. Kim, K. Lee, G.T. Kim, G.S. Duesberg, T. Hallam, J.J. Boland, J.J. Wang, J.F. Donegan, J.C. Grunlan, G. Moriarty, A. Shmeliov, R.J. Nicholls, J.M. Perkins, E.M. Grieveson, K. Theuwissen, D.W. McComb, P.D. Nellist, V. Nicolosi, Two-Dimensional Nanosheets Produced by Liquid Exfoliation of Layered Materials, Science 331 (2011) 568–571. https://doi.org/10.1126/science.1194975

[6] M. Gao, J. Jiang, S. Yu, Solution-Based Synthesis and Design of Late Transition Metal Chalcogenide Materials for Oxygen Reduction Reaction (ORR), Small 8 (2012) 13–27. https://doi.org/10.1002/smll.201101573

[7] M.H. Chung, B.R. Park, E.J. Choi, Y.J. Choi, C. Lee, J. Hong, H.U. Cho, J.H. Cho, J.W. Moon, Performance level criteria for semi-transparent photovoltaic windows based on dye-sensitized solar cells, Solar Energy Materials and Solar Cells 217 (2020) 110683. https://doi.org/10.1016/j.solmat.2020.110683

[8] A.B. Muñoz-García, I. Benesperi, G. Boschloo, J.J. Concepcion, J.H. Delcamp, E.A. Gibson, G.J. Meyer, M. Pavone, H. Pettersson, A. Hagfeldt, M. Freitag, Dye-sensitized solar cells strike back, Chem. Soc. Rev. 50 (2021) 12450–12550. https://doi.org/10.1039/D0CS01336F

[9] G. Richhariya, A. Kumar, P. Tekasakul, B. Gupta, Natural dyes for dye sensitized solar cell: A review, Renewable and Sustainable Energy Reviews 69 (2017) 705–718. https://doi.org/10.1016/j.rser.2016.11.198

[10] S.N. Karthick, K.V. Hemalatha, Suresh Kannan Balasingam, F. Manik Clinton, S. Akshaya, H. Kim, Dye-Sensitized Solar Cells: History, Components, Configuration, and Working Principle, in: A. Pandikumar, K. Jothivenkatachalam, K. Bhojanaa (Eds.), Interfacial Engineering in Functional Materials for Dye-Sensitized Solar Cells, 1st ed., Wiley, 2019: pp. 1–16. https://doi.org/10.1002/9781119557401.ch1

[11] P. Kulkarni, S.K. Nataraj, R.G. Balakrishna, D.H. Nagaraju, M.V. Reddy, Nanostructured binary and ternary metal sulfides: synthesis methods and their application in energy conversion and storage devices, J. Mater. Chem. A 5 (2017) 22040–22094. https://doi.org/10.1039/C7TA07329A

[12] H. Yuan, J. Liu, H. Li, Y. Li, X. Liu, D. Shi, Q. Wu, Q. Jiao, Graphitic carbon nitride quantum dot decorated three-dimensional graphene as an efficient metal-free electrocatalyst for triiodide reduction, J. Mater. Chem. A 6 (2018) 5603–5607. https://doi.org/10.1039/C8TA00205C

[13] M. Hosseinnezhad, S. Nasiri, M. Fathi, M. Ghahari, K. Gharanjig, Introduction of new configuration of dyes contain indigo group for dye-sensitized solar cells: DFT and photovoltaic study, Optical Materials 124 (2022) 111999. https://doi.org/10.1016/j.optmat.2022.111999

[14] J. Gong, K. Sumathy, Q. Qiao, Z. Zhou, Review on dye-sensitized solar cells (DSSCs): Advanced techniques and research trends, Renewable and Sustainable Energy Reviews 68 (2017) 234–246. https://doi.org/10.1016/j.rser.2016.09.097

[15] A.N.B. Zulkifili, T. Kento, M. Daiki, A. Fujiki, The Basic Research on the Dye-Sensitized Solar Cells (DSSC), JOCET 3 (2015) 382–387. https://doi.org/10.7763/JOCET.2015.V3.228

Materials Research Forum LLC
https://doi.org/10.21741/9781644903711-7

[16]  K. Wu, J. Ma, W. Cui, B. Ruan, M. Wu, The Impact of Metal Ion Doping on the Performance of Flexible Poly(3,4-ethylenedioxythiophene) (PEDOT) Cathode in Dye-Sensitized Solar Cells, Journal of Photochemistry and Photobiology A: Chemistry 340 (2017) 29–34. https://doi.org/10.1016/j.jphotochem.2017.02.023

[17]  F.C. Krebs, M. Hösel, M. Corazza, B. Roth, M.V. Madsen, S.A. Gevorgyan, R.R. Søndergaard, D. Karg, M. Jørgensen, Freely available OPV—The fast way to progress, Energy Tech 1 (2013) 378–381. https://doi.org/10.1002/ente.201300057

[18]  Y. Wang, Recent research progress on polymer electrolytes for dye-sensitized solar cells, Solar Energy Materials and Solar Cells 93 (2009) 1167–1175. https://doi.org/10.1016/j.solmat.2009.01.009

[19]  D. Wei, Dye Sensitized Solar Cells, IJMS 11 (2010) 1103–1113. https://doi.org/10.3390/ijms11031103

[20]  P.P. Kumavat, P. Sonar, D.S. Dalal, An overview on basics of organic and dye sensitized solar cells, their mechanism and recent improvements, Renewable and Sustainable Energy Reviews 78 (2017) 1262–1287. https://doi.org/10.1016/j.rser.2017.05.011

[21]  S.E. Sheela, R. Sekar, D.K. Maurya, M. Paulraj, S. Angaiah, Progress in transition metal chalcogenides-based counter electrode materials for dye-sensitized solar cells, Materials Science in Semiconductor Processing 156 (2023) 107273. https://doi.org/10.1016/j.mssp.2022.107273

[22]  S. Suhaimi, M.M. Shahimin, Z.A. Alahmed, J. Chyský, A.H. Reshak, Materials for Enhanced Dye-sensitized Solar Cell Performance: Electrochemical Application, International Journal of Electrochemical Science 10 (2015) 2859–2871. https://doi.org/10.1016/S1452-3981(23)06503-3

[23]  D. Zhang, M. Stojanovic, Y. Ren, Y. Cao, F.T. Eickemeyer, E. Socie, N. Vlachopoulos, J.-E. Moser, S.M. Zakeeruddin, A. Hagfeldt, M. Grätzel, A molecular photosensitizer achieves a Voc of 1.24 V enabling highly efficient and stable dye-sensitized solar cells with copper(II/I)-based electrolyte, Nat Commun 12 (2021) 1777. https://doi.org/10.1038/s41467-021-21945-3

[24]  M. Freitag, J. Teuscher, Y. Saygili, X. Zhang, F. Giordano, P. Liska, J. Hua, S.M. Zakeeruddin, J.-E. Moser, M. Grätzel, A. Hagfeldt, Dye-sensitized solar cells for efficient power generation under ambient lighting, Nature Photon 11 (2017) 372–378. https://doi.org/10.1038/nphoton.2017.60

[25]  J. Gong, J. Liang, K. Sumathy, Review on dye-sensitized solar cells (DSSCs): Fundamental concepts and novel materials, Renewable and Sustainable Energy Reviews 16 (2012) 5848–5860. https://doi.org/10.1016/j.rser.2012.04.044

[26]  N. Huang, F. Zheng, J. Xu, H. Huang, G. Li, Z. Xia, P. Sun, X. Sun, Solution-Based in–situ Synthesis of Transition Metal Sulfides as Efficient Counter Electrodes for Dye-Sensitized Solar Cells, ChemistrySelect 1 (2016) 4613–4619. https://doi.org/10.1002/slct.201600903

[27]  S. Wang, X. Wang, L. Yu, M. Sun, Progress and trends of photodynamic therapy: From traditional photosensitizers to AIE-based photosensitizers, Photodiagnosis and Photodynamic Therapy 34 (2021) 102254. https://doi.org/10.1016/j.pdpdt.2021.102254

[28]    A. Carella, F. Borbone, R. Centore, Research Progress on Photosensitizers for DSSC, Front. Chem. 6 (2018) 481. https://doi.org/10.3389/fchem.2018.00481

[29]    K. Sharma, V. Sharma, S.S. Sharma, Dye-Sensitized Solar Cells: Fundamentals and Current Status, Nanoscale Res Lett 13 (2018) 381. https://doi.org/10.1186/s11671-018-2760-6

[30]    L. Li, D.L. Jacobs, B.R. Bunes, H. Huang, X. Yang, L. Zang, Anomalous high photovoltages observed in shish kebab-like organic p–n junction nanostructures, Polym. Chem. 5 (2014) 309–313. https://doi.org/10.1039/C3PY01026K

[31]    S. Aghazada, P. Gao, A. Yella, G. Marotta, T. Moehl, J. Teuscher, J.-E. Moser, F. De Angelis, M. Grätzel, M.K. Nazeeruddin, Ligand Engineering for the Efficient Dye-Sensitized Solar Cells with Ruthenium Sensitizers and Cobalt Electrolytes, Inorg. Chem. 55 (2016) 6653–6659. https://doi.org/10.1021/acs.inorgchem.6b00842

[32]    F. Gao, Y. Wang, D. Shi, J. Zhang, M. Wang, X. Jing, R. Humphry-Baker, P. Wang, S.M. Zakeeruddin, M. Grätzel, Enhance the Optical Absorptivity of Nanocrystalline TiO$_2$ Film with High Molar Extinction Coefficient Ruthenium Sensitizers for High Performance Dye-Sensitized Solar Cells, J. Am. Chem. Soc. 130 (2008) 10720–10728. https://doi.org/10.1021/ja801942j

[33]    M. Bavarian, S. Nejati, K.K.S. Lau, D. Lee, M. Soroush, Theoretical and Experimental Study of a Dye-Sensitized Solar Cell, Ind. Eng. Chem. Res. 53 (2014) 5234–5247. https://doi.org/10.1021/ie4016914

[34]    M.-E. Yeoh, K.-Y. Chan, Recent advances in photo-anode for dye-sensitized solar cells: a review: Recent advances in photo-anode for DSSCs: A review, Int J Energy Res 41 (2017) 2446–2467. https://doi.org/10.1002/er.3764

[35]    S.H. Ahn, W.S. Chi, D.J. Kim, S.Y. Heo, J.H. Kim, Honeycomb-Like Organized TiO$_2$ Photoanodes with Dual Pores for Solid-State Dye-Sensitized Solar Cells, Adv Funct Materials 23 (2013) 3901–3908. https://doi.org/10.1002/adfm.201203851

[36]    J.-W. Choi, H. Kang, M. Lee, J.S. Kang, S. Kyeong, J.-K. Yang, J. Kim, D.H. Jeong, Y.-S. Lee, Y.-E. Sung, Plasmon-enhanced dye-sensitized solar cells using SiO2 spheres decorated with tightly assembled silver nanoparticles, RSC Adv. 4 (2014) 19851. https://doi.org/10.1039/c4ra00596a

[37]    S. Anandan, Recent improvements and arising challenges in dye-sensitized solar cells, Solar Energy Materials and Solar Cells 91 (2007) 843–846. https://doi.org/10.1016/j.solmat.2006.11.017

[38]    N.A. Ludin, A.M. Al-Alwani Mahmoud, A. Bakar Mohamad, Abd.A.H. Kadhum, K. Sopian, N.S. Abdul Karim, Review on the development of natural dye photosensitizer for dye-sensitized solar cells, Renewable and Sustainable Energy Reviews 31 (2014) 386–396. https://doi.org/10.1016/j.rser.2013.12.001

[39]    Z. Tang, J. Wu, M. Zheng, J. Huo, Z. Lan, A microporous platinum counter electrode used in dye-sensitized solar cells, Nano Energy 2 (2013) 622–627. https://doi.org/10.1016/j.nanoen.2013.07.014

[40]    M.A.M. Al-Alwani, A.B. Mohamad, N.A. Ludin, Abd.A.H. Kadhum, K. Sopian, Dye-sensitised solar cells: Development, structure, operation principles, electron kinetics,

characterisation, synthesis materials and natural photosensitisers, Renewable and Sustainable Energy Reviews 65 (2016) 183–213. https://doi.org/10.1016/j.rser.2016.06.045

[41]    I. Benesperi, H. Michaels, M. Freitag, The researcher's guide to solid-state dye-sensitized solar cells, J. Mater. Chem. C 6 (2018) 11903–11942. https://doi.org/10.1039/C8TC03542C

[42]    M.Z. Iqbal, S.R. Ali, S. Khan, Progress in dye sensitized solar cell by incorporating natural photosensitizers, Solar Energy 181 (2019) 490–509. https://doi.org/10.1016/j.solener.2019.02.023

[43]    M. Freitag, G. Boschloo, The revival of dye-sensitized solar cells, Current Opinion in Electrochemistry 2 (2017) 111–119. https://doi.org/10.1016/j.coelec.2017.03.011

[44]    King-Chuen Lin, Chun-Li Chang, Photo-Induced Electron Transfer from Dye or Quantum Dot to TiO$_2$ Nanoparticles at Single Molecule Level, INTECH Open Access Publisher, 2011

[45]    F. Liu, J. Zhu, L. Hu, B. Zhang, J. Yao, Md.K. Nazeeruddin, M. Grätzel, S. Dai, Low-temperature, solution-deposited metal chalcogenide films as highly efficient counter electrodes for sensitized solar cells, J. Mater. Chem. A 3 (2015) 6315–6323. https://doi.org/10.1039/C5TA00028A

[46]    L.-D. Zhao, S.-H. Lo, Y. Zhang, H. Sun, G. Tan, C. Uher, C. Wolverton, V.P. Dravid, M.G. Kanatzidis, Ultralow thermal conductivity and high thermoelectric figure of merit in SnSe crystals, Nature 508 (2014) 373–377. https://doi.org/10.1038/nature13184

[47]    A.J. Jackson, D. Tiana, A. Walsh, A universal chemical potential for sulfur vapours, Chem. Sci. 7 (2016) 1082–1092. https://doi.org/10.1039/C5SC03088A

[48]    S.K. Swami, N. Chaturvedi, A. Kumar, R. Kapoor, V. Dutta, J. Frey, T. Moehl, M. Grätzel, S. Mathew, M.K. Nazeeruddin, Investigation of electrodeposited cobalt sulphide counter electrodes and their application in next-generation dye sensitized solar cells featuring organic dyes and cobalt-based redox electrolytes, Journal of Power Sources 275 (2015) 80–89. https://doi.org/10.1016/j.jpowsour.2014.11.003

[49]    S.Y. Shajaripour Jaberi, A. Ghaffarinejad, Z. Khajehsaeidi, A. Sadeghi, The synthesis, properties, and potential applications of CoS2 as a transition metal dichalcogenide (TMD), International Journal of Hydrogen Energy 48 (2023) 15831–15878. https://doi.org/10.1016/j.ijhydene.2023.01.056

[50]    Z. Wang, C. Zhao, R. Gui, H. Jin, J. Xia, F. Zhang, Y. Xia, Synthetic methods and potential applications of transition metal dichalcogenide/graphene nanocomposites, Coordination Chemistry Reviews 326 (2016) 86–110. https://doi.org/10.1016/j.ccr.2016.08.004

[51]    X. Qian, H. Liu, Y. Huang, Z. Ren, Y. Yu, C. Xu, L. Hou, Co-Ni-MoSx yolk-shell nanospheres as superior Pt-free electrode catalysts for highly efficient dye-sensitized solar cells, Journal of Power Sources 412 (2019) 568–574. https://doi.org/10.1016/j.jpowsour.2018.11.067

[52]    Q. Han, Z. Hu, H. Wang, Y. Sun, J. Zhang, L. Gao, M. Wu, High performance metal sulfide counter electrodes for organic sulfide redox couple in dye-sensitized solar cells, Materials Today Energy 8 (2018) 1–7. https://doi.org/10.1016/j.mtener.2018.02.004

Materials Research Forum LLC
https://doi.org/10.21741/9781644903711-7

[53] F. Gong, H. Wang, X. Xu, G. Zhou, Z.-S. Wang, In Situ Growth of Co $_{0.85}$ Se and Ni $_{0.85}$ Se on Conductive Substrates as High-Performance Counter Electrodes for Dye-Sensitized Solar Cells, J. Am. Chem. Soc. 134 (2012) 10953–10958. https://doi.org/10.1021/ja303034w

[54] Z. Jin, M. Zhang, M. Wang, C. Feng, Z.-S. Wang, Metal Selenides as Efficient Counter Electrodes for Dye-Sensitized Solar Cells, Acc. Chem. Res. 50 (2017) 895–904. https://doi.org/10.1021/acs.accounts.6b00625

[55] G. Liu, M. Wang, H. Wang, R.E.A. Ardhi, H. Yu, D. Zou, J.K. Lee, Hierarchically structured photoanode with enhanced charge collection and light harvesting abilities for fiber-shaped dye-sensitized solar cells, Nano Energy 49 (2018) 95–102. https://doi.org/10.1016/j.nanoen.2018.04.037

[56] L. Qiu, Q. Wu, Z. Yang, X. Sun, Y. Zhang, H. Peng, Freestanding Aligned Carbon Nanotube Array Grown on a Large-Area Single-Layered Graphene Sheet for Efficient Dye-Sensitized Solar Cell, Small 11 (2015) 1150–1155. https://doi.org/10.1002/smll.201400703

[57] H. Hu, B. Dong, B. Chen, X. Gao, D. Zou, High performance fiber-shaped perovskite solar cells based on lead acetate precursor, Sustainable Energy Fuels 2 (2018) 79–84. https://doi.org/10.1039/C7SE00462A

[58] A. Ali, K. Shehzad, F. Ur-Rahman, S.M. Shah, M. Khurram, M. Mumtaz, R.U.R. Sagar, Flexible, Low Cost, and Platinum-Free Counter Electrode for Efficient Dye-Sensitized Solar Cells, ACS Appl. Mater. Interfaces 8 (2016) 25353–25360. https://doi.org/10.1021/acsami.6b08826

[59] D.P. Joseph, S. Ganesan, M. Kovendhan, S.A. Suthanthiraraj, P. Maruthamuthu, C. Venkateswaran, Fabrication of dye sensitized solar cell using Cr doped Cu-Zn-Se type chalcopyrite thin film, Physica Status Solidi (a) 208 (2011) 2215–2219. https://doi.org/10.1002/pssa.201026368

[60] Z. He, Y. Yang, J.-W. Liu, S.-H. Yu, Emerging tellurium nanostructures: controllable synthesis and their applications, Chem. Soc. Rev. 46 (2017) 2732–2753. https://doi.org/10.1039/C7CS00013H

[61] J. Guo, Y. Shi, Y. Chu, T. Ma, Highly efficient telluride electrocatalysts for use as Pt-free counter electrodes in dye-sensitized solar cells, Chem. Commun. 49 (2013) 10157. https://doi.org/10.1039/c3cc45698f

[62] K. Wan, F. Wu, Y. Dou, L. Fang, C. Mao, Enhance the performance of dye-sensitized solar cells by Bi$_2$Te$_3$ nanosheet/ZnO nanoparticle composite photoanode, Journal of Alloys and Compounds 680 (2016) 373–380. https://doi.org/10.1016/j.jallcom.2016.04.124

[63] S. Hussain, S.A. Patil, D. Vikraman, N. Mengal, H. Liu, W. Song, K.-S. An, S.H. Jeong, H.-S. Kim, J. Jung, Large area growth of MoTe$_2$ films as high performance counter electrodes for dye-sensitized solar cells, Sci Rep 8 (2018) 29. https://doi.org/10.1038/s41598-017-18067-6

[64] S. Hussain, S.A. Patil, D. Vikraman, I. Rabani, A.A. Arbab, S.H. Jeong, H.-S. Kim, H. Choi, J. Jung, Enhanced electrocatalytic properties in MoS$_2$/MoTe$_2$ hybrid heterostructures for dye-sensitized solar cells, Applied Surface Science 504 (2020) 144401. https://doi.org/10.1016/j.apsusc.2019.144401

[65]   J. Zhang, Z. Wang, X. Li, J. Yang, C. Song, Y. Li, J. Cheng, Q. Guan, B. Wang, Flexible Platinum-Free Fiber-Shaped Dye Sensitized Solar Cell with 10.28% Efficiency, ACS Appl. Energy Mater. 2 (2019) 2870–2877. https://doi.org/10.1021/acsaem.9b00207

[66]   X. Wen, Z. Lu, X. Sun, Y. Xiang, Z. Chen, J. Shi, I. Bhat, G.-C. Wang, M. Washington, T.-M. Lu, Epitaxial CdTe Thin Films on Mica by Vapor Transport Deposition for Flexible Solar Cells, ACS Appl. Energy Mater. 3 (2020) 4589–4599. https://doi.org/10.1021/acsaem.0c00265

[67]   J.A. Castillo-Robles, E. Rocha-Rangel, J.A. Ramírez-de-León, F.C. Caballero-Rico, E.N. Armendáriz-Mireles, Advances on Dye-Sensitized Solar Cells (DSSCs) Nanostructures and Natural Colorants: A Review, J. Compos. Sci. 5 (2021) 288. https://doi.org/10.3390/jcs5110288

[68]   H. Yuan, L. Kong, T. Li, Q. Zhang, A review of transition metal chalcogenide/graphene nanocomposites for energy storage and conversion, Chinese Chemical Letters 28 (2017) 2180–2194. https://doi.org/10.1016/j.cclet.2017.11.038

[69]   S. Yun, A. Hagfeldt, T. Ma, Pt-Free Counter Electrode for Dye-Sensitized Solar Cells with High Efficiency, Advanced Materials 26 (2014) 6210–6237. https://doi.org/10.1002/adma.201402056

[70]   M. Heydari Gharahcheshmeh, K.K. Gleason, Device Fabrication Based on Oxidative Chemical Vapor Deposition (oCVD) Synthesis of Conducting Polymers and Related Conjugated Organic Materials, Adv Materials Inter 6 (2019) 1801564. https://doi.org/10.1002/admi.201801564

[71]   P.P. Sanap, S.P. Gupta, S.S. Kahandal, J.L. Gunjakar, C.D. Lokhande, B.R. Sankapal, Z. Said, R.N. Bulakhe, J. Man Kim, A.B. Bhalerao, Exploring vanadium-chalcogenides toward solar cell application: A review, Journal of Industrial and Engineering Chemistry 129 (2024) 124–142. https://doi.org/10.1016/j.jiec.2023.09.004

[72]   M.C. Mathpal, P. Kumar, F.H. Aragón, M.A.G. Soler, H.C. Swart, Basic Concepts, Engineering, and Advances in Dye-Sensitized Solar Cells, in: S.K. Sharma, K. Ali (Eds.), Solar Cells, Springer International Publishing, Cham, 2020: pp. 185–233. https://doi.org/10.1007/978-3-030-36354-3_8

[73]   H.S. Nalwa, Encyclopedia of nanoscience and nanotechnology, American scientific publishers, North Lewis Way (Calif.), 2004

[74]   K.W. Shah, S.-X. Wang, Y. Zheng, J. Xu, Solution-Based Synthesis and Processing of Metal Chalcogenides for Thermoelectric Applications, Applied Sciences 9 (2019) 1511. https://doi.org/10.3390/app9071511

[75]   W. Yao, S. Yu, Synthesis of Semiconducting Functional Materials in Solution: From II-VI Semiconductor to Inorganic–Organic Hybrid Semiconductor Nanomaterials, Adv Funct Materials 18 (2008) 3357–3366. https://doi.org/10.1002/adfm.200800672

[76]   Y.X. Gan, A.H. Jayatissa, Z. Yu, X. Chen, M. Li, Hydrothermal Synthesis of Nanomaterials, Journal of Nanomaterials 2020 (2020) 1–3. https://doi.org/10.1155/2020/8917013

Materials Research Forum LLC
https://doi.org/10.21741/9781644903711-7

[77]   H. Yu, J. Pan, Y. Bai, X. Zong, X. Li, L. Wang, Hydrothermal Synthesis of a Crystalline Rutile TiO $_2$ Nanorod Based Network for Efficient Dye-Sensitized Solar Cells, Chemistry A European J 19 (2013) 13569–13574. https://doi.org/10.1002/chem.201300999

[78]   M. Andersson, L. Österlund, S. Ljungström, A. Palmqvist, Preparation of Nanosize Anatase and Rutile $TiO_2$ by Hydrothermal Treatment of Microemulsions and Their Activity for Photocatalytic Wet Oxidation of Phenol, J. Phys. Chem. B 106 (2002) 10674–10679. https://doi.org/10.1021/jp025715y

[79]   D. Jamwal, S.K. Mehta, Metal Telluride Nanomaterials: Facile Synthesis, Properties and Applications for Third Generation Devices., ChemistrySelect 4 (2019) 1943–1963. https://doi.org/10.1002/slct.201803680

[80]   N. Chaudhary, M. Khanuja, Abid, S.S. Islam, Hydrothermal synthesis of $MoS_2$ nanosheets for multiple wavelength optical sensing applications, Sensors and Actuators A: Physical 277 (2018) 190–198. https://doi.org/10.1016/j.sna.2018.05.008

[81]   M. Li, D. Wang, J. Li, Z. Pan, H. Ma, Y. Jiang, Z. Tian, Facile hydrothermal synthesis of MoS $_2$ nano-sheets with controllable structures and enhanced catalytic performance for anthracene hydrogenation, RSC Adv. 6 (2016) 71534–71542. https://doi.org/10.1039/C6RA16084K

[82]   K. Byrappa, M. Yoshimura, Handbook of hydrothermal technology: a technology for crystal growth and materials processing, Noyes Publications, Norwich, N.Y, 2001

[83]   J. Li, Q. Wu, J. Wu, Synthesis of Nanoparticles via Solvothermal and Hydrothermal Methods, in: M. Aliofkhazraei (Ed.), Handbook of Nanoparticles, Springer International Publishing, Cham, 2016: pp. 295–328. https://doi.org/10.1007/978-3-319-15338-4_17

[84]   A. Kagkoura, T. Skaltsas, N. Tagmatarchis, Transition-Metal Chalcogenide/Graphene Ensembles for Light-Induced Energy Applications, Chemistry A European J 23 (2017) 12967–12979. https://doi.org/10.1002/chem.201700242

[85]   M.-R. Gao, Y.-F. Xu, J. Jiang, S.-H. Yu, Nanostructured metal chalcogenides: synthesis, modification, and applications in energy conversion and storage devices, Chem. Soc. Rev. 42 (2013) 2986. https://doi.org/10.1039/c2cs35310e

[86]   S.D. Henam, F. Ahmad, M.A. Shah, S. Parveen, A.H. Wani, Microwave synthesis of nanoparticles and their antifungal activities, Spectrochimica Acta Part A: Molecular and Biomolecular Spectroscopy 213 (2019) 337–341. https://doi.org/10.1016/j.saa.2019.01.071

[87]   A.M. Schwenke, S. Hoeppener, U.S. Schubert, Synthesis and Modification of Carbon Nanomaterials utilizing Microwave Heating, Advanced Materials 27 (2015) 4113–4141. https://doi.org/10.1002/adma.201500472

[88]   A. De La Hoz, A. Loupy, eds., Microwaves in Organic Synthesis, 1st ed., Wiley, 2012. https://doi.org/10.1002/9783527651313

[89]   L.T. Gularte, C.D. Fernandes, M.L. Moreira, C.W. Raubach, P.L.G. Jardim, S.S. Cava, In situ microwave-assisted deposition of CoS counter electrode for dye-sensitized solar cells, Solar Energy 198 (2020) 658–664. https://doi.org/10.1016/j.solener.2020.01.052

[90]   D. Tonelli, E. Scavetta, I. Gualandi, Electrochemical Deposition of Nanomaterials for Electrochemical Sensing, Sensors 19 (2019) 1186. https://doi.org/10.3390/s19051186

[91]   S. Domínguez-Domínguez, J. Arias-Pardilla, Á. Berenguer-Murcia, E. Morallón, D. Cazorla-Amorós, Electrochemical deposition of platinum nanoparticles on different carbon supports and conducting polymers, J Appl Electrochem 38 (2008) 259–268. https://doi.org/10.1007/s10800-007-9435-9

[92]   X. Shi, Z.-G. Chen, W. Liu, L. Yang, M. Hong, R. Moshwan, L. Huang, J. Zou, Achieving high Figure of Merit in p-type polycrystalline $Sn_{0.98}Se$ via self-doping and anisotropy-strengthening, Energy Storage Materials 10 (2018) 130–138. https://doi.org/10.1016/j.ensm.2017.08.014

[93]   E.J. Menke, Q. Li, R.M. Penner, Bismuth Telluride ($Bi_2Te_3$) Nanowires Synthesized by Cyclic Electrodeposition/Stripping Coupled with Step Edge Decoration, Nano Lett. 4 (2004) 2009–2014. https://doi.org/10.1021/nl048627t

[94]   W.-L. Wang, C.-C. Wan, Y.-Y. Wang, Composition-dependent characterization and optimal control of electrodeposited $Bi_2Te_3$ films for thermoelectric application, Electrochimica Acta 52 (2007) 6502–6508. https://doi.org/10.1016/j.electacta.2007.04.037

[95]   L. Li, Y. Yang, X. Huang, G. Li, L. Zhang, Pulsed electrodeposition of single-crystalline $Bi_2Te_3$ nanowire arrays, Nanotechnology 17 (2006) 1706–1712. https://doi.org/10.1088/0957-4484/17/6/027

[96]   X.-H. Li, B. Zhou, L. Pu, J.-J. Zhu, Electrodeposition of $Bi_2Te_3$ and $Bi_2Te_3$ Derived Alloy Nanotube Arrays, Crystal Growth & Design 8 (2008) 771–775. https://doi.org/10.1021/cg7006759

[97]   M. Yang, Z. Shen, X. Liu, W. Wang, Electrodeposition and Thermoelectric Properties of Cu-Se Binary Compound Films, Journal of Elec Materi 45 (2016) 1974–1981. https://doi.org/10.1007/s11664-016-4344-5

[98]   M. Biçer, İ. Şişman, Electrodeposition and growth mechanism of SnSe thin films, Applied Surface Science 257 (2011) 2944–2949. https://doi.org/10.1016/j.apsusc.2010.10.096

[99]   V.H.V. Quy, J.-H. Park, S.-H. Kang, H. Kim, K.-S. Ahn, Improved electrocatalytic activity of electrodeposited $Ni_3S_4$ counter electrodes for dye- and quantum dot-sensitized solar cells, Journal of Industrial and Engineering Chemistry 70 (2019) 322–329. https://doi.org/10.1016/j.jiec.2018.10.032

[100]  J. Huo, M. Zheng, Y. Tu, J. Wu, L. Hu, S. Dai, A high performance cobalt sulfide counter electrode for dye-sensitized solar cells, Electrochimica Acta 159 (2015) 166–173. https://doi.org/10.1016/j.electacta.2015.01.214

[101]  K. Hachem, M.J. Ansari, R.O. Saleh, H.H. Kzar, M.E. Al-Gazally, U.S. Altimari, S.A. Hussein, H.T. Mohammed, A.T. Hammid, E. Kianfar, Methods of Chemical Synthesis in the Synthesis of Nanomaterial and Nanoparticles by the Chemical Deposition Method: A Review, BioNanoSci. 12 (2022) 1032–1057. https://doi.org/10.1007/s12668-022-00996-w

[102]  I.R. S., X. Xu, W. Yang, F. Yang, L. Hou, Y. Li, Highly active and reflective MoS2 counter electrode for enhancement of photovoltaic efficiency of dye sensitized solar cells, Electrochimica Acta 212 (2016) 614–620. https://doi.org/10.1016/j.electacta.2016.07.059

[103] A. Sharma, S. Masoumi, D. Gedefaw, S. O'Shaughnessy, D. Baran, A. Pakdel, Flexible solar and thermal energy conversion devices: Organic photovoltaics (OPVs), organic thermoelectric generators (OTEGs) and hybrid PV-TEG systems, Applied Materials Today 29 (2022) 101614. https://doi.org/10.1016/j.apmt.2022.101614

[104] G. Cao, Nanostructures & nanomaterials: synthesis, properties & applications, Imperial College Press, London ; Hackensack, NJ, 2004

[105] Q. Wang, H. Wu, M. Qin, Z. Li, B. Jia, A. Chu, X. Qu, Study on influencing factors and mechanism of high-quality tungsten carbide nanopowders synthesized via carbothermal reduction, Journal of Alloys and Compounds 867 (2021) 158959. https://doi.org/10.1016/j.jallcom.2021.158959

[106] S.M. George, Atomic Layer Deposition: An Overview, Chem. Rev. 110 (2010) 111–131. https://doi.org/10.1021/cr900056b

[107] M. Salavati-Niasari, M. Bazarganipour, F. Davar, A.A. Fazl, Simple routes to synthesis and characterization of nanosized tin telluride compounds, Applied Surface Science 257 (2010) 781–785. https://doi.org/10.1016/j.apsusc.2010.07.065

[108] L. Zhang, Z. Ai, F. Jia, L. Liu, X. Hu, J.C. Yu, Controlled Hydrothermal Synthesis and Growth Mechanism of Various Nanostructured Films of Copper and Silver Tellurides, Chemistry A European J 12 (2006) 4185–4190. https://doi.org/10.1002/chem.200501404

[109] A. Paca, P. Ajibade, Synthesis, Optical, and Structural Studies of Iron Sulphide Nanoparticles and Iron Sulphide Hydroxyethyl Cellulose Nanocomposites from Bis-(Dithiocarbamato)Iron(II) Single-Source Precursors, Nanomaterials 8 (2018) 187. https://doi.org/10.3390/nano8040187

[110] Z.Y. Aydın, S. Abacı, Characterization of CuTe nanofilms grown by underpotential deposition based on an electrochemical codeposition technique, J Solid State Electrochem 21 (2017) 1417–1430. https://doi.org/10.1007/s10008-016-3496-9

[111] C.-L. Chen, Y.-Y. Chen, S.-J. Lin, J.C. Ho, P.-C. Lee, C.-D. Chen, S.R. Harutyunyan, Fabrication and Characterization of Electrodeposited Bismuth Telluride Films and Nanowires, J. Phys. Chem. C 114 (2010) 3385–3389. https://doi.org/10.1021/jp909926z

[112] K.O. Ighodalo, B.N. Ezealigo, A. Agbogu, A.C. Nwanya, D. Obi, S.L. Mammah, S. Botha, R. Bucher, M. Maaza, F.I. Ezema, The effect of deposition cycles on intrinsic and electrochemical properties of metallic cobalt sulfide by Simple chemical route, Materials Science in Semiconductor Processing 101 (2019) 16–27. https://doi.org/10.1016/j.mssp.2019.05.015

[113] Q. Yao, Y. Zhu, L. Chen, Z. Sun, X. Chen, Microwave-assisted synthesis and characterization of $Bi_2Te_3$ nanosheets and nanotubes, Journal of Alloys and Compounds 481 (2009) 91–95. https://doi.org/10.1016/j.jallcom.2009.03.001

[114] H. Rojas-Chávez, H. Cruz-Martínez, E. Flores-Rojas, J.M. Juárez-García, J.L. González-Domínguez, N. Daneu, J. Santoyo-Salazar, The mechanochemical synthesis of PbTe nanostructures: following the Ostwald ripening effect during milling, Phys. Chem. Chem. Phys. 20 (2018) 27082–27092. https://doi.org/10.1039/C8CP04915G

[115] F. Haque, T. Daeneke, K. Kalantar-zadeh, J.Z. Ou, Two-Dimensional Transition Metal Oxide and Chalcogenide-Based Photocatalysts, Nano-Micro Lett. 10 (2018) 23. https://doi.org/10.1007/s40820-017-0176-y

[116] Y. Xiao, C. Xiong, M.-M. Chen, S. Wang, L. Fu, X. Zhang, Structure modulation of two-dimensional transition metal chalcogenides: recent advances in methodology, mechanism and applications, Chem. Soc. Rev. 52 (2023) 1215–1272. https://doi.org/10.1039/D1CS01016F

[117] M. Dai, R. Wang, Synthesis and Applications of Nanostructured Hollow Transition Metal Chalcogenides, Small 17 (2021) 2006813. https://doi.org/10.1002/smll.202006813

[118] Q. Gao, W. Zhang, Z. Shi, L. Yang, Y. Tang, Structural Design and Electronic Modulation of Transition-Metal-Carbide Electrocatalysts toward Efficient Hydrogen Evolution, Advanced Materials 31 (2019) 1802880. https://doi.org/10.1002/adma.201802880

[119] X. Wang, A. Chen, X. Wu, J. Zhang, J. Dong, L. Zhang, Synthesis and Modulation of Low-Dimensional Transition Metal Chalcogenide Materials via Atomic Substitution, Nano-Micro Lett. 16 (2024) 163. https://doi.org/10.1007/s40820-024-01378-5

[120] Y. Guo, J. Li, MoS2 quantum dots: synthesis, properties and biological applications, Materials Science and Engineering: C 109 (2020) 110511. https://doi.org/10.1016/j.msec.2019.110511

[121] E. Singh, K.S. Kim, G.Y. Yeom, H.S. Nalwa, Atomically Thin-Layered Molybdenum Disulfide (MoS2) for Bulk-Heterojunction Solar Cells, ACS Appl. Mater. Interfaces 9 (2017) 3223–3245. https://doi.org/10.1021/acsami.6b13582

[122] P. Acevedo-Peña, G. Vázquez, D. Laverde, J.E. Pedraza-Rosas, J. Manríquez, I. González, Electrochemical Characterization of TiO2 Films Formed by Cathodic EPD in Aqueous Media, J. Electrochem. Soc. 156 (2009) C377. https://doi.org/10.1149/1.3208009

[123] J. Li, L. Li, Y. Du, X. Liu, L. Wang, Carbon spheres decorated with SnS2 nanosheets as a low-cost counter-electrode material for dye-sensitized solar cell, Materials Letters 363 (2024) 136239. https://doi.org/10.1016/j.matlet.2024.136239

[124] X. Wang, Y. Xie, B. Bateer, K. Pan, X. Zhang, J. Wu, H. Fu, CoSe 2 /N-Doped Carbon Hybrid Derived from ZIF-67 as High-Efficiency Counter Electrode for Dye-Sensitized Solar Cells, ACS Sustainable Chem. Eng. 7 (2019) 2784–2791. https://doi.org/10.1021/acssuschemeng.8b05995

[125] S. Hussain, K. Akbar, D. Vikraman, I. Rabani, W. Song, K.-S. An, H.-S. Kim, S.-H. Chun, J. Jung, Experimental and theoretical insights to demonstrate the hydrogen evolution activity of layered platinum dichalcogenides electrocatalysts, Journal of Materials Research and Technology 12 (2021) 385–398. https://doi.org/10.1016/j.jmrt.2021.02.097

[126] S. Rashidi, S. Rashidi, R.K. Heydari, S. Esmaeili, N. Tran, D. Thangi, W. Wei, WS 2 and MoS 2 counter electrode materials for dye-sensitized solar cells, Progress in Photovoltaics 29 (2021) 238–261. https://doi.org/10.1002/pip.3350

[127] M.Z. Iqbal, S. Alam, M.M. Faisal, S. Khan, Recent advancement in the performance of solar cells by incorporating transition metal dichalcogenides as counter electrode and photoabsorber, Int J Energy Res 43 (2019) 3058–3079. https://doi.org/10.1002/er.4375

Materials Research Forum LLC
https://doi.org/10.21741/9781644903711-7

[128] V. Diez-Cabanes, S. Fantacci, M. Pastore, First-principles modeling of dye-sensitized solar cells: From the optical properties of standalone dyes to the charge separation at dye/$TiO_2$ interfaces, in: Theoretical and Computational Photochemistry, Elsevier, 2023: pp. 215–245. https://doi.org/10.1016/B978-0-323-91738-4.00013-0

[129] Electro-Materials Research Laboratory, Centre for Nanoscience and Technology, Pondicherry University, Puducherry-605014, India, R. Kottayi, Department of Physics, Kanchi Mamunivar Govt. Institute for PG Studies and Research, Puducherry-605008, India, D.K. Maurya, Electro-Materials Research Laboratory, Centre for Nanoscience and Technology, Pondicherry University, Puducherry-605014, India, R. Sittaramane, Department of Physics, Kanchi Mamunivar Govt. Institute for PG Studies and Research, Puducherry-605008, India, S. Angaiah, Electro-Materials Research Laboratory, Centre for Nanoscience and Technology, Pondicherry University, Puducherry-605014, India, Recent developments in Metal Chalcogenides based Quantum Dot Sensitized Solar Cells, ES Energy Environ. (2022). https://doi.org/10.30919/esee8c754

[130] M. Chhowalla, H.S. Shin, G. Eda, L.-J. Li, K.P. Loh, H. Zhang, The chemistry of two-dimensional layered transition metal dichalcogenide nanosheets, Nature Chem 5 (2013) 263–275. https://doi.org/10.1038/nchem.1589

[131] G. Eda, S.A. Maier, Two-Dimensional Crystals: Managing Light for Optoelectronics, ACS Nano 7 (2013) 5660–5665. https://doi.org/10.1021/nn403159y

[132] C. Tan, X. Cao, X.-J. Wu, Q. He, J. Yang, X. Zhang, J. Chen, W. Zhao, S. Han, G.-H. Nam, M. Sindoro, H. Zhang, Recent Advances in Ultrathin Two-Dimensional Nanomaterials, Chem. Rev. 117 (2017) 6225–6331. https://doi.org/10.1021/acs.chemrev.6b00558

[133] H. Zhou, Q. Chen, G. Li, S. Luo, T. Song, H.-S. Duan, Z. Hong, J. You, Y. Liu, Y. Yang, Interface engineering of highly efficient perovskite solar cells, Science 345 (2014) 542–546. https://doi.org/10.1126/science.1254050

[134] W. Zhang, Z. Huang, W. Zhang, Y. Li, Two-dimensional semiconductors with possible high room temperature mobility, Nano Res. 7 (2014) 1731–1737. https://doi.org/10.1007/s12274-014-0532-x

[135] P. Vijayakumar, M. Senthil Pandian, S.P. Lim, A. Pandikumar, N.M. Huang, S. Mukhopadhyay, P. Ramasamy, Investigations of tungsten carbide nanostructures treated with different temperatures as counter electrodes for dye sensitized solar cells (DSSC) applications, J Mater Sci: Mater Electron 26 (2015) 7977–7986. https://doi.org/10.1007/s10854-015-3452-y

[136] N. Yu, L. Wang, M. Li, X. Sun, T. Hou, Y. Li, Molybdenum disulfide as a highly efficient adsorbent for non-polar gases, Phys. Chem. Chem. Phys. 17 (2015) 11700–11704. https://doi.org/10.1039/C5CP00161G

[137] M. Grätzel, Dye-sensitized solar cells, Journal of Photochemistry and Photobiology C: Photochemistry Reviews 4 (2003) 145–153. https://doi.org/10.1016/S1389-5567(03)00026-1

[138] W. Wei, K. Sun, Y.H. Hu, An efficient counter electrode material for dye-sensitized solar cells—flower-structured 1T metallic phase $MoS_2$, J. Mater. Chem. A 4 (2016) 12398–12401. https://doi.org/10.1039/C6TA04743B

[139] E. Singh, K.S. Kim, G.Y. Yeom, H.S. Nalwa, Two-dimensional transition metal dichalcogenide-based counter electrodes for dye-sensitized solar cells, RSC Adv. 7 (2017) 28234–28290. https://doi.org/10.1039/C7RA03599C

[140] L. Su, Y. Xiao, G. Han, L. Lu, H. Li, M. Zhu, Performance enhancement of perovskite solar cells using trimesic acid additive in the two-step solution method, Journal of Power Sources 426 (2019) 11–15. https://doi.org/10.1016/j.jpowsour.2019.04.024

[141] G. Zhuang, J. Yan, Y. Wen, Z. Zhuang, Y. Yu, Two-Dimensional Transition Metal Oxides and Chalcogenides for Advanced Photocatalysis: Progress, Challenges, and Opportunities, Solar RRL 5 (2021) 2000403. https://doi.org/10.1002/solr.202000403

[142] N.N. Rosman, R. Mohamad Yunus, L. Jeffery Minggu, K. Arifin, M.N.I. Salehmin, M.A. Mohamed, M.B. Kassim, Photocatalytic properties of two-dimensional graphene and layered transition-metal dichalcogenides based photocatalyst for photoelectrochemical hydrogen generation: An overview, International Journal of Hydrogen Energy 43 (2018) 18925–18945. https://doi.org/10.1016/j.ijhydene.2018.08.126

[143] Y.L. Huang, W. Chen, A.T.S. Wee, Two-dimensional magnetic transition metal chalcogenides, Smart Mat 2 (2021) 139–153. https://doi.org/10.1002/smm2.1031

[144] W. Chen, Y. Qu, L. Yao, X. Hou, X. Shi, H. Pan, Electronic, magnetic, catalytic, and electrochemical properties of two-dimensional Janus transition metal chalcogenides, J. Mater. Chem. A 6 (2018) 8021–8029. https://doi.org/10.1039/C8TA01202D

[145] Q. Chen, Q. Ding, Y. Wang, Y. Xu, J. Wang, Electronic and Magnetic Properties of a Two-Dimensional Transition Metal Phosphorous Chalcogenide TMPS $_4$, J. Phys. Chem. C 124 (2020) 12075–12080. https://doi.org/10.1021/acs.jpcc.0c02432

[146] J.-F. Yang, B. Parakash, J. Hardell, Q.-F. Fang, Tribological properties of transition metal di-chalcogenide based lubricant coatings, Front. Mater. Sci. 6 (2012) 116–127. https://doi.org/10.1007/s11706-012-0155-7

[147] D. Voiry, A. Mohite, M. Chhowalla, Phase engineering of transition metal dichalcogenides, Chem. Soc. Rev. 44 (2015) 2702–2712. https://doi.org/10.1039/C5CS00151J

[148] F. Özel, A. Sarılmaz, B. İstanbullu, A. Aljabour, M. Kuş, S. Sönmezoğlu, Penternary chalcogenides nanocrystals as catalytic materials for efficient counter electrodes in dye-synthesized solar cells, Sci Rep 6 (2016) 29207. https://doi.org/10.1038/srep29207

[149] U. Ahmed, M. Alizadeh, N.A. Rahim, S. Shahabuddin, M.S. Ahmed, A.K. Pandey, A comprehensive review on counter electrodes for dye sensitized solar cells: A special focus on Pt-TCO free counter electrodes, Solar Energy 174 (2018) 1097–1125. https://doi.org/10.1016/j.solener.2018.10.010

[150] S.-H. Chang, M.-D. Lu, Y.-L. Tung, H.-Y. Tuan, Gram-Scale Synthesis of Catalytic $Co_9S_8$ Nanocrystal Ink as a Cathode Material for Spray-Deposited, Large-Area Dye-Sensitized Solar Cells, ACS Nano 7 (2013) 9443–9451. https://doi.org/10.1021/nn404272j

[151] J. Theerthagiri, R.A. Senthil, K. Susmitha, M. Raghavender, J. Madhavan, Synthesis of Efficient $Ni_{0.9}X_{0.1}Se_2$ (X=Cd, Co, Sn and Zn) Based Ternary Selenides for Dye-Sensitized Solar Cells, MSF 832 (2015) 61–71. https://doi.org/10.4028/www.scientific.net/MSF.832.61

[152]  Y. Zeng, S. Zhang, X. Li, J. Ao, Y. Sun, W. Liu, F. Liu, P. Gao, Y. Zhang, Two-step growth of VSe $_2$ films and their photoelectric properties*, Chinese Phys. B 28 (2019) 058101. https://doi.org/10.1088/1674-1056/28/5/058101

[153]  X. Liu, G. Yue, H. Zheng, A promising vanadium sulfide counter electrode for efficient dye-sensitized solar cells, RSC Adv. 7 (2017) 12474–12478. https://doi.org/10.1039/C6RA28667D

[154]  X. Sun, J. Dou, F. Xie, Y. Li, M. Wei, One-step preparation of mirror-like NiS nanosheets on ITO for the efficient counter electrode of dye-sensitized solar cells, Chem. Commun. 50 (2014) 9869. https://doi.org/10.1039/C4CC03798G

[155]  N. Mahuli, S.K. Sarkar, Atomic layer deposition of NiS and its application as cathode material in dye sensitized solar cell, Journal of Vacuum Science & Technology A: Vacuum, Surfaces, and Films 34 (2016) 01A142. https://doi.org/10.1116/1.4938078

175

Applications for Earth-Abundant Transition Metals
Materials Research Foundations 179 (2025) 176-205

Materials Research Forum LLC
https://doi.org/21741/9781644903711-8

Chapter 8

# Earth Abundant Transition Metals-Based Materials in Chemical Sensors

Melkamu Biyana Regasa, Shibiru Yadeta Ejeta

Department of Chemistry, College of Natural and Computational Sciences, P.O. Box 395, Nekemte, Ethiopia

**Abstract**

Sensors are devices that detect and measure specific chemical substances in their environment and are useful in various fields. The recent advancement in materials, synthesis techniques, and signal processing yields highly sensitive and selective chemical sensors. Recently, using earth-abundant transition metals got attention in different research particularly, the first row of transition metals and their compounds, owning to their abundance, low cost, and tunable properties. Pure metals, their oxides as well as their mixture offer additional benefits in the development of chemical sensors, with better merits. Therefore, discussed in this chapter are the general idea and the progress made in utilizing transition metals-based materials for the development of sensors in the context environmental monitoring, biomedical diagnostics, and industrial processes.

**Keywords**

Chemical Sensors, Electrochemical Sensors, Optical Sensor, Thermal Sensor, Transition Metal

## Contents

## 1.   Introduction

Chemical sensors are analytical devices designed to interact with a target analyte, causing a change in some chemical or physical property. This change is then converted into measurable physical and chemical signals using transducer elements [1]. It is bonding or intermolecular attraction that facilitate the attachment of the analyte [2]. This characteristic of a chemical sensor is essential for producing high-quality signal output and ensuring reliable detection methods.

Most chemical sensors consist of sensing systems and transduction systems. The preparation of chemical sensors encompasses the integration of recognition layers on top of the exterior of working electrodes [3]. Materials like metallic and semiconducting materials, diverse-walled nanotubes of carbon, organic nanomaterial, and hybrid composites with tailored functionalities are commonly used for these purposes. The smaller size and penetrability makes them suitable for chemical and biological sensor with higher degree of sensitivity and selectivity as compared to their bulk counterparts [4]. Recently, earth-abundant transition metals are gaining extensive attention in the arena of chemical devices owing to their availability, cost-effectiveness, and desirable possessions [5]. Unique chemical and functional properties of earth-abundant transition metals are used in a variety of uses such as well as detecting and identifying, fuel cells, solar cells, $H_2O$ electrolyzers, photo-electrochemical water splitting, adsorbents, drug delivery, and energy storage systems. [6]. Thus, this chapter brings an insight into the applications of earth-abundant transition metals in the growth of numerous kinds of biochemical sensors.

## 2.   Earth abundant transition elements

Earth abundant transition elements such as nickel (Ni), copper (Cu), manganese (Mn), iron (Fe), and zinc (Zn) are very important in many industrial applications. The earth's crust mass fraction (kg/kg) of some transition metals is $5.63 \times 10^{-2}$ for Fe, $7.0 \times 10^{-5}$ for Zn, $84.0 \times 10^{-6}$ for Ni, $25.0 \times 10^{-6}$ for Co, and $95.0 \times 10^{-5}$ for Mn [7]. Compared to the varied applications of expensive transition metals, they (e.g. first-row) become much more attractive for various applications. This might be payable to their noticeable benefits, including great quantity in earth, cheap, negligible harmfulness, and exceptional catalytic features [8]. These materials offer promising alternatives to rare and expensive metals like platinum, palladium, and gold [9].

During the past decades; palladium, rhodium, ruthenium, and iridium based transition metal catalysts have proven to be efficient for numerous organic transformations reactions [10]. However, these metals bear inherent drawbacks, like less abundance, expensive, and toxicity. In this regard, abundant transition metals like first-row (Mn, Fe, Co, Ni, Zn, and Cu) play important role due to their economic and ecological benefits [11].

Earth-abundant transition metals have been emerging to overcome the aforementioned problems of precious-metal such as low natural abundance. Therefore, materials based on earth-abundant transition metals are emerging area of chemical sensors, catalysis, drug delivery, energy systems, and adsorbents [7, 12, 13] as shown in Figure 1. These materials are very important in line for their sole features like good catalytic activity, low toxicity, and more labile environment [6, 10]. Moreover, they are sustainable due to their cheapness, accessibility, reasonably not harmful, broader interaction. Lately, the pure and oxide forms of these metals are attracting interest from various bodies in arrears to their extraordinary discrimination, strength, feasibility, and humble production methods [14]. Furthermore, due to some attractive properties, they are used in catalyst for sensing various emergent biomarkers [15]. The understanding of the advanced chemical sensing practices is very important at present time [5, 16].

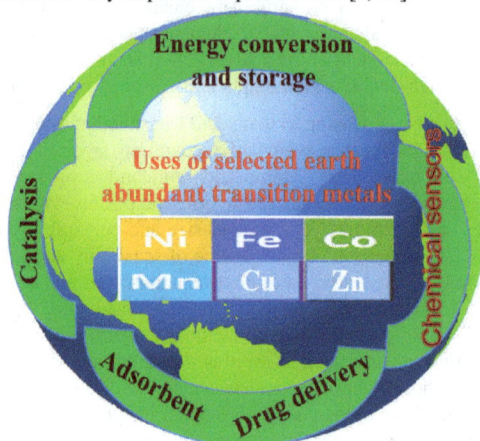

*Figure 1. Applications of selected earth abundant transition metals*

## 3. Biochemical devices (Sensors)

They are electronic operation that remain used to sense and measure the presence and amount of specific chemical species in a given environment [17]. They are classified into some groups based on the transducer element employed to convert any change into a measurable signal as: gravimetric sensors, thermal sensors, electrochemical sensors (potentiometry, conductivity), amperometry, and optical sensors (Figure 2) [18]. These different types of chemical sensors have their own unique characteristics and applications [19].

## 3.1 Electrochemical sensors (ECS)

ECS comprises of conducting substances employed as reference, indicator and counter electrodes. The electrical potential, current, or impedance of a sensing element is used to measure concentration changes or molecular bonding mode changes that are caused by interaction with a target. The devices generate responses (signals) including resistance, capacitance, impedance, and conductance are used to study the kinetic and thermodynamic properties of the analyte [20]. These sensors measure the electrical properties (e.g. current, voltage, resistance) of a chemical reaction to sense and quantify the presence of specific chemical species. Examples include potentiometric sensors (e.g. pH sensors, ion-selective electrodes), amperometric sensors (e.g. oxygen sensors, glucose sensors), and conductometric sensors (e.g., gas sensors).

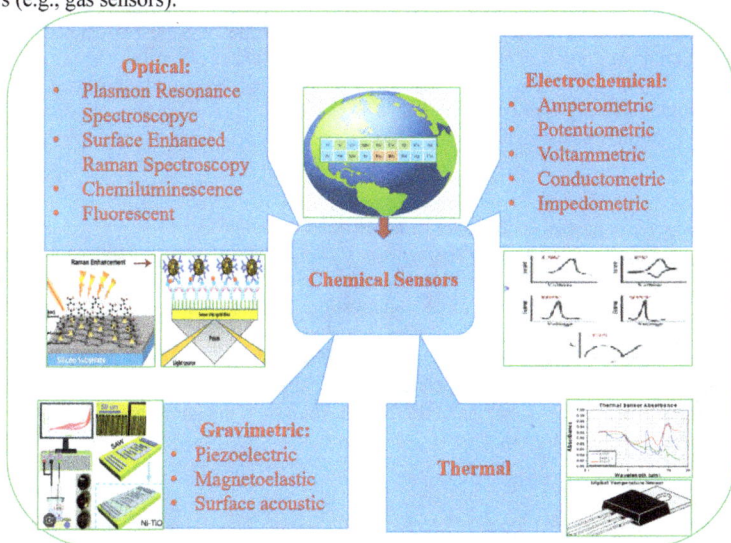

*Figure 2. Chemical sensors-based earth-abundant transition metals materials*

Moreover, in redox reactions, electron exchange between species with opposite charges generates a signal sensor. Numerous modified nanoparticle integrated with electrochemical detection using surface-functionalized electrodes have been employed in ECS [20, 21]. According to the signals generated, ECS can be grouped into conductometric, potentiometric, amperometric, voltammetric, and impedometric. The ECS encompasses the usage of modified electrodes aimed at the recognition of the targets and transformation of the response [22, 23].

## 3.2 Optical sensors

The optical sensors perform depending on the principle of the interaction of light with matter. Both organic and inorganic substances can be detected using this types of sensor based on the light absorption, emission, refraction etc. [24] These sensors use light-based techniques to sense and qualify or quantity the presence of chemical species. Examples include absorption spectroscopy, fluorescence-based sensors, and surface plasmon resonance sensors. Absorption-based sensors (e.g. infrared, ultraviolet), surface Plasmon resonance sensors, and fluorescence-based sensors (e.g., pH-sensitive dyes) are examples. Surface Plasmon Resonance Spectroscopy

(SPRS) works based on the changes in the refractive indexes when analytes are bound to the surface of the sensor [25]. Surface Enhanced Raman Spectroscopy (SERS) is grounded on change in frequencies from scattered radiation produced from light-matter interactions [26]. Fluorescence based sensors are other types of optical sensors that are extensively used in biomedicine [27], bioimaging [28], environmental detection [29], catalysis [30].

### 3.3 Thermal sensors

Thermal sensors measure the heat of reaction or phase change associated with the communication among the sensing material and the analyte chemical species. Examples include calorimetric sensors and thermometric devices. It is a device that stands designed to quantity hotness or coldness changes upon interaction with the analyte [18, 31]. Also, they are proficient in determining numerous thermal indicators like flow of heat, speeding up and angular swiftness [32].

Thermal sensors are mainly developed different applications ranging from industrial process monitoring to environmental control. For example, the thermal properties of the gas that can alter the heat transfer may be used to test the gas [33]. Various types of thermal sensors exist, such as silicon sensors, radiation thermometers, resistance thermometers, and thermocouples. Thermoelectric sensors has a widespread applications such as automotive, optical, thermal-cycling, telecom, power industries, biomedical, consumer factories [34]. Wearable producers are also of great interest because they can harvest human body heat to automated sensors [35]. Interestingly, the development nanotechnology has facilitated the development of thermal sensors that are more sensitive, stable, and efficient, leading to enhanced accuracy and precision in temperature measurement [34].

### 3.4 Gravimetric sensors

Gravimetric sensors are the type of sensors that works based on the mass change upon interaction with a target chemical species [36]. The most common examples are surface acoustic wave (SAW) sensors and quartz crystal microbalance (QCM) sensors [37, 38]. These two types of sensors are widely used in innumerable fields due to their practicality in measuring gases and liquids. The functioning QCM as a bulk audile sensor influences electromechanical resonance, which is very sensitive to changes in machine-driven properties like toughness, mass, and curbing. These changes are converted into electrical signals to detect differences in the motorized possessions. Nowadays, QCM is used in health diagnosis, air pollutants, water pollutants, microorganisms, moisture content, biomolecules, and food quality among others [39].

### 4. Chemical sensors based on earth abundant transition metals

### 4.1 Synthesis methods

The design and preparation of chemical sensors mainly depends on the preparation of sensing materials that are considered as heart of sensors. The preparation of sensing materials from earth-abundant transition metals can be achieved through two main approaches [40]. Bottom-up and top-down approaches are commonly utilized for synthesizing micro- and nano-sized materials. In top-down methods, these materials are produced by contravention of large materials, while in bottom-up techniques, chemical or physical forces at the atomic level assist in assembling basic units into larger structures [41]. This approach may include mechanical exfoliation, liquid aided exfoliation (physical and chemical) means, scotch-tape-aided machine-driven exfoliation methods [42-44].

Conversely, bottom-up approaches encompass techniques such as hydrothermal method, electrochemical and gas coatings, and metal-organic CVD (MOCVD). Among these, hydrothermal method stands out as a highly efficient and environmentally friendly method for producing nanomaterials [45-47]. Furthermore, inkjet printing of nanomaterials also reported. The merit of these method is overcoming the requirement for straight expensive lithography and minimize price [48]. Though, transition metals liquefied in an appropriate dissolving agent to prepare ink and excess solvent can hinder the final shows. This implies that cheap materials in high demand for the future [49].

## 4.2 Applications of chemical sensors

Chemical sensors works based on an immobilization of the target receptors into the indicator electrode surface in order to detect important targets [50]. Earth-abundant transition metals and their compounds offer a sustainable and cost-effective solution for developing chemical sensors with enhanced performance and reliability [9, 51]. This might be payable to abundance, inexpensive, and exceptional features for detecting analytes with high sensitivity, selectivity, and stability. The methodical enactment can be willingly adjusted through modifying the constructions, subdivision magnitude, apparent area, and electron transmission possessions of the transition metal based materials [6, 52]. Furthermore, integrating nanocomposite dendrimers provides a flexible framework for electrode surface modification, increasing necessary connections, and enhancing strength, thus boosting the general healthiness of sensors' recital [9, 51, 53]. The application of some earth-abundant transition metals-based materials in chemical sensors is presented in the next subsections.

### 4.2.1 Iron based chemical sensors

Iron nanomaterials possess great potential in the preparation of biochemical sensors due to their exclusive possessions like extraordinary external area, outstanding conduction of electron, and catalytic activity [54]. These properties make iron-based materials as an ideal candidate for detecting numerous analytes with high selectivity and sensitivity. Iron and its oxide-based materials such as iron oxides (hematite ($\alpha$-$Fe_2O_3$), magnetite ($Fe_3O_4$) and maghemite ($\gamma$-$Fe_2O_3$) nanoparticles), are widely used in the production of chemical devices owing to their excellent sensing performance, low cost, and abundance [55].

Iron oxide nanoparticles, nanowires, and thin films exhibit high sensitivity towards various gases, including hydrogen, carbon monoxide, nitrogen dioxide, and VOCs [50]. Chemical sensors based on magnetic nanoparticles are attracting attention because of their easy synthesis, good strength, less hazardous, and environmental-friendliness. Moreover, they have biocompatibility, manageable dimensions and form, ability to be coated or modified, possess larger surface area t, enhanced electron transfer kinetics between electrodes, and superior catalytic efficiencies [56]. Various Fe nanomaterials were engaged in the production of chemical devices with enhanced performances for the recognition of different analytes of interest (Table 1).

**Table 1.** *Selected application of iron-based nanomaterials in chemical sensors*

| Materials | Type | Target | LOD | Reference |
|---|---|---|---|---|
| α-Fe$_2$O$_3$ | Electrochemical | Phenyl hydrazine | 97 μM | [57] |
| Fe$_3$O$_4$ | | Aminoaldehydes | 750 nM | [58] |
| Fe$_3$O$_4$ | | Glucose | 0.9 μM | [59] |
| Fe$_3$O$_4$ | | Polyphenols | 81 nM | [60] |
| Fe$_3$O$_4$ Composites | | Chloramphenicol | 0.09 mM | [61] |
| Fe$_2$O$_3$/graphene | | Zn$^{2+}$, Cd$^{2+}$, Pb$^{2+}$ | 0.11 μg/ L, 0.08 μg/L, 0.07 μg/L | [62] |
| Fe$_3$O$_4$@-SiO$_2$-MIP | | Gram-negative bacterial quorum | 0.80 nM | [63] |
| Bi$_{0.9}$Er$_{0.1}$FeO$_3$ | Thermal | acetone | 100 ppm | [64] |
| Pd@α-Fe$_2$O$_3$ | | Acetone | 50 ppb | [65] |
| Pd@Fe$_3$O$_4$ | | H$_2$ | 250 ppb | [66] |
| Fe$_3$O$_4$NPs/chitosan | | Urea | 0.50 mg/dL | [67] |
| Ag–Fe$_3$O$_4$–graphene oxide | | Nitrite | 0.17 μM | [68] |

Surface active maghemite nanoparticles were used to cultivate chemical devices for the detection of glucose [59], phenolic compounds [58], serotonin [69], dopamine and uric acid [70], and aminoaldehydes [58]. Similarly, iron oxide nanoparticles were used for phenyl hydrazine [57], Erbium-doped Bismuth ferrite nanoparticles for Acetone [64], Fe$_3$O$_4$@-SiO$_2$-molecularly imprinted polymer (MIP) for gram-negative bacterial quorum [63], and magnetite Fe$_3$O$_4$ nanoparticles for chloramphenicol [61].

Silver-doped iron oxide (Ag–Fe$_2$O$_3$) load on to polyaniline (PANI) was coated on to the glassy carbon electrode (GCE) (Ag–Fe$_2$O$_3$@PANI/GCE) was used in the electrochemical sensor for the detection of uric acid (UA). This sensor can measure the UA in the linear dynamic range of 0.001–0.900 μM with the LOD of 102 pM showing the decent discrimination, precision, and reusability of the device [71]. Moreover, iron oxides were successfully employed in the preparation of gas sensors [65, 66]. Urea [67], nitrite [68], and metal ions (Zn$^{2+}$, Cd$^{2+}$ and Pb$^{2+}$) [62] sensors were developed by using iron-based nanomaterials and composites.

**4.2.2 Copper based chemical sensors**

Copper and its compounds, such as copper oxides (CuO, Cu$_2$O), have shown promising sensing properties for gases like carbon monoxide, ammonia, hydrogen, and volatile organic compounds. Such sensors are eye-catching due to their great indifference, rapid response, and low operating temperatures [72].

CuO is widely utilized in the development of gas sensors. Pd-decorated Cu$_2$O/CuO hydrogen sulfide (H$_2$S) sensors have been suggested, with the idea that Pd amendment can enhance the interaction among oxygen adsorbed and H$_2$S. This modification helps to prevent the formation of Cu$_2$S, thereby increasing the steadiness of the device's reaction [73]. Nevertheless, Pd-decorated Cu$_2$O/CuO sensors are by far better in performance than that made of CuO nanoplates [74].

Hydrogen peroxide ($H_2O_2$) is a powerful oxidant and lightening instrument commonly utilized in biomedicine, households, and industries. It also serves as a key reactive oxygen species (ROS) in numerous biological and neurotic progressions. $H_2O_2$ is associated with several anthropological ailments, including diabetes, circulatory disorders, neurodegenerative conditions, and Huntington's diseases, as well as digestive sicknesses. Therefore, accurate detection of $H_2O_2$ is critical for both engineering and educational research applications. Consequently, the advance of $H_2O_2$ sensors that are inexpensive, fast, delicate, and discriminatory is essential [75].

CuO nanoparticles were prepared for sensing $H_2O_2$ [76], and palladium-copper oxide nanoparticles for CuO nanoparticles decorated with carbon spheres for glucose [77, 78] sensing to achieve the LOD of 0.5 µM, 30 nM, 0.1 µM, respectively. Synchronized recognition of ascorbic acid, acetaminophen, and caffeine was achieved through the development of CuO-graphene nanocomposite based electrochemical sensors with the LOD of 0.011, 0.008, and 0.010 µmol $L^{-1}$, respectively [79]. In another works, $H_2O_2$ and cholesterol sensors were developed using hollow spheres made of highly porous copper oxide (CuO) [80] and using thionine and $Cu_2O$ nanomaterials [81].

Cu-coordinated molecularly imprinted polymer (Cu-MIP) grounded electrochemical sensor was invented for monitoring the overdose adverse effects dextromethorphan (DXM) drug that is widely utilized central antitussive and recreational agent. $Cu^{2+}$ frolicked a part in simplifying electron transmission and tying DXM so that good electrocatalytic properties and recognition will be achieved [82].

Nanocoposites such as copper-graphene oxide (Cu-GO) and copper nanoparticles CuNPs) were developed as dual SERS and colorimetric sensing devices. The work shown that colorimetric sensor (Cu-GO) exhibited elevated sensitivity towards amoxicillin within the linear range of 5 to 50 µM and LOD of 1.71 µM when compared to that based on CuNPs (2.17 µM) [83]. In another work, CuO nanoneedles based metal-organic framework synthesized to obtain electrochemiluminescence (ECL) sensing device for ultrasensitive recognition of catechol and luteolin by means of extensive undeviating ranges from 3 nM to 0.10 mM for catechol and 0.1 nM to 0.2 mM for luteolin, with corresponding LOD of 1.5 nM for catechol and 5.3 nM for luteolin [84]. Moreover, $Cu_2O$ NPs and polyalizarin yellow R (PYAR) on bare GCE were employed for the electrochemical determination of $H_2O_2$ by differential pulse voltammetry (DPV) with the linear range and LOD of 0.1-140 µM, and 0.03 µM [85]. Generally, different Cu based nanomaterials were prepared and integrated into various sensor types to determine different targets as shown in Table 2.

*Table 2.* Cu constructed nanomaterials based sensing devices for detecting various analytes

| Materials | Method | Target | LOD | Reference |
|---|---|---|---|---|
| CuO | Electrochemical | Glucose | 0.06 µM | [86] |
| | | | 0.3 µM | [68] |
| | | | 0.04 µM | [87] |
| CuO | | $H_2O_2$ | 0.5 µM | [76] |
| $Co_3O_4$/CuO | | Glucose | 0.38 µM | [88] |
| $Cu_2O$/Pd | | | 0.1 µM | [89] |
| rGO/Pd/CuO | | Glucose | 30 nM | [77] |
| CuO/Graphene | | Acetaminophen, ascorbic acid and caffeine | 0.008, 0.011 and 0.010 µM | [79] |
| CuONPs-CSs | | Glucose | 0.1 µM | [78] |
| CuO/g-$C_3N_4$ nanocomposites | | Dopamine | 0.10 nM | [90] |
| $Cu_2O$–TiNTs | | Eugenol | 1.3 µM | [91] |
| CuO NPs on Carbon ceramic electrode | | Tyrosine | 160 nM | [92] |
| $Cu^{2+}$Y/ZMCPE | | Paracetamol, Mefenamic acid | 0.1 µM | [93] |
| MWCNTs/ CTS–Cu | | Paracetamol | 0.024µM | [93] |
| Cu nanomaterial | Optical | Nitrite | - | [94] |
| | | Temperature | - | |

### 4.2.3   Nickel based chemical sensors

Nickel created materials look like to be the utmost widespread contestant for the production of enzyme free chemical sensors (e.g. glucose sensors) ensuing a redox pair of Ni (III)/Ni(II) in basic solution [95].

Materials derived from Ni, particularly NiO, are extensively investigated for gas sensing applications. NiO nanostructures exhibit excellent sensitivity towards different gases (e.g. ammonia, hydrogen, carbon monoxide, and volatile organic compounds. Nickel-based sensors provide benefits such as extraordinary stability, power efficiency, and compatibility with microfabrication techniques [96].

Pt@NiO NPs manufactured by sol–gel method [97] and Pt/NiO skinny print prepared by radio frequency sputtering [98] were used for $H_2$ and $NH_3$ sensing, respectively. Furthermore, ascorbic acid electrochemical sensor based on the $NiCo_2O_4$ nanostructures was reported [99]. The prepared enzyme free ascorbic acid sensing device achieved the LOD of 0.085 µM in working range of 5.0 µM–4.4 mM and specificity.

Durai and the co-workers [100] also informed diverse NiCoZnO hybrids to functionalize the GCE surface to detect dopamine in biological samples in the linear range from 1.0 nM to 0.5 µM with a LOD of 0.01 nM. In this work, NiO offered high catalytic vigorous locates with manifold oxidation numbers ($Ni^{2+}$/$Ni^{3+}$), and CoO and ZnO offers materials stability. Other

Materials Research Forum LLC
https://doi.org/21741/9781644903711-8

reliable sensing technology founded on variegated $Al-Mn_{0.65}-Cr_{1.76}$-oxide functional materials were reported for the same analyte by Alam and co-workers [101].

Carbohydrate antigen 125 (CA125) is important for the prompt identification and medical controlling of ovarian malignance. Thus, $CdS/Bi_2S_3/NiS$ ternary sulfide was integrated into electrochemical system to develop photoelectrochemical sensor for this analyte. The sensor was able to obtain 0.85 pg/mL detection limit in the linear range of 1 pg /mL$-$50 ng/mL. In this report, a modest, firm and sufficient profound method for CA125 checking was attained [102]. The nanocomposite made from Ni NPs, chitosan and nitrogen-doped carbon nanoshee (NiNP/NCN/CS) with three-dimensional hierarchical cylinder-like structure was employed for the GCE surface modification for the detection bisphenol A (BPA). DPV was used to detect BPA in the 0.1–2.5 µM and 2.5–15.0 µM ranges with two peaks. The LOD of 45 nM (S/N = 3) was calculated for this sensor [103]. In another work, Ni NPs supported nitrogen-doped carbon materials (NiNPs/N–C) was employed to prepare a sensors to measure vanillin down to 0.01 µM (S/N = 3) (Nie et al., 2020). Also, square wave anodic stripping voltammetry (SWASV) was applied in the detection of copper Cu(II) at trace using the sensors constructed through the application of NiO NPs to modify the carbon graphite electrode (GrE) [104]. Similarly, GCE was surface functionalized with NiCo-P@NiCo-LDH core-shell in order to sense isoprenaline analyte. The sensing layer showed extraordinary conductivity, small particle size, and extra interaction area amongst the materials and analyte, and abundant energetic sites for improving electrochemical response to give LOD (0.17 µM), widespread ranges (0.5–2110 µM), storage and operational stability, cheap, and lightness [105]. Nickel based nanomaterials have been widely castoff in the growth of different chemical sensors for the recognition of countless analytes in different matrices as shown in Table 3.

*Table 3. Ni nanomaterials based chemical sensors aimed at the recognition of several targets.*

| Materials | Method | Target | LR | LOD | Reference |
|---|---|---|---|---|---|
| $ERG/Ni_2O_3$–NiO | Electrochemical | | 0.04-100µM | 0.02µM | [106] |
| Ni/Cu MOF | | Glucose | 0.51 µM | 1 µM–20 mM | [107] |
| NiO/PANINS | | | 0.06 µM | 1-3000 µM | [108] |
| β-nickel sulfide (β-NiS) nanobelts | | | 1–45 µM | 0.173 µM | [109] |
| NiO | | | 0.5–20 µM | 0.00 | [110] |
| NiO | | | 0.5nM-9 mM | 7 nM | [111] |
| $NiMn_2O_4$ | | | 1–1000; 1000 –3500 µM | - | [112] |
| NiO@Au | | Lactic acid | 11.6 µM | 100.0 µM–0.5 M | [113] |
| NiS/S-g-C3N4 nanoparticles | | Glucose | 0.001–2.1 mM | 1.5 µM | [114] |
| NC-NiS@NS-NiS nanoparticles | | | 0.02–5.0 | 0.0083 µM | [115] |
| β-NiS nanoparticles | | | 0.005–0.06 mM | 0.052 µM | [116] |

#### 4.2.4 Zinc based chemical sensors

Zinc nanomaterials have shown great potential in the manufacturing of biochemical sensors due to their exceptional characteristics such as large surface area, tremendous catalytic activity, and good electrical conductivity. These properties make Zn nanomaterials ideal for sensing several injurious gases and VOCs with high selectivity and sensitivity [117].

Materials derived oxides of Zn show various interesting and versatile properties and wide applications. Its wide range of applications is primarily attributed to its tall definite surface space, less hazardous, electrical and optical activities, good material stability, excellent electron communication possessions, non-toxicity, and simplicity of production [118]. ZnO is widely used in chemical sensors in arrears to its incomparable characteristics, together with extraordinary sensitivity, huge surface area, and adjustable morphology. Chemical sensors based on ZnO materials showed superior performance for detecting gases such as VOCs, nitrogen dioxide, carbon monoxide, and hydrogen [119].

ZnO is well-thought-out as a appropriate material for the provision of various chemical sensors for the reason that its features such as large isoelectric point (IEP) and broad band gap (3.37eV) [120] (Table 4). As a result, ZnO has been used to develop sensing technology for various targets chemicals and cancer cells. The detection of naproxen using square wave voltammetry sensor was reported for the sensors produced through the application of ZnO NPs and multiwalled carbon nanotubes to modify carbon paste electrode [121]. The sensor was able to measure the target in the concentration series of 1.0 µM to 0.20 mM and the LOD of 0.23 µM. Besides, Ag–ZnO graphene oxide were reported for the recognition of *E. coli* bacteria [122], and ZnO coated carbon electrode to detect *para*-nitrophenol [123]. After a year, 3-D ZnO sensors were developed to detect glucose with the LOD of 0.02 mM [124]. Interestingly, ZnO base biosensors developed based on carbon cloth showed excellent performance with low LOD in the concentration series of 0.5 nM to 5 µM, finely tuned compassion, and improved storage stability [125].

On the other hand, Enzyme-free ZnO/CuO nanoleaves oxides-based sensors were established for concurrent sensing of acetylcholine and ascorbic acid [126]. The sensor presented an innovative technology for the coincident recognition of acetylcholine and ascorbic acid, which is crucial for advancing biological analysis, clinical diagnostics, and healthcare. The proposed sensor confirmed a low LOD and broad linear range. LOD and linear ranges are 14.7 nM and for acetylcholine as well as 12.0 pM for ascorbic acid, respectively.

Nanorods [127] and nanowires [128] are important one dimensional ZNO materials to prepare improved gas sensors established on Au-decorated ZnO [129]. For example, ethanol sensors were developed based on the Au-functionalized ZnO nanorods [130, 131]. In these works, it was claimed that Au nanoparticles were used as sensitizers to improve selectivity and response time. In addition to this, Au/ZnO sensors are also used to $C_2H_2$ at 3 ppb [132] and ethanol [133] successfully.

*Table 4.* Zn materials based chemical sensors used for detection of various analytes

| Materials | Method | Target | LR | LOD | Reference |
|---|---|---|---|---|---|
| AC-ZnO | Electroche mical | Acetaminophen | 0.05-1380 µM | 0.83 µM | [134] |
| ZnO-rGO | | Urea | 0.02 - 7.2 µM | 0.012 µM | [135] |
| ZnO nanorods-rGO | | Dopamine | 0.01– 6 µM and 6 – 80 µM | 3.6 nM | [136] |
| ZnO quantum dots | | Uric acid | 1-10 mM | 22.97 µM | [137] |
| ZnO nanoparticles | | | 50–500 µM | 1.65 µM | [138] |
| ZnO-Ag$_2$O-Co$_3$O4 nanoparticles | | | 0.1 nM–0.01 mM | 89.14±4.46 pM | [139] |
| Co doped ZnO nanoparticles | | H$_2$O$_2$ | 5 – 30 mM | 14.3 µM | [140] |
| Cu doped ZnO nanoparticles | | Myoglobin | 3 - 15 nM | 0.46 nM | [141] |
| ZnO-rGO | | Phenol | 2 – 15 µM 15 – 40 µM | 1.94 µM | [142] |
| ZnO nanowires | | Hydrazine | 500 - 1200 nM | 84.7 nM | [143] |
| ZnO-MoS$_2$ | | Trinitrotoluene | $1 \times 10-4$ - 1 ppm | $5.2 \times 10-5$ ppm | [144] |

## 4.2.5 Manganese based chemical sensors

Manganese and its oxides, such as manganese dioxide (MnO$_2$), are favorable materials aimed at gas sensing applications since the have interesting properties required for sensor fabrication. Sensors based on MnO$_2$ demonstrate outstanding performance in detecting a range of gases, including hydrogen, carbon monoxide, ammonia, and VOCs [145].

The use of MnO$_2$ in sensor applications has attracted significant interest owing to its distinctive properties and its potent for integration with biological systems. Manganese oxides has various oxidation states that enable them to support oxidation-reduction reactions and process that require facilitated electron transmission which are essential for sensing functions [146]. This unique characteristic prompted the investigation of MnO$_2$-based biosensors in multiple fields. MnO$_2$ demonstrates exceptional catalytic property, making it a strong aspirant for electrochemical sensing technology. Its natural capability to facilitate electron transferal at the interface has enabled the creation of delicate and effective biosensing platforms [145]. The adjustable electrocatalytic activities of MnO$_2$, along with its suitability with natural sensing elements, offer significant potential for detecting a extensive range of analytes [147]. Additionally, MnO$_2$ nanostructures are important to improve biomolecule adherence onto the electrode surface and enhance signal intensification. These materials are incorporated into chemical sensors to realize greater sensitivity and better LODs [148, 149].

MnO$_2$ has shown its capacity to enable straight electron transferal at the interface (between electrode and enzymes), eliminating the necessity for supplementary redox intermediaries. This simplifies sensor strategy and enhances steadiness. The use MnO$_2$ surfaces for the immobilization of enzymes on the electrode surface maintains the bioactivity of sensing layer, ensuring steadfast and precise sensor fabrication. $MnO_2$-based biosensors have diverse applications, including the recognition of several analytes such as heavy metals, $H_2O_2$, ecological pollutants and glucose [149].

*Table 5. Different $MnO_2$ based nanomaterial chemical sensors.*

| Materials | Method | Target | LR | LOD | Reference |
|---|---|---|---|---|---|
| $MnO_2$ | Electroche mical | Salmonella | $3.0 \times 101$ -3.0 $\times 106$ | 19 CFU mL$^{-1}$ | [150] |
| $MnO_2$@GNR composites | | Glucose | 0.1-1.4 mM | 0.05 mM | [151] |
| M13-E4 @$MnO_2$ | | | 5 µM-2 mM | 1.8 µM | [152] |
| $MnO_2$/Polythiophene composite | | Dopamine | 0.04-9.0 µM | 41 nM | [153] |
| $MnO_2$/Ta Electrode | | $H_2O_2$ | 1-2 µM | 0.06 µM | [154] |
| $MnO_2$/Ag@C shell nanocomposites | | | 0.5 µM–5.7 mM | 0.17 µM | [155] |
| Au/$MnO_2$ | | Histamine | 0.3-5.1 µM | 0.08 µM | [156] |
| Lucigenin/$MnO_2$ | | Glutathione | 10-2000 nM | 3.7µM | [157] |
| MWCNT-$MnO_2$/rGO | | Acetylcholi ne | 0.1-100 µM | 0.1 µM | [158] |
| $MnO_2$/3D-RGO @Ni foam | | Ractopamin e | 17-962 nM | 1.6 nM | [159] |
| $MnO_2$ nanofibrous-mesh | | Ascorbic Acid | 0.20-10 mM | 1.33 µM | [160] |

MnO, $MnO_2$, and $Mn_3O_4$ are three distinct manganese oxides that have been extensively studied and effectively used as electrode materials in various biosensors. These oxides are not hazardous, environmentally friendly, cheerfully accessible, and simple to synthesize. Their high energy density and activity in alkaline environments make them ideal materials for developing biosensors for various analytes [161, 162]. Moreover, manganese oxides can exist in four diverse sizes, ranging from zero dimensions (0-D) to three dimensions (3-D), similar to zinc oxides. Compared to 0-D and 1-D nanostructures, 3-D nanoparticles offer a greater surface area both externally and internally providing more reaction sites. The fabricated Al–$Mn_{0.65}$–$Cr_{1.76}$-oxide nanomaterial sensing devices showed broad working range from 0.1 nM to 0.01 mM, low LOD (~96.8 pM), enhanced sensitivity (~55.8 µA µM$^{-1}$cm$^{-2}$), better analytical figures of merit against the measurement of dopamine.

$MnO_2$ single bond nanoparticles modified GCE was applied in the manufacturing of sensing device to detect olanzapine (OLZ) in the working series of 10 µM to 0.23 µM and obtained LOD is 2.27 nM [163]. In conclusion, various chemical sensors based on Mn constituents are described by many investigators and summarized in Table 5.

### 4.2.6 Cobalt based chemical sensors

Cobalt Oxide ($Co_3O_4$) has garnered attention for gas sensors preparation, demonstrating notable strength and exceptional enactment in the detection of gases like $H_2$, CO, NO, $NH_3$, and VOCs [164]. The decoration of a single Pd atom in the $Pd^{4+}$ state at $Co^{3+}$ positions on the surface of

Materials Research Forum LLC
https://doi.org/21741/9781644903711-8

$Co_3O_4$ nanoparticles (NPs) contributes electrons to the $Co_3O_4$ empty orbitals. This increases the concentration of free electrons, enhancing the likelihood of oxygen adsorbing these electrons to form ion-adsorbed oxygen. This method of Pd single-atom catalysis successfully decreases the response time and increases the sensitivity of the sensor. Additionally, Pd nanoparticles-supported $Co_3O_4$ porous materials, prepared by means of metal-organic framework templates, were used for acetone sensing [165]. Here the Pd was applied to advance the catalytic properties of the sensing layers towards acetone at various concentrations in the existence of gases such as $H_2S$, $NH_3$, ethanol, pentane, and toluene as potential interferents. This work demonstrated the capability of the sensors towards exhaled acetone analyzers.

$Co_3O_4$ is a cubic spinel structure p-type semiconductor metal oxide (SMO), whose surface could adsorb more oxygen [27]. In another work, the SMOs-based Au NPs adsorbed $Co_3O_4$ NPs sensors were employed for $NO_2$ detection. The prepared SMOs-based Au/$Co_3O_4$-nanoparticles devices showed the maximum performance (33%) to 100 ppb $NO_2$ and extra sensitivity to $NO_2$ when compared to other gases (CO, $NH_3$, $SO_2$, $CO_2$) at 136 °C [166]. Such technologies are proper for wearable devices and are favorably for detecting networks in the internet. So far, different chemical sensors were established built on the application of Co derived materials as depicted in Table 6.

*Table 6. Co materials based chemical sensors used for the detection of different targets.*

| Materials | Method | Target | LR | LOD | Reference |
|---|---|---|---|---|---|
| $Co_3O_4$/N doped CNT | Electroche mical | Glucose | $5–1.1 \times 10^4\,\mu M$ | $1.0\ \mu M$ | [167] |
| $Co_3O_4$ | | | $10^{-4} \times 10^3\ \mu M$ | $4.4\ \mu M$ | [168] |
| $NiCo_2O_4$/N doped graphene | | | $1–5^{10}\ \mu M$ | $0.14\ \mu M$ | [169] |
| $SiO_2$/C/$Co_3O_4$ | | Dopamine | $10–240\ \mu M$ | $0.02\ \mu M$ | [170] |
| Co/ $SiO_2$/C | | | $38.9–460\ \mu M$ | $0.58\ \mu M$ | [170] |
| Co/N-doped graphene | | Uric acid | $0.4–41,950\ \mu M$ | $33.30\ nM$ | [171] |
| $Co_3O_4$ | | | $5–3000\ \mu M$ | $2.4\ \mu M$ | [172] |
| Cobalt-based metal-organic framework | | Glucose | $5\ \mu M$ to $900\ \mu M$ | $1.6\ \mu M$ | [173] |
| Co hydroxide nanosheets | | | $200\ nM$-$1.94\ mM$ | $76.5\ nM$ | [174] |
| Cobalt(II) sulfide | | | $8\ \mu M$-$1.5\ mM$ | $5\ \mu M$ | [175] |
| Co(DBrTPA)(NMP)]n (CoMOF-1) | | 3-Nitrotyrosi ne | - | $23.6\ ng/mL$ | [176] |
| $Co_3O_4$/$TiO_2$ | | $H_2O_2$ | - | $0.8\ M$ | [177] |

**Conclusion and future perspectives**

Earth abundant transition metals in general, and first row-based materials in particular are attracting a huge interest from medical, food industry, and environmental fields. Compared to precious metals, first row transition metals in the periodic table of elements are earth abundant elements and become an alternative to the widely used metals which are rare in nature and expensive to obtain. These earth-abundant transition metals and their compounds offer a sustainable and cost-effective solution for developing chemical sensors with enhanced performance and reliability. So far, electrochemical, optical, thermal and acoustic wave sensors are prepared successfully for environmental and food quality monitoring, health, and industrial applications. The quality of the sensors output directly depends on the materials selection, composition and engineering to address the current demand from various sectors. As such, future works need to continue to explore novel synthesis methods, device architectures, and surface functionalization techniques to advance the recognition capabilities of these materials and expand their applications in environmental monitoring, industrial safety, healthcare, and homeland security.

## References

[1] A. Hulanicki, S. Glab, F. Ingman, Chemical sensors: definitions and classification, Pure and applied chemistry, 63 (1991) 1247-1250. https://doi.org/10.1351/pac199163091247

[2] L. Thobakgale, S. Ombinda-Lemboumba, P. Mthunzi-Kufa, Chemical sensor nanotechnology in pharmaceutical drug research, Nanomaterials, 12 (2022) 2688. https://doi.org/10.3390/nano12152688

[3] C. Karunakaran, R. Rajkumar, K. Bhargava, Introduction to biosensors, Biosensors and bioelectronics, Elsevier2015, pp. 1-68. https://doi.org/10.1016/B978-0-12-803100-1.00001-3

[4] R. Abdel-Karim, Y. Reda, A. Abdel-Fattah, Nanostructured materials-based nanosensors, Journal of The Electrochemical Society, 167 (2020) 037554. https://doi.org/10.1149/1945-7111/ab67aa

[5] E. Fazio, S. Spadaro, C. Corsaro, G. Neri, S.G. Leonardi, F. Neri, N. Lavanya, C. Sekar, N. Donato, G. Neri, Metal-oxide based nanomaterials: Synthesis, characterization and their applications in electrical and electrochemical sensors, Sensors, 21 (2021) 2494. https://doi.org/10.3390/s21072494

[6] G. Maduraiveeran, M. Sasidharan, W. Jin, Earth-abundant transition metal and metal oxide nanomaterials: Synthesis and electrochemical applications, Progress in Materials Science, 106 (2019) 100574. https://doi.org/10.1016/j.pmatsci.2019.100574

[7] K.M.P. Wheelhouse, R.L. Webster, G.L. Beutner, Advances and applications in catalysis with earth-abundant metals, ACS Publications, 2023, pp. 1677-1679. https://doi.org/10.1021/acs.organomet.3c00292

[8] B. Su, Z.-C. Cao, Z.-J. Shi, Exploration of earth-abundant transition metals (Fe, Co, and Ni) as catalysts in unreactive chemical bond activations, Accounts of chemical research, 48 (2015) 886-896. https://doi.org/10.1021/ar500345f

[9] S. Malik, J. Singh, R. Goyat, Y. Saharan, V. Chaudhry, A. Umar, A.A. Ibrahim, S. Akbar, S. Ameen, S. Baskoutas, Nanomaterials-based biosensor and their applications: A review, Heliyon, (2023). https://doi.org/10.1016/j.heliyon.2023.e19929

[10] L.p. Cui, Y.w. Wang, K. Yu, Y.j. Ma, B.b. Zhou, A Multi-Active Site Subnano Heterostructures Catalyst Grown In situ POM and Fe0. 2Ni0. 8Co2O4 onto Nickel Foam Toward Efficient Electrocatalytic Overall Water Splitting, Advanced Functional Materials, 2408968.

[11] P. Chirik, R. Morris, Getting down to earth: the renaissance of catalysis with abundant metals, ACS Publications, 2015, pp. 2495-2495. https://doi.org/10.1021/acs.accounts.5b00385

[12] N. Kaplaneris, L. Ackermann, Earth-abundant 3d transition metals on the rise in catalysis, Beilstein-Institut, 2022, pp. 86-88. https://doi.org/10.3762/bjoc.18.8

[13] M. Albrecht, R. Bedford, B. Plietker, Catalytic and organometallic chemistry of earth-abundant metals, ACS Publications, 2014, pp. 5619-5621. https://doi.org/10.1021/om5010379

[14] M.S. Krishna, S. Singh, M. Batool, H.M. Fahmy, K. Seku, A.E. Shalan, S. Lanceros-Mendez, M.N. Zafar, A review on 2D-ZnO nanostructure based biosensors: from materials to devices, Materials Advances, 4 (2023) 320-354. https://doi.org/10.1039/D2MA00878E

[15] P. Kannan, G. Maduraiveeran, Metal oxides nanomaterials and nanocomposite-based electrochemical sensors for healthcare applications, Biosensors, 13 (2023) 542. https://doi.org/10.3390/bios13050542

[16] A. Huang, Y. He, Y. Zhou, Y. Zhou, Y. Yang, J. Zhang, L. Luo, Q. Mao, D. Hou, J. Yang, A review of recent applications of porous metals and metal oxide in energy storage, sensing and catalysis, Journal of Materials Science, 54 (2019) 949-973. https://doi.org/10.1007/s10853-018-2961-5

[17] T.M. Swager, K.A. Mirica, Introduction: chemical sensors, ACS Publications, 2019, pp. 1-2. https://doi.org/10.1021/acs.chemrev.8b00764

[18] H. Rostamzad, 14 - Active and intelligent biodegradable films and polymers, in: S. Mavinkere Rangappa, J. Parameswaranpillai, S. Siengchin, M. Ramesh (Eds.) Biodegradable Polymers, Blends and Composites, Woodhead Publishing2022, pp. 415-430. https://doi.org/10.1016/B978-0-12-823791-5.00023-5

[19] R. Wills, J. Farhi, P. Czabala, S. Shahin, J.M. Spangle, M. Raj, Chemical sensors for imaging total cellular aliphatic aldehydes in live cells, Chemical Science, 14 (2023) 8305-8314. https://doi.org/10.1039/D3SC02025H

[20] V. Naresh, N. Lee, A review on biosensors and recent development of nanostructured materials-enabled biosensors, Sensors, 21 (2021) 1109. https://doi.org/10.3390/s21041109

[21] D. Grieshaber, R. MacKenzie, J. Vörös, E. Reimhult, Electrochemical biosensors-sensor principles and architectures, Sensors, 8 (2008) 1400-1458. https://doi.org/10.3390/s80314000

[22] H. Vovusha, R.G. Amorim, H. Bae, S. Lee, T. Hussain, H. Lee, Sensing of sulfur containing toxic gases with double transition metal carbide MXenes, Materials Today Chemistry, 30 (2023) 101543. https://doi.org/10.1016/j.mtchem.2023.101543

[23] H. Louis, M. Patrick, I.O. Amodu, I. Benjamin, I.J. Ikot, G.E. Iniama, A.S. Adeyinka, Sensor behavior of transition-metals (X= Ag, Au, Pd, and Pt) doped Zn11-XO12 nanostructured materials for the detection of serotonin, Materials Today Communications, 34 (2023) 105048. https://doi.org/10.1016/j.mtcomm.2022.105048

[24] S. Firdous, S. Anwar, R. Rafya, Development of surface plasmon resonance (SPR) biosensors for use in the diagnostics of malignant and infectious diseases, Laser physics letters, 15 (2018) 065602. https://doi.org/10.1088/1612-202X/aab43f

[25] C.M. Miyazaki, F.M. Shimizu, M. Ferreira, Surface plasmon resonance (SPR) for sensors and biosensors, Nanocharacterization techniques, Elsevier2017, pp. 183-200. https://doi.org/10.1016/B978-0-323-49778-7.00006-0

[26] J.I.S.d.S.d. Jesus, R. Löbenberg, N.A. Bou-Chacra, Raman spectroscopy for quantitative analysis in the pharmaceutical industry, Journal of Pharmacy and Pharmaceutical Sciences, 23 (2020) 24-46. https://doi.org/10.18433/jpps30649

[27] D. Zhang, J. Hu, X.-y. Yang, Y. Wu, W. Su, C.-y. Zhang, Target-initiated synthesis of fluorescent copper nanoparticles for the sensitive and label-free detection of bleomycin, Nanoscale, 10 (2018) 11134-11142. https://doi.org/10.1039/C8NR02780C

[28] A.S. Satyvaldiev, Z.K. Zhasnakunov, E. Omurzak, T.D. Doolotkeldieva, S.T. Bobusheva, G.T. Orozmatova, Z. Kelgenbaeva, Copper nanoparticles: synthesis and biological activity, IOP Publishing, pp. 012075. https://doi.org/10.1088/1757-899X/302/1/012075

[29] S. Li, T. Wei, M. Tang, F. Chai, F. Qu, C. Wang, Facile synthesis of bimetallic Ag-Cu nanoparticles for colorimetric detection of mercury ion and catalysis, Sensors and Actuators B: Chemical, 255 (2018) 1471-1481. https://doi.org/10.1016/j.snb.2017.08.159

[30] N. Wang, L. Ga, J. Ai, Synthesis of novel fluorescent copper nanomaterials and their application in detection of iodide ions and catalysis, Analytical Methods, 11 (2019) 44-48. https://doi.org/10.1039/C8AY01871E

[31] M.B. Kulkarni, S. Rajagopal, B. Prieto-Simón, B.W. Pogue, Recent advances in smart wearable sensors for continuous human health monitoring, Talanta, 272 (2024) 125817. https://doi.org/10.1016/j.talanta.2024.125817

[32] X. Zuo, X. Zhang, L. Qu, J. Miao, Smart fibers and textiles for personal thermal management in emerging wearable applications, Advanced Materials Technologies, 8 (2023) 2201137. https://doi.org/10.1002/admt.202201137

[33] E.L.W. Gardner, J.W. Gardner, F. Udrea, Micromachined thermal gas sensors-A review, Sensors, 23 (2023) 681. https://doi.org/10.3390/s23020681

[34] F.J. Tovar-Lopez, Recent progress in micro-and nanotechnology-enabled sensors for biomedical and environmental challenges, Sensors, 23 (2023) 5406. https://doi.org/10.3390/s23125406

[35] A. Kumar, G. Gupta, K. Bapna, D.D. Shivagan, Semiconductor-metal-oxide-based nano-composites for humidity sensing applications, Materials Research Bulletin, 158 (2023) 112053. https://doi.org/10.1016/j.materresbull.2022.112053

[36] S. Lama, B.-G. Bae, S. Ramesh, Y.-J. Lee, N. Kim, J.-H. Kim, Nano-Sheet-like Morphology of Nitrogen-Doped Graphene-Oxide-Grafted Manganese Oxide and Polypyrrole Composite for Chemical Warfare Agent Simulant Detection, Nanomaterials, 2022. https://doi.org/10.3390/nano12172965

[37] J. Kim, E. Kim, J. Kim, J.-H. Kim, S. Ha, C. Song, W.J. Jang, J. Yun, Four-Channel Monitoring System with Surface Acoustic Wave Sensors for Detection of Chemical Warfare Agents, Journal of Nanoscience and Nanotechnology, 20 (2020) 7151-7157. https://doi.org/10.1166/jnn.2020.18851

[38] Y.-J. Lee, J.-G. Kim, J.-H. Kim, J. Yun, W.J. Jang, Detection of Dimethyl Methylphosphonate (DMMP) Using Polyhedral Oligomeric Silsesquioxane (POSS), Journal of Nanoscience and Nanotechnology, 18 (2018) 6565-6569. https://doi.org/10.1166/jnn.2018.15698

[39] V. Tramonti, C. Lofrumento, M.R. Martina, G. Lucchesi, G. Caminati, Graphene Oxide/Silver Nanoparticles Platforms for the Detection and Discrimination of Native and Fibrillar Lysozyme: A Combined QCM and SERS Approach, Nanomaterials, 2022. https://doi.org/10.3390/nano12040600

[40] I. Amorim, F. Bento, Electrochemical Sensors Based on Transition Metal Materials for Phenolic Compound Detection, Sensors, 2024. https://doi.org/10.3390/s24030756

[41] A.K. Mia, M. Meyyappan, P.K. Giri, Two-Dimensional Transition Metal Dichalcogenide Based Biosensors: From Fundamentals to Healthcare Applications, Biosensors, 2023. https://doi.org/10.3390/bios13020169

[42] P. Budania, P.T. Baine, J.H. Montgomery, D.W. McNeill, S.J. Neil Mitchell, M. Modreanu, P.K. Hurley, Comparison between Scotch tape and gel-assisted mechanical exfoliation techniques for preparation of 2D transition metal dichalcogenide flakes, Micro & Nano Letters, 12 (2017) 970-973. https://doi.org/10.1049/mnl.2017.0280

[43] H. Li, G. Lu, Y. Wang, Z. Yin, C. Cong, Q. He, L. Wang, F. Ding, T. Yu, H. Zhang, Mechanical Exfoliation and Characterization of Single- and Few-Layer Nanosheets of WSe2, TaS2, and TaSe2, Small, 9 (2013) 1974-1981. https://doi.org/10.1002/smll.201202919

[44] S. Thomas, V.K. Greenacre, D.E. Smith, Y.J. Noori, N.M. Abdelazim, A.L. Hector, C.H. de Groot, W. Levason, P.N. Bartlett, G. Reid, Tungsten disulfide thin films via electrodeposition from a single source precursor, Chemical Communications, 57 (2021) 10194-10197. https://doi.org/10.1039/D1CC03297F

[45] D.Ö. Özgür, G. Özkan, O. Atakol, H. Çelikkan, Facile Ion-Exchange Method for Zn Intercalated MoS2 As an Efficient and Stable Catalyst toward Hydrogen Evaluation Reaction, ACS Applied Energy Materials, 4 (2021) 2398-2407. https://doi.org/10.1021/acsaem.0c02899

[46] B.A. Sperling, B. Kalanyan, J.E. Maslar, Atomic Layer Deposition of Al2O3 Using Trimethylaluminum and H2O: The Kinetics of the H2O Half-Cycle, The Journal of Physical Chemistry C, 124 (2020) 3410-3420. https://doi.org/10.1021/acs.jpcc.9b11291

[47] B. Yorulmaz, A. Özden, H. Şar, F. Ay, C. Sevik, N.K. Perkgöz, CVD growth of monolayer WS2 through controlled seed formation and vapor density, Materials Science in Semiconductor Processing, 93 (2019) 158-163. https://doi.org/10.1016/j.mssp.2018.12.035

[48] J. Li, M.M. Naiini, S. Vaziri, M.C. Lemme, M. Östling, Inkjet Printing of MoS2, Advanced Functional Materials, 24 (2014) 6524-6531. https://doi.org/10.1002/adfm.201400984

[49] Y. Yoon, P.L. Truong, D. Lee, S.H. Ko, Metal-Oxide Nanomaterials Synthesis and Applications in Flexible and Wearable Sensors, ACS Nanoscience Au, 2 (2022) 64-92. https://doi.org/10.1021/acsnanoscienceau.1c00029

[50] M. Adampourezare, M. Hasanzadeh, M.-A. Hoseinpourefeizi, F. Seidi, Iron/iron oxide-based magneto-electrochemical sensors/biosensors for ensuring food safety: recent progress

and challenges in environmental protection, RSC Advances, 13 (2023) 12760-12780. https://doi.org/10.1039/D2RA07415J

[51] R.M. Bullock, J.G. Chen, L. Gagliardi, P.J. Chirik, O.K. Farha, C.H. Hendon, C.W. Jones, J.A. Keith, J. Klosin, S.D. Minteer, R.H. Morris, A.T. Radosevich, T.B. Rauchfuss, N.A. Strotman, A. Vojvodic, T.R. Ward, J.Y. Yang, Y. Surendranath, Using nature's blueprint to expand catalysis with Earth-abundant metals, Science, 369 (2020) eabc3183. https://doi.org/10.1126/science.abc3183

[52] A. Bhardwaj, I.-h. Kim, J.-w. Hong, A. Kumar, S.-J. Song, Transition metal oxide (Ni, Co, Fe)-tin oxide nanocomposite sensing electrodes for a mixed-potential based NO2 sensor, Sensors and Actuators B: Chemical, 284 (2019) 534-544. https://doi.org/10.1016/j.snb.2019.01.003

[53] D.J.L. Golding, N. Carter, D. Robinson, A.J. Fitzpatrick, Crystallisation-Induced Emission Enhancement in Zn(II) Schiff Base Complexes with a Tuneable Emission Colour, Sustainability, 2020. https://doi.org/10.3390/su12229599

[54] M.A. El-Shal, S.M. Azab, H.A.M. Hendawy, A facile nano-iron oxide sensor for the electrochemical detection of the anti-diabetic drug linagliptin in the presence of glucose and metformin, Bulletin of the National Research Centre, 43 (2019) 95. https://doi.org/10.1186/s42269-019-0132-8

[55] M.M. Vinay, Y. Arthoba Nayaka, Iron oxide (Fe2O3) nanoparticles modified carbon paste electrode as an advanced material for electrochemical investigation of paracetamol and dopamine, Journal of Science: Advanced Materials and Devices, 4 (2019) 442-450. https://doi.org/10.1016/j.jsamd.2019.07.006

[56] M. Hasanzadeh, N. Shadjou, M. de la Guardia, Iron and iron-oxide magnetic nanoparticles as signal-amplification elements in electrochemical biosensing, TrAC Trends in Analytical Chemistry, 72 (2015) 1-9. https://doi.org/10.1016/j.trac.2015.03.016

[57] S.W. Hwang, A. Umar, G.N. Dar, S.H. Kim, R.I. Badran, Synthesis and Characterization of Iron Oxide Nanoparticles for Phenyl Hydrazine Sensor Applications, Sensor Letters, 12 (2014) 97-101. https://doi.org/10.1166/sl.2014.3224

[58] M. Magro, D. Baratella, G. Miotto, J. Frömmel, M. Šebela, M. Kopečná, E. Agostinelli, F. Vianello, Enzyme self-assembly on naked iron oxide nanoparticles for aminoaldehyde biosensing, Amino Acids, 51 (2019) 679-690. https://doi.org/10.1007/s00726-019-02704-7

[59] D. Baratella, M. Magro, G. Sinigaglia, R. Zboril, G. Salviulo, F. Vianello, A glucose biosensor based on surface active maghemite nanoparticles, Biosensors and Bioelectronics, 45 (2013) 13-18. https://doi.org/10.1016/j.bios.2013.01.043

[60] M. Magro, D. Baratella, V. Colò, F. Vallese, C. Nicoletto, S. Santagata, P. Sambo, S. Molinari, G. Salviulo, A. Venerando, C.R. Basso, V.A. Pedrosa, F. Vianello, Electrocatalytic nanostructured ferric tannate as platform for enzyme conjugation: Electrochemical determination of phenolic compounds, Bioelectrochemistry, 132 (2020) 107418. https://doi.org/10.1016/j.bioelechem.2019.107418

[61] K. Giribabu, S.-C. Jang, Y. Haldorai, M. Rethinasabapathy, S.Y. Oh, A. Rengaraj, Y.-K. Han, W.-S. Cho, C. Roh, Y.S. Huh, Electrochemical determination of chloramphenicol using

a glassy carbon electrode modified with dendrite-like $Fe_3O_4$ nanoparticles, Carbon letters, 23 (2017) 38-47.

[62] S. Lee, J. Oh, D. Kim, Y. Piao, A sensitive electrochemical sensor using an iron oxide/graphene composite for the simultaneous detection of heavy metal ions, Talanta, 160 (2016) 528-536. https://doi.org/10.1016/j.talanta.2016.07.034

[63] H. Jiang, D. Jiang, J. Shao, X. Sun, Magnetic molecularly imprinted polymer nanoparticles based electrochemical sensor for the measurement of Gram-negative bacterial quorum signaling molecules (N-acyl-homoserine-lactones), Biosensors and Bioelectronics, 75 (2016) 411-419. https://doi.org/10.1016/j.bios.2015.07.045

[64] X. Liu, J. Li, L. Guo, G. Wang, Highly Sensitive Acetone Gas Sensors Based on Erbium-Doped Bismuth Ferrite Nanoparticles, Nanomaterials, 2022. https://doi.org/10.3390/nano12203679

[65] H. Yang, Q. Lei, Z. Zhao, Y. Sun, P. Li, S. Zhuiykov, J. Hu, Electrospinning Encapsulation of Pd Nanoparticles into α-Fe2O3 Nanofibers Windows Enhanced Acetone Sensing, IEEE Sensors Journal, 21 (2021) 15944-15951. https://doi.org/10.1109/JSEN.2021.3076216

[66] B. Sharma, J.-S. Sung, A.A. Kadam, J.-h. Myung, Adjustable n-p-n gas sensor response of Fe3O4-HNTs doped Pd nanocomposites for hydrogen sensors, Applied Surface Science, 530 (2020) 147272. https://doi.org/10.1016/j.apsusc.2020.147272

[67] A. Kaushik, P.R. Solanki, A.A. Ansari, G. Sumana, S. Ahmad, B.D. Malhotra, Iron oxide-chitosan nanobiocomposite for urea sensor, Sensors and Actuators B: Chemical, 138 (2009) 572-580. https://doi.org/10.1016/j.snb.2009.02.005

[68] Z. Li, Y. Chen, Y. Xin, Z. Zhang, Sensitive electrochemical nonenzymatic glucose sensing based on anodized CuO nanowires on three-dimensional porous copper foam, Scientific Reports, 5 (2015) 16115. https://doi.org/10.1038/srep16115

[69] G. Ran, X. Chen, Y. Xia, Electrochemical detection of serotonin based on a poly(bromocresol green) film and Fe3O4 nanoparticles in a chitosan matrix, RSC Advances, 7 (2017) 1847-1851. https://doi.org/10.1039/C6RA25639B

[70] Z. Cai, Y. Ye, X. Wan, J. Liu, S. Yang, Y. Xia, G. Li, Q. He, Morphology-Dependent Electrochemical Sensing Properties of Iron Oxide-Graphene Oxide Nanohybrids for Dopamine and Uric Acid, Nanomaterials, 2019. https://doi.org/10.3390/nano9060835

[71] S.K. Ponnaiah, P. Periakaruppan, B. Vellaichamy, New Electrochemical Sensor Based on a Silver-Doped Iron Oxide Nanocomposite Coupled with Polyaniline and Its Sensing Application for Picomolar-Level Detection of Uric Acid in Human Blood and Urine Samples, The Journal of Physical Chemistry B, 122 (2018) 3037-3046. https://doi.org/10.1021/acs.jpcb.7b11504

[72] G. Ashraf, M. Asif, A. Aziz, Z. Wang, X. Qiu, Q. Huang, F. Xiao, H. Liu, Nanocomposites consisting of copper and copper oxide incorporated into MoS4 nanostructures for sensitive voltammetric determination of bisphenol A, Microchimica Acta, 186 (2019) 337. https://doi.org/10.1007/s00604-019-3406-9

[73] K. Mikami, Y. Kido, Y. Akaishi, A. Quitain, T. Kida, Synthesis of Cu2O/CuO Nanocrystals and Their Application to H2S Sensing, Sensors, 2019. https://doi.org/10.3390/s19010211

[74] H.T. Nha, P. Van Tong, N. Van Duy, C.M. Hung, N.D. Hoa, Facile Synthesis of Pd-CuO Nanoplates with Enhanced SO2 and H2 Gas-Sensing Characteristics, Journal of Electronic Materials, 50 (2021) 2767-2778. https://doi.org/10.1007/s11664-021-08799-7

[75] S. Malik, J. Singh, R. Goyat, Y. Saharan, V. Chaudhry, A. Umar, A.A. Ibrahim, S. Akbar, S. Ameen, S. Baskoutas, Nanomaterials-based biosensor and their applications: A review, Heliyon, 9 (2023) e19929. https://doi.org/10.1016/j.heliyon.2023.e19929

[76] J. Ping, S. Ru, K. Fan, J. Wu, Y. Ying, Copper oxide nanoparticles and ionic liquid modified carbon electrode for the non-enzymatic electrochemical sensing of hydrogen peroxide, Microchimica Acta, 171 (2010) 117-123. https://doi.org/10.1007/s00604-010-0420-3

[77] K. Dhara, R. Thiagarajan, B.G. Nair, G.S.B. Thekkedath, Highly sensitive and wide-range nonenzymatic disposable glucose sensor based on a screen printed carbon electrode modified with reduced graphene oxide and Pd-CuO nanoparticles, Microchimica Acta, 182 (2015) 2183-2192. https://doi.org/10.1007/s00604-015-1549-x

[78] J. Zhang, J. Ma, S. Zhang, W. Wang, Z. Chen, A highly sensitive nonenzymatic glucose sensor based on CuO nanoparticles decorated carbon spheres, Sensors and Actuators B: Chemical, 211 (2015) 385-391. https://doi.org/10.1016/j.snb.2015.01.100

[79] Z.M. Khoshhesab, Simultaneous electrochemical determination of acetaminophen, caffeine and ascorbic acid using a new electrochemical sensor based on CuO-graphene nanocomposite, RSC Advances, 5 (2015) 95140-95148. https://doi.org/10.1039/C5RA14138A

[80] D. Cheng, J. Qin, Y. Feng, J. Wei, Synthesis of Mesoporous CuO Hollow Sphere Nanozyme for Paper-Based Hydrogen Peroxide Sensor, Biosensors, 2021. https://doi.org/10.3390/bios11080258

[81] Q. Yan, R. Wu, H. Chen, H. Wang, W. Nan, Highly sensitive cholesterol biosensor based on electron mediator thionine and cubic-shaped Cu2O nanomaterials, Microchemical Journal, 185 (2023) 108201. https://doi.org/10.1016/j.microc.2022.108201

[82] J. Yuan, S. Wang, S. Cheng, Y. Liu, F. Zhao, B. Zeng, A novel electrochemical sensor based on a Cu-coordinated molecularly imprinted polymer and MoS2 modified chitin-derived carbon for selective detection of dextromethorphan, Analytical Methods, 16 (2024) 3278-3286. https://doi.org/10.1039/D4AY00549J

[83] N.T. Anh, N.X. Dinh, H. Van Tuan, M.Q. Doan, N.H. Anh, N.T. Khi, V.T. Trang, D.Q. Tri, A.-T. Le, Eco-friendly copper nanomaterials-based dual-mode optical nanosensors for ultrasensitive trace determination of amoxicillin antibiotics residue in tap water samples, Materials Research Bulletin, 147 (2022) 111649. https://doi.org/10.1016/j.materresbull.2021.111649

[84] B. Wang, W. Zhao, L. Wang, K. Kang, X. Li, D. Zhang, J. Ren, X. Ji, Binary-amplifying electrochemiluminescence sensor for sensitive assay of catechol and luteolin based on HKUST-1 derived CuO nanoneedles as a novel luminophore, Talanta, 273 (2024) 125836. https://doi.org/10.1016/j.talanta.2024.125836

[85] N. Amini, B. Rashidzadeh, N. Amanollahi, A. Maleki, J.-K. Yang, S.-M. Lee, Application of an electrochemical sensor using copper oxide nanoparticles/polyalizarin yellow R nanocomposite for hydrogen peroxide, Environmental Science and Pollution Research, 28 (2021) 38809-38816. https://doi.org/10.1007/s11356-021-13299-6

[86] A. Rahim, Z.U. Rehman, S. Mir, N. Muhammad, F. Rehman, M.H. Nawaz, M. Yaqub, S.A. Siddiqi, A.A. Chaudhry, A non-enzymatic glucose sensor based on CuO-nanostructure modified carbon ceramic electrode, Journal of Molecular Liquids, 248 (2017) 425-431. https://doi.org/10.1016/j.molliq.2017.10.087

[87] S. Ayaz, S. Karakaya, G. Emir, D.G. Dilgin, Y. Dilgin, A novel enzyme-free FI-amperometric glucose biosensor at Cu nanoparticles modified graphite pencil electrode, Microchemical Journal, 154 (2020) 104586. https://doi.org/10.1016/j.microc.2019.104586

[88] S. Cheng, S. DelaCruz, C. Chen, Z. Tang, T. Shi, C. Carraro, R. Maboudian, Hierarchical Co3O4/CuO nanorod array supported on carbon cloth for highly sensitive non-enzymatic glucose biosensing, Sensors and Actuators B: Chemical, 298 (2019) 126860. https://doi.org/10.1016/j.snb.2019.126860

[89] L. Tang, K. Huan, D. Deng, L. Han, Z. Zeng, L. Luo, Glucose sensor based on Pd nanosheets deposited on Cu/Cu2O nanocomposites by galvanic replacement, Colloids and Surfaces B: Biointerfaces, 188 (2020) 110797. https://doi.org/10.1016/j.colsurfb.2020.110797

[90] J. Zou, S. Wu, Y. Liu, Y. Sun, Y. Cao, J.-P. Hsu, A.T. Shen Wee, J. Jiang, An ultra-sensitive electrochemical sensor based on 2D g-C3N4/CuO nanocomposites for dopamine detection, Carbon, 130 (2018) 652-663. https://doi.org/10.1016/j.carbon.2018.01.008

[91] S.-Z. Kang, H. Liu, X. Li, M. Sun, J. Mu, Electrochemical behavior of eugenol on TiO2 nanotubes improved with Cu2O clusters, RSC Advances, 4 (2014) 538-543. https://doi.org/10.1039/C3RA44895A

[92] H. Razmi, H. Nasiri, R. Mohammad-Rezaei, Amperometric determination of L-tyrosine by an enzymeless sensor based on a carbon ceramic electrode modified with copper oxide nanoparticles, Microchimica Acta, 173 (2011) 59-64. https://doi.org/10.1007/s00604-010-0527-6

[93] A. Babaei, B. Khalilzadeh, M. Afrasiabi, A new sensor for the simultaneous determination of paracetamol and mefenamic acid in a pharmaceutical preparation and biological samples using copper(II) doped zeolite modified carbon paste electrode, Journal of Applied Electrochemistry, 40 (2010) 1537-1543. https://doi.org/10.1007/s10800-010-0131-9

[94] N. Wang, L. Ga, J. Ai, Y. Wang, Fluorescent Copper Nanomaterials for Sensing NO2 and Temperature, Frontiers in Chemistry, 9 (2022) 805205. https://doi.org/10.3389/fchem.2021.805205

[95] X. Niu, X. Li, J. Pan, Y. He, F. Qiu, Y. Yan, Recent advances in non-enzymatic electrochemical glucose sensors based on non-precious transition metal materials: opportunities and challenges, RSC Advances, 6 (2016) 84893-84905. https://doi.org/10.1039/C6RA12506A

[96] J. Fu, C. Zhao, J. Zhang, Y. Peng, E. Xie, Enhanced Gas Sensing Performance of Electrospun Pt-Functionalized NiO Nanotubes with Chemical and Electronic Sensitization,

ACS Applied Materials & Interfaces, 5 (2013) 7410-7416.
https://doi.org/10.1021/am4017347

[97] C.-H. Wu, Z. Zhu, H.-M. Chang, Z.-X. Jiang, C.-Y. Hsieh, R.-J. Wu, Pt@NiO core-shell
nanostructure for a hydrogen gas sensor, Journal of Alloys and Compounds, 814 (2020)
151815. https://doi.org/10.1016/j.jallcom.2019.151815

[98] H.-I. Chen, C.-Y. Hsiao, W.-C. Chen, C.-H. Chang, T.-C. Chou, I.P. Liu, K.-W. Lin, W.-C.
Liu, Characteristics of a Pt/NiO thin film-based ammonia gas sensor, Sensors and Actuators
B: Chemical, 256 (2018) 962-967. https://doi.org/10.1016/j.snb.2017.10.032

[99] S.I.B. T, G. Karuppasamy, Assessing the electrochemical sensing behavior of manganese
based inverse spinel towards ascorbic acid detection, Materials Today Communications, 33
(2022) 104607. https://doi.org/10.1016/j.mtcomm.2022.104607

[100] L. Durai, A. Gopalakrishnan, S. Badhulika, One-pot hydrothermal synthesis of NiCoZn a
ternary mixed metal oxide nanorod based electrochemical sensor for trace level recognition of
dopamine in biofluids, Materials Letters, 298 (2021) 130044.
https://doi.org/10.1016/j.matlet.2021.130044

[101] M.M. Alam, A.M. Asiri, M.M. Rahman, M.A. Islam, Fabrication of dopamine sensor
based on ternary AlMn0.645Cr1.76O7.47 nanoparticles, Materials Chemistry and Physics,
244 (2020) 122740. https://doi.org/10.1016/j.matchemphys.2020.122740

[102] S. Wang, J. Yuan, C. Wang, T. Wang, F. Zhao, B. Zeng, CdS/Bi2S3/NiS ternary
heterostructure-based photoelectrochemical immunosensor for the sensitive detection of
carbohydrate antigen 125, Analytica Chimica Acta, 1312 (2024) 342765.
https://doi.org/10.1016/j.aca.2024.342765

[103] Y. Wang, C. Yin, Q. Zhuang, An electrochemical sensor modified with nickel
nanoparticle/nitrogen-doped carbon nanosheet nanocomposite for bisphenol A detection,
Journal of Alloys and Compounds, 827 (2020) 154335.
https://doi.org/10.1016/j.jallcom.2020.154335

[104] S. Moussaoui, F. Smaili, S.E. Berrabah, A. Manseri, Fabrication of novel electrochemical
sensor based on NiO-nanoparticles for copper detection in drinking water, Inorganic
Chemistry Communications, 158 (2023) 111563.
https://doi.org/10.1016/j.inoche.2023.111563

[105] S. Farokhi, M. Roushani, H. Hosseini, Advanced core-shell nanostructures based on
porous NiCo-P nanodiscs shelled with NiCo-LDH nanosheets as a high-performance
electrochemical sensing platform, Electrochimica Acta, 362 (2020) 137218.
https://doi.org/10.1016/j.electacta.2020.137218

[106] G.-T. Liu, H.-F. Chen, G.-M. Lin, P.-p. Ye, X.-P. Wang, Y.-Z. Jiao, X.-Y. Guo, Y. Wen,
H.-F. Yang, One-step electrodeposition of graphene loaded nickel oxides nanoparticles for
acetaminophen detection, Biosensors and Bioelectronics, 56 (2014) 26-32.
https://doi.org/10.1016/j.bios.2014.01.005

[107] B. Wang, Y. Luo, L. Gao, B. Liu, G. Duan, High-performance field-effect transistor
glucose biosensors based on bimetallic Ni/Cu metal-organic frameworks, Biosensors and
Bioelectronics, 171 (2021) 112736. https://doi.org/10.1016/j.bios.2020.112736

[108] S. Kailasa, B.G. Rani, M.S. Bhargava Reddy, N. Jayarambabu, P. Munindra, S. Sharma, K. Venkateswara Rao, NiO nanoparticles -decorated conductive polyaniline nanosheets for amperometric glucose biosensor, Materials Chemistry and Physics, 242 (2020) 122524. https://doi.org/10.1016/j.matchemphys.2019.122524

[109] H. Lin, C. Peng, J. Shi, B. Zheng, H. Lee, P. Wu, M. Lee, The Slight Adjustment in the Weight of Sulfur Sheets to Synthesize $\beta$-NiS Nanobelts for Maintaining Detection of Lower Concentrations of Glucose through a Long-Term Storage Test, Nanomaterials, 2023. https://doi.org/10.3390/nano13162371

[110] L. Zhang, N. Wang, P. Cao, M. Lin, L. Xu, H. Ma, Electrochemical non-enzymatic glucose sensor using ionic liquid incorporated cobalt-based metal-organic framework, Microchemical Journal, 159 (2020) 105343. https://doi.org/10.1016/j.microc.2020.105343

[111] N. Singer, R.G. Pillai, A.I.D. Johnson, K.D. Harris, A.B. Jemere, Nanostructured nickel oxide electrodes for non-enzymatic electrochemical glucose sensing, Microchimica Acta, 187 (2020) 196. https://doi.org/10.1007/s00604-020-4171-5

[112] M. Dong, H. Hu, S. Ding, C. Wang, L. Li, Fabrication of NiMn2O4 nanosheets on reduced graphene oxide for non-enzymatic detection of glucose, Materials Technology, 36 (2021) 203-211. https://doi.org/10.1080/10667857.2020.1740861

[113] G. Maduraiveeran, A. Chen, Design of an enzyme-mimicking NiO@Au nanocomposite for the sensitive electrochemical detection of lactic acid in human serum and urine, Electrochimica Acta, 368 (2021) 137612. https://doi.org/10.1016/j.electacta.2020.137612

[114] S. Vinoth, P.M. Rajaitha, A. Venkadesh, K.S. Shalini Devi, S. Radhakrishnan, A. Pandikumar, Nickel sulfide-incorporated sulfur-doped graphitic carbon nitride nanohybrid interface for non-enzymatic electrochemical sensing of glucose, Nanoscale Advances, 2 (2020) 4242-4250. https://doi.org/10.1039/D0NA00172D

[115] M. Arivazhagan, Y. Manova Santhosh, G. Maduraiveeran, Non-Enzymatic Glucose Detection Based on NiS Nanoclusters@NiS Nanosphere in Human Serum and Urine, Micromachines, 2021. https://doi.org/10.3390/mi12040403

[116] A. Singh, A. Sharma, A. Ahmed, A.K. Sundramoorthy, H. Furukawa, S. Arya, A. Khosla, Recent Advances in Electrochemical Biosensors: Applications, Challenges, and Future Scope, Biosensors, 2021. https://doi.org/10.3390/bios11090336

[117] K.G. Krishna, G. Umadevi, S. Parne, N. Pothukanuri, Zinc oxide based gas sensors and their derivatives: a critical review, Journal of Materials Chemistry C, 11 (2023) 3906-3925. https://doi.org/10.1039/D2TC04690C

[118] S. Chaudhary, A. Umar, K.K. Bhasin, S. Baskoutas, Chemical Sensing Applications of ZnO Nanomaterials, Materials, 2018. https://doi.org/10.3390/ma11020287

[119] A. Ulyankina, I. Leontyev, M. Avramenko, D. Zhigunov, N. Smirnova, Large-scale synthesis of ZnO nanostructures by pulse electrochemical method and their photocatalytic properties, Materials Science in Semiconductor Processing, 76 (2018) 7-13. https://doi.org/10.1016/j.mssp.2017.12.011

[120] P.-X. Li, A.-Y. Yang, L. Xin, B. Xue, C.-H. Yin, Photocatalytic Activity and Mechanism of Cu2+ Doped ZnO Nanomaterials, Science of Advanced Materials, 14 (2022) 1599-1604. https://doi.org/10.1166/sam.2022.4363

[121] J. Tashkhourian, B. Hemmateenejad, H. Beigizadeh, M. Hosseini-Sarvari, Z. Razmi, ZnO nanoparticles and multiwalled carbon nanotubes modified carbon paste electrode for determination of naproxen using electrochemical techniques, Journal of Electroanalytical Chemistry, 714-715 (2014) 103-108. https://doi.org/10.1016/j.jelechem.2013.12.026

[122] E. Roy, S. Patra, A. Tiwari, R. Madhuri, P.K. Sharma, RETRACTED: Single cell imprinting on the surface of Ag-ZnO bimetallic nanoparticle modified graphene oxide sheets for targeted detection, removal and photothermal killing of E. Coli, Biosensors and Bioelectronics, 89 (2017) 620-626. https://doi.org/10.1016/j.bios.2015.12.085

[123] R.M. Bashami, A. Hameed, M. Aslam, I.M.I. Ismail, M.T. Soomro, The suitability of ZnO film-coated glassy carbon electrode for the sensitive detection of 4-nitrophenol in aqueous medium, Analytical Methods, 7 (2015) 1794-1801. https://doi.org/10.1039/C4AY02857K

[124] L. Fang, B. Liu, L. Liu, Y. Li, K. Huang, Q. Zhang, Direct electrochemistry of glucose oxidase immobilized on Au nanoparticles-functionalized 3D hierarchically ZnO nanostructures and its application to bioelectrochemical glucose sensor, Sensors and Actuators B: Chemical, 222 (2016) 1096-1102. https://doi.org/10.1016/j.snb.2015.08.032

[125] A. Fallatah, N. Kuperus, M. Almomtan, S. Padalkar, Sensitive Biosensor Based on Shape-Controlled ZnO Nanostructures Grown on Flexible Porous Substrate for Pesticide Detection, Sensors, 2022. https://doi.org/10.3390/s22093522

[126] M. Musarraf Hussain, A.M. Asiri, M.M. Rahman, Non-enzymatic simultaneous detection of acetylcholine and ascorbic acid using ZnO·CuO nanoleaves: Real sample analysis, Microchemical Journal, 159 (2020) 105534. https://doi.org/10.1016/j.microc.2020.105534

[127] C. Chen, Q. Zhang, G. Xie, M. Yao, H. Pan, H. Du, H. Tai, X. Du, Y. Su, Enhancing visible light-activated NO2 sensing properties of Au NPs decorated ZnO nanorods by localized surface plasmon resonance and oxygen vacancies, Materials Research Express, 7 (2020) 015924. https://doi.org/10.1088/2053-1591/ab6b64

[128] A. Kaiser, E. Torres Ceja, Y. Liu, F. Huber, R. Müller, U. Herr, K. Thonke, H2S sensing for breath analysis with Au functionalized ZnO nanowires, Nanotechnology, 32 (2021) 205505. https://doi.org/10.1088/1361-6528/abe004

[129] G. Korotcenkov, Current Trends in Nanomaterials for Metal Oxide-Based Conductometric Gas Sensors: Advantages and Limitations. Part 1: 1D and 2D Nanostructures, Nanomaterials, 2020. https://doi.org/10.3390/nano10071392

[130] J. Guo, J. Zhang, M. Zhu, D. Ju, H. Xu, B. Cao, High-performance gas sensor based on ZnO nanowires functionalized by Au nanoparticles, Sensors and Actuators B: Chemical, 199 (2014) 339-345. https://doi.org/10.1016/j.snb.2014.04.010

[131] J. Li, Y. Yang, Q. Wang, X. Cheng, Y. Luo, B. An, J. Bai, Y. Wang, E. Xie, Design of size-controlled Au nanoparticles loaded on the surface of ZnO for ethanol detection, CrystEngComm, 23 (2021) 783-792. https://doi.org/10.1039/D0CE01318H

[132] J. Miao, J.Y.S. Lin, Nanometer-Thick Films of Aligned ZnO Nanowires Sensitized with Au Nanoparticles for Few-ppb-Level Acetylene Detection, ACS Applied Nano Materials, 3 (2020) 9174-9184. https://doi.org/10.1021/acsanm.0c01807

[133] N.M. Vuong, L.H. Than, T.H. Phan, H.N. Hieu, N. Van Nghia, N. Tu, Ultra Responsive and Highly Selective Ethanol Gas Sensor Based on Au Nanoparticles Embedded ZnO Hierarchical Structures, Journal of The Electrochemical Society, 168 (2021) 027503. https://doi.org/10.1149/1945-7111/abdde3

[134] M. Sivakumar, M. Sakthivel, S.-M. Chen, Activated Carbon -ZnO Nanocomposite for Electrochemical Sensing of Acetaminophen, International Journal of Electrochemical Science, 11 (2016) 8363-8373. https://doi.org/10.20964/2016.10.51

[135] K.B. Babitha, P.S. Soorya, A. Peer Mohamed, R.B. Rakhi, S. Ananthakumar, Development of ZnO@rGO nanocomposites for the enzyme free electrochemical detection of urea and glucose, Materials Advances, 1 (2020) 1939-1951. https://doi.org/10.1039/D0MA00445F

[136] F. Li, B. Ni, Y. Zheng, Y. Huang, G. Li, A simple and efficient voltammetric sensor for dopamine determination based on ZnO nanorods/electro-reduced graphene oxide composite, Surfaces and Interfaces, 26 (2021) 101375. https://doi.org/10.1016/j.surfin.2021.101375

[137] M. Ali, I. Shah, S.W. Kim, M. Sajid, J.H. Lim, K.H. Choi, Quantitative detection of uric acid through ZnO quantum dots based highly sensitive electrochemical biosensor, Sensors and Actuators A: Physical, 283 (2018) 282-290. https://doi.org/10.1016/j.sna.2018.10.009

[138] B. Ramya, P.G. Priya, Rapid phytochemical microwave-assisted synthesis of zinc oxide nano flakes with excellent electrocatalytic activity for non-enzymatic electrochemical sensing of uric acid, Journal of Materials Science: Materials in Electronics, 32 (2021) 21406-21424. https://doi.org/10.1007/s10854-021-06644-5

[139] M.M. Alam, A.M. Asiri, M.T. Uddin, M.A. Islam, M.R. Awual, M.M. Rahman, Detection of uric acid based on doped ZnO/Ag2O/Co3O4 nanoparticle loaded glassy carbon electrode, New Journal of Chemistry, 43 (2019) 8651-8659. https://doi.org/10.1039/C9NJ01287G

[140] S.B. Khan, M.M. Rahman, A.M. Asiri, S.A.B. Asif, S.A.S. Al-Qarni, A.G. Al-Sehemi, S.A. Al-Sayari, M.S. Al-Assiri, Fabrication of non-enzymatic sensor using Co doped ZnO nanoparticles as a marker of H2O2, Physica E: Low-dimensional Systems and Nanostructures, 62 (2014) 21-27. https://doi.org/10.1016/j.physe.2014.04.007

[141] M. Haque, H. Fouad, H.K. Seo, O.Y. Alothman, Z.A. Ansari, Cu-Doped ZnO Nanoparticles as an Electrochemical Sensing Electrode for Cardiac Biomarker Myoglobin Detection, IEEE Sensors Journal, 20 (2020) 8820-8832. https://doi.org/10.1109/JSEN.2020.2982713

[142] R. Sha, S.K. Puttapati, V.V.S.S. Srikanth, S. Badhulika, Ultra-sensitive phenol sensor based on overcoming surface fouling of reduced graphene oxide-zinc oxide composite electrode, Journal of Electroanalytical Chemistry, 785 (2017) 26-32. https://doi.org/10.1016/j.jelechem.2016.12.001

[143] A. Umar, M.M. Rahman, Y.-B. Hahn, Ultra-sensitive hydrazine chemical sensor based on high-aspect-ratio ZnO nanowires, Talanta, 77 (2009) 1376-1380. https://doi.org/10.1016/j.talanta.2008.09.020

[144] T. Yang, Y. Cui, M. Chen, R. Yu, S. Luo, W. Li, K. Jiao, Uniform and Vertically Oriented ZnO Nanosheets Based on Thin-Layered MoS2: Synthesis and High-Sensing Ability, ACS Sustainable Chemistry & Engineering, 5 (2017) 1332-1338. https://doi.org/10.1021/acssuschemeng.6b01699

[145] W. Yang, Y. Peng, Y. Wang, Y. Wang, H. Liu, Z.a. Su, W. Yang, J. Chen, W. Si, J. Li, Controllable redox-induced in-situ growth of MnO2 over Mn2O3 for toluene oxidation: Active heterostructure interfaces, Applied Catalysis B: Environmental, 278 (2020) 119279. https://doi.org/10.1016/j.apcatb.2020.119279

[146] N. Sohal, B. Maity, N.P. Shetti, S. Basu, Biosensors Based on MnO2 Nanostructures: A Review, ACS Applied Nano Materials, 4 (2021) 2285-2302. https://doi.org/10.1021/acsanm.0c03380

[147] S. Zhou, Q. Wang, J. Chen, Y. Shen, L. Liu, C. Wang, Preparation and Optimization of MnO2 Nanoparticles, Science of Advanced Materials, 14 (2022) 927-933. https://doi.org/10.1166/sam.2022.4255

[148] C. Battilocchio, J.M. Hawkins, S.V. Ley, Mild and Selective Heterogeneous Catalytic Hydration of Nitriles to Amides by Flowing through Manganese Dioxide, Organic Letters, 16 (2014) 1060-1063. https://doi.org/10.1021/ol403591c

[149] W. Xiao, D. Wang, X.W. Lou, Shape-Controlled Synthesis of MnO2 Nanostructures with Enhanced Electrocatalytic Activity for Oxygen Reduction, The Journal of Physical Chemistry C, 114 (2010) 1694-1700. https://doi.org/10.1021/jp909386d

[150] L. Xue, R. Guo, F. Huang, W. Qi, Y. Liu, G. Cai, J. Lin, An impedance biosensor based on magnetic nanobead net and MnO2 nanoflowers for rapid and sensitive detection of foodborne bacteria, Biosensors and Bioelectronics, 173 (2021) 112800. https://doi.org/10.1016/j.bios.2020.112800

[151] V. Vukojević, S. Djurdjić, M. Ognjanović, M. Fabián, A. Samphao, K. Kalcher, D.M. Stanković, Enzymatic glucose biosensor based on manganese dioxide nanoparticles decorated on graphene nanoribbons, Journal of Electroanalytical Chemistry, 823 (2018) 610-616. https://doi.org/10.1016/j.jelechem.2018.07.013

[152] L. Han, C. Shao, B. Liang, A. Liu, Genetically Engineered Phage-Templated MnO2 Nanowires: Synthesis and Their Application in Electrochemical Glucose Biosensor Operated at Neutral pH Condition, ACS Applied Materials & Interfaces, 8 (2016) 13768-13776. https://doi.org/10.1021/acsami.6b03266

[153] Y. Shoja, A.A. Rafati, J. Ghodsi, Polythiophene supported MnO2 nanoparticles as nano-stabilizer for simultaneously electrostatically immobilization of d-amino acid oxidase and hemoglobin as efficient bio-nanocomposite in fabrication of dopamine bi-enzyme biosensor, Materials Science and Engineering: C, 76 (2017) 637-645. https://doi.org/10.1016/j.msec.2017.03.155

[154] K. Vijayalakshmi, A. Renitta, K. Alagusundaram, A. Monamary, Novel two-step process for the fabrication of MnO2 nanostructures on tantalum for enhanced electrochemical H2O2 detection, Materials Chemistry and Physics, 214 (2018) 431-439. https://doi.org/10.1016/j.matchemphys.2018.04.108

[155] S. Zhang, J. Zheng, Synthesis of single-crystal α-MnO2 nanotubes-loaded Ag@C core-shell matrix and their application for electrochemical sensing of nonenzymatic hydrogen peroxide, Talanta, 159 (2016) 231-237. https://doi.org/10.1016/j.talanta.2016.06.014

[156] S. Knežević, M. Ognjanović, N. Nedić, J.F.M.L. Mariano, Z. Milanović, B. Petković, B. Antić, S.V. Djurić, D. Stanković, A single drop histamine sensor based on AuNPs/MnO2 modified screen-printed electrode, Microchemical Journal, 155 (2020) 104778. https://doi.org/10.1016/j.microc.2020.104778

[157] W. Gao, Z. Liu, L. Qi, J. Lai, S.A. Kitte, G. Xu, Ultrasensitive Glutathione Detection Based on Lucigenin Cathodic Electrochemiluminescence in the Presence of MnO2 Nanosheets, Analytical Chemistry, 88 (2016) 7654-7659. https://doi.org/10.1021/acs.analchem.6b01491

[158] N. Chauhan, S. Balayan, U. Jain, Sensitive biosensing of neurotransmitter: 2D material wrapped nanotubes and MnO2 composites for the detection of acetylcholine, Synthetic Metals, 263 (2020) 116354. https://doi.org/10.1016/j.synthmet.2020.116354

[159] M.Y. Wang, W. Zhu, L. Ma, J.J. Ma, D.E. Zhang, Z.W. Tong, J. Chen, Enhanced simultaneous detection of ractopamine and salbutamol - Via electrochemical-facial deposition of MnO2 nanoflowers onto 3D RGO/Ni foam templates, Biosensors and Bioelectronics, 78 (2016) 259-266. https://doi.org/10.1016/j.bios.2015.11.062

[160] B. Tehseen, A. Rehman, M. Rahmat, H.N. Bhatti, A. Wu, F.K. Butt, G. Naz, W.S. Khan, S.Z. Bajwa, Solution growth of 3D MnO2 mesh comprising 1D nanofibres as a novel sensor for selective and sensitive detection of biomolecules, Biosensors and Bioelectronics, 117 (2018) 852-859. https://doi.org/10.1016/j.bios.2018.06.061

[161] J. Zhang, W. Chu, J. Jiang, X.S. Zhao, Synthesis, characterization and capacitive performance of hydrous manganese dioxide nanostructures, Nanotechnology, 22 (2011) 125703. https://doi.org/10.1088/0957-4484/22/12/125703

[162] S. Devaraj, N. Munichandraiah, Effect of Crystallographic Structure of MnO2 on Its Electrochemical Capacitance Properties, The Journal of Physical Chemistry C, 112 (2008) 4406-4417. https://doi.org/10.1021/jp7108785

[163] J.I. Gowda, R.M. Hanabaratti, P.D. Pol, R.C. Sheth, P.P. Joshi, S.T. Nandibewoor, Manganese oxide nanoparticles modified electrode for electrosensing of antipsychotic drug olanzapine, Chemical Data Collections, 38 (2022) 100824. https://doi.org/10.1016/j.cdc.2021.100824

[164] G. Yuan, Y. Zhong, Y. Chen, Q. Zhuo, X. Sun, Highly sensitive and fast-response ethanol sensing of porous Co3O4 hollow polyhedra via palladium reined spillover effect, RSC Advances, 12 (2022) 6725-6731. https://doi.org/10.1039/D1RA09352E

[165] W.-T. Koo, S. Yu, S.-J. Choi, J.-S. Jang, J.Y. Cheong, I.-D. Kim, Nanoscale PdO Catalyst Functionalized Co3O4 Hollow Nanocages Using MOF Templates for Selective Detection of

Acetone Molecules in Exhaled Breath, ACS Applied Materials & Interfaces, 9 (2017) 8201-82 https://doi.org/10.1021/acsami.7b01284

[166] T.-J. Hsueh, S.-S. Wu, Highly sensitive Co3O4 nanoparticles/MEMS NO2 gas sensor with the adsorption of the Au nanoparticles, Sensors and Actuators B: Chemical, 329 (2021) 129201. https://doi.org/10.1016/j.snb.2020.129201

[167] Y. Qin, Y. Sun, Y. Li, C. Li, L. Wang, S. Guo, MOF derived Co3O4/N-doped carbon nanotubes hybrids as efficient catalysts for sensitive detection of H2O2 and glucose, Chinese Chemical Letters, 31 (2020) 774-778. https://doi.org/10.1016/j.cclet.2019.09.016

[168] J. Mu, L. Zhang, M. Zhao, Y. Wang, Co3O4 nanoparticles as an efficient catalase mimic: Properties, mechanism and its electrocatalytic sensing application for hydrogen peroxide, Journal of Molecular Catalysis A: Chemical, 378 (2013) 30-37. https://doi.org/10.1016/j.molcata.2013.05.016

[169] Z. Lu, L. Wu, J. Zhang, W. Dai, G. Mo, J. Ye, Bifunctional and highly sensitive electrochemical non-enzymatic glucose and hydrogen peroxide biosensor based on NiCo2O4 nanoflowers decorated 3D nitrogen doped holey graphene hydrogel, Materials Science and Engineering: C, 102 (2019) 708-717. https://doi.org/10.1016/j.msec.2019.04.072

[170] A.R. Younus, J. Iqbal, N. Muhammad, F. Rehman, M. Tariq, A. Niaz, S. Badshah, T.A. Saleh, A. Rahim, Nonenzymatic amperometric dopamine sensor based on a carbon ceramic electrode of type SiO2/C modified with Co3O4 nanoparticles, Microchimica Acta, 186 (2019) 471. https://doi.org/10.1007/s00604-019-3605-4

[171] F.X. Hu, T. Hu, S. Chen, D. Wang, Q. Rao, Y. Liu, F. Dai, C. Guo, H.B. Yang, C.M. Li, Single-Atom Cobalt-Based Electrochemical Biomimetic Uric Acid Sensor with Wide Linear Range and Ultralow Detection Limit, Nano-Micro Letters, 13 (2020) 7. https://doi.org/10.1007/s40820-020-00536-9

[172] V. Nagal, T. Tuba, V. Kumar, S. Alam, A. Ahmad, M.B. Alshammari, A.K. Hafiz, R. Ahmad, A non-enzymatic electrochemical sensor composed of nano-berry shaped cobalt oxide nanostructures on a glassy carbon electrode for uric acid detection, New Journal of Chemistry, 46 (2022) 12333-12341. https://doi.org/10.1039/D2NJ01961B

[173] Y. Zhang, Y.-Q. Liu, Y. Bai, W. Chu, J. Sh, Confinement preparation of hierarchical NiO-N-doped carbon@reduced graphene oxide microspheres for high-performance non-enzymatic detection of glucose, Sensors and Actuators B: Chemical, 309 (2020) 127779. https://doi.org/10.1016/j.snb.2020.127779

[174] P. Balasubramanian, S.-B. He, H.-H. Deng, H.-P. Peng, W. Chen, Defects engineered 2D ultrathin cobalt hydroxide nanosheets as highly efficient electrocatalyst for non-enzymatic electrochemical sensing of glucose and l-cysteine, Sensors and Actuators B: Chemical, 320 (2020) 128374. https://doi.org/10.1016/j.snb.2020.128374

[175] J. Li, Y. Liu, X. Tang, L. Xu, L. Min, Y. Xue, X. Hu, Z. Yang, Multiwalled carbon nanotubes coated with cobalt(II) sulfide nanoparticles for electrochemical sensing of glucose via direct electron transfer to glucose oxidase, Microchimica Acta, 187 (2020) 80. https://doi.org/10.1007/s00604-019-4047-8

[176] W. Li, W. Li, X. Liu, D. Zhao, L. Liu, J. Yin, X. Li, G. Zhang, L. Fan, Two Chemorobust Cobalt(II) Organic Frameworks as High Sensitivity and Selectivity Sensors for Efficient Detection of 3-Nitrotyrosine Biomarker in Serum, Crystal Growth & Design, 23 (2023) 7716-7724. https://doi.org/10.1021/acs.cgd.3c00478

[177] R. Ullah, M.A. Rasheed, S. Abbas, K.-u. Rehman, A. Shah, K. Ullah, Y. Khan, M. Bibi, M. Ahmad, G. Ali, Electrochemical sensing of H2O2 using cobalt oxide modified TiO2 nanotubes, Current Applied Physics, 38 (2022) 40-48. https://doi.org/10.1016/j.cap.2022.02.008

Applications for Earth-Abundant Transition Metals
Materials Research Foundations 179 (2025) 206-258

Materials Research Forum LLC
https://doi.org/21741/9781644903711-9

# Chapter 9

# Earth-Abundant Transition Metals for Water Treatment

A. Ayub[1], M. Rizwan[2*], A. Shoukat[1], M. A. Waris[1], K. Zaman[1,3]

[1] Department of Physics, University of Gujrat, Gujrat, Pakistan

[2] School of Physical Sciences, University of the Punjab, Lahore, Pakistan

[3] Institute of Modren Physics, Northwest University, Xi'an, P. R. China

*rizwan.sps@pu.edu.pk

**Abstract**

This comprehensive analysis explores the potential of earth-abundant transition metals (EATMs) as sustainable alternatives for water treatment. Starting with an overview of water scarcity and traditional treatment limitations, the chapter delves into the types of water contaminants and the challenges associated with their removal. It then introduces EATMs, discussing their unique properties and potential applications. The analysis continues by examining the catalytic reactions involved in water treatment using EATMs, highlighting their advantages in terms of cost-effectiveness, environmental sustainability, and adjustability. Finally, the paper addresses the challenges associated with using EATMs, such as durability, regeneration, and long-term stability, and outlines future research directions to overcome these obstacles.

**Keywords**

Water Treatment, Earth-Abundant Transition Metals, Sustainability, Cost-Effectiveness, Durability, Regeneration, Scalability, Adsorption, Ion Exchange, Membrane Technology, Catalytic Properties, Pollutant Degradation, Disinfection

## Contents

## 1.    Introduction

Water is regarded as an essential resource for the survival of all living organisms on Earth [1]. Human utilizes water for diverse purposes like cooking, washing, drinking, agriculture, and transportation as well as different industrial processes [2]. The demand for freshwater have been rising quickly due to the rise of the human population. Water scarcity not only affects human nutrition supplies but also drastically lowers biodiversity in terrestrial and aquatic ecosystems. Water, however, ought to meet the minimal requirements for quality set forth by the World Health Organization (WHO) [3]. Water purity is influenced by various physical, chemical, and biological transformations. Only 2.8% of the water on Earth is suitable for human consumption, out of the Earth's total three quarters as Shown in Figure 1 [4]. The remaining 97.2% of water comes from salty oceans, and it is very expensive to remove salts to produce pure water. The remaining freshwater sources, such as glaciers and icebergs, contain frozen water [4].

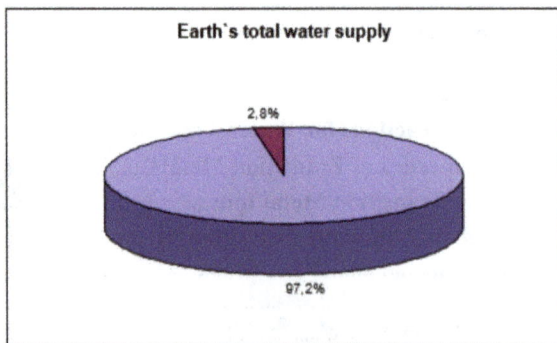

*Figure 1: Overall Earth water supply [4]*

Water is essential to all living things and performs many essential functions in the human body, some of which are listed in the following order: It is a biological solvent that aids in the body's mineral and vitamin transportation and dissolution; it plays a crucial role in regulating the body's temperature; it helps kidneys and other body part function; it serves as a cushion and shield.; it is essential for skin-moisturizing, eliminating pollutants, and cleaning the body; it also aids in nutrient absorption and food conversion into energy. Lastly, water is the primary component of proteins, fats, and carbs in the body. Aside from all of the above, water is essential for carrying out several essential functions including elimination, circulation, and reproduction. It also makes up 80–90% of our blood and 75–80% of our muscles. We require water to survive, and even a little period of dehydration makes us uncomfortable [2, 5]. We come to understand how valuable water is when we lack really important blessings, such as our access to water.

Numerous chemical pollutants possess long-term impact, some have serious health implications. At present, more than 40,000 water bodies in the United States are classified as 'impaired' according to the EPA's definition indicating their failure to support a thriving ecosystem or meet water quality standards. Water contamination refers to the presence of a substance that is overly prevalent in water and can adversely impact people, the environment, or both. [6]. Contaminated water is also called polluted water or wastewater [7]. Different industrial and municipal water waste sources are displayed in Figure 2 [8].

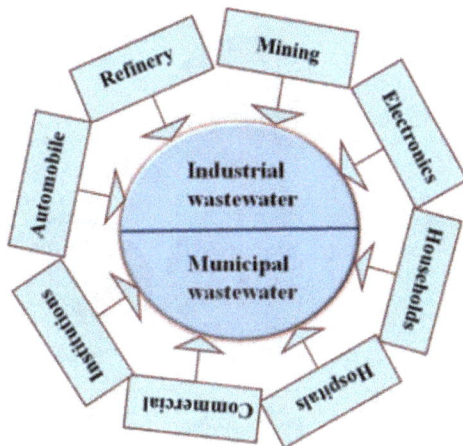

*Figure 2: Sources of wastewater [8]*

Municipal wastewater resources include Households, commercial, hospitals as well as Institutions [9] while industrial wastewater sources include refineries, mining, electronics, and automobiles. In Industries fresh water is utilized in various processes from refining as a result form producing the final product a lot of contaminated water is generated [9-12]. Industrial wastewater contains complicated organic compounds, poisonous heavy metals, and suspended atoms, which are harmful to both human health and the environment [9]. Thus, it is essential to create a system that is fast, simple, effective, and efficient for managing industrial and municipal water contaminants.

The objective of wastewater treatment (WWT) is to eradicate contaminants, remove toxic chemicals, refine the water by killing pathogens, etc. to produce effluents that can be reutilized for beneficial purposes of the environment [6, 13]. On average about 770 million of the world's population lack availability of potable water for home use as well as drinking purposes [13]. WWT correspondingly aims to increase the strain on natural water resources and boost the quantity of water that is accessible for usage [13]. Due to the increasing demand for clean water, the WWT can be a very useful and sustainable solution for obtaining clean water [14]. Al-Juaidi et al. [15], predict as the human population grows, the demand for natural resources, including clean water for industrial and residential use, will rise. Without a sustainable water supply, we may encounter severe water shortages and health issues related to contaminated water.

This chapter elaborates on diverse types of contaminated water pollutants, and the challenges that are faced in the treatment of the contaminated water also provides a comprehensive view of how the earth's abundant transition metals can be a sustainable solution to wastewater treatment and make it useful for drinking and other purposes. Also, different applications of earth-abundant transition metals are discussed for the treatment of contaminated water.

## 2. Challenges in Water Treatment

Wastewater treatments are facing a large number of challenges as shown in Figure 3, some of which are elaborated here.

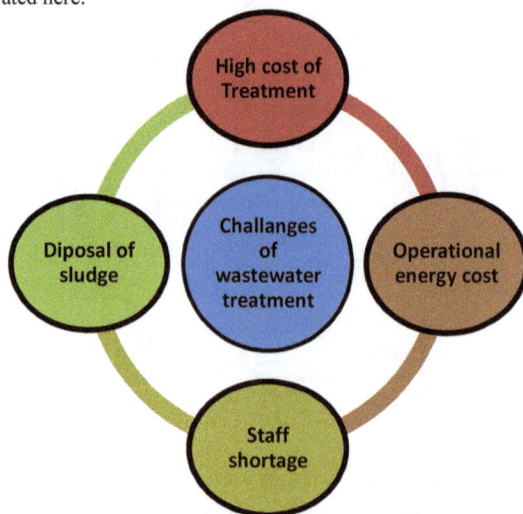

*Figure 3: Wastewater Treatment (WWT) Challenges*

The Leading obstacle in wastewater treatment (WWT) is the expensive cost associated with the establishment of treatment. The typical cost of setting up a functional WWT is about GBP700,00 to GBP.5illion, relative to the facilities as stated by Manning et al. [16]. Costly wastewater treatment makes it inaccessible to low-income areas, due to which these communities cannot get benefits from the sustainable WWT [17, 18]. Costly Installation of setup makes it unavailable for the low-cost regions and communities to install it, as a result, they can't get facilities for clear water as well as energy generation [19-21]. A further barrier to many prospective communities

Materials Research Forum LLC
https://doi.org/21741/9781644903711-9

developing small wastewater treatment facilities is the high cost of implementation [22, 23]. As the expense of these initiatives may be prohibitive for the taxpayers, and low-income economies, the government mainly depends on donations for proper function. Wastewater treatment is an essential requirement that cannot be disregarded, even though the cost is prohibitive for certain administrations. To improve water quality and hygiene, the communities in addition to cities need to collect more funds for the pare plantation of wastewater treatment [24-26].

Another challenge faced by wastewater treatment is the operational energy cost of WWT facilities. WWT utilizes almost 2-3% of the energy in industrialized nations. WWT amid to considered one of the largest sources of energy consumption [27]. While producing biomethane to supplement electrical energy sources is one way to overcome this obstacle, most wastewater treatment facilities are not equipped to produce enough electricity [28]. Presently, 50% to 60% of the facilities can be powered by the electrical energy produced by WWT plants [28, 29]. Its revenues around 40% of it needs to remain driven by other sources. Due to this, the administrators need to spend a large amount on the electricity. Wastewater sludges can be turned into beneficial products like fertilizers to get around this problem. The revenue produced by the different operations in the WWT plants helps to cover certain ongoing expenses and guarantee the proper functionality of the plant [30].

Another obstacle in the wastewater treatment is Staff shortage, which can stop further plant of the wastewater treatment. The staff of WWT is highly trained and professional in operating the WWT facilities and are certified by appropriate agencies for the specified operation [31]. Staff hiring cost for a particular operation is also a noteworthy challenge. To meet this obstacle, some facilities trained their internal staff for operating particular operations as they required [32-34]. It can be accomplished by enlisting people with limited knowledge of wastewater treatment and training them to become fully qualified specialists capable of offering the necessary services [29, 35, 36].

Disposal of sludge is another significant challenge in WWT plants. Huge amounts of sludge formed by both the primary and secondary treatment processes and these must be removed for treatment of additional wastewater [27, 34, 37]. Sludge occupies a large amount of nutrients which can be utilized as a fertilizer for crop growth [38]. Nevertheless, it presents a significant challenge to the organization when the amount of sludge exceeds what can be processed into fertilizer and needs a sizable plot of land to be disposed of safely [39-41]. Some municipal administrations might not have access to the amount of land required for sludge production, which is one of the numerous difficulties that activated sludge may encounter [40, 42, 43]. A considerable amount of land is needed for the installation of multiple tanks that serve as aeration basins for primary in addition to secondary treatment procedures. To guarantee that There is sufficient room on even tiny pieces of land for the production of sludge, creative solutions must be developed [43]. Increasing biomass concentration through utilizing cutting-edge technology, such as a membrane aerated biofilm reactor (MABR), is one of the creative solutions. Concentrated biomass is an essential part of the solution, as suggested by Manning et al. [16]. Higher concentrated biomass can be utilized as a significant strategy for accomplishing the objective. Furthermore, to lessen the strain on natural resources, sustainable resource management depends on creative solutions [25, 44].

## 3.    Types of Water Contaminants

The water contaminants are basically of four types.

- Organic Pollutants
- Inorganic Pollutants
- Biological Pollutants
- Radiological Pollutants

### 3.1    Organic Pollutants

The major cause of organic contamination includes households as well as industrial wastes etc. These pollutants can lead to severe health issues like hormone imbalance, cancer, and neurological disorders [45, 46]. The formation of trihalomethanes (THMs) occurs when organic materials and chlorine in treated water are mixed.

Pesticides pollute both public and agricultural hygiene sources. Pesticides used in public health and agriculture have a detrimental impact on the environment because they are handled and applied improperly [47]. The chemistry of a pest's living body interacts with different chemical processes that pesticides are intended to engage with. Regretfully, by doing this, all pesticides have the potential to affect the metabolism of unintended living organisms. Pesticides primarily harm the liver and nervous system and can also lead to liver tumor [48]. Some of the maximum contamination levels as well as Low contamination level materials through literature are given in Table 1 [49].

*Table 1: Organic pollutants [49]*

| Pesticides | Maximum Contamination level (MCL), µg/L |
|---|---|
| Dinoseb | 7 |
| Carofuran | 40 |
| Dalapon | 200 |
| Dibromochloropopane | 0.2 |
| Dioxin | 0.0003 |
| Endothall | 100 |
| Glyphosate | 700 |
| Diquat | 20 |
| Ethylene dibromide | 0.05 |
| Oxamyl | 200 |
| **Pesticides** | **Low Contamination level (LCL) µg/L** |
| Sulfamethoxazole | 20,000 |
| Naproxen | 37,000 |
| Ciprofloxacin | 20 |
| Diclonofenac | 10,000 |
| Atenolol | $10 \times 10^6$ |
| Ibuprofen | 5000 |

Materials Research Forum LLC
https://doi.org/21741/9781644903711-9

- Solvents, organic compounds such as styrene, chloride, benzene, etc. fuel, and petrol additives are instances of volatile organic compounds (VOCs). These VOCs have a great influence on chronic health which can cause cancer, birth defects, abnormal central nervous system, as well as injury to the liver and kidney, etc. [50, 51].

- Dyes are among the most common kinds of organic compounds that are becoming more and more harmful to the environment. The waste-contaminated water through the process of oxidation, hydrolysis, reduction, and other reaction processes is a great source of non-esthetic contaminants which leads to initiate the harmful by-products [52, 53].

- Moreover, the substances present in water may also have unfavorable impacts on human health or the environmental ecosystem. We refer to these substances as "Emerging Organic Contaminants (EOCs)." It includes industrial compounds, personal care products, and water treatment by utilization of products. Generally, it comprises carcinogenic, petroleum hydrocarbons, endocrine, etc. [54-57].

- The organic pollutants source can be of industrial as well as agricultural means. Hospital effluents as well as chemicals obtained from chemical means for the production of Pharmaceuticals like antibiotics, analgesics, etc. Some of newly EOCs have low contamination levels with no effect given in Table 1[49].

## 3.2    Inorganic Pollutants

Its chemical properties can also be utilized to measure the level of pollutants present. Drinking water hardness is a naturally occurring contaminant. According to the molecules' hardness belonging (Calcium or magnesium content), classified as carbonate or non-carbonate. Carbonate hardness is defined as molecules' combination with the carbonate ions, while in the case of non-carbonate hardness, the molecules are combined with other ions. The general water hardness level is between 300 and 400 mg/L which can be utilized for drinking. If the hardness level exceeds the form 500mg/L , then the contaminated water caused kidney stones, etc. [49].

Other inorganic pollutants are heavy metals, cations, and radioactive materials [58]. In aqueous systems, inorganic contaminants take longer to break down and cause further damage. Groundwater in many places of the world, like India, has been discovered to have high amounts of metals (especially heavy metals) and other harmful substances, such as fluoride and nitrate, over the cutoff point, rendering it unsafe for human consumption [59]. In Industries, heavy metals like lead, mercury, zinc, etc. are the main cause of wastewater. The inorganic pollutants are listed in Table 2 [60].

*Table 2: Inorganic pollutants [60]*

| Heavy metals | Nature | Anthropogenic | MCL (mg/L) | Health issues |
|---|---|---|---|---|
| Arsenic | Geogenic | Industrial leakage | 0.05 | Skin problems |
| Cadmium | Geogenic | Batteries discharge | 0.005 | Kidney problem |
| Chromium | Geogenic | Paper mills | 0.1 | Allergy |
| Copper | Geogenic | Pipe corrosion, Wood preservation | 1.3 | Stomach problem |
| Lead | Geogenic | Pipe corrosion | 0.015 | Kidney disorder |

### 3.3    Biological Pollutants

Water becomes contaminated biologically when it comprises live things like bacteria, viruses, algae, or protozoa. Algae are tiny, unicellular creatures. They depend on the phosphorus (or other nutrients) in the water [61, 62]. The majority of the nutrients originate from residential runoff or industrial pollution. Overgrowth of algae not only gives water a bad taste and smell, but it Additionally, blocks filters and leads to undesirable slime growth on the carriers. Blue-green algae like Anabaena, Aphanizomenon, and Microcystis can sometimes release toxins that are harmful to the skin, liver, and nervous system [63, 64].

Furthermore, Bacteria are tiny, unicellular organisms. Various harmful microorganisms can be harmful and make water contaminated. They may cause diarrhea, typhoid, etc. While not hazardous, certain non-pathogenic bacteria (such as sulfur and phenothrin iron bacteria) can lead to issues with taste and odor [65-67].

Protozoans (Giardia and Cryptosporidium) are tiny, unicellular organisms that are mostly found in rivers, streams, etc., and contaminated as a result of animal wastes. It leads to headaches, dehydration, diarrhea, and cramping in the stomach. The tiniest living things that can spread illness and cause infection are viruses. There are frequent reports of polio and hepatitis viruses in tainted water [49].

### 3.4    Radiological Pollutants

Radioactive elements are the basis of radiological pollutants. Radiological pollutants can cause cancer. Its origins include both industrial waste and soil or rocks which allow water to pass. Erosion may cause the alpha and beta radiation emission of radioactive elements. The radioactive elements cause great trouble for the underground water [68, 69].

### 4.    Limitations of Existing Water Treatments

Reducing or eliminating hazardous materials, pathogenic microbes, and organic and inorganic components are the main objectives of wastewater treatment [34, 58, 70-74]. As a consequence, the treated water's quality is raised to meet WHO standards. Municipal and industrial wastewater contains different contaminants [75-77]. As a consequence, kinds of WWT are based on features of effluent and the required post-treatment water quality.

Three stages of primary, secondary, and tertiary treatments are typically included in wastewater treatment as represented in Figure 4 [78]. Most big particles and organic materials are removed by the first and secondary treatment processes, respectively. Tertiary treatment, which functions as an enhancing unit, is used to eradicate different undesired materials that are left in the treated water after the primary and secondary treatments. These therapies often include several physical, chemical, and biological procedures [77].

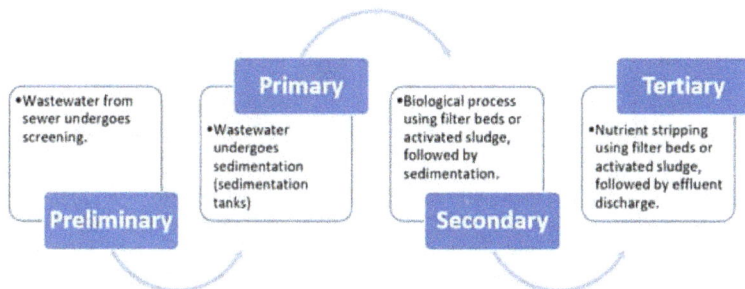

*Figure 4: Contaminated water treatment processes [78]*

Membrane filtration may be employed even at elevated concentrations, as it is a simple and rapid physical process. They are said to be appropriate for practically all kinds of pollutants, such as colors, mineral ions granules in suspension, etc. Nevertheless, due to the occurrence of fast membrane clogging at elevated concentrations, these systems entail significant energy, maintenance, and operational expenses, rendering them impractical for minor businesses.

Coagulation and flocculation are fundamental physiochemical processes. However, effectively controlling the formation of large-sized flocs, which results in the production of sludge, can be challenging and increases the overall cost of this process.

Chemical methods are simple, rapid, and efficient processes that use a range of oxidant-related methodologies. Chemical processes are typically conducted on a lab scale and are often not financially viable for small businesses to meet their energy needs, in addition to all these other qualities.

The biological treatment is also attractive for adaptation since it is a simple and financially attractive process. But it breaks down slowly, takes a long time, and requires the ideal environment for bacteria to thrive.

Adsorption is a non-destructive, cost-effective, and technically simple method of treating wastewater. It was found by looking into a variety of WWT techniques. It efficiently purges water and wastewater of a broad spectrum of pollutants like metal ions, minerals, dyes, and other impurities [79].

Wastewater treatment processes have some limitations. The conventional methods used for water purification like filtration, and chlorination have limited scope, and they cannot address and remove all types of pollutants. Other methods like biological treatment, advanced oxidation, and adsorption are employed to eliminate the contaminants from water but they still have barriers of cost, environmental factors, etc. [80]. Moreover, some techniques produce secondary pollutants which affect the water quality as well as the ecosystem. For removal of these secondary pollutants different adsorbent materials are being employed in wastewater treatment processes, but their effectiveness has to be improved, and the problem of their safe disposal and ecologically responsible manner needs to be addressed [81]. Even though anaerobic wastewater treatment is good at breaking down organic molecules and creating biogas, rates of nutrient removal, stability, and flexibility still need to be increased [82]. The broad scope of traditional methods, the limits of alternative methods, the creation of secondary pollutants, the effectiveness and disposal of adsorbents, and the requirement for advancements in anaerobic treatment are among the general constraints of wastewater treatment approaches[81, 82].

## 5.   Earth-Abundant Transition Metal s(EATMs): A Sustainable Alternative

Access to clean water is both a fundamental human right and essential for maintaining health. The Major wastewater contamination is Nitrate [83-85]. Increased $NO^{3-}$ Levels in drinking water have been associated with thyroid issues, cancer, and harmful respiratory consequences [86-88]. Throughout the last hundred years, there has been a notable rise in nitrate concentrations in both surface and groundwater through the use of nitrogen fertilizers by humans [89]. About 45 million Americans depend on uncontrolled private groundwater wells that contain nitrate levels exceeding permitted thresholds  [90]. Furthermore, nitrate affects a lot of municipal water sources, and expensive ion exchange technologies are frequently used to remedy them.

Therefore, effective nitrate removal technologies are desperately needed for both extensive water treatment systems and small-scale point-of-use (POU) treatment systems installed in homes [91, 92]. It is reported that how electrocatalytic reduction of nitrate (ERN) works as a practical drinking water treatment method by selectively reducing $NO^{3-}$ to safe dinitrogen (N2) [93]. ERN, as electrode cost accounts for the majority of the capital cost of electrochemical systems [93-95]. Platinoid materials (Pt and Pd) were used in most ERN articles [96-102].  In the water-energy nexus, platinum electrodes are very efficient electrocatalysts for nitrate reduction, primarily because of their exceptional corrosion resistance [103-105]. Unfortunately, because of price and restricted accessibility as resources on Earth, platinum group elements (PGEs) are categorized as expensive and rare elements [106]. The mining and cleansing of these extinct components have been linked to negative environmental repercussions, as identified by lifecycle analysis [107, 108] [109, 110]. Finding substitute electrocatalysts based on highly efficient earth-abundant elements is a scientific and engineering issue.

Earth-abundant transition metals can be utilized in replacement of the PGE-based electrodes. ETMs can be catalysts that help to degrade nitrate a reduce the purifying of the water. The First-row transition metals (e. g. , Ti, Fe, Co, Ni, Cu, Zn) as well as Carbon and tin can also be utilized for killing water pollutants. The ETMS-based electrodes are a good and sustainable alternative because they reduce the cost of the electrodes, are less harmful to the environment, and have good catalytic and redox activity [106].

## 6.   Unique Properties of Transition Metals Relevant to Water Treatment

Water is being contaminated by different sources like industrial, pharmaceutical, organic dyes, household waste, etc. The contaminated water has detrimental consequences for human health and ecological systems [111, 112]. To overcome this, different wastewater treatment processes are being utilized like adsorption [113], coagulation-flocculation[114], advanced oxidation process (AOP) [115, 116] etc.

Some water treatment systems, however, can only concentrate target contaminants into several phases and are unable to totally "eliminate" or "destroy" them. Similarly, high operational costs and the possibility of secondary contaminants entering the environment are associated with

Applications for Earth-Abundant Transition Metals          Materials Research Forum LLC
Materials Research Foundations 179 (2025) 206-258          https://doi.org/21741/9781644903711-9

membrane separation, adsorption, and sedimentation technologies [117]. AOPs are considered the most effective and environmentally benign of these technologies [118].

AOPs offer extremely responsive oxygen species $(OH, SO_4^-, 1O_2, O_2^-)$ offering outstanding potential for the effective breakdown and acceptable mineralization of hazardous and bio-refractory contaminants removed from the dirty water system. These treatment methods have several appealing features, including ease of use, exceptional efficiency, low energy consumption, the ability to treat many pollutants at once, affordability, and environmental friendliness as compared to other treatments [119-124].

A wide range of transition metal compounds like phosphides, sulfides[125], metal alloys, etc.[126, 127] are being employed in the oxidation degradation process. In particular, transition metal sulfides (TMSs) and their derivatives-based heterogeneous catalytic oxidation play a significant role [128-130].

TMSs included among earth-abundant transition metals and sulfides are combined with other metals to form the cations of stoichiometries $MS, M_2S, M_3S_4, MS_2$ [131] . Similarly, $A_{1-x}B_xS_y$ is a concise description of the creation of typical bimetallic sulfides (where A and B are distinct metals and x and y are integers) [132]. Numerous TMS kinds have been documented, including pyrite, tetrahedral crystal structure, anti-fluorite, simple and symmetrical sodium chloride, and fluorite structural type [133-135]. Unlike other commonly used transition metal catalysts, TMSs have superior visual properties, remarkable mechanical and thermal stabilities, strong catalytic activity, and richer redox sites with greater electrical conductivity.

TMSs are widely used in light-emitting devices [136], water splitting solar cells, energy conversion and storage [137], cathode materials[138], sensors, and environmental cleanup because of these benefits[138, 139]. Because of their distinct physicochemical characteristics, AOPs utilizing TMSs and their derivatives have attracted a lot of interest.

Heterogeneous catalysis for water cleansing using TMSs-based catalysts may be easily achieved in these oxidation processes. As far as we are aware, there have only been a few prior reviews of TMS-based catalysts. These reviews primarily concentrate on photocatalysis and environmental cleanup using electrochemical energy storage, with a focus on methods for synthesizing catalysts, characterization techniques, and control of morphology and structure. Research of TMSs-based AOPs and application of heterogeneous catalysts depending on TMSs to clean dirty water using AOPs are still in their infancy, regardless of heterogeneous catalysis advancements for water decontamination [140].

## 6.1    Redox Activity

The term Redox comprises two words, reduction and oxidation, in case of reduction gain of electrons is done while in case of oxidation loss of electrons is done moreover decrease and increase of oxidation number occur respectively. These two reactions occurred consequently. Redox reactions, which encompass all transformations in which the interacting species experience alterations in their oxidation state, are among the most fundamental reactions in chemistry. The transition metals have the unique property of the redox activity[140-142].

## 6.2 Catalytic Abilities

The transition metals have catalytic abilities. Because of their small optical bandgaps, charge transport properties, and capacity to absorb visible and/or UV light, Transition metal sulfides (TMSs) and their derivatives are thought to be viable options for photocatalytic water purification. Due to the significant optical characteristics of TMSs-based photocatalysts can be used for the treatment of wastewater. Due to excellent properties and a low bandgap of 2.26eV, the CdS (~0.5 µm) in size shows its potential for photodegradation of acetaminophen [143]. The mechanism of the Photo-degradation is presented in figure 5 **[140]**.

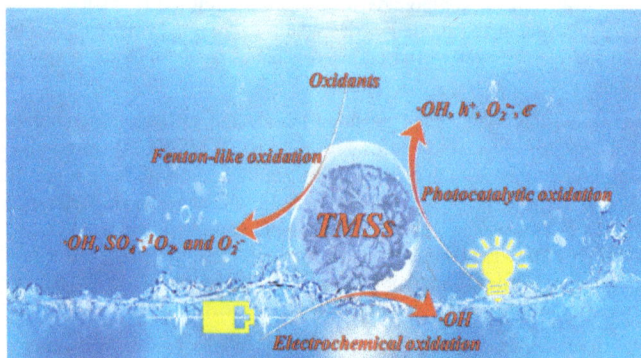

*Figure 5: Transition metals mechanism for water purification[140]*

## 7. Properties of Earth Abundant Transition Metals (EATMs)

The physiochemical properties of EATMs Photocatalysts are closely linked to their structure and morphology. These physiochemical characteristics can be tuned by monitoring size, shape, and composition or via doping and manufacturing heterostructures.

### 7.1 Tuning of Shape, Size and Composition

Their physiochemical properties can be regulated by varying the size, shape, and composition of EATMs. CdS PL spectra showed a blue shift owed to variation in different sizes of CdS nanocrystals, the varying sizes are achieved by manipulating the concentration of $Cd^{+2}$ in do decanethiol solvent. In samples that had very small nanocrystals(NCs), PL showed minimum intensity due to increases in defects[144]. In another study, it was revealed that tuning the morphology of MoS2, which are active centers in catalysis, leads to exposure to more edge sites due to improved surface area. Sometimes, tuning the size and shape of these EATMs could lead to the disappearance of quantum effects and instability. The peak shift in CdS NCs is owed to quantum size effects. The stability of these EATMs is dependent on synthesis parameters such as precursors, capping ligands, and reaction temperature. The introduction of different metals in these EATMs can lead to unique structures with tunable properties[145]. In conclusion, the properties of TMSs can be tailored for various applications by controlling their structure, composition, and synthesis conditions [140].

## 7.2 Doping

Tunable EATMs can also be achieved via doping. The effect of doping on EATMs as reported by many studies is summarized in Table 03.

*Table 3: Doping of Transition Metal Sulfides (TMSs)*

| Doping Method | Effect on Properties | Example | Reference |
|---|---|---|---|
| Substitutional Doping | Changes in morphology, Raman spectra, PL quenching | Fe-doped WS2 | [146] |
| Ion Implantation | Introduces impurities, affects carrier concentration and mobility | Fe-doped NiCo2S4 | [147] |
| Plasma Treatment | Introduces impurities, which can alter the crystal structure | Fe-doped WS2 | [146] |
| Vacancy Formation | Creates free holes, p-type doping, LSPR | Cu2-xS NCs | [148] |

## 7.3 Heterostructures

EATMs heterostructures are of great importance in water treatment due to their improved physiochemical properties. Heterostructures EATMs combine different EATMS to create synergistic effects and improve their properties. The quantum efficiency (QE) of NiS/CdS is reported to be 51.3%, the HER was increased by 10 times in comparison to pure CdS[140].

Table 04, summarizes some studies of heterostructure EATMS for pollutant degradation.

*Table 4: catalytic Degradation of Organic Pollutants by Transition Metal Sulfides*

| Organic Contaminant | TM-Photocatalyst | Reaction Parameters | Degradation Efficiency | Major Reactive Oxygen Species | Reference |
|---|---|---|---|---|---|
| DEP (20 mg/L) | FeS | [PMS] = 2 mM , pH=7.0, 0.5 g/L | 100%, 120 minutes, | $SO_4^{\cdot-}$ | [149] |
| ACR (0.074 mM) | FeS | [$H_2O_2$] = 0.8 mM , pH =6.2, 0.5 g/L, | >99%, 60 minutes, | $OH^{\cdot}$ | [150] |
| p-ASA (1 mg/L) | FeS | [$H_2O_2$] = 1.5 mM , pH =3.0, 5 g/L, | 100%, 120 minutes, | $OH^{\cdot}$ | [151] |
| PCA (0.2 mM) | CuS | [PS] = 4 mM , pH =7.0, 0.35 g/L, | 99%, 30 minutes, | - | [152] |
| ATZ (50 μM) | CuS | 25 mM, pH =7.0 | 91.6%, 20 minutes, | $OH^{\cdot}$ $SO_4^{\cdot-}$ | [153] |
| MB (10 mg/L) | CuS/rGO | [$H_2O_2$] = 0.08 M , pH =9.0, 20 mg/L, | 93.1%, 90 minutes, | $OH^{\cdot}$ | [154] |
| BPA (20 mg/L) | CuFeS2 | [PMS] = 0.4 mM , pH =6.2, 2 g/L, | 83%, 30 minutes, | $OH^{\cdot}$, $SO_4^{\cdot-}$ | [153] |
| BPS (25 μM) | CuFeS2 | [PMS] = 0.3 mM , pH =6.0, 0.1 g/L, | 99.7%, 20 minutes, | $OH^{\cdot}$, $SO_4^{\cdot-}$ | [155] |
| 2,4-DCP (10 mg/L) | CoSx | [$NaHCO_3$] = 15 mM, , pH =8.5, [$H_2O_2$] = 0.08 M, 50 mg/L, | 91.6%,120 minutes, | $OH^{\cdot}$, $O_2$, $CO_3^{\cdot-}$ | [156] |
| TC (30 mg/L) | CoSx | [PMS] = 0.3 g/L , pH= 5.0, 0.2 g/L, | 100%, 30 minutes, | $OH^{\cdot}$, $SO_4^{\cdot-}$, $O_2$ | [157] |

Materials Research Forum LLC
https://doi.org/21741/9781644903711-9

## 8. Applications of Earth-Abundant Transition Metals in Water Treatment

Water pollution and scarcity have raised a serious threat to environmental health, thus requiring novel sustainable solutions. Water treatment is one of the steps to eradicate hazardous contaminants and render them safe for human consumption and industrial use. Traditional pollutant degradants have been effective, but more efficient, cost-effective solutions are required. Earth-abundant transition metals (EATMs) are considered potential contenders for water treatment as pollutant degradants. EATMs include Fe, Mn, Co, Ni, Cu, and Zn, and offer many advantages over traditional precious metals used in catalysis for wastewater management. EATMs are easily available, inexpensive, and have better catalytic activity than traditional alternatives[158].

The application of these materials in the field of water treatment consists of many processes such as

Adsorption: EATMs can be utilized in the production of adsorbents that have a high attraction for a variety of pollutants such as dyes, organic pollutants, or heavy metals such as lead. Transition metal's surface chemistry and porous nature play a vital role in their adsorption of water-polluting contaminants.

Catalysis: These transition metals can be employed as a catalyst to treat polluted water via Oxidation and reduction reactions, and pollutant degradation mechanisms. Fe has been widely used as a pollutant remover via Oxidation reactions.

Membrane Technology: Transition metals can be introduced in membrane materials for better selectivity and antifouling performance in water treatment.

Disinfection: Transition metals such as copper with antimicrobic properties, can be employed for water clarification.

By utilizing these above-mentioned properties of EATMS, we can develop novel water treatment strategies that can solve the water scarcity issues[159]. Fig Figure 06 dictates how d-block transition metals can be utilized for organic pollutant photodegradation.

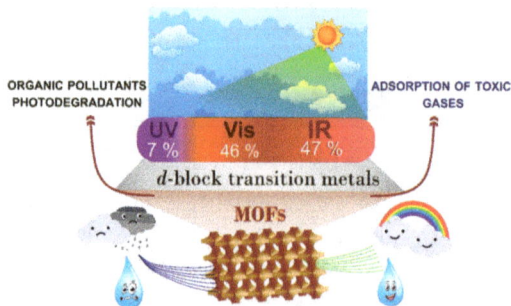

*Figure 06: Transition metal-based MOFs for environmental remediation. These materials excel in photodegrading organic pollutants and capturing toxic gases due to their broad light absorption and high porosity[160]*

## 9. Adsorbents for Contaminant Removal

Transition metal-based oxide frameworks are used for contaminant removal in water via photocatalysis, these MOFs are used due to their high porosity and superior optical properties. MOFs that have transition metal nodes have garnered significant attention due to their structural and morphological variability. MOFs that contain EATM such as Cd, Co, Cu, Fe, and Zn centers are suggested as photocatalysts/absorbents in the context of degradation of water pollutants under light the transition metal centers within the MOFs contain at least one weakly coordinated ligand, such as water molecules or other solvents, which can be removed without compromising the crystal structure. When the ligand is removed creating a coordination vacancy, the TM center acts as Lewis acids, therefore receiving electrons from a light-sensitive molecule that is present in the medium. Transition metals have a huge role in MOF photocatalytic activity, as well as the porous space and organic linker which contribute to redox reactions.[161] The photocatalytic activity can be improved via the enclosure of photocatalytic active molecules in the MOF pores. Charge transfer is enhanced by the porous nature of the polymers, which results in a reduced recombination rate of $e^-$-$h^+$ pairs. The organic linker in MOF absorbs light and produces charge separation via $e^-$-transfer between the ligand and the TM node. Aromatic polycarboxylates, 2-amino terephthalic acid, and porphyrins are preferred ligands because they exhibit strong conjugation with metal ions, facilitating efficient electronic transitions. Introducing additional metal elements into MOFs can improve their porosity and adsorption capacity [162].

## 10. Processes Involved in Water Treatment

### 10.1 Mechanisms of Adsorption

Adsorption is a process where a solute/adsorbate is attached to an adsorbent via attractive forces as shown in figure 08. The material used in adsorption has high porosity and surface and is characterized by the parameters shown in Figure 07.

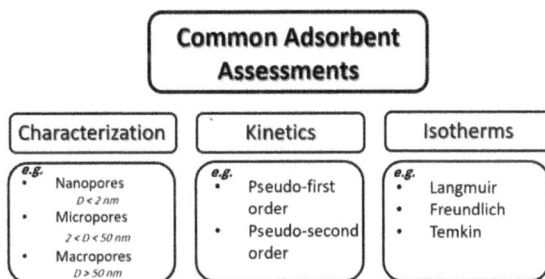

*Figure 7: Common parameters for describing adsorption systems[163]*

The adsorption process can also be defined as an intermolecular transfer of pollutants onto the adsorbent. The adsorption relies on the composition and chemical nature of the adsorbent and adsorbate. Adsorption can involve van der Waals forces, transfer via electrostatic attraction, $\pi$-$\pi$ interactions, and chemisorption. The physical adsorption can occur via hydrophobicity, dipole-

dipole interaction, steric interaction, and polarity. Chemical adsorption involves charge transfer between pollutant and adsorbent by making a chemical bond. Chemical adsorption most likely happens between metallic ions and adsorbents with functional groups. It is suggested that the main mechanism of removal of organic pollutants is physical sorption. In a solid-liquid system, the pollutant/solute is removed via diffusion layer[163].

To comprehend the absorption of water contaminants, characterizing the adsorbents is necessary. A good adsorbent must have a large surface area and an internal network of pores and pore size between 2 nm- 50 nm.

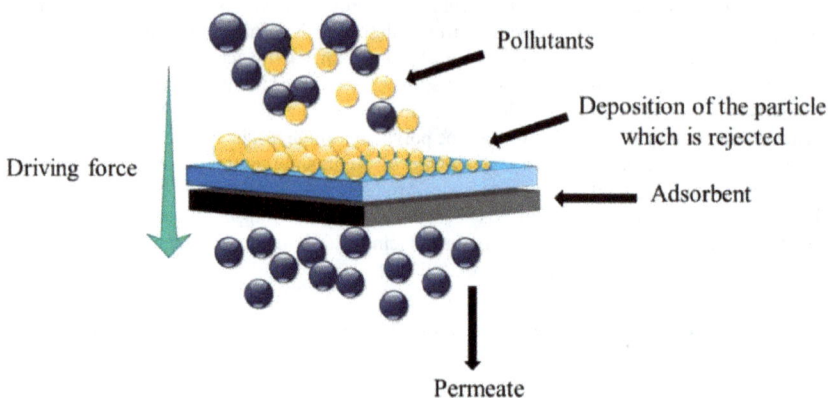

*Figure 8: Adsorption of pollutants on the transition metal absorbents [164]*

## 10.2 Ion Exchange

The charge state and density of transition metals can vary. The electrostatic attraction allows for the separation of untouched EATM ions in ion exchange processes. Due to the negative charge of most EATMs, anion exchange is frequently utilized[165].

## 10.3 Membrane Technology

This is another mechanism to remove pollutants from water, in this porous membrane is treated as a barrier that stops particular particles while allowing others. This separation technique, usually involves adsorption, the Donnan effect, ion exchange, and the Seieve effect. Depending on the pollutants and background substances present in water, membrane technology can be applied through various methods such as microfiltration and nanofiltration. The accompanying Figure 09 illustrates the membrane technology process*[161]*.

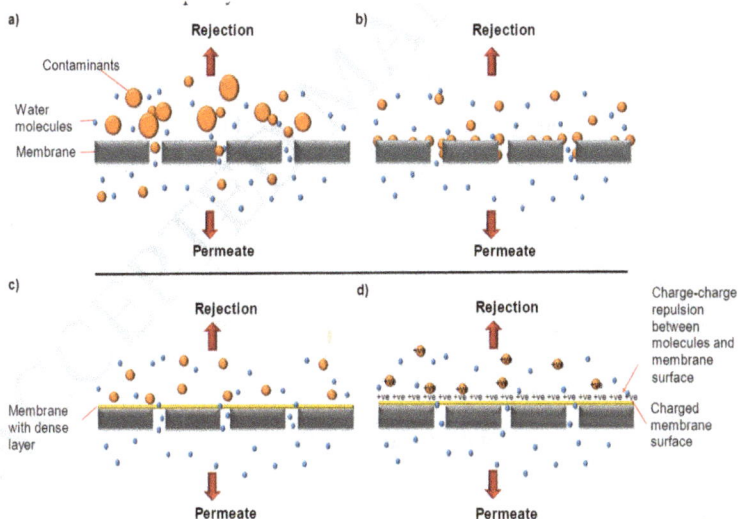

*Figure 09: (a) Steric Hindrance Mechanism in a Low-Pressure Membrane, (b) Adsorption Apparatus in a Low-Pressure Membrane, (c) Size-Based Separation in a Low-Pressure Membrane[161]*

## 11. Examples of Transition Metal-Based Adsorbents

Transition metal oxides(TMOs) and transition metal Chalcogenides have garnered much attention due to their tunable composition, band gap, earth abundance superior optical properties, and catalytic activity for water treatment. TMOs can be prepared through various synthesis methods such as solvothermal, hydrothermal, electrospinning, and refluxing methods. TMOs are explored as heterogenous photocatalysts owed to their biocompatible nature, stability in various environments, generation of electron-hole pairs upon incident light absorption, and electronic structure. The performance of these TMOs depends on their properties such as size, morphology, and surface defects. The TMOs can be synthesized in various ways, and these synthesis methods characterize their optical properties. TiO2 and ZnO have drawn various attention for the purification of water. ZrO2 is a photocatalyst as well. These TMOs are used to eradicate hazardous contaminants in water, but there are still limitations imposed. (TMOs) are employed in various water treatment technologies, including adsorption and Fenton-like processes. The table showcases several TM-based frameworks utilized for pollutant degradation in water. Table 05 characterizes the type of pollutants, the photocatalysts degrade, their efficiency in doing so, and the operating conditions.

Materials Research Forum LLC
https://doi.org/21741/9781644903711-9

*Table 5: TM and TMO-based Photocatalysts and their Pollutant degradation*

| Photocatalyst | Contaminants | Experimental parameters | Oxidizing Setup | Efficiency (TOC %) | References |
|---|---|---|---|---|---|
| MIL-53 (Fe) | Rhodamine B(RhB) | pH: 180, Time: 0.40 min, C: 10 mg/L | $H_2O_2$/Xe lamp | 90 | [166] |
| AgI/MIL-53(Fe) | RhB | pH: 200, Time: 0.30 min, C: 5 mg/L | Xe lamp | 100 | [167] |
| CdS@NH$_2$.MIL-125(Ti) | RhB | pH: 120, Time: 0.15 min, C: 180 mg/L | Xe lamp | 97 | [168] |
| BiOBr/NH-MIL-125(Ti) | RhB | pH: 100, Time: 0.20 min, C: 20 mg/L | Xe lamp | 80 | [169] |
| BiOI@MIL-88A(Fe)@g-CN | RhB | pH: 180, Time: 0.10 min, C: 10 mg/L | Xe lamp | 75 | [170] |
| CuS/UiO-66 | RhB | pH: 60, Time: 0.15 min, C: 10 mg/L | $H_2O_2$/Xe lamp | 90 | [171] |
| 1-NO$_3$-OH-20H$_2$O | RhB | pH: 120, Time: 0.40 min, C: 10 mg/L | $H_2O_2$/Xe lamp | 92 | [172] |
| Prussian blue | RhB | pH: 120, Time: 7.5 min, C: - mg/L | $H_2O_2$ | 80 | [173] |
| HPU-5 | RhB | pH: 150, Time: 0.07 min, C: 10 mg/L | $H_2O_2$/Hg lamp | 99 | |
| HPU-6 | Methyl orange | pH: -, Time: -, C: - mg/L | - | 83 | |
| HPU-6 | RhB | pH: -, Time: -, C: - mg/L | - | 78 | |
| HPU-6 | Methyl orange | pH: -, Time: -, C: - mg/L | - | 31 | [174] |
| α-Fe$_2$O$_3$/UiO-66 | Methylene blue | pH: 60, Time: 1.00 min, C: 12.8 mg/L | Xe lamp | 100 | [175] |
| Fe$_3$O$_4$/Cu$_3$(BTC)$_2$ | Methylene blue | pH: 4, Time: 120 min, C: 0.6 mg/L, C: 100 mg/L | $H_2O_2$ | 85 | [176] |
| TiO$_2$@NH$_2$-MIL-88B(Fe) | Methylene blue | pH: 7.0, Time: 150 min, C: 0.20 mg/L, C: 100 mg/L | $H_2O$/LED lamp | 100 | [177] |
| Mn(3,4-dimethylthieno[2,3-b]thiophene-2,5-dicarboxylic acid)(DMF) | Methylene blue | pH: 70, Time: 1.00 min, C: 10 mg/L | Hg lamp | 91 | [178] |
| BiOI/ZnFe$_2$O$_4$/MIL-88B(Fe) | Methylene blue | pH: 120, Time: 0.10 min, C: 10 mg/L | LED | 65 | [179] |
| BiOI/ZnFe$_2$O$_4$/MIL-88B(Fe) | Acid Blue 92 | pH: -, Time: -, C: - mg/L | LED | 73 | [180] |
| CoNi(μ$_3$-bdc)$_2$(H$_2$O)$_2$ | Methylene blue | pH: 135, Time: 0.10 min, C: 10 mg/L | LED | 98 | [181] |

## 12. Catalysts for Degradation of Pollutants

Transition metals are extensively used as photocatalysts in the field of wastewater treatment. Transition metals possess the versatility of being employed as photocatalysts or coupled with other compounds to act as heterogeneous catalysts [182]. The photocatalytic activity of these materials hugely depends on their crystal structure, phase pureness, morphology, band gap, and electronic

Materials Research Forum LLC
https://doi.org/21741/9781644903711-9

behavior. Zinc Oxide, $Co_3O_4$, Nickel oxide, $Cr_2O_3$, and copper oxide are prepared through the same synthesis method but all showed different morphology and 80-88% degradation efficiency, this is due to the charge recombination, which affects their photocatalytic efficiency[182]. One way to address this via metal/non-metal doping is to produce heterostructures. TM-doped $TiO_2$ efficiency eradicated Rhodamine B, due to the enhanced surface area and reduced band gap. The photocatalytic performance of these TM-based frameworks can also be enhanced via enzyme incorporation. Laccase incorporation in zinc oxide and $MnO_2$ produced 95 and 85% degradation of red S dye[183]. Ternary metal oxides $A_xB_{3-x}O_4$ (A, B = Cd, Co, Cu, Fe, Mn, Ni, Zn,) have garnered devotion as photocatalysts owed to their high stability, high photocatalytic activity, low cost, effortless synthesis, and selectivity over morphology. $CoFe_2O_4$, as a ternary photocatalyst is reported to degrade MB[184]. $ZnFe_2O_3$ is another example of a spinel structure ternary photocatalyst, suggested for the degradation of MB, solo phenyl red 3BL, and tetracycline hydrochloride.

Transition metal-based heterostructures with two-dimensional materials have proved the synergy between adsorption and photocatalysis. Table 06, summarizes transition metal-based photocatalysts, their degradation performance, and the pollutant, they are used to eradicate.

*Table 6: TMOs and their pollutant degradation efficiency*

| Transition Metal Oxide | Pollutant | Degradation Efficiency | Reference |
|---|---|---|---|
| ZnO, Co3O4, NiO, Cr2O3, CuO | Alizarin Red S, Methylene Blue | 80-88% in 180 min | [182] |
| Mo-TiO2/rGO | Rhodamine B | 100% adsorption and degradation | [185] |
| Lac-ZnO, Lac-MnO2 | Alizarin Red S | 95% and 85% respectively | [183] |
| CoFe2O4 | Methylene Blue | High | [184] |
| ZnFe2O3 | Methylene Blue, Tetracycline hydrochloride, Solophenyl Red 3BL | Varies | [186, 187] [188] |
| CuBi2O4 | Methylene Blue, Diclofenac | Lower than the crystalline counterpart | [189, 190] |
| ZnO/ML-Ti3C2, SnO2/ML-Ti3C2, ZBi2MoO6/ML-Ti3C2 | Methylene Blue, 4-chlorophenol | Complete degradation in 4 h | [191] |
| ZnFe2O4@ZnO | Congo Red | High | [192] |

## 12.1 Mn-Based Photocatalysts

Mn is the 12th most widely available element in the earth's crust. Mn and Mn-based catalysts are reported to exhibit oxidation reactions of substrates such as carbon monoxide, alcohol, etc. Mn photocatalysts are mainly MnO species that have exhibited photocatalytic activity for carbon monoxide owed to their high surface area as shown in Table 07. Mn-based catalysts have shown catalytic performance in oxidation reactions involving pollutants[158].

*Table 7: Mn-Based Nano catalysts and Their Applications*

| Catalyst | Synthesis Method | Application | Substrate | References |
|---|---|---|---|---|
| $MnO_2$ NPs | Microwave-assisted solution | CO oxidation | CO | [193] |
| $Mn_3O_4$ NPs | Thermal treatment | Alcohol oxidation | Benzylic alcohols, allylic alcohols, aliphatic alcohols | [194] |
| $Mn_3O_4$ NPs | Surfactant-free strategy | Organic pollutant degradation | Methyl orange, Congo red | [195] |
| Mn(III) meso-tetraphenyl porphyrin NPs | Host-guest strategy | Olefin epoxidation, hydrocarbon oxygenation | Cyclooctene, alkenes | [196] |
| Mn(phox)2(CH3O)@NH2 | Immobilization on Fe3O4 NPs | Thiol oxidation | Thiols | [197] |
| Mn(TPP)Cl@Im | Immobilization on Fe3O4 NPs | Biomass-derived compound oxidation | 5-Hydroxymethylfurfural | [198] |
| $Mn_3O_4$ NPs | Immobilization on Fe3O4 NPs | Olefin epoxidation | Alkenes | [194] |

## 12.2 Fe-Based Photocatalyst

Fe is the 4th most widely available element that is inexpensive and eco-friendly. Fe-based catalysts have received attention due to their catalytic abilities. FeO-based nanomaterials have been employed as catalysts in various reactions such as the reduction of $N_2O$, and decomposition of $H_2O_2$, $NH_3$ as displayed in Table 08[158].

*Table 8: Fe-Based Nano catalysts and Their Applications*

| Catalyst | Synthesis Method | Application | Substrate | References |
|---|---|---|---|---|
| $Fe_2O_3$ NPs | - | Alcohol oxidation | Alcohols | [199] |
| FeOx/N-doped graphene | - | Oxidative dehydrogenation | N-heterocycles | [200, 201] |
| $Fe_2O_3$ NPs | Solvothermal | CO oxidation | CO | [202] |
| $Fe_3O_4$ NPs | Aqueous FeCl3 solution and A. annua extract | Three-component reaction | Benzoxazinone, benzthioxazinone derivatives | [203] |
| $Fe_2O_3$ NPs | In situ generation | Water oxidation | Water | [204] |
| g-$Fe_2O_3$ nanorods | Fabrication from {110} and {001} planes | Nitrogen oxide reduction | NO | [205] |
| FeO NPs | Reduction of Fe salts | Hydrogenation | Alkenes, alkynes | [206] |

## 12.3 Co-Based Photocatalyst

Cobalt-based photocatalysts have been utilized in various reactions such as oxidation reaction, reduction, and water splitting. Table 09, discussed some Cobalt-based photocatalyst and their properties.

*Table 9: Co-Based Nano catalysts and Their Applications*

| Catalyst | Synthesis Method | Application | Substrate | References |
|---|---|---|---|---|
| $Co_3O_4$ nanorods | Calcination of cobalt hydroxide carbonate | CO oxidation | CO | [207] |
| $Co_3O_4$ NPs | Impregnation of activated carbon | CO oxidation | CO | [208] |
| Mesoporous $Co_3O_4$ NPs | Nanocasting | Ethylene oxidation | Ethylene | [209] |
| $Co_3O_4$ NPs | Wetness impregnation | Alcohol oxidation | Alcohols | [210] |
| N-doped graphite-coated Co NPs | Nano-coating | Alcohol oxidation to esters | Alcohols | [211] |
| $Co_3O_4$ NPs | Nano-coating | Amine oxidation to nitriles | Amines | [212] |
| N-graphene-modified $Co_3O_4$ NPs | Nano-coating | Hydrogenation of heteroarenes | Heteroarenes | [213] |

## 12.4 Ni-Based Catalysts

Ni is quite abundant in Earth's crust with 84 g present in every one ton. Nickel-based nanomaterials act as carrier or support materials in various reactions as exhibited in the table 10.

*Table 10: Ni-Based Nano catalysts and Their Applications*

| Catalyst | Synthesis Method | Application | Substrate | References |
|---|---|---|---|---|
| Ni NPs (PEG-stabilized) | Reduction of Ni oxalate with hydrazine | Reduction of p-nitrophenol | p-nitrophenol | [214] |
| Ni NPs (Ionic liquid-stabilized) | Reduction of nickel(II) salt with hydrazine | Hydrogenation of CQC double bonds | CQC double bonds | [215] |
| Ni NPs (MOF-supported) | Impregnation or synthesis within MOF | Reduction of nitroarenes | Nitroarenes | [216, 217] |
| Ni NPs ($Fe_3O_4$-supported) | Impregnation | CO oxidation | CO | [218] |
| Ni NPs (P-doped) | Chemical vapor deposition | Hydrogenation | Alkenes | [219] |
| Ni NPs (N-doped) | Chemical vapor deposition | Oxygen reduction | Oxygen | [220] |

## 12.5 Cu-Based Catalyst

Cu is quite abundant in earth crust with 60 g being present in every ton. Owed to their environment-friendly behavior, wide availability, efficiency, and inexpensiveness, Cu-based photocatalysts are used in various reactions as displayed in table 11.

*Table 11: Cu-based Nano catalysts and Their Applications*

| Catalyst | Synthesis Method | Application | Substrate | References |
|---|---|---|---|---|
| $Cu_2ONPs@PVP$ | Complexation-reduction | Click reaction (CuAAC) | Benzyl azide, terminal alkynes | [221] |
| $Cu_3N/Fe_3N@SiO_2$ | Infiltration and annealing | Click reaction (CuAAC) | Benzyl azide, terminal alkynes | [222] |
| Cu NPs (supported on various materials) | Impregnation or synthesis | C–X coupling reactions | Thiols, aryl halides, phenols | [223, 224] |
| $Cu_2O$ nanocubes | Synthesis from $Cu(OAc)2$ | C–O cross-coupling | Aryl halides, phenols | [225] |
| CuI NPs | Synthesis from CuI | C–N cross-coupling | Aryl halides, amines | [226] |

## 13. Types of Catalytic Reactions for Water Treatment

Many catalytic methods are used in water treatments for the degradation of pollutants, they can involve oxidation and reduction, but the main goal is to convert the harmful contaminants in water to harmless products. The main hurdles in water purification with TM-based heterogeneous catalysts are

- To ensure that catalytic reactions occur at the catalyst's surface, not via direct contact of oxidizing/reducing agent with water or via homogenous catalysis.

- The catalyst active sites don't require external regeneration.

- No toxic compounds are formed.

The main oxidation-based water treatment is wet air oxidation (WAO) and advanced oxidation processes(AOPs). The operating conditions for WAO are severe such as high temperatures (125-320°C) and high pressure (0.5-20MPa) and WAO is suitable for treatments that involve eradicating high pollutant concentrations. WAO utilizes organic radicals such as R from RH and inorganic radicals (OH) generated via heat[227].

On the other hand, AOPs work at ambient temperature and pressure and involve the production of OH radicals by chemical or photo-chemical decomposition of $O_3$ or $H_2O_2$. AOP was then further extended to photocatalysis, electrolysis, and other techniques that include strong oxidizing agents[228]. Figure 10 gives an overview of the classification of AOPs.

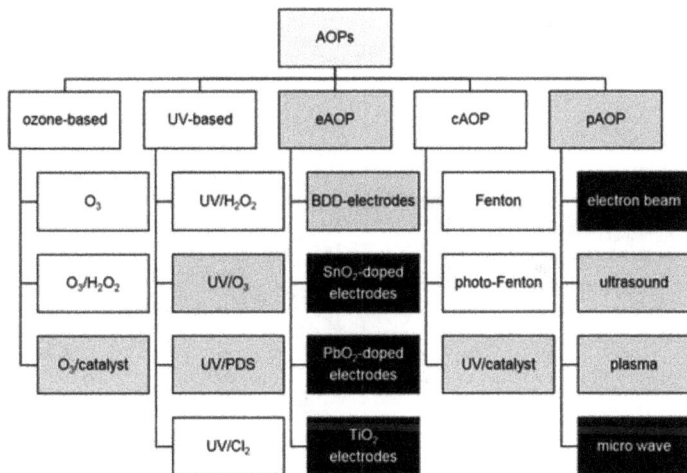

Figure 10: Classification of Different Advanced Oxidation Processes[229]

The Redox potentials associated with Oxidizing agents in the wet oxidation and advanced oxidation are reported in table 12.

Table 12: Redox potentials of the Oxidants in advanced oxidation processes. [230, 231]

| Oxidant | Half-Reaction | Redox Potential (V vs NHE) |
|---|---|---|
| $S_2O_8^{2-}$ | $S_2O_8^{2-} + 2H^+ + 2e^- \rightarrow 2HSO^{4-}$ | +2.08 |
| $O_3$ | $O^3 + 2H^+ + 2e^- \rightarrow O_2 + H_2O$ | +2.07 |
| $HSO^{5-}$ | $HSO_5^- + 2H^+ + 2e^- \rightarrow HSO_4^- + H_2O$ | +1.81 |
| $H_2O_2$ | $H_2O_2 + 2H^+ + 2e^- \rightarrow 2H_2O$ | +1.78 |
| $O_2$ | $O_2 + 4H^+ + 4e^- \rightarrow 2H_2O$ | +1.23 |
| $HO^\bullet$ | $HO^\bullet + e^- \rightarrow OH^-$ | +2.73 |
| $SO_4^{\bullet-}$ | $SO_4^{\bullet-} + e^- \rightarrow SO_4^{2-}$ | +2.43 |
| $ClO^\bullet$ | $ClO^\bullet + e^- \rightarrow Cl^-$ | +2.43 |

Water treatment with reduction is less common and is used for the conversion of particular pollutants such as nitrate into nitrogen or to eradicate Cl. Nitrate removal from water is carried out at room temperature and pressure with a noble metal catalyst[232].

## 14. Antimicrobial Properties of Transition Metals

Certain TMs such as Cu, Zn, Ag, and Fe have antimicrobial capabilities making them a contender for water treatment. The antimicrobial action of these materials is associated with their tendency to react with the microbial cell components. The transition metal-based nanomaterials employed in the eradication of pharmaceuticals, and dyes have antimicrobial properties. The mechanism behind the antimicrobial degradation of contaminants I still terse, but it is suggested that in the early stage, reactive oxygen species are generated, that interact with the proteins of microbes, therefore resulting in the denaturation of microbes. The microbe wall is destroyed giving access to oxygen species to disturb the respiratory system of the microbe, ultimately leading to its death[233]. Figure 11 explains this mechanism.

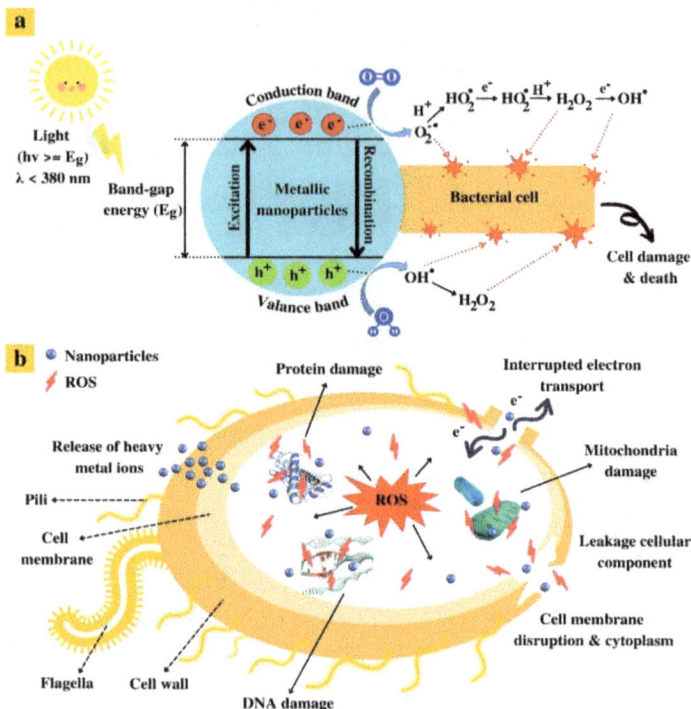

Figure 11: (a) Proposed Apparatus for the Generation of Reactive Oxygen Species (ROS) by Nanoparticles(b) Proposed Apparatus for the Antibacterial Activity of Nanoparticle-Generated Reactive Oxygen Species [233].

## 15. Disinfection Using Transition Metal Ions

The eradication of viruses via photocatalysts could potentially overcome the existing limitations of traditional water treatments. In this method, the photocatalyst destroys the cell membrane thus eliminating the proteins, and RNA via the production of ROS, in this way, photocatalysts inhibit virus growth or kill the virus cells. Titanium dioxide is a well-known photocatalyst, which can also be used for the deactivation of various viruses and bacteria such as norovirus, and bacteriophage.

Fenton reactions have been explored with Fe and Cu-based catalysts to kill $MS_2$ coliphage. The oxidant that inactivates the virus is hydroxyl which is generated via reaction between oxidant and hydrogen peroxide. In this study, the concentration of catalyst, as well as sunlight dependence on virus activation was explored[234].

Applications for Earth-Abundant Transition Metals                    Materials Research Forum LLC
Materials Research Foundations 179 (2025) 206-258          https://doi.org/21741/9781644903711-9

## 16.    Advantages of Earth-Abundant Transition Metal for Water Treatment

### 16.1    Cost-Effectiveness

Efficiency is one of the main advantages of employing transition metals such as iron, manganese, and copper in water treatment procedures. This metal is less expensive than other options, particularly precious metals or synthetic compounds, because of several variables and the first advantage is the cost-effectiveness[158].

*The Earth's Crust's Richness*

As we know the earth's crust contains a vast variety of transition metals, and there are significant natural reserves all around the planet. So, in the Earth's crust, iron, for instance, is the fourth most common metal; copper and manganese are also present in a significant amount. So, in rare commodities, this widespread supply reduces price volatility and offers stability. The cost of treating water during operation is directly reduced as a result of this metal's lower costs and processing expenses[235].

*Lower Material Expenses*

As compared to precious metals used in catalytic processes like gold, palladium, and platinum, substitute metals are far less expensive. So the scarcity, difficult extraction procedures, and strong demand from sectors like jewelry and electronics are the main causes of precious metals' high prices and the large quantities of metals are required for industrial water treatment, and bulk metals are especially helpful because of their generally cheaper market pricing[236].

*Minimizing workers*

The usage of plentiful metal alternatives can increase consistency in work in addition to lowering the investment. So, since they need to be replaced or reprocessed more frequently than other metals, these metals are typically robust and are durable. The operating costs can be further reduced by constantly remanufacturing and reusing metal-based catalysts, which are utilized in processes like the Fenton reactions. Throughout the therapy, its resilience and reusability contribute to a decrease in expenses[237].

*Financial Scalability*

Now, another factor that makes scalability easier is the low cost of various metal substitutes available at this time. So, measuring the effectiveness of water treatment is essential since the demand for clean water around the world keeps rising. So, water treatment plants are able to grow in response to rising population or corporate demand because of the ubiquitous and affordable use of steel, which keeps prices from rising as the work is done.

Therefore, transition metals are a useful option for water treatment due to their affordability. And the worldwide water crisis can be solved with these metals because of their low consumption and capacity to lower capital and operating expenses. Water treatment facilities can reach high levels of efficacy and efficiency by utilizing these benefits without having to deal with the financial strain of alternative, costlier solutions[238].

## 16.2    Environmental Sustainability

IAs, it has grown crucial as the globe faces more and more difficulties like pollution, hunger, and climate change. So, transition earth metals, such as iron, manganese, copper, and nickel, have numerous benefits for the environment when used in water treatment activities. So these advantages contribute to the development of more secure and safe water solutions[239].

*Reduced Environmental Impact and Abundance Extracted*

The crust of the Earth has significant concentrations of transition metals in the planet. One of the most common metals is iron, which comprises around 5% of the Earth's crust. And since these metals are more common than rare or valuable metals, which frequently need more intervention and result in considerable environmental damage, their extraction poses less of a threat to the outer environment. So these metals' low environmental impact during extraction and processing adds to the overall sustainability of the water treatment methods that employ them[240].

*Decreased Toxicology and Eco-Friendly Features*

Compared to the rare metals or synthetic chemicals utilized in some water systems, many transition metals found throughout the world are less harmful to ecosystems and to humans. For instance, at the right levels, iron and manganese, which are crucial components of biological systems, are not harmful to the system and the danger of disease in groundwater and the water supply can be decreased when certain metals are employed in the water treatment. So these metals can also be biocompatible and are less likely to cause environmental issues because they are found in nature[241].

*Lowering the level of secondary pollutants*

Preventing secondary contamination, which occurs when the treatment process releases more contaminants into the environment, is one of the primary concerns in water treatment. And alternative metals can help lower this danger because they are widely available in the world. So, to target and eradicate dangerous germs, metal-based coagulants and catalysts are frequently manufactured under specific pressure. So furthermore, most of these metals' byproducts are non-toxic, which lessens their possible impact on the environment

Therefore, transition metals provide many environmental benefits when used in water treatment around the world. So, these benefits include their availability, non-toxicity, adherence to green chemistry principles, recyclability, and compatibility with renewable energy sources. These metals offer a technique to treat water more sustainably, minimizing its negative effects on the environment while preserving or even enhancing the removal of pathogens. Using the sustainability advantages of these metals is essential to creating water treatment technologies that are advantageous to the economy and environment as the demand for clean water around the whole world rises[242].

## 16.3    Adjustability

Therefore, adjustable qualities of transition metals, which include iron, manganese, copper, and nickel, are among the main justifications for their application in the industry of water treatment and these metals are adaptable and helpful since they may be chemically altered and processed to enhance their performance in various water treatment procedures. So because of this metal's tunability, individual diseases and treatment goals can be addressed, enhancing the overall advantages and improvements of cleaning the water[243].

*Various States of Oxidation*

So, the ability of transition metals rich in soil to live in many oxidation states is a crucial characteristic that can be applied to the treatment of the water. For instance, iron has two states: Fe III and Fe all and the transition between these states is useful in the Fenton reaction, wherein Fe II catalyzes the formation of hydroxyl radical peroxides from hydrogen, which degrades the organic contaminants. So these metals have the capacity to alter their oxidation state, which enables them to take part in a variety of redox reactions and treat a wide range of illnesses, including those caused by bacteria, heavy metals, and organic chemicals[244].

*Surface alteration and functionality*

As many transition metals can have their characteristics changed to increase their stability, selectivity, and reactivity in the treatment of the water and for instance, to increase the dispersion of metal nanoparticles in water, inhibit aggregation, and optimize their interaction with targets, their surfaces can be coated with an organic or inorganic layer. The adsorption capability of these metals toward certain pollutants, such as arsenic, fluoride, or heavy metals, can be increased by working with particular ligands or polymers. Maintaining this chemical environment enables the creation of metal products with desirable qualities that can address issues related to water treatment[245].

*Doped and alloy process*

The doping or alloying of various metals in the soil with other elements can further tailor their properties to meet the specific needs of water for example, doping a metal oxide with nonmetals such as sulfur or carbon can improve its electrical conductivity and catalytic activity, greatly improving its performance in the high-oxidation processes. Alloying copper with zinc to make brass can enhance its anticorrosive and antibacterial properties, making it more effective in antibacterial applications these modifications allow metal products to be produced with specialized properties that can target specific diseases or increase the durability and performance of treatment processes for water.

So, the earth-rich transition metals have tunable features that make them adaptable to a wide range of water applications. So, these metals can be alloyed or doped to change their size, shape, oxidation state, surface characteristics, and doping to achieve our desired results. Therefore, enhancing their catalytic activity, stability, and redox potential can expand their use, boost stability, and efficiency, and advance water technology. So this change not only improves the effectiveness of water treatment but also makes it possible to create solutions for a range of issues related to pollution of water[246].

## 17.  Challenges for Earth-Abundant Transition Metal for Water Treatment

The earth-abundant transition metals present a number of significant and varied challenges as firstly, these metals must be robust in the face of complex water matrices, which may contain a variety of organic and inorganic contaminants. So over time, these matrices may cause the metals to degrade or become inactive, which would reduce their usefulness in long-term applications. Now secondly, regeneration of these metals is a challenge because repeated cycles may cause a loss of catalytic activity, which would definitely reduce their efficacy. Furthermore, a number of issues need to be addressed in order to develop more reliable and sustainable water treatment methods that use earth-abundant transition metals[241]. Now these issues include the stability of transition metals in varying pH levels and under different environmental conditions, as fluctuations can affect their performance; additionally, the possibility of metals leaching into the treated water raises environmental and health concerns, requiring closely monitored management.

## 17.1   The Durability

Now, even though metals like iron, manganese, and copper are abundant and beneficial, there are still significant challenges with their application in the treatment of water so when exposed to water and different diseases, these metals are prone to oxidation and corrosion, which can lead to a decline in performance and subsequent the illnesses. For instance, iron will transform into iron oxides, which will lessen the potency of the germs it kills and complicate the cleaning. So over time, things like organic matter buildup, catalytic activity, and ions will deteriorate so despite its ability to oxidize contaminants, manganese loses its reactivity as manganese oxides form, rendering the catalytic site useless. In addition, these metals may interact with organic matter NOM in water to produce scaling, which lowers the metal's active surface and promotes the regeneration of the coating standard and an accumulation of dirt can clog the pores of a metal catalyst or adsorbent, decreasing its ability to eradicate microorganisms and necessitating regular equipment replacement or cleaning so this not only drives up operational expenses but also prompts questions about the waste's potential effects on our environment because materials must be resistant to various chemicals and microorganisms found in various water sources, including biofilms in wastewater, durability presents an extra problem[247]. So, to overcome these obstacles, robust designs that can survive real-world conditions, enhanced catalytic processes, and corrosion-resistant materials are needed. Then it will take more investigation into metal degradation mechanisms and mitigation techniques to enable these materials to be used for long-term, extensive treatment of water.

## 17.2   Regeneration

Therefore, sustainability and affordability of the water treatment process are directly impacted by the competitiveness of recycled materials because of their catalytic qualities and capacity to remove a wide range of contaminants, metals including iron, manganese, and copper are frequently utilized in the water treatment but as time passes, the filth, scale, and debris that accumulate on their surfaces prevents them from being as effective in catalyzing reactions or absorbing the noxious air. So reconditioning entails repurposing the metal, which means scraping off this substance and reworking its surface but there are a lot of issues with this approach. Cleaning, heat treatment, and pickling are some of the most recycling techniques that are popular. These processes can be harsh on the metal and cause partial breaking, which lowers the metal's quality and workability over time. In addition, these techniques frequently call for the use of acids or bases, which have the potential to harm the environment and need to be handled and disposed of very carefully. So, damage to the metal or the substance supporting it during the recycling process is another significant challenge. So, for instance, heat regeneration may alter the crystal structure of the metal, lowering its catalytic activity. In a similar way, mechanical deterioration and product loss may result from chemical renewal weakening the metal support matrix. Now this reduces the metal's lifespan and raises operational expenses because more metal needs to be recycled or

Applications for Earth-Abundant Transition Metals                    Materials Research Forum LLC
Materials Research Foundations 179 (2025) 206-258          https://doi.org/21741/9781644903711-9

replaced. So, in conclusion, regeneration poses major issues even though it is crucial for limiting the amount of metal in the treatment of water. Therefore, harsh circumstances necessary for regeneration may result in the loss of metal, structural damage, and environmental harm. The process's sustainability and security may also be impacted by energy and financial considerations so to ensure the long-term potential of metal treatment systems in water treatment, it is imperative to design a more economical, sustainable, and efficient solution[248].

## 17.3   Long-Term Stability of Transition Metal in Complex Aqueous Matrices

Now a significant obstacle to water utilization is the long-term stability of transition metals that are common in soil, such as iron, manganese, and copper, in the complex aqueous matrices and these metals are valuable because of their efficiency, availability, and catalytic qualities; nevertheless, when the water system undergoes changes, their performance may be significantly impacted. So, the presence and longevity of metals are greatly impacted by the range of organic and inorganic pollutants, pH fluctuations, and varying ionic compositions seen in the complex water matrices. Therefore, the development of different ions and compounds in water, or the synthesis of non-metals, is one of the main issues for instance, transition metals like iron can create anhydrous metal phosphates or carbonates in high-phosphate or calcium environments and this can be catalytic and restrict the availability of metals in the area. Waste has the potential to obstruct vision or seep out of the water, delaying the treatment process in its entirety and normal pH levels of aqueous matrices and variations in redox circumstances provide further challenges to the chemical stability of these metals[248]. Metals like iron and manganese will dissolve in acidic environments or change their oxidation state, which will lead to the creation of fewer or inactive species for instance  Iron hydroxide, is created when iron is oxidized and is less reactive and soluble in bacteria that can catalyze their destruction and on the other hand, these metals will be lowered to a lower oxidation state in a reducing environment, making them less effective in oxidation processes like the destruction of the contaminants. So, another level of complexity is introduced by competing ions in complicated water matrices and these ions have the capacity to attach themselves to the metal's active sites and prevent the metal from catalyzing the process or killing the microorganisms for instance, ions such as chloride or sulfate can compete with bacteria for active sites, lowering the amount of treatment needed overall and necessitating higher iron levels or more frequent regeneration. In summary, a variety of chemical, biological, and mechanical variables significantly impede the long-term stability of the Earth's many transition metals in intricate aquatic matrices however, the advantages of these metals may be progressively diminished by the onset of water disputes, biofouling, competition with other ions, and degradation of product. So strong materials, innovative building techniques, and a thorough understanding of the interactions between these metals and other minerals in water systems are required to address these issues so ensuring the long-term and effective use of iron in water treatment worldwide requires taking this crucial step[249].

## 18.   Future direction and outlook

The research on scalability, cost-effectiveness, and integration with current technologies is essential to advancing the use of transition metals in water treatment globally and scalability is the ability to use these metals on a commercial scale while keeping them consistent and taking economic and environmental issues into account. The goal of discounting is to reduce the cost of these metals by improving their efficiency, recyclable nature, and efficiency so using current technology to include these metals in water treatment systems will increase their efficacy without requiring significant changes. Therefore, to enable the world to move toward a more effective and efficient solution to the global water crisis, all of these initiatives are so essential[250].

## 18.1   Scalability

Therefore, scalability is critical to the continued research and application of metal gradients in water treatment across the world so the problem of controlling efficacy, efficiency, and the environment on a broader scale arises when science transitions from lab research to practical applications. A tiny quantity of flexible metal can accomplish great control performance in the laboratory but still, a lot of things need to be taken into account to make sure the process keeps working as it goes along. So, the availability and stability of certain metals are important considerations and these metals are more common than precious metals, but if their widespread use is not strictly controlled, it may have a negative effect on the environment. Therefore, the primary focus of research should be on developing environmentally friendly mining and refining methods that minimize environmental damage. As in small-scale manufacturing, the size, surface area, and purity of materials are typically strictly regulated. However, there are difficulties in achieving these qualities on a large scale of the industry[251].

So, the availability and stability of certain metals are important considerations and these metals are more common than precious metals, but if their widespread use is not strictly controlled, it may have negative effects on the environment. Therefore, the primary focus of research should be on developing environmentally friendly mining and refining methods that minimize environmental damage so in small-scale manufacturing, the size, surface area, and purity of materials are typically strictly regulated. However, there are difficulties in achieving these qualities on a big industrial scale as it is essential to look into cost-cutting measures like maximizing mix preparation, recycling metals, and reducing energy use during production. In addition, the development of isolated and adaptable treatment systems that can be efficiently scaled up or down in response to the various districts' water treatment requirements appears to improve the overall cost viability of using earth-abundant move metals. In the last, the long-term operational solidity and solidness of these metals in actual water treatment scenarios must also be taken into account for adaptability, and metals used in large-scale water treatment must be protected from fouling and erosion[252].

## 18.2   Cost reduction

The reduced toll could be a crucial question in the development of earth-abundant move metals for water treatment, as it directly affects the feasibility and widespread use of these developments and these metals are more plentiful and less expensive than their rare or valuable counterparts, although the general cost of using them in large-scale water treatment systems and can still be significant. A careful oversight of the expenses associated with extraction, preparation, union, and shipping is necessary to ensure that these commodities remain economically viable because it directly impacts the viability and widespread application of these advancements, reduced toll may be a critical question in the development of earth-abundant mobile metals for treatment of water. While these metals are more accessible and less costly than their rare or valuable counterparts, employing them in large-scale water treatment systems can still come with a hefty price tag to keep these commodities economically viable, close monitoring of the costs related to extraction, preparation, union, and shipment is also required.

In addition, it is crucial to reduce the operational expenses associated with transporting these metals via the current water treatment systems and this includes reducing the energy needed for popular applications of move metals in water treatment, such as catalysis or adsorption. By increasing these forms' proficiency, energy expenses can be reduced, leading to a generally more conservative treatment plan and isolated designs that allow for easy scaling up or down in accordance with the specific requirements of a water treatment office can help optimize expenses by avoiding unnecessary hardware and foundation use. Ultimately, achieving a reduction in the use of earth-abundant move metals for water treatment may be a complex task requiring a combination of creative blending techniques, persuasive reusing tactics, and energy-efficient processes. Addressing these problems won't exactly make these metals more sensible, but it will increase their maintainability and common sense in global water treatment projects. So as research continues, hitting critical fetched drops will be crucial to the widespread use of these promising materials in addressing global issues of water treatment[252].

## 18.3    Combining with Current Technologies

Integrating earth-rich metals into current water treatment technologies may be a fundamental research topic that has a significant impact on their practical use and the primary difficulty is ensuring that these metals can be reliably integrated into existing systems without necessitating extensive changes or sacrificing the functionality. So, for the integration to be successful and economical, it must be compatible with current forms, such as filtration, adsorption, and catalysis. Therefore, research has to focus on comprehending the relationships that exist between these metals and other materials that are frequently used in the treatment of water, like films, activated carbon, and chemical coagulants and this knowledge will guide the development of hybrid frameworks that boost treatment efficiency by fusing the best aspects of traditional materials with the unique characteristics of mobile metals. Another important aspect of integration is designing flexible frameworks that can accommodate the shifting characteristics of moveable metals. For example, for certain metals to function at their best, specific operating parameters, like pH or temperature regulation, may be necessary[253].

Furthermore, it is necessary to take coordinate framework adaptability into account and move metals to be compatible with both large-scale, centralized offices and small-scale, decentralized water treatment units. So, to do this, a thorough comprehension of the behavior of these metals on various scales is necessary, as is the development of defined usage guidelines. The integration of movable metals into current innovation can be made more practical and wider by creating flexible frameworks that can be scaled up or down as needed. In summary, the successful incorporation of earth-abundant metals into current water treatment advances depends on establishing flexible frameworks, ensuring flexibility, and having an intuitive understanding of fabric. By addressing these research issues, it will be feasible to fully utilize these metals, enhancing the efficacy and sustainability of water treatment systems globally[253].

## 18.4    Potential Role of EATMs in Global Water Problems

The earth-abundant metals have significant potential for addressing global water issues, particularly in providing affordable and sustainable solutions for water treatment and filtration and these metals, which include press, manganese, copper, and zinc, are more widely available and less expensive than their valuable counterparts, which attracts them for use in large-scale applications in areas with the limited resources. Therefore, they are effective at eliminating a variety of toxins, including heavy metals, organic substances, and pathogens, thanks to their intriguing chemical characteristics, which include the ability to catalyze redox reactions and adsorb pollutants. As an example, nanoscale press particles have shown amazing promise in cleaning up contaminated groundwater by reducing toxic metals like arsenic and chromium to less harmful forms. Additionally, the ability of manganese-based catalysts to degrade various naturally occurring toxins, including pesticides and medications, which are increasingly turning up in water supplies across the globe, is being researched[254].

Furthermore, using earth-abundant move metals aligns well with green chemistry and maintainability standards. In conclusion, earth-abundant move metals have enormous potential to address global water challenges. Therefore, they are ideal for developing creative and long-lasting water treatment arrangements due to their affordability, adaptability, and accessibility and as research in this area advances, these metals are probably going to become increasingly important in ensuring that communities worldwide have access to safe and clean water, helping to alleviate one of the most pressing environmental and public health concerns of our day. So, by utilizing the intriguing qualities of these metals, we are prepared to advance toward a time when everyone will have access to cost-effective and productive water treatment, regardless of geographical or financial constraints.

## Conclusion

Earth-abundant transition metals offer a promising solution for water treatment due to their cost-effectiveness, environmental sustainability, and adjustability. However, challenges such as durability, regeneration, and long-term stability in complex water matrices need to be addressed. Continued research and development are essential to improve their efficiency and integrate them effectively with current technologies. By overcoming these challenges, EATMs can play a significant role in addressing global water scarcity and pollution issues.

## References

[1] Jéquier, Constant.Water as an Essential Nutrient: The Physiological Basis of Hydration,European journal of clinical nutrition,115-123.64 (2010) https://doi.org/10.1038/ejcn.2009.111

[2] Kılıç.The Importance of Water and Conscious Use of Water,International Journal of Hydrology,239-241.4 (2020) https://doi.org/10.15406/ijh.2020.04.00250

[3] Mitchell, Frisbie.A Comprehensive Survey and Analysis of International Drinking Water Regulations for Inorganic Chemicals with Comparisons to the World Health Organization's Drinking-Water Guidelines,Plos one,e0287937.18 (2023) https://doi.org/10.1371/journal.pone.0287937

[4] Viman, Oroian, Fleşeriu.Types of Water Pollution: Point Source and Nonpoint Source,Aquaculture, Aquarium, Conservation & Legislation,393-397.3 (2010)

[5] Akın, Akın.Suyun Önemi, Türkiye'de Su Potansiyeli, Su Havzalari Ve Su Kirliliği,Ankara Üniversitesi Dil ve Tarih-Coğrafya Fakültesi Dergisi,105-118.47 (2007) https://doi.org/10.1501/Dtcfder_0000000992

[6] Dwivedi.Researches in Water Pollution: A Review,International Research Journal of Natural and Applied Sciences,118-142.4 (2017)

[7] Chan, Chong, Law, Hassell.A Review on Anaerobic-Aerobic Treatment of Industrial and Municipal Wastewater,Chemical engineering journal,1-18.155 (2009) https://doi.org/10.1016/j.cej.2009.06.041

[8] Gedda, Balakrishnan, Devi, Shah, Gandhi, Gandh, Shah.Introduction to Conventional Wastewater Treatment Technologies: Limitations and Recent Advances,Mater. Res. Found,1-36.91 (2021) https://doi.org/10.21741/9781644901151-1

[9] Nugent.The Impact of Urban Agriculture on the Household and Local Economies,Bakker N., Dubbeling M., Gündel S., Sabel-Koshella U., de Zeeuw H. Growing cities, growing food. Urban agriculture on the policy agenda. Feldafing, Germany: Zentralstelle für Ernährung und Landwirtschaft (ZEL),67-95(2000)

[10] Ankush, Ritambhara, Lamba, Deepika, Prakash.Cadmium in Environment-an Overview,Cadmium Toxicity in Water: Challenges and Solutions,3-20(2024) https://doi.org/10.1007/978-3-031-54005-9_1

[11] Majumdar.The Blue Baby Syndrome: Nitrate Poisoning in Humans,Resonance,20-30.8 (2003) https://doi.org/10.1007/BF02840703

[12] Bryan, Alexander, Coughlin, Milkowski, Boffetta.Ingested Nitrate and Nitrite and Stomach Cancer Risk: An Updated Review,Food and Chemical Toxicology,3646-3665.50 (2012) https://doi.org/10.1016/j.fct.2012.07.062

[13] Villarín, Merel.Paradigm Shifts and Current Challenges in Wastewater Management,Journal of hazardous materials,122139.390 (2020) https://doi.org/10.1016/j.jhazmat.2020.122139

[14] Tortajada.Contributions of Recycled Wastewater to Clean Water and Sanitation Sustainable Development Goals,NPJ Clean Water,22.3 (2020) https://doi.org/10.1038/s41545-020-0069-3

[15] Al-Juaidi, Kaluarachchi, Mousa.Hydrologic-Economic Model for Sustainable Water Resources Management in a Coastal Aquifer,Journal of Hydrologic Engineering,04014020.19 (2014) https://doi.org/10.1061/(ASCE)HE.1943-5584.0000960

[16] Melin, Jefferson, Bixio, Thoeye, De Wilde, De Koning, van der Graaf, Wintgens.Membrane Bioreactor Technology for Wastewater Treatment and Reuse,Desalination,271-282.187 (2006) https://doi.org/10.1016/j.desal.2005.04.086

[17] Vymazal.Constructed Wetlands for Wastewater Treatment,Water,530-549.2 (2010) https://doi.org/10.3390/w2030530

[18] Yoshida, Mønster, Scheutz.Plant-Integrated Measurement of Greenhouse Gas Emissions from a Municipal Wastewater Treatment Plant,Water research,108-118.61 (2014) https://doi.org/10.1016/j.watres.2014.05.014

[19] Yenkie.Integrating the Three E's in Wastewater Treatment: Efficient Design, Economic Viability, and Environmental Sustainability,Current Opinion in Chemical Engineering,131-138.26 (2019) https://doi.org/10.1016/j.coche.2019.09.002

[20] Jodar-Abellan, López-Ortiz, Melgarejo-Moreno.Wastewater Treatment and Water Reuse in Spain. Current Situation and Perspectives,Water,1551.11 (2019) https://doi.org/10.3390/w11081551

[21] Chaúque, Rott.Solar Disinfection (Sodis) Technologies as Alternative for Large-Scale Public Drinking Water Supply: Advances and Challenges,Chemosphere,130754.281 (2021) https://doi.org/10.1016/j.chemosphere.2021.130754

[22] Zheng, Zhang, Yu, Chen, Cheng, Min, Wang, Xiao, Wang.Overview of Membrane Technology Applications for Industrial Wastewater Treatment in China to Increase Water Supply,Resources, Conservation and Recycling,1-10.105 (2015) https://doi.org/10.1016/j.resconrec.2015.09.012

[23] Kivaisi.The Potential for Constructed Wetlands for Wastewater Treatment and Reuse in Developing Countries: A Review,Ecological engineering,545-560.16 (2001) https://doi.org/10.1016/S0925-8574(00)00113-0

[24] Herrero, Stuckey.Bioaugmentation and Its Application in Wastewater Treatment: A Review,Chemosphere,119-128.140 (2015) https://doi.org/10.1016/j.chemosphere.2014.10.033

[25] Stottmeister, Wießner, Kuschk, Kappelmeyer, Kaestner, Bederski, Müller, Moormann.Effects of Plants and Microorganisms in Constructed Wetlands for Wastewater Treatment,Biotechnology advances,93-117.22 (2003) https://doi.org/10.1016/j.biotechadv.2003.08.010

[26] Yan, Xu.Improving Winter Performance of Constructed Wetlands for Wastewater Treatment in Northern China: A Review,Wetlands,243-253.34 (2014) https://doi.org/10.1007/s13157-013-0444-7

[27] Vesilind, Wastewater Treatment Plant Design, IWA publishing2003.

[28] Zhang, Yang, Ngo, Guo, Jin, Dzakpasu, Yang, Wang, Wang, Ao.Current Status of Urban Wastewater Treatment Plants in China,Environment international,11-22.92 (2016) https://doi.org/10.1016/j.envint.2016.03.024

[29] Brame, Li, Alvarez.Nanotechnology-Enabled Water Treatment and Reuse: Emerging Opportunities and Challenges for Developing Countries,Trends in Food Science & Technology,618-624.22 (2011) https://doi.org/10.1016/j.tifs.2011.01.004

[30] Zeng, Chen, Dong, Liu.Efficiency Assessment of Urban Wastewater Treatment Plants in China: Considering Greenhouse Gas Emissions,Resources, Conservation and Recycling,157-165.120 (2017) https://doi.org/10.1016/j.resconrec.2016.12.005

[31] Sun, Chen, Wu, Wu, Zhang, Niu, Hu.Characteristics of Water Quality of Municipal Wastewater Treatment Plants in China: Implications for Resources Utilization and Management,Journal of Cleaner Production,1-9.131 (2016) https://doi.org/10.1016/j.jclepro.2016.05.068

[32] Kesari, Soni, Jamal, Tripathi, Lal, Jha, Siddiqui, Kumar, Tripathi, Ruokolainen.Wastewater Treatment and Reuse: A Review of Its Applications and Health Implications,Water, Air, & Soil Pollution,1-28.232 (2021) https://doi.org/10.1007/s11270-021-05154-8

[33] López-Morales, Rodríguez-Tapia.On the Economic Analysis of Wastewater Treatment and Reuse for Designing Strategies for Water Sustainability: Lessons from the Mexico Valley Basin,Resources, Conservation and Recycling,1-12.140 (2019) https://doi.org/10.1016/j.resconrec.2018.09.001

[34] Sonune, Ghate.Developments in Wastewater Treatment Methods,Desalination,55-63.167 (2004) https://doi.org/10.1016/j.desal.2004.06.113

[35] Sobsey, Stauber, Casanova, Brown, Elliott.Point of Use Household Drinking Water Filtration: A Practical, Effective Solution for Providing Sustained Access to Safe Drinking Water in the Developing World,Environmental science & technology,4261-4267.42 (2008) https://doi.org/10.1021/es702746n

[36] Ali, Khan, Peng, Naz, Sultan, Ali, Mahmood, Niaz.Identification and Elucidation of the Designing and Operational Issues of Trickling Filter Systems for Wastewater Treatment,Polish Journal of Environmental Studies.26 (2017) https://doi.org/10.15244/pjoes/70627

[37] Sardana, Cottrell, Soulsby, Aziz.Dissolved Organic Matter Processing and Photoreactivity in a Wastewater Treatment Constructed Wetland,Science of the total environment,923-934.648 (2019) https://doi.org/10.1016/j.scitotenv.2018.08.138

[38] Meena, Kannah, Sindhu, Ragavi, Kumar, Gunasekaran, Banu.Trends and Resource Recovery in Biological Wastewater Treatment System,Bioresource Technology Reports,100235.7 (2019) https://doi.org/10.1016/j.biteb.2019.100235

[39] Maurer, Rothenberger, Larsen.Decentralised Wastewater Treatment Technologies from a National Perspective: At What Cost Are They Competitive?,Water Science and Technology: Water Supply,145-154.5 (2005) https://doi.org/10.2166/ws.2005.0059

[40] Choudhary, Kumar, Sharma.Constructed Wetlands: An Approach for Wastewater Treatment,Elixir Pollut,3666-3672.37 (2011)

[41] Hamed, Khalafallah, Hassanien.Prediction of Wastewater Treatment Plant Performance Using Artificial Neural Networks,Environmental Modelling & Software,919-928.19 (2004) https://doi.org/10.1016/j.envsoft.2003.10.005

[42] Ashraf, Ramamurthy, Rene.Wastewater Treatment and Resource Recovery Technologies in the Brewery Industry: Current Trends and Emerging Practices,Sustainable Energy Technologies and Assessments,101432.47 (2021) https://doi.org/10.1016/j.seta.2021.101432

[43] Libhaber, Orozco-Jaramillo, Sustainable Treatment and Reuse of Municipal Wastewater: For Decision Makers and Practising Engineers, Iwa publishing2012. https://doi.org/10.2166/9781780400631

[44] Melo, da Costa, Pinto, Barroso, Oliveira.Adequacy Analysis of Drinking Water Treatment Technologies in Regard to the Parameter Turbidity, Considering the Quality of Natural Waters Treated by Large-Scale Wtps in Brazil,Environmental monitoring and assessment,1-12.191 (2019) https://doi.org/10.1007/s10661-019-7526-9

[45] Ram, Significance and Treatment of Volatile Organic Compounds in Water Supplies, CRC Press1990.

[46] Harvey, Smith, George.Effect of Organic Contamination Upon Microbial Distributions and Heterotrophic Uptake in a Cape Cod, Mass., Aquifer,Applied and Environmental Microbiology,1197-1202.48 (1984) https://doi.org/10.1128/aem.48.6.1197-1202.1984

[47] Organization, International Code of Conduct on the Distribution and Use of Pesticides: Guidelines for the Registration of Pesticides, World Health Organization, 2010.

[48] Bolognesi.Genotoxicity of Pesticides: A Review of Human Biomonitoring Studies,Mutation Research/Reviews in Mutation Research,251-272.543 (2003) https://doi.org/10.1016/S1383-5742(03)00015-2

[49] Sharma, Bhattacharya.Drinking Water Contamination and Treatment Techniques,Applied water science,1043-1067.7 (2017) https://doi.org/10.1007/s13201-016-0455-7

[50] Brown, Bishop, Rowan.The Role of Skin Absorption as a Route of Exposure for Volatile Organic Compounds (Vocs) in Drinking Water,American journal of public health,479-484.74 (1984) https://doi.org/10.2105/AJPH.74.5.479

[51] Wehrmann.Ground-Water Contamination by Volatile Organic Compounds: Site Characterization, Spatial and Temporal Variability,ISWS Contract Report CR 591,(1996)

[52] Pagga, Brown.The Degradation of Dyestuffs: Part Ii Behaviour of Dyestuffs in Aerobic Biodegradation Tests,Chemosphere,479-491.15 (1986) https://doi.org/10.1016/0045-6535(86)90542-4

[53] Bianco Prevot, Baiocchi, Brussino, Pramauro, Savarino, Augugliaro, Marci, Palmisano.Photocatalytic Degradation of Acid Blue 80 in Aqueous Solutions Containing Tio2 Suspensions,Environmental science & technology,971-976.35 (2001) https://doi.org/10.1021/es000162v

[54] Pal, He, Jekel, Reinhard, Gin.Emerging Contaminants of Public Health Significance as Water Quality Indicator Compounds in the Urban Water Cycle,Environment international,46-62.71 (2014) https://doi.org/10.1016/j.envint.2014.05.025

[55] Pal, Gin, Lin, Reinhard.Impacts of Emerging Organic Contaminants on Freshwater Resources: Review of Recent Occurrences, Sources, Fate and Effects,Science of the total environment,6062-6069.408 (2010) https://doi.org/10.1016/j.scitotenv.2010.09.026

[56] Stuart, Lapworth, Crane, Hart.Review of Risk from Potential Emerging Contaminants in Uk Groundwater,Science of the Total Environment,1-21.416 (2012) https://doi.org/10.1016/j.scitotenv.2011.11.072

[57] Lapworth, Baran, Stuart, Ward.Emerging Organic Contaminants in Groundwater: A Review of Sources, Fate and Occurrence,Environmental pollution,287-303.163 (2012) https://doi.org/10.1016/j.envpol.2011.12.034

[58] Mohan, Sarswat, Ok, Pittman Jr.Organic and Inorganic Contaminants Removal from Water with Biochar, a Renewable, Low Cost and Sustainable Adsorbent-a Critical Review,Bioresource technology,191-202.160 (2014) https://doi.org/10.1016/j.biortech.2014.01.120

[59] Sundaram, Viswanathan, Meenakshi.Defluoridation Chemistry of Synthetic Hydroxyapatite at Nano Scale: Equilibrium and Kinetic Studies,Journal of Hazardous Materials,206-215.155 (2008) https://doi.org/10.1016/j.jhazmat.2007.11.048

[60] Srivastav, Ranjan, Inorganic Water Pollutants, Inorganic Pollutants in Water, Elsevier2020, pp. 1-15. https://doi.org/10.1016/B978-0-12-818965-8.00001-9

[61] Daschner, Rüden, Simon, Clotten.Microbiological Contamination of Drinking Water in a Commercial Household Water Filter System,European Journal of Clinical Microbiology and Infectious Diseases,233-237.15 (1996) https://doi.org/10.1007/BF01591360

[62] Ashbolt.Microbial Contamination of Drinking Water and Disease Outcomes in Developing Regions,Toxicology,229-238.198 (2004) https://doi.org/10.1016/j.tox.2004.01.030

[63] BC.Cyanobacterial Toxins: Removal During Drinking Water Treatment, and Human Risk Assessment,Environ Health Perspect,113-122.108 (2000) https://doi.org/10.1289/ehp.00108s1113

[64] Rao, Gupta, Bhaskar, Jayaraj.Toxins and Bioactive Compounds from Cyanobacteria and Their Implications on Human Health,Journal of Environmental Biology,215-224.23 (2002)

[65] Inamori, Fujimoto.Water Quality and Standards-Vol. Ii, Microbial/Biological Contamination of Water,Encyclopaedia of Life support systems (EOLSS),(2009)

[66] Nwachcuku, Gerba.Emerging Waterborne Pathogens: Can We Kill Them All?,Current opinion in biotechnology,175-180.15 (2004) https://doi.org/10.1016/j.copbio.2004.04.010

[67] Rusin, Rose, Haas, Gerba.Risk Assessment of Opportunistic Bacterial Pathogens in Drinking Water,Reviews of Environmental Contamination and Toxicology: Continuation of Residue Reviews,57-83(1997) https://doi.org/10.1007/978-1-4612-1964-4_2

[68] Binesh, Mohammadi, Mowavi, Parvaresh.Measurement of Heavy Radioactive Pollution: Radon and Radium in Drinking Water Samples of Mashhad,Int J Curr Res,54-58.10 (2010)

[69] Hakl, Hunyadi, Varga, Csige.Determination of Radon and Radium Content of Water Samples by Ssntd Technique,Radiation measurements,657-658.25 (1995) https://doi.org/10.1016/1350-4487(95)00214-Y

[70] Zhou, Ying, Liu, Zhou, Chen, Peng.Simultaneous Removal of Inorganic and Organic Compounds in Wastewater by Freshwater Green Microalgae,Environmental Science: Processes & Impacts,2018-2027.16 (2014) https://doi.org/10.1039/C4EM00094C

[71] Gedda, Lee, Lin, Wu.Green Synthesis of Carbon Dots from Prawn Shells for Highly Selective and Sensitive Detection of Copper Ions,Sensors and Actuators B: Chemical,396-403.224 (2016) https://doi.org/10.1016/j.snb.2015.09.065

[72] Gedda, Abdelhamid, Khan, Wu.Zno Nanoparticle-Modified Polymethyl Methacrylate-Assisted Dispersive Liquid-Liquid Microextraction Coupled with Maldi-Ms for Rapid Pathogenic Bacteria Analysis,RSC advances,45973-45983.4 (2014) https://doi.org/10.1039/C4RA03391D

[73] Bhaisare, Gedda, Khan, Wu.Fluorimetric Detection of Pathogenic Bacteria Using Magnetic Carbon Dots,Analytica chimica acta,63-71.920 (2016) https://doi.org/10.1016/j.aca.2016.02.025

[74] Gedda, Pandey, Lin, Wu.Antibacterial Effect of Calcium Oxide Nano-Plates Fabricated from Shrimp Shells,Green Chemistry,3276-3280.17 (2015) https://doi.org/10.1039/C5GC00615E

[75] Reemtsma, Miehe, Duennbier, Jekel.Polar Pollutants in Municipal Wastewater and the Water Cycle: Occurrence and Removal of Benzotriazoles,Water research,596-604.44 (2010) https://doi.org/10.1016/j.watres.2009.07.016

[76] Panizza, Cerisola.Removal of Organic Pollutants from Industrial Wastewater by Electrogenerated Fenton's Reagent,Water research,3987-3992.35 (2001) https://doi.org/10.1016/S0043-1354(01)00135-X

[77] Wang, Hu, Chen, Wu.Total Concentrations and Fractions of Cd, Cr, Pb, Cu, Ni and Zn in Sewage Sludge from Municipal and Industrial Wastewater Treatment Plants,Journal of hazardous materials,245-249.119 (2005) https://doi.org/10.1016/j.jhazmat.2004.11.023

[78] Silva.Wastewater Treatment and Reuse for Sustainable Water Resources Management: A Systematic Literature Review,Sustainability,10940.15 (2023) https://doi.org/10.3390/su151410940

[79] Crini, Lichtfouse.Advantages and Disadvantages of Techniques Used for Wastewater Treatment,Environmental chemistry letters,145-155.17 (2019) https://doi.org/10.1007/s10311-018-0785-9

[80] Amil Usmani, Khan, H Bhat, S Pillai, Ahmad, K Mohamad Haafiz, Oves.Current Trend in the Application of Nanoparticles for Waste Water Treatment and Purification: A Review,Current Organic Synthesis,206-226.14 (2017) https://doi.org/10.2174/1570179413666160928125328

[81] Younas, Mustafa, Farooqi, Wang, Younas, Mohy-Ud-Din, Ashir Hameed, Mohsin Abrar, Maitlo, Noreen.Current and Emerging Adsorbent Technologies for Wastewater Treatment: Trends, Limitations, and Environmental Implications,Water,215.13 (2021) https://doi.org/10.3390/w13020215

[82] Zielinski, Kazimierowicz, Debowski.Advantages and Limitations of Anaerobic Wastewater Treatment-Technological Basics, Development Directions, and Technological Innovations,Energies.16 (2023) https://doi.org/10.3390/en16010083

[83] Mohammadpour, Gharehchahi, Badeenezhad, Parseh, Khaksefidi, Golaki, Dehbandi, Azhdarpoor, Derakhshan, Rodriguez-Chueca.Nitrate in Groundwater Resources of Hormozgan Province, Southern Iran: Concentration Estimation, Distribution and Probabilistic Health Risk Assessment Using Monte Carlo Simulation,Water,564.14 (2022) https://doi.org/10.3390/w14040564

[84] Zhai, Lei, Wu, Teng, Wang, Zhao, Pan.Does the Groundwater Nitrate Pollution in China Pose a Risk to Human Health? A Critical Review of Published Data,Environmental Science and Pollution Research,3640-3653.24 (2017) https://doi.org/10.1007/s11356-016-8088-9

[85] Allaire, Wu, Lall.National Trends in Drinking Water Quality Violations,Proceedings of the National Academy of Sciences,2078-2083.115 (2018) https://doi.org/10.1073/pnas.1719805115

[86] Schullehner, Hansen, Thygesen, Pedersen, Sigsgaard.Nitrate in Drinking Water and Colorectal Cancer Risk: A Nationwide Population-Based Cohort Study,International journal of cancer,73-79.143 (2018) https://doi.org/10.1002/ijc.31306

[87] Wolfe, Patz.Reactive Nitrogen and Human Health: Acute and Long-Term Implications,Ambio: A journal of the human environment,120-125.31 (2002) https://doi.org/10.1579/0044-7447-31.2.120

[88] Jones, Weyer, DellaValle, Inoue-Choi, Anderson, Cantor, Krasner, Robien, Freeman, Silverman.Nitrate from Drinking Water and Diet and Bladder Cancer among Postmenopausal Women in Iowa,Environmental health perspectives,1751-1758.124 (2016) https://doi.org/10.1289/EHP191

Materials Research Forum LLC
https://doi.org/21741/9781644903711-9

[89] Ward, Jones, Brender, De Kok, Weyer, Nolan, Villanueva, Van Breda.Drinking Water Nitrate and Human Health: An Updated Review,International journal of environmental research and public health,1557.15 (2018) https://doi.org/10.3390/ijerph15071557

[90] Pennino, Compton, Leibowitz.Trends in Drinking Water Nitrate Violations across the United States,Environmental science & technology,13450-13460.51 (2017) https://doi.org/10.1021/acs.est.7b04269

[91] Garcia-Segura, Nienhauser, Fajardo, Bansal, Conrad, Fortner, Marcos-Hernández, Rogers, Villagran, Wong.Disparities between Experimental and Environmental Conditions: Research Steps toward Making Electrochemical Water Treatment a Reality,Current Opinion in Electrochemistry,9-16.22 (2020) https://doi.org/10.1016/j.coelec.2020.03.001

[92] Liu, Lopez, Dueñas-Osorio, Stadler, Xie, Alvarez, Li.The Importance of System Configuration for Distributed Direct Potable Water Reuse,Nature Sustainability,548-555.3 (2020) https://doi.org/10.1038/s41893-020-0518-5

[93] Garcia-Segura, Lanzarini-Lopes, Hristovski, Westerhoff.Electrocatalytic Reduction of Nitrate: Fundamentals to Full-Scale Water Treatment Applications,Applied Catalysis B: Environmental,546-568.236 (2018) https://doi.org/10.1016/j.apcatb.2018.05.041

[94] Chaplin.The Prospect of Electrochemical Technologies Advancing Worldwide Water Treatment,Accounts of chemical research,596-604.52 (2019) https://doi.org/10.1021/acs.accounts.8b00611

[95] Stirling, Walker, Westerhoff, Garcia-Segura.Techno-Economic Analysis to Identify Key Innovations Required for Electrochemical Oxidation as Point-of-Use Treatment Systems,Electrochimica Acta,135874.338 (2020) https://doi.org/10.1016/j.electacta.2020.135874

[96] Nishimura, Machida, Enyo.On-Line Mass Spectroscopy Applied to Electroreduction of Nitrite and Nitrate Ions at Porous Pt Electrode in Sulfuric Acid Solutions,Electrochimica acta,877-880.36 (1991) https://doi.org/10.1016/0013-4686(91)85288-I

[97] Gootzen, Peeters, Dukers, Lefferts, Visscher, Van Veen.The Electrocatalytic Reduction of No3− on Pt, Pd and Pt+ Pd Electrodes Activated with Ge,Journal of electroanalytical chemistry,171-183.434 (1997) https://doi.org/10.1016/S0022-0728(97)00093-4

[98] Ureta-Zañartu, Yáñez.Electroreduction of Nitrate Ion on Pt, Ir and on 70: 30 Pt: Ir Alloy,Electrochimica acta,1725-1731.42 (1997) https://doi.org/10.1016/S0013-4686(96)00372-6

[99] Dima, De Vooys, Koper.Electrocatalytic Reduction of Nitrate at Low Concentration on Coinage and Transition-Metal Electrodes in Acid Solutions,Journal of Electroanalytical Chemistry,15-23.554 (2003) https://doi.org/10.1016/S0022-0728(02)01443-2

[100] De Groot, Koper.The Influence of Nitrate Concentration and Acidity on the Electrocatalytic Reduction of Nitrate on Platinum,Journal of Electroanalytical Chemistry,81-94.562 (2004) https://doi.org/10.1016/j.jelechem.2003.08.011

[101] Dima, Beltramo, Koper.Nitrate Reduction on Single-Crystal Platinum Electrodes,Electrochimica acta,4318-4326.50 (2005) https://doi.org/10.1016/j.electacta.2005.02.093

[102] Lacasa, Canizares, Llanos, Rodrigo.Effect of the Cathode Material on the Removal of Nitrates by Electrolysis in Non-Chloride Media,Journal of hazardous materials,478-484.213 (2012) https://doi.org/10.1016/j.jhazmat.2012.02.034

[103] Cao, Wu, Cao.Recent Advances in the Stabilization of Platinum Electrocatalysts for Fuel-Cell Reactions,ChemCatChem,26-45.6 (2014) https://doi.org/10.1002/cctc.201300647

[104] Sánchez-Sánchez, Souza-Garcia, Herrero, Aldaz.Electrocatalytic Reduction of Carbon Dioxide on Platinum Single Crystal Electrodes Modified with Adsorbed Adatoms,Journal of Electroanalytical Chemistry,51-59.668 (2012) https://doi.org/10.1016/j.jelechem.2011.11.002

[105] Kettler.Platinum Group Metals in Catalysis: Fabrication of Catalysts and Catalyst Precursors,Organic process research & development,342-354.7 (2003) https://doi.org/10.1021/op034017o

[106] Fajardo, Westerhoff, Sanchez-Sanchez, Garcia-Segura.Earth-Abundant Elements a Sustainable Solution for Electrocatalytic Reduction of Nitrate,Applied Catalysis B: Environmental,119465.281 (2021) https://doi.org/10.1016/j.apcatb.2020.119465

[107] Chirik, Morris, Getting Down to Earth: The Renaissance of Catalysis with Abundant Metals, ACS Publications, 2015, pp. 2495-2495. https://doi.org/10.1021/acs.accounts.5b00385

[108] Choe, Bergquist, Jeong, Guest, Werth, Strathmann.Performance and Life Cycle Environmental Benefits of Recycling Spent Ion Exchange Brines by Catalytic Treatment of Nitrate,Water research,267-280.80 (2015) https://doi.org/10.1016/j.watres.2015.05.007

[109] Chen, Huo, Liu, Wang, Werth, Strathmann.Exploring Beyond Palladium: Catalytic Reduction of Aqueous Oxyanion Pollutants with Alternative Platinum Group Metals and New Mechanistic Implications,Chemical Engineering Journal,745-752.313 (2017) https://doi.org/10.1016/j.cej.2016.12.058

[110] Lèbre, Owen, Corder, Kemp, Stringer, Valenta.Source Risks as Constraints to Future Metal Supply,Environmental science & technology,10571-10579.53 (2019) https://doi.org/10.1021/acs.est.9b02808

[111] Peng, Xia, Kong, Hu, Wang.Uv Light Irradiation Improves the Aggregation and Settling Performance of Metal Sulfide Particles in Strongly Acidic Wastewater,Water research,114860.163 (2019) https://doi.org/10.1016/j.watres.2019.114860

[112] Yang, Lu, Jiang, Ma, Liu, Cao, Liu, Li, Pang, Kong.Degradation of Sulfamethoxazole by Uv, Uv/H2o2 and Uv/Persulfate (Pds): Formation of Oxidation Products and Effect of Bicarbonate,Water research,196-207.118 (2017) https://doi.org/10.1016/j.watres.2017.03.054

[113] Tian, Zhang, Sun, Tadé, Wang.One-Step Synthesis of Flour-Derived Functional Nanocarbons with Hierarchical Pores for Versatile Environmental Applications,Chemical Engineering Journal,432-439.347 (2018) https://doi.org/10.1016/j.cej.2018.04.139

[114] Ly, Lee, Hur.Using Fluorescence Surrogates to Track Algogenic Dissolved Organic Matter (Aom) During Growth and Coagulation/Flocculation Processes of Green Algae,Journal of Environmental Sciences,311-320.79 (2019) https://doi.org/10.1016/j.jes.2018.12.006

[115] Ji, Aleisa, Duan, Zhang, Yin, Xing.Metallic Active Sites on Moo2 (110) Surface to Catalyze Advanced Oxidation Processes for Efficient Pollutant Removal,Iscience.23 (2020) https://doi.org/10.1016/j.isci.2020.100861

[116] Chen, Fang, Xia, Huang, Huang.Selective Transformation of B-Lactam Antibiotics by Peroxymonosulfate: Reaction Kinetics and Nonradical Mechanism,Environmental science & technology,1461-1470.52 (2018) https://doi.org/10.1021/acs.est.7b05543

[117] Herney-Ramirez, Vicente, Madeira.Heterogeneous Photo-Fenton Oxidation with Pillared Clay-Based Catalysts for Wastewater Treatment: A Review,Applied Catalysis B: Environmental,10-26.98 (2010) https://doi.org/10.1016/j.apcatb.2010.05.004

[118] Tian, Dong, Chen, Li, Xie.Amorphous Co3o4 Nanoparticles-Decorated Biochar as an Efficient Activator of Peroxymonosulfate for the Removal of Sulfamethazine in Aqueous Solution,Separation and Purification Technology,117246.250 (2020) https://doi.org/10.1016/j.seppur.2020.117246

[119] Duan, Sun, Shao, Wang.Nonradical Reactions in Environmental Remediation Processes: Uncertainty and Challenges,Applied Catalysis B: Environmental,973-982.224 (2018) https://doi.org/10.1016/j.apcatb.2017.11.051

[120] Gerrity, Stanford, Trenholm, Snyder.An Evaluation of a Pilot-Scale Nonthermal Plasma Advanced Oxidation Process for Trace Organic Compound Degradation,Water Research,493-504.44 (2010) https://doi.org/10.1016/j.watres.2009.09.029

[121] Kim, Kim, Cha, Yu.Degradation of Sulfamethoxazole by Ionizing Radiation: Identification and Characterization of Radiolytic Products,Chemical Engineering Journal,556-566.313 (2017) https://doi.org/10.1016/j.cej.2016.12.080

[122] Demetrio, Nikolaos, Elefteria, Dionissions.Photocatalytic Degradation of Reactive Black 5 in Aqueous Solution: Effect of Operating Conditions and Coupling with Ultrasound Irradiation,Water Research,2236-2246.41 (2007) https://doi.org/10.1016/j.watres.2007.01.048

[123] Li, Zhao, Yan, Yan, Pan, Zhang, Lai.Enhanced Sulfamethoxazole Degradation by Peroxymonosulfate Activation with Sulfide-Modified Microscale Zero-Valent Iron (S-Mfe0): Performance, Mechanisms, and the Role of Sulfur Species,Chemical Engineering Journal,121302.376 (2019) https://doi.org/10.1016/j.cej.2019.03.178

[124] Yang, Yu, Shan, Gao, Pan.Enhanced Fe (Iii)-Mediated Fenton Oxidation of Atrazine in the Presence of Functionalized Multi-Walled Carbon Nanotubes,Water research,37-46.137 (2018) https://doi.org/10.1016/j.watres.2018.03.006

[125] Guan, Han, Asakura, Wang, Chen, Yan, Guan, Keenan, Hayama, A van Spronsen.Angewandte Chemie International Edition Angewandte Chemie International Edition Research Article Open Access Subsurface Single-Atom Catalyst Enabled by Mechanochemical Synthesis for Oxidation Chemistry,Angewandte Chemie International Edition,(2024) https://doi.org/10.1002/anie.202410457

[126] Xin, Liu, Ma, Gong, Ma, Yan, Chen, Ma, Zhang, Gao.High Efficiency Heterogeneous Fenton-Like Catalyst Biochar Modified Cufeo2 for the Degradation of Tetracycline: Economical Synthesis, Catalytic Performance and Mechanism,Applied Catalysis B: Environmental,119386.280 (2021) https://doi.org/10.1016/j.apcatb.2020.119386

Materials Research Forum LLC
https://doi.org/21741/9781644903711-9

[127] Dong, Jiang, Deng, Zhang, Cheng, Hou, Zhang, Tang, Zeng.Physicochemical Transformation of Fe/Ni Bimetallic Nanoparticles During Aging in Simulated Groundwater and the Consequent Effect on Contaminant Removal,Water research,51-57.129 (2018) https://doi.org/10.1016/j.watres.2017.11.002

[128] Dalui, Thupakula, Khan, Ghosh, Satpati, Acharya.Mechanism of Versatile Catalytic Activities of Quaternary Cuznfes Nanocrystals Designed by a Rapid Synthesis Route,Small,1829-1839.11 (2015) https://doi.org/10.1002/smll.201402837

[129] Ye, Shi, Yang, Fu, Chen.P-Doped Znxcd1− Xs Solid Solutions as Photocatalysts for Hydrogen Evolution from Water Splitting Coupled with Photocatalytic Oxidation of 5-Hydroxymethylfurfural,Applied Catalysis B: Environmental,70-79.233 (2018) https://doi.org/10.1016/j.apcatb.2018.03.060

[130] Yi, Tan, Liu, Lu, Xing, Zhang.Peroxymonosulfate Activation by Three-Dimensional Cobalt Hydroxide/Graphene Oxide Hydrogel for Wastewater Treatment through an Automated Process,Chemical Engineering Journal,125965.400 (2020) https://doi.org/10.1016/j.cej.2020.125965

[131] Chandrasekaran, Yao, Deng, Bowen, Zhang, Chen, Lin, Peng, Zhang.Recent Advances in Metal Sulfides: From Controlled Fabrication to Electrocatalytic, Photocatalytic and Photoelectrochemical Water Splitting and Beyond,Chemical Society Reviews,4178-4280.48 (2019) https://doi.org/10.1039/C8CS00664D

[132] Zhang, Xu, Wang.Ultrathin Single Crystal Zns Nanowires,Chemical Communications,8941-8943.46 (2010) https://doi.org/10.1039/C0CC02549F

[133] Zhang, Zhou, Wu, Xu, Lou.Unusual Formation of Single-Crystal Manganese Sulfide Microboxes Co-Mediated by the Cubic Crystal Structure and Shape,Angewandte Chemie (International ed. in English),7267-7270.51 (2012) https://doi.org/10.1002/anie.201202877

[134] Zhao, Yan, Chen, Chen.Spinels: Controlled Preparation, Oxygen Reduction/Evolution Reaction Application, and Beyond,Chemical reviews,10121-10211.117 (2017) https://doi.org/10.1021/acs.chemrev.7b00051

[135] Huang, Lv, Zhang, Ding, Lai, Wu, Wang, Li, Cai, Ma.Co-Fe Bimetallic Sulfide with Robust Chemical Adsorption and Catalytic Activity for Polysulfides in Lithium-Sulfur Batteries,Chemical Engineering Journal,124122.387 (2020) https://doi.org/10.1016/j.cej.2020.124122

[136] Kwon, Kim, Le, Gim, Jeon, Ham, Lee, Jang, Kim.Synthesis of Atomically Thin Transition Metal Disulfides for Charge Transport Layers in Optoelectronic Devices,ACS nano,4146-4155.9 (2015) https://doi.org/10.1021/acsnano.5b01504

[137] Yilmaz, Yam, Zhang, Fan, Ho.In Situ Transformation of Mofs into Layered Double Hydroxide Embedded Metal Sulfides for Improved Electrocatalytic and Supercapacitive Performance,Advanced Materials,1606814.29 (2017) https://doi.org/10.1002/adma.201606814

[138] Xing, Tan, Yuan, Wang, Ma, Xie, Li, Wu, Ren, Shahbazian-Yassar.Consolidating Lithiothermic-Ready Transition Metals for Li2s-Based Cathodes,Advanced Materials,2002403.32 (2020) https://doi.org/10.1002/adma.202002403

[139] Li, Su, Rao, Wu, Rudolf, Blake, de Groot, Besenbacher, Palstra.Band Gap Narrowing of Sns 2 Superstructures with Improved Hydrogen Production,Journal of Materials Chemistry A,209-216.4 (2016) https://doi.org/10.1039/C5TA07283B

[140] Li, Dong, Li, Tang, Tian, Li, Chen, Xie, Jin, Xiao.Recent Advances in Waste Water Treatment through Transition Metal Sulfides-Based Advanced Oxidation Processes,Water Research,116850.192 (2021) https://doi.org/10.1016/j.watres.2021.116850

[141] Taube, Mechanisms of Redox Reactions of Simple Chemistry, Advances in Inorganic Chemistry and Radiochemistry, Elsevier1959, pp. 1-53. https://doi.org/10.1016/S0065-2792(08)60251-4

[142] Cheung, Williams.Separation of Transition Metals and Chelated Complexes in Wastewaters,Environmental Progress & Sustainable Energy,761-783.34 (2015) https://doi.org/10.1002/ep.12065

[143] Al Balushi, Al Marzouqi, Al Wahaibi, Kuvarega, Al Kindy, Kim, Selvaraj.Hydrothermal Synthesis of Cds Sub-Microspheres for Photocatalytic Degradation of Pharmaceuticals,Applied Surface Science,559-565.457 (2018) https://doi.org/10.1016/j.apsusc.2018.06.286

[144] Zhuang, Lu, Peng, Li.A Facile "Dispersion-Decomposition" Route to Metal Sulfide Nanocrystals,Chemistry-A European Journal,10445-10452.17 (2011) https://doi.org/10.1002/chem.201101145

[145] Chakraborty, Thangavel, Komninou, Zhou, Gupta.Nanospheres and Nanoflowers of Copper Bismuth Sulphide (Cu3bis3): Colloidal Synthesis, Structural, Optical and Electrical Characterization,Journal of Alloys and Compounds,142-148.776 (2019) https://doi.org/10.1016/j.jallcom.2018.10.151

[146] Zhang, Fujisawa, Zhang, Liu, Lucking, Gontijo, Lei, Liu, Crust, Granzier-Nakajima.Universal in Situ Substitutional Doping of Transition Metal Dichalcogenides by Liquid-Phase Precursor-Assisted Synthesis,ACS nano,4326-4335.14 (2020) https://doi.org/10.1021/acsnano.9b09857

[147] Lai, Feng, Heil, Tian, Schmidt, Wang, Oschatz.Partially Delocalized Charge in Fe-Doped Nico 2 S 4 Nanosheet-Mesoporous Carbon-Composites for High-Voltage Supercapacitors,Journal of Materials Chemistry A,19342-19347.7 (2019) https://doi.org/10.1039/C9TA06250E

[148] Elimelech, Liu, Plonka, Frenkel, Banin.Size Dependence of Doping by a Vacancy Formation Reaction in Copper Sulfide Nanocrystals,Angewandte Chemie,10471-10476.129 (2017) https://doi.org/10.1002/ange.201702673

[149] Zhou, Wang, Zhu, Dionysiou, Zhao, Fang, Zhou.New Insight into the Mechanism of Peroxymonosulfate Activation by Sulfur-Containing Minerals: Role of Sulfur Conversion in Sulfate Radical Generation,Water research,208-216.142 (2018) https://doi.org/10.1016/j.watres.2018.06.002

[150] Liu, Wang, Ai, Zhang.Hydrothermal Synthesis of Fes2 as a High-Efficiency Fenton Reagent to Degrade Alachlor Via Superoxide-Mediated Fe (Ii)/Fe (Iii) Cycle,ACS applied materials & interfaces,28534-28544.7 (2015) https://doi.org/10.1021/acsami.5b09919

[151] Zhao, Pan, Ye, Zhang, Pan.Fes2/H2o2 Mediated Water Decontamination from P-Arsanilic Acid Via Coupling Oxidation, Adsorption and Coagulation: Performance and Mechanism,Chemical Engineering Journal,122667.381 (2020) https://doi.org/10.1016/j.cej.2019.122667

[152] Yuan, Tao, Fan, Ma.Degradation of P-Chloroaniline by Persulfate Activated with Ferrous Sulfide Ore Particles,Chemical Engineering Journal,38-46.268 (2015) https://doi.org/10.1016/j.cej.2014.12.092

[153] Peng, Zhou, Liu, Ao, Ji, Liu, Su, Yao, Lai.Insights into Heterogeneous Catalytic Activation of Peroxymonosulfate by Natural Chalcopyrite: Ph-Dependent Radical Generation, Degradation Pathway and Mechanism,Chemical Engineering Journal,125387.397 (2020) https://doi.org/10.1016/j.cej.2020.125387

[154] Qian, Wang, Guan, Li, Xu, Liu, Liu, Qiu.Enhanced Wet Hydrogen Peroxide Catalytic Oxidation Performances Based on Cus Nanocrystals/Reduced Graphene Oxide Composites,Applied surface science,633-640.288 (2014) https://doi.org/10.1016/j.apsusc.2013.10.086

[155] Nie, Mao, Ding, Hu, Tang.Highly Efficient Catalysis of Chalcopyrite with Surface Bonded Ferrous Species for Activation of Peroxymonosulfate toward Degradation of Bisphenol A: A Mechanism Study,Journal of Hazardous Materials,59-68.364 (2019) https://doi.org/10.1016/j.jhazmat.2018.09.078

[156] Xu, Tang, Cai, Xi, Zhang, Pi, Mao.Heterogeneous Activation of Peroxymonocarbonate by Chalcopyrite (Cufes2) for Efficient Degradation of 2, 4-Dichlorophenol in Simulated Groundwater,Applied Catalysis B: Environmental,273-282.251 (2019) https://doi.org/10.1016/j.apcatb.2019.03.080

[157] Wu, Zhao, Huang, Zhang.A Mechanistic Study of Amorphous Cosx Cages as Advanced Oxidation Catalysts for Excellent Peroxymonosulfate Activation Towards Antibiotics Degradation,Chemical Engineering Journal,122768.381 (2020) https://doi.org/10.1016/j.cej.2019.122768

[158] Wang, Astruc.The Recent Development of Efficient Earth-Abundant Transition-Metal Nanocatalysts,Chemical Society Reviews,816-854.46 (2017) https://doi.org/10.1039/C6CS00629A

[159] Maduraiveeran, Sasidharan, Jin.Earth-Abundant Transition Metal and Metal Oxide Nanomaterials: Synthesis and Electrochemical Applications,Progress in Materials Science,100574.106 (2019) https://doi.org/10.1016/j.pmatsci.2019.100574

[160] Lopez, Viltres, Gupta, Acevedo-Pena, Leyva, Ghaffari, Gupta, Kim, Bae, Kim.Transition Metal-Based Metal-Organic Frameworks for Environmental Applications: A Review,Environmental Chemistry Letters,1295-1334.19 (2021) https://doi.org/10.1007/s10311-020-01119-1

[161] Abdullah, Yusof, Lau, Jaafar, Ismail.Recent Trends of Heavy Metal Removal from Water/Wastewater by Membrane Technologies,Journal of Industrial and Engineering Chemistry,17-38.76 (2019) https://doi.org/10.1016/j.jiec.2019.03.029

[162] Vu, Tran, Le, Nguyen, Vu, Vu.Synthesis and Application of Novel Fe-Mil-53/Go Nanocomposite for Photocatalytic Degradation of Reactive Dye from Aqueous

Solution,Vietnam Journal of Chemistry,681-685.57 (2019)
https://doi.org/10.1002/vjch.201900055

[163] Yousef, Qiblawey, El-Naas.Adsorption as a Process for Produced Water Treatment: A
Review,Processes,1657.8 (2020) https://doi.org/10.3390/pr8121657

[164] Yadav, Yashas, Shivaraju.Transitional Metal Chalcogenide Nanostructures for
Remediation and Energy: A Review,Environmental Chemistry Letters,3683-3700.19 (2021)
https://doi.org/10.1007/s10311-021-01269-w

[165] Li, Song, Han, Wang, Li.Efficient Removal of Cu (Ii) and Citrate Complexes by
Combined Permanent Magnetic Resin and Its Mechanistic Insights,Chemical Engineering
Journal,1-10.366 (2019) https://doi.org/10.1016/j.cej.2019.02.070

[166] Liu, Jing, Li, Huang, Gao, You, Zhang.Effect of Synthesis Conditions on the
Photocatalytic Degradation of Rhodamine B of Mil-53 (Fe),Materials Letters,92-95.237
(2019) https://doi.org/10.1016/j.matlet.2018.11.079

[167] Han, Zhang, Bai, Wu, Meng, Xu, Liang, Wang, Zhang.Fabrication of Agi/Mil-53 (Fe)
Composites with Enhanced Photocatalytic Activity for Rhodamine B Degradation under
Visible Light Irradiation,Applied Organometallic Chemistry,e4325.32 (2018)
https://doi.org/10.1002/aoc.4325

[168] Wang, Cui, Shi, Tan, Zhang, Zhang, Zhang.Controllable Self-Assembly of Cds@ Nh2-
Mil-125 (Ti) Heterostructure with Enhanced Photodegradation Efficiency for Organic
Pollutants through Synergistic Effect,Materials Science in Semiconductor Processing,91-
100.97 (2019) https://doi.org/10.1016/j.mssp.2019.03.016

[169] Zhu, Liu, Wu, Zhao, Li, Tao, Yi, Han.Enhanced Photocatalytic Performance of Biobr/Nh
2-Mil-125 (Ti) Composite for Dye Degradation under Visible Light,Dalton
Transactions,17521-17529.45 (2016) https://doi.org/10.1039/C6DT02912D

[170] Khasevani, Gholami.Engineering a Highly Dispersed Core@ Shell Structure for Efficient
Photocatalysis: A Case Study of Ternary Novel Bioi@ Mil-88a (Fe)@ G-C3n4
Nanocomposite,Materials Research Bulletin,93-102.106 (2018)
https://doi.org/10.1016/j.materresbull.2018.05.024

[171] Chen, Chao, Ma, Zhu, Jiang, Ren, Guo, Lou.Synthesis of Flower-Like Cus/Uio-66
Composites with Enhanced Visible-Light Photocatalytic Performance,Inorganic Chemistry
Communications,223-228.104 (2019) https://doi.org/10.1016/j.inoche.2019.04.022

[172] Zheng, Qian, Li, Wang, Li, Zhang, Li, Wu.A Bifunctional Cationic Metal-Organic
Framework Based on Unprecedented Nonanuclear Copper (Ii) Cluster for High Dichromate
and Chromate Trapping and Highly Efficient Photocatalytic Degradation of Organic Dyes
under Visible Light Irradiation,Dalton Transactions,9103-9113.47 (2018)
https://doi.org/10.1039/C8DT01685B

[173] Liu, Cheng, Luo, Cheng, Wang, Lou.Degradation of Dye Rhodamine B under Visible
Irradiation with Prussian Blue as a Photo-Fenton Reagent,Environmental Chemistry
Letters,31-35.9 (2011) https://doi.org/10.1007/s10311-009-0242-x

[174] Li, Li, He, Xu, Tang.Two Novel Porous Mofs with Square-Shaped Cavities for the
Removal of Toxic Dyes: Adsorption or Degradation?,New Journal of Chemistry,15204-
15209.41 (2017) https://doi.org/10.1039/C7NJ02904G

[175] Bao, Li, Ning, Peng, Jin, Tang.Highly Effective Removal of Mercury and Lead Ions from Wastewater by Mercaptoamine-Functionalised Silica-Coated Magnetic Nano-Adsorbents: Behaviours and Mechanisms,Applied Surface Science,457-466.393 (2017) https://doi.org/10.1016/j.apsusc.2016.09.098

[176] Sun, Zhang, Mao, Yu, Han, Bhat.Facile Synthesis of the Magnetic Metal-Organic Framework Fe 3 O 4/Cu 3 (Btc) 2 for Efficient Dye Removal,Environmental Chemistry Letters,1091-1096.17 (2019) https://doi.org/10.1007/s10311-018-00833-1

[177] Li, Jiang, Fang, Cao, Chen, Li, Xu, Lu.Tio2 Nanoparticles Anchored onto the Metal-Organic Framework Nh2-Mil-88b (Fe) as an Adsorptive Photocatalyst with Enhanced Fenton-Like Degradation of Organic Pollutants under Visible Light Irradiation,ACS Sustainable Chemistry & Engineering,16186-16197.6 (2018) https://doi.org/10.1021/acssuschemeng.8b02968

[178] Fang, Yang, Dou, Liu, Yao, Xu, Zhu.Synthesis, Crystal Structure and Photocatalytic Properties of a Mn (Ii) Metal-Organic Framework Based on a Thiophene-Functionalized Dicarboxylate Ligand,Inorganic Chemistry Communications,124-127.96 (2018) https://doi.org/10.1016/j.inoche.2018.08.017

[179] Khasevani, Gholami.Synthesis of Bioi/Znfe2o4-Metal-Organic Framework and G-C3n4-Based Nanocomposites for Applications in Photocatalysis,Industrial & Engineering Chemistry Research,9806-9818.58 (2019) https://doi.org/10.1021/acs.iecr.8b05871

[180] Ramezanalizadeh, Manteghi.Synthesis of a Novel Mof/Cuwo4 Heterostructure for Efficient Photocatalytic Degradation and Removal of Water Pollutants,Journal of Cleaner Production,2655-2666.172 (2018) https://doi.org/10.1016/j.jclepro.2017.11.145

[181] Liu, Zhang, Wang, Guo, Muhler, Wang, Lin, Chen, Yang.Highly Efficient Photocatalytic Degradation of Dyes by a Copper-Triazolate Metal-Organic Framework,Chemistry-A European Journal,16804-16813.24 (2018) https://doi.org/10.1002/chem.201803306

[182] Shanker, Jassal, Rani.Catalytic Removal of Organic Colorants from Water Using Some Transition Metal Oxide Nanoparticles Synthesized under Sunlight,RSC advances,94989-94999.6 (2016) https://doi.org/10.1039/C6RA17555D

[183] Rani, Shanker, Chaurasia.Catalytic Potential of Laccase Immobilized on Transition Metal Oxides Nanomaterials: Degradation of Alizarin Red S Dye,Journal of environmental chemical engineering,2730-2739.5 (2017) https://doi.org/10.1016/j.jece.2017.05.026

[184] Velayutham, Parvathiraja, Anitha, Mahalakshmi, Jenila, Alasmary, Almalki, Iqbal, Lai.Photocatalytic and Antibacterial Activity of Cofe2o4 Nanoparticles from Hibiscus Rosa-Sinensis Plant Extract,Nanomaterials,3668.12 (2022) https://doi.org/10.3390/nano12203668

[185] Thalgaspitiya, Kapuge, He, Deljoo, Meguerdichian, Aindow, Suib.Multifunctional Transition Metal Doped Titanium Dioxide Reduced Graphene Oxide Composites as Highly Efficient Adsorbents and Photocatalysts,Microporous and Mesoporous Materials,110521.307 (2020) https://doi.org/10.1016/j.micromeso.2020.110521

[186] Yadav, Chaudhary, Sakhare, Dongale, Patil, Sheikh.Impact of Collected Sunlight on Znfe2o4 Nanoparticles for Photocatalytic Application,Journal of colloid and interface science,289-297.527 (2018) https://doi.org/10.1016/j.jcis.2018.05.051

[187] Cao, Lei, Chen, Kang, Li, Liu.Enhanced Photocatalytic Degradation of Tetracycline Hydrochloride by Novel Porous Hollow Cube Znfe2o4,Journal of Photochemistry and Photobiology A: Chemistry,794-800.364 (2018) https://doi.org/10.1016/j.jphotochem.2018.07.023

[188] Boutra, Trari, Nassrallah, Bellal.Adsorption and Photodegradation of Solophenyl Red 3bl on Nanosized Znfe 2 O 4 under Solar Light,Theoretical and Experimental Chemistry,303-309.52 (2016) https://doi.org/10.1007/s11237-016-9482-6

[189] Chen, Dai, Guo.Hydrothermal Synthesis of Well-Distributed Spherical Cubi2o4 with Enhanced Photocatalytic Activity under Visible Light Irradiation,Materials Letters,251-254.161 (2015) https://doi.org/10.1016/j.matlet.2015.08.118

[190] Chen, Yao, Chen, Li, Yu, Zhang, Lai.Hydrothermal Synthesis of Dendritic Cubi2o4 and Its Photocatalytic Performance Towards Tetracycline Degradation under Different Light Conditions,Materials Science in Semiconductor Processing,106503.142 (2022) https://doi.org/10.1016/j.mssp.2022.106503

[191] Zhang, Ma, Dai, Bu, Li, Yu, Cao, Guan.Removal of Pollutants Via Synergy of Adsorption and Photocatalysis over Mxene-Based Nanocomposites,Chemical Engineering Journal Advances,100285.10 (2022) https://doi.org/10.1016/j.ceja.2022.100285

[192] Nguyen, Nguyen, Nguyen, Nguyen, Nguyen, Nguyen, Van Tran.Green Synthesis of Znfe2o4@ Zno Nanocomposites Using Chrysanthemum Spp. Floral Waste for Photocatalytic Dye Degradation,Journal of Environmental Management,116746.326 (2023) https://doi.org/10.1016/j.jenvman.2022.116746

[193] Ching, Kriz, Luthy, Njagi, Suib.Self-Assembly of Manganese Oxide Nanoparticles and Hollow Spheres. Catalytic Activity in Carbon Monoxide Oxidation,Chemical Communications,8286-8288.47 (2011) https://doi.org/10.1039/c1cc11764e

[194] Sun, Hua, Guo, Wang, Huang.Selective Aerobic Oxidation of Alcohols by Using Manganese Oxide Nanoparticles as an Efficient Heterogeneous Catalyst,Advanced Synthesis & Catalysis,569-573.354 (2012) https://doi.org/10.1002/adsc.201100666

[195] Mansournia, Azizi, Rakhshan.A Novel Ammonia-Assisted Method for the Direct Synthesis of Mn3o4 Nanoparticles at Room Temperature and Their Catalytic Activity During the Rapid Degradation of Azo Dyes,Journal of Physics and Chemistry of Solids,91-97.80 (2015) https://doi.org/10.1016/j.jpcs.2015.01.001

[196] Rezaeifard, Soltani, Jafarpour.Nanoaggregates of Simple Mn Porphyrin Complexes as Catalysts for the Selective Oxidation of Hydrocarbons,European Journal of Inorganic Chemistry,2657-2664.2013 (2013) https://doi.org/10.1002/ejic.201201437

[197] Bagherzadeh, Haghdoost, Moghaddam, Foroushani, Saryazdi, Payab.Mn (Iii) Complex Supported on Fe3o4 Nanoparticles: Magnetically Separable Nanocatalyst for Selective Oxidation of Thiols to Disulfides,Journal of Coordination Chemistry,3025-3036.66 (2013) https://doi.org/10.1080/00958972.2013.821699

[198] Bagherzadeh, Mortazavi-Manesh.Nanoparticle Supported, Magnetically Separable Manganese Porphyrin as an Efficient Retrievable Nanocatalyst in Hydrocarbon Oxidation Reactions,RSC advances,41551-41560.6 (2016) https://doi.org/10.1039/C6RA02123A

[199] Shi, Tse, Pohl, Brückner, Zhang, Beller.Tuning Catalytic Activity between Homogeneous and Heterogeneous Catalysis: Improved Activity and Selectivity of Free Nano-Fe2o3 in Selective Oxidations,Angewandte Chemie International Edition,8866-8868.46 (2007) https://doi.org/10.1002/anie.200703418

[200] Cui, Li, Bachmann, Scalone, Surkus, Junge, Topf, Beller.Synthesis and Characterization of Iron-Nitrogen-Doped Graphene/Core-Shell Catalysts: Efficient Oxidative Dehydrogenation of N-Heterocycles,Journal of the American Chemical Society,10652-10658.137 (2015) https://doi.org/10.1021/jacs.5b05674

[201] Jagadeesh, Natte, Junge, Beller.Nitrogen-Doped Graphene-Activated Iron-Oxide-Based Nanocatalysts for Selective Transfer Hydrogenation of Nitroarenes,Acs Catalysis,1526-1529.5 (2015) https://doi.org/10.1021/cs501916p

[202] Zheng, Cheng, Wang, Bao, Zhou, Wei, Zhang, Zheng.Quasicubic A-Fe2o3 Nanoparticles with Excellent Catalytic Performance,The Journal of Physical Chemistry B,3093-3097.110 (2006) https://doi.org/10.1021/jp056617q

[203] Basavegowda, Magar, Mishra, Lee.Green Fabrication of Ferromagnetic Fe 3 O 4 Nanoparticles and Their Novel Catalytic Applications for the Synthesis of Biologically Interesting Benzoxazinone and Benzthioxazinone Derivatives,New Journal of Chemistry,5415-5420.38 (2014) https://doi.org/10.1039/C4NJ01155D

[204] Chen, Chen, Ng, Man, Lau.Chemical and Visible-Light-Driven Water Oxidation by Iron Complexes at Ph 7-9: Evidence for Dual-Active Intermediates in Iron-Catalyzed Water Oxidation,Angewandte Chemie International Edition,1789-1791.52 (2013) https://doi.org/10.1002/anie.201209116

[205] Mou, Zhang, Li, Yao, Wei, Su, Shen.Rod-Shaped Fe2o3 as an Efficient Catalyst for the Selective Reduction of Nitrogen Oxide by Ammonia,Angewandte Chemie International Edition,2989-2993.51 (2012) https://doi.org/10.1002/anie.201107113

[206] Jagadeesh, Stemmler, Surkus, Junge, Junge, Beller.Hydrogenation Using Iron Oxide-Based Nanocatalysts for the Synthesis of Amines,Nature Protocols,548-557.10 (2015) https://doi.org/10.1038/nprot.2015.025

[207] Xie, Li, Liu, Haruta, Shen.Low-Temperature Oxidation of Co Catalysed by Co3o4 Nanorods,Nature,746-749.458 (2009) https://doi.org/10.1038/nature07877

[208] Dangwal Pandey, Jia, Schmidt, Leoni, Schwickardi, Schüth, Weidenthaler.Size-Controlled Synthesis and Microstructure Investigation of Co3o4 Nanoparticles for Low-Temperature Co Oxidation, The Journal of Physical Chemistry C,19405-19412.116 (2012) https://doi.org/10.1021/jp306166g

[209] Ma, Mu, Li, Jin, Cheng, Lu, Hao, Qiao.Mesoporous Co3o4 and Au/Co3o4 Catalysts for Low-Temperature Oxidation of Trace Ethylene,Journal of the American Chemical Society,2608-2613.132 (2010) https://doi.org/10.1021/ja906274t

[210] Zhu, Kailasam, Fischer, Thomas.Supported Cobalt Oxide Nanoparticles as Catalyst for Aerobic Oxidation of Alcohols in Liquid Phase,Acs Catalysis,342-347.1 (2011) https://doi.org/10.1021/cs100153a

[211] Chen, Surkus, He, Pohl, Radnik, Topf, Junge, Beller.Selective Catalytic Hydrogenation of Heteroarenes with N-Graphene-Modified Cobalt Nanoparticles (Co3o4-Co/Ngr@ A-

Al2o3),Journal of the American Chemical Society,11718-11724.137 (2015)
https://doi.org/10.1021/jacs.5b06496

[212] Natte, Jagadeesh, Sharif, Neumann, Beller.Synthesis of Nitriles from Amines Using Nanoscale Co 3 O 4-Based Catalysts Via Sustainable Aerobic Oxidation,Organic & Biomolecular Chemistry,3356-3359.14 (2016) https://doi.org/10.1039/C6OB00184J

[213] Jagadeesh, Banerjee, Arockiam, Junge, Junge, Pohl, Radnik, Brückner, Beller.Highly Selective Transfer Hydrogenation of Functionalised Nitroarenes Using Cobalt-Based Nanocatalysts,Green Chemistry,898-902.17 (2015) https://doi.org/10.1039/C4GC00731J

[214] Liu, Wang, Chen, Zhong, Liu, Li, Wang, Wang, Lu, Wang.Noncrystalline Nickel Phosphide Decorated Poly (Vinyl Alcohol-Co-Ethylene) Nanofibrous Membrane for Catalytic Hydrogenation of P-Nitrophenol,Applied Catalysis B: Environmental,223-231.196 (2016) https://doi.org/10.1016/j.apcatb.2016.05.059

[215] Hu, Yu, Hou, Yang, Feng, Li, Qiao, Wang, Hua, Pan.Ionic Liquid Immobilized Nickel (0) Nanoparticles as Stable and Highly Efficient Catalysts for Selective Hydrogenation in the Aqueous Phase,Chemistry-An Asian Journal,1178-1184.5 (2010) https://doi.org/10.1002/asia.200900628

[216] Kalbasi, Zamani.Synthesis and Characterization of Ni Nanoparticles Incorporated into Hyperbranched Polyamidoamine-Polyvinylamine/Sba-15 Catalyst for Simple Reduction of Nitro Aromatic Compounds,RSC advances,7444-7453.4 (2014) https://doi.org/10.1039/c3ra44662j

[217] Marakatti, Peter.Nickel-Antimony Nanoparticles Confined in Sba-15 as Highly Efficient Catalysts for the Hydrogenation of Nitroarenes,New Journal of Chemistry,5448-5457.40 (2016) https://doi.org/10.1039/C5NJ03479E

[218] Gawande, Rathi, Branco, Nogueira, Velhinho, Shrikhande, Indulkar, Jayaram, Ghumman, Bundaleski.Regio-and Chemoselective Reduction of Nitroarenes and Carbonyl Compounds over Recyclable Magnetic Ferrite-Nickel Nanoparticles (Fe3o4-Ni) by Using Glycerol as a Hydrogen Source,Chemistry-a European Journal,12628.18 (2012) https://doi.org/10.1002/chem.201202380

[219] Tong, Gu, Zhang, Tang, Wang, Tu.Thermal Growth of Nio on Interconnected Ni-P Tube Network for Electrochemical Oxidation of Methanol in Alkaline Medium,International Journal of Hydrogen Energy,6342-6352.41 (2016) https://doi.org/10.1016/j.ijhydene.2016.03.018

[220] Zhuang, Giles, Zheng, Jenness, Caratzoulas, Vlachos, Yan.Nickel Supported on Nitrogen-Doped Carbon Nanotubes as Hydrogen Oxidation Reaction Catalyst in Alkaline Electrolyte,Nature communications,10141.7 (2016) https://doi.org/10.1038/ncomms10141

[221] Tornøe, Christensen, Meldal.Peptidotriazoles on Solid Phase:[1, 2, 3]-Triazoles by Regiospecific Copper (I)-Catalyzed 1, 3-Dipolar Cycloadditions of Terminal Alkynes to Azides,The Journal of organic chemistry,3057-3064.67 (2002) https://doi.org/10.1021/jo011148j

[222] Molteni, Bianchi, Marinoni, Santo, Ponti.Cu/Cu-Oxide Nanoparticles as Catalyst in the "Click" Azide-Alkyne Cycloaddition,New Journal of Chemistry,1137-1139.30 (2006) https://doi.org/10.1039/B604297J

[223] Nasrollahzadeh, Jaleh, Fakhri, Zahraei, Ghadery.Synthesis and Catalytic Activity of Carbon Supported Copper Nanoparticles for the Synthesis of Aryl Nitriles and 1, 2, 3-Triazoles,RSC Advances,2785-2793.5 (2015) https://doi.org/10.1039/C4RA09935D

[224] Sharghi, Khalifeh, Doroodmand.Copper Nanoparticles on Charcoal for Multicomponent Catalytic Synthesis of 1, 2, 3-Triazole Derivatives from Benzyl Halides or Alkyl Halides, Terminal Alkynes and Sodium Azide in Water as a "Green" Solvent,Advanced Synthesis & Catalysis,207-218.351 (2009) https://doi.org/10.1002/adsc.200800612

[225] Jayaramulu, Suresh, Maji.Stabilization of Cu 2 O Nanoparticles on a 2d Metal-Organic Framework for Catalytic Huisgen 1, 3-Dipolar Cycloaddition Reaction,Dalton Transactions,83-86.44 (2015) https://doi.org/10.1039/C4DT02661F

[226] Zeng, Yang, Hudson, Song, Moores, Li.Fe3o4 Nanoparticle-Supported Copper (I) Pybox Catalyst: Magnetically Recoverable Catalyst for Enantioselective Direct-Addition of Terminal Alkynes to Imines,Organic letters,442-445.13 (2011) https://doi.org/10.1021/ol102759w

[227] Bhargava, Tardio, Prasad, Föger, Akolekar, Grocott.Wet Oxidation and Catalytic Wet Oxidation,Industrial & engineering chemistry research,1221-1258.45 (2006) https://doi.org/10.1021/ie051059n

[228] Glaze, Kang, Chapin.The Chemistry of Water Treatment Processes Involving Ozone, Hydrogen Peroxide and Ultraviolet Radiation,(1987) https://doi.org/10.1080/01919518708552148

[229] Miklos, Remy, Jekel, Linden, Drewes, Hübner.Evaluation of Advanced Oxidation Processes for Water and Wastewater Treatment-a Critical Review,Water research,118-131.139 (2018) https://doi.org/10.1016/j.watres.2018.03.042

[230] Armstrong, Huie, Koppenol, Lymar, Merényi, Neta, Ruscic, Stanbury, Steenken, Wardman.Standard Electrode Potentials Involving Radicals in Aqueous Solution: Inorganic Radicals (Iupac Technical Report),Pure and Applied Chemistry,1139-1150.87 (2015) https://doi.org/10.1515/pac-2014-0502

[231] Zhang, Zhang, Teng, Fan.Sulfate Radical and Its Application in Decontamination Technologies,Critical Reviews in Environmental Science and Technology,1756-1800.45 (2015) https://doi.org/10.1080/10643389.2014.970681

[232] Parvulescu, Epron, Garcia, Granger.Recent Progress and Prospects in Catalytic Water Treatment,Chemical Reviews,2981-3121.122 (2021) https://doi.org/10.1021/acs.chemrev.1c00527

[233] Nguyen, Van Tran, Nguyen, Nguyen, Alhassan, Lee.New Frontiers of Invasive Plants for Biosynthesis of Nanoparticles Towards Biomedical Applications: A Review,Science of The Total Environment,159278.857 (2023) https://doi.org/10.1016/j.scitotenv.2022.159278

[234] Nieto-Juarez, Pierzchła, Sienkiewicz, Kohn.Inactivation of Ms2 Coliphage in Fenton and Fenton-Like Systems: Role of Transition Metals, Hydrogen Peroxide and Sunlight,Environmental science & technology,3351-3356.44 (2010) https://doi.org/10.1021/es903739f

[235] Li, Sun, Yao, Han.Earth-Abundant Transition-Metal-Based Electrocatalysts for Water Electrolysis to Produce Renewable Hydrogen,Chemistry-A European Journal,18334-18355.24 (2018) https://doi.org/10.1002/chem.201803749

[236] Wang, Yue, Yang, Sirisomboonchai, Wang, Ma, Abudula, Guan.Earth-Abundant Transition-Metal-Based Bifunctional Catalysts for Overall Electrochemical Water Splitting: A Review,Journal of Alloys and Compounds,153346.819 (2020) https://doi.org/10.1016/j.jallcom.2019.153346

[237] Yu, Le, Tran, Lee.Earth-Abundant Transition-Metal-Based Bifunctional Electrocatalysts for Overall Water Splitting in Alkaline Media,Chemistry-A European Journal,6423-6436.26 (2020) https://doi.org/10.1002/chem.202000209

[238] Zhou, Zhou, Zhang, Sun, Wen, Yuan.Upgrading Earth-Abundant Biomass into Three-Dimensional Carbon Materials for Energy and Environmental Applications,Journal of Materials Chemistry A,4217-4229.7 (2019) https://doi.org/10.1039/C8TA12159A

[239] Faber, Jin.Earth-Abundant Inorganic Electrocatalysts and Their Nanostructures for Energy Conversion Applications,Energy & Environmental Science,3519-3542.7 (2014) https://doi.org/10.1039/C4EE01760A

[240] Li, Wang, Priest, Li, Xu, Wu.Advanced Electrocatalysis for Energy and Environmental Sustainability Via Water and Nitrogen Reactions,Advanced Materials,2000381.33 (2021) https://doi.org/10.1002/adma.202000381

[241] Sable, Kumar, Singh, Rustagi, Chahal, Chaudhary.Strategically Engineering Advanced Nanomaterials for Heavy-Metal Remediation from Wastewater,Coordination Chemistry Reviews,216079.518 (2024) https://doi.org/10.1016/j.ccr.2024.216079

[242] Zhang, Nai, Yu, Lou.Metal-Organic-Framework-Based Materials as Platforms for Renewable Energy and Environmental Applications,Joule,77-107.1 (2017) https://doi.org/10.1016/j.joule.2017.08.008

[243] Li, Zheng.One-Dimensional Earth-Abundant Nanomaterials for Water-Splitting Electrocatalysts,Advanced Science,1600380.4 (2017) https://doi.org/10.1002/advs.201600380

[244] Sanati, Morsali, Garcia.First-Row Transition Metal-Based Materials Derived from Bimetallic Metal-Organic Frameworks as Highly Efficient Electrocatalysts for Electrochemical Water Splitting,Energy & Environmental Science,3119-3151.15 (2022) https://doi.org/10.1039/D1EE03614A

[245] Danish, Bhattacharya, Stepanova, Mikhaylov, Grilli, Khosravy, Senjyu.A Systematic Review of Metal Oxide Applications for Energy and Environmental Sustainability,Metals,1604.10 (2020) https://doi.org/10.3390/met10121604

[246] Luo, Fu, Yu, Hristovski, Westerhoff, Crittenden.Review of Advances in Engineering Nanomaterial Adsorbents for Metal Removal and Recovery from Water: Synthesis and Microstructure Impacts,ACS ES&T Engineering,623-661.1 (2021) https://doi.org/10.1021/acsestengg.0c00174

[247] An, Lv, Jiang, Wang, Shi, Hang, Pang.The Stability of Mofs in Aqueous Solutions-Research Progress and Prospects,Green Chemical Engineering,187-204.5 (2024) https://doi.org/10.1016/j.gce.2023.07.004

[248] Singh, Roy.Evolution in the Design of Water Oxidation Catalysts with Transition-Metals: A Perspective on Biological, Molecular, Supramolecular, and Hybrid Approaches,ACS omega,9886-9920.9 (2024) https://doi.org/10.1021/acsomega.3c07847

[249] Rocky, Rahman, Endo, Hasegawa.Comprehensive Insights into Aqua Regia-Based Hybrid Methods for Efficient Recovery of Precious Metals from Secondary Raw Materials,Chemical Engineering Journal,153537(2024) https://doi.org/10.1016/j.cej.2024.153537

[250] Choi, Lee, Jang.Interconnection between Renewable Energy Technologies and Water Treatment Processes,Water Research,122037(2024) https://doi.org/10.1016/j.watres.2024.122037

[251] Oarga-Mulec, Luin, Valant.Back to the Future with Emerging Iron Technologies,RSC advances,20765-20779.14 (2024) https://doi.org/10.1039/D4RA03565H

[252] Akash, Shovon, Rahman, Rahman, Chakraborty, Prasetya, Monir.Advancements in Ceramic Membrane Technology for Water and Wastewater Treatment: A Comprehensive Exploration of Current Utilizations and Prospective Horizons,Desalination and Water Treatment,100569(2024) https://doi.org/10.1016/j.dwt.2024.100569

[253] Al-Juboori, Ahmed, Khanzada, Khatri, Al-shaeli, Ibrahim, Hilal.Burgeoning Innovation and Scalability Activities for Phosphorus Recovery from Wastewater Treatment Facilities,Sustainable Materials and Technologies,e00907.40 (2024) https://doi.org/10.1016/j.susmat.2024.e00907

[254] Vilanova, Dias, Lopes, Mendes.The Route for Commercial Photoelectrochemical Water Splitting: A Review of Large-Area Devices and Key Upscaling Challenges,Chemical Society Reviews,(2024) https://doi.org/10.1039/D1CS01069G

Applications for Earth-Abundant Transition Metals    Materials Research Forum LLC
Materials Research Foundations 179 (2025) 259-319    https://doi.org/21741/9781644903711-10

Chapter 10

# Earth-Abundant Metals as Catalysts for |Hydrogen Production

Guocai Tian[1]*

[1] State Key Laboratory of Complex Nonferrous Metal Resources Clean Utilization, Faculty of Metallurgical and Energy Engineering, Kunming University of Science and Technolgy, Kunming, Yunnan Province, China, 650093

*tiangc@kust.edu.cn, tiangc01@gmail.com

## Abstract

The development of highly active, low-cost, and durable earth rich metal oxidants is the key to achieving electro-catalytic and photo-catalytic water splitting for hydrogen production. For this purpose, hydrogen evolution catalysts such as transition metal elements, transition metal carbides, phosphides, sulfides, selenides, nitrides, oxides and hydroxides, and transition metal hybrid materials have been developed and achieved good results. In this chapter, we summarized and analyzed the research progress of using single elements, compounds, and complexes of transition metals, which are abundant on Earth, as catalysts in fields such as electrolysis and photolysis of water in recent years. We also discussed the existing problems and future development directions.

## Keywords

Hydrogen Production, Earth's Abundant Metals, Catalysts, Water Electrolysis, Water Photolysis, Transition Metals, Transition Metal Compounds, Nanomaterials

## Contents

## 1.   Introduction

The increasingly serious energy crisis and environmental problems caused by excessive energy consumption have made it an urgent need for future social development to find clean energy sources that can replace fossil fuels [1, 2]. Hydrogen energy is an ideal energy carrier, with high-energy density and wide sources, as well as advantages such as cleanliness and sustainability. It is considered an important alternative to fossil fuels. Since John Bockris proposed the hydrogen economy in the 1970s, hydrogen had been regarded as an efficient zero carbon energy source with enormous potential for future development and is considered an important driving force for the replacement of new energy sources in the context of carbon neutrality [2].

Applications for Earth-Abundant Transition Metals         Materials Research Forum LLC
Materials Research Foundations 179 (2025) 259-319      https://doi.org/21741/9781644903711-10

Nowadays, the technologies for industrial production of hydrogen mainly include methane reforming for hydrogen production, coal gasification, and water splitting [3, 5]. Water splitting technology also includes electro-catalytic cracking water, photocatalytic cracking water, and thermal catalytic cracking water. Steam methane reforming and coal gasification produce hydrogen through the reaction of steam with methane or coal at high temperatures, during which by-products such as $CO_2$ are produced. Although these two methods account for over 95% of global hydrogen fuel production, the process consumes a large amount of fossil fuels, causing environmental pollution and greenhouse-gas emissions [6-7]. Electro-catalytic water splitting for hydrogen production, using water as the sole raw material, can achieve zero carbon emissions and is the greenest and the most sustainable method [8]. Currently, electro-catalytic water splitting for hydrogen production (also known as electrolysis of water for hydrogen production) has advantages of high efficiency, green environmental protection, and sustainability, and is considered the most promising technology for development. Electricity bills are the biggest obstacle to the development of an electro-catalytic water splitting hydrogen production process, accounting for 70% of the total process cost. However, with the vigorous development of clean and renewable energy, the cost of this process will continue to decrease in the future, providing a solid foundation for the development of large-scale and low-cost electro-catalytic water splitting hydrogen production technology.

At present, the main electro-catalytic water splitting (electrolysis of water to produce hydrogen) technologies include alkaline water electrolysis (AWE), proton exchange membrane water electrolysis (PEMWE), solid oxide electrolysis cell (SOEC), and anion exchange membrane water electrolysis (AEMWE) [6-8]. The AWE method uses potassium hydroxide (KOH) aqueous solution as the electrolyte and resin or asbestos as the separator to electrolyze water to produce hydrogen and oxygen, with an efficiency typically ranging from 63% to 70%. PEMWE replaces the membrane and liquid electrolyte in AWE with polymer, and protons exchange membrane. Solid oxide SOE electrolysis technology is a high-temperature electrolysis technology that utilizes YSZ oxygen ion conductor (Yttria stabilized zirconia) as the electrolyte, with an operating temperature range of up to 600-1000 °C. It has advantages of high-energy conversion efficiency and does not require the use of precious metal catalysts. Alkaline water electrolysis AWE technology is currently the most mature electrolysis technology and is commonly used in large-scale projects both domestically and internationally. The electro-catalytic water splitting reaction is divided into two and a half reactions: hydrogen evolution reaction (HER) and oxygen evolution reaction (OER). One of the key issues hindering the practical application of electro-catalytic water splitting for hydrogen production is the high over-potential and slow reaction kinetics of OER and HER [9-10]. Therefore, the selection of catalysts plays an important role in both OER and HER, and efficient catalysts are needed to minimize the over-potential of OER and HER, thereby efficiently producing $H_2$ and $O_2$. In addition, photo-catalytic decomposition of water is an effective strategy for directly utilizing solar energy to produce hydrogen fuel. The photo-catalytic water splitting process can be divided into three basic steps: light absorption, charge separation and migration, and surface redox reaction. Photocatalytic water splitting for hydrogen production is milder and more effective compared to other hydrogen production methods and has received widespread attention [11].

Design efficient catalyst to improve the reaction kinetics of a catalytic process. It is very important to enhance the hydrogen evolution reaction efficiency of catalytic cracking water and reduce energy consumption. At present, noble metal catalysts such as metal oxides of ruthenium and iridium, platinum-based materials and so on are all reaction catalysts with good activity. However, the scarcity and high cost of precious metals have greatly hindered their large-scale application. Therefore, it is of profound significance to develop non noble metal catalysts with high activity and low cost. The earth's rich metals generally include transition metal elements, such as titanium (Tl), vanadium (V), chromium (Cr), manganese (Mn), iron (Fe), cobalt (Co), nickel (Ni), copper (Cu), zirconium (Zr), niobium (Nb), molybdenum (Mo), hafnium (Hf), tantalum (Ta), tungsten (W), etc. Through a lot of research work, researchers found that the earth's rich metals (transition metal based compounds) such as carbides, selenides, phosphates, sulfides, nitrides, etc. not only have good conductivity and semiconductor properties [2], but also have high stability and large specific surface area, which is expected to replace precious metal based electro-catalysts in an all-round way. However, at present, the appearance of transition metal based catalyst is mainly powder. In order to be coated on the conductive current collector, a certain amount of polymer (naphthol, etc.) has to be added as a binder. This process will not only increase the resistance, but also affect the distribution of active sites and reduce the transfer charge rate. Therefore, how to improve the stability and efficiency of catalysts and decrease the cost is the current problem to be solved. According to the different ways of hydrogen production, this paper mainly introduces the research progress of the earth's rich transition metal elements in hydrogen production from alkaline water, hydrogen production from proton exchange membrane electrolysis of water and photo-catalytic hydrogen production [3-16].

## 2. Hydrogen production through alkaline water electrolysis using Earth's abundant transition metal elements

### 2.1 HER transition metal catalysts

### 2.1.1 Transition metal elements

Transition metal elements (such as pure nickel mesh, foam nickel, etc.) generally have the advantages of low cost, high stability, simple preparation process and convenient large-scale industrial application. However, the intrinsic catalytic activity of transition metal elements is low, and the specific surface area of their electrochemical activity can be increased by nano-structure modification and other methods. Ma et al. [17] obtained a practical nickel mesh electrode with a 3D ordered surface structure by electroetching polycrystalline nickel. The nickel atoms on the surface of the nickel mesh can be orderly stripped along the crystal plane direction by anodic etching, forming an ordered surface structure, accelerating the detachment of bubbles and mass transfer of electrolyte during electrolysis, and reducing most of the polarization loss caused by bubbles. The etched nickel mesh electrode exhibited low HER over-potential in electrochemical testing, reaching a current density of 400 mA.cm$^{-2}$ at only 0.413 mV over-potential, and maintaining high HER catalytic activity after 600 start stop cycles of testing.

### 2.1.2 Transition metal carbides (TMCs)

TMCs have attracted widespread attention in the development of non-precious metal based electro-catalysts. For example, molybdenum carbide ($Mo_2C$) and tungsten carbide (WC) exhibit high catalytic activity for HER. In addition to superior conductivity, their hydrogen adsorption characteristics and d-band electron density states (similar to Pt) exhibit the best combination, which is considered the main reason for the observed high HER activity. In 1973, Levy et al. [18]

first discovered that WC has a d-band electron density state similar to Pt, exhibiting catalytic behavior similar to Pt. In addition, Kitchin et al. [19] further utilized density functional theory (DFT) calculations to study the physical, chemical, and electronic structural properties of a series of transition metal carbides. The theoretical calculation results were same as the experimental data in 2012, which indicates that adding carbon atoms into the lattice gap, results in d-band electron density state similar to Pt. Vrubel et al. [20] reported that commercially available $Mo_2C$ particles exhibited well HER catalytic activity in both acidic and alkaline media, but their over-potential is relatively high (190-230 mV) at a cathode current density of 10 mA/cm$^2$. Inspired by this, researchers have adopted different methods to optimize $Mo_2C$ catalysts, such as exposing more active sites through nanotechnology. Xiao et al. [21] successfully synthesized $Mo_2C$ nano-rods with porous structure by carburizing molybdate aniline in hydrogen gas. Through field emission scanning electron microscopy (FE-SEM) and transmission electron microscopy (TEM). It can be seen that the surface of the nano-rods is smooth and has a porous structure. Linear sweep voltammetry (LSV) tests were conducted in a 0.5 mol/L $H_2SO_4$ solution, and the results showed that $Mo_2C$-R nano-rods exhibited better catalytic activity than commercial $Mo_2C$. The stability test of $Mo_2C$-R is shown that after 2000 cycles, it is similar to the first cycle, proving that it has a reliable cycle life. At the same time, the activity of $Mo_2C$-R nano-rods in 1 mol/L KOH was also studied, which showed superior performance compared to commercial $Mo_2C$. $Mo_2C$-R nano-rods exhibit excellent HER performance in both acidic and alkaline media, thanks to their high conductivity and good porous morphology.

Yang et al. [22] prepared a series of transition metal carbide catalysts including VC, $Cr_3C_2$, $Mo_2C$, and $Fe_3C$, and introduced Ni atoms to regulate their electro-catalytic activity. And Ni GF/VC nanocomposites were prepared. Research has shown that Ni GF/VC exhibits excellent HER performance under both acidic and alkaline conditions. When the current density is 10 mA.cm$^{-2}$, the over-potential and Tafel's slope of NiGF/VC in 1 mol. L$^{-1}$ KOH and 0.5 mol. L$^{-1}$ $H_2SO_4$ are 128 mV, 80 mV.dec$^{-1}$, and 111 mV, 86 mV.dec$^{-1}$, respectively. Compared to other transition metal carbides, the introduction and activation of Ni atoms significantly enhance the activity of transition metal carbides in both acidic and alkaline solutions.

### 2.1.3 Transition metal phosphides

Transition metal phosphides have good mechanical strength, chemical stability, and conductivity. Therefore, it was widely used in the field of electrochemical non-precious metal hydrogen evolution as catalysts. Due to the large radius of phosphorus atoms in transition metal phosphides, reaching 0.109 nm, most phosphides are comprised of triangular prisms, which are then stacked to form anisotropic growth structures. These structures will expose more unsaturated atomic sites on the surface of transition metal phosphides, thereby forming higher catalytic activity internally. Generally speaking, hydrogen strongly binds to TMPs, so doping heteroatoms into TMPs were widely adopted as an effective method to regulate the electronic structure, optimize hydrogen adsorption strength, and ultimately improve their intrinsic HER activity. At the same time, structural regulation can also be utilized to enhance the catalytic activity of TMPs. Transition metal phosphides (TMPs) exhibit high catalytic activity and stability in both acidic and alkaline electrolytes. Owing to its excellent conductivity and unique electronic structure, P atoms play an important role in TMPs. It is precisely this unique structure and physicochemical properties that have made transition metal phosphides a hot topic of concern in recent years.

*(1) Nickel phosphide*

The discovery of $Ni_2P$ catalyst as one of the best practical catalysts for HER can be traced back to 2005. Liu and Rodriguze [23] calculated through density functional theory that Ni2P exhibits superior activity compared to pure Pt and pure Ni. The presence of P atoms reduces the concentration of highly active Ni sites, resulting in moderate binding strength between intermediates and products, referred to as the "overall effect." The discovery that $Ni_2P$ is a promising hydrogen evolution catalyst has greatly stimulated the exploration of $Ni_2P$ in the field of hydrogen evolution catalysts. Popczun et al. [24] provided direct experimental evidence supporting the catalytic synergy mentioned above, demonstrating that $Ni_2P$ nanoparticles deposited on a Ti foil substrate exhibit excellent HER activity with an exchange current density of $3.3 \times 10^{-5}$ mA/cm$^2$ and a Tafel's slope of 46mV/dec. However, the stability of $Ni_2P$/Ti electrodes in alkaline electrolytes is not good. Zhang et al. [25] further investigated the bimetallic structure of phosphide electro-catalysts ($NiCo_2P_x$), which exhibit excellent durability and long-term stability indifferent electrolytes. Lin et al. [26] synthesized a nanoscale CoP/NiCoP heterojunction catalyst using solid-phase transition method. This catalyst is primarily formed by utilizing the difference in surface binding energy between CoP and NiCoP. The electron distribution of the prepared catalyst is changed due to the heterojunction structure, making its catalytic activity superior to similar catalysts.

In order to improve the overall water decomposition efficiency, Chen and Duan [27] prepared FeRh-$Ni_2P$/N by anchoring the Fe and Rh co-doped $Ni_2P$ nano-sheet array on the three-dimensional foam nickel in situ under hydrothermal conditions, and by continuous phosphating treatment, its unique nano-sheet array effectively enriched the active sites, which is easy to combine with hydrogen. Under alkaline conditions, the hydrogen evolution over-potential of the material was just 73 mV at a current density of 10 mA/cm$^2$, with RHZ (reversible hydrogen electrode) as the reference electrode and a Tafel's slope of 117.3 mV/dec Moreover, the performance of the material did not significantly decrease after 24 hours of continuous operation. Zhang et al. [28] first prepared $Ni(OH)_2$@NC/NF, and then phosphorized it with sodium hypophosphite as the phosphorus source to obtain $Ni_2P$@NC/NF. However, its process is relatively complicated, and the conditions are harsh, which is unsuitable for large-scale production applications. In recent years, an increasing number of research reports have indicated that $Ni_2P$ is a promising catalyst that offers the opportunity to replace precious metal catalysts for water electrolysis and hydrogen production. Theoretical and experimental evidence shows that doping with heteroatoms is an effective method for regulating and enhancing the catalytic activity of $Ni_2P$. Li et al. [29] modified the surface of metal $Ni_2P$ nano-sheet catalyst by adding oxygen, and compared with pure $Ni_2P$, its OER performance was significantly improved. At a current density of 10 mA/cm$^2$, this oxygen-containing $Ni_2P$ nano-sheet generates an over-potential of 347 mV and a Tafel's slope of 63 mV/dec. Compared with pure $Ni_2P$, the catalytic current at 1.6 V is increased by about 25 times.

*(2) Cobalt phosphide*

Liu et al. [30] prepared a novel hollow polyhedral cobalt phosphide using ZIF-67 as a template. This material has a large specific surface area and high porosity, thus providing abundant catalytic active sites. In the HER test, the open-circuit voltage is 35 mV, and the Tafel's slope is 59 mV/dec, and the over-potential at a current density of 10 mA/cm$^2$ is 159 mV. In the OER test, the open-circuit voltage is 300 mV, where the Tafel's slope is 57 mV/dec, and the over-potential at a current density of 10 mA/cm$^2$ is 400 mV. Its performance is significantly better than granular CoP and comparable to commercial precious metal catalysts. Pan et al. [31] successfully synthesized a series of cobalt phosphide electro-catalysts through a simple thermal decomposition method, and obtained different phases of cobalt phosphide catalysts by changing the type of phosphorus source. The synthesized CoP/NCNTs catalyst exhibited the best HER electro-catalytic activity. In 0.5 M H$_2$SO$_4$, the open-circuit voltage is 32 mV and the Tafel's slope is 49 mV/dec. The HER catalytic activity order of the prepared series of catalysts is as follows: CoP/NCNTs>Co$_2$P/NCNTs>CoP/CNTs>Co$_2$P/CNTs>CoP>Co$_2$P [32].

*(3) Copper phosphide and metal phosphides*

In recent years, due to the metallic properties and good conductivity of Cu$_3$P, it has proven to be active against HER [33-37]. Yang et al. [38] directionally synthesized layered copper phosphide nano-arrays (Cu$_3$P/NF). At a current density of 10 mA/cm$^2$, the hydrogen evolution over-potential was 115 mV, and the performance did not significantly decrease after 24 hours. Due to its rich redox properties, copper-based materials are increasingly being designed as bi-functional catalysts in scientific research. Wang et al. [39] designed and fabricated S and Co co-doped self-supporting Cu$_3$P metal nanowires (S@Co-Cu$_3$P NWs/CF) for hydrogen evolution reaction under neutral conditions. In seawater electrolytes with near neutral pH values, the over-potential of S@Co-Cu$_3$P NWs/CF at 10 mA/cm$^2$ is 127.2 mV, indicating excellent activity of S@Co-Cu$_3$P NWs/CF towards neutral HER. At present, there is relatively little research on Cu$_3$P as a hydrogen evolution catalyst, possibly due to the lower stability and activity of copper compared to other metals. However, it also has advantages such as abundant reserves and low prices. It is believed that there will be more and more research on Cu$_3$P in the future.

*(4) Polymetallic phosphides*

Although many single metal phosphides have demonstrated excellent OER or HER catalytic performance, their inherent limitations limit further improvement of catalytic activity, such as poor conductivity and adsorption energy issues of intermediates [40]. To overcome these limitations, some researchers modify single TMPs by doping them with foreign metal atoms. Sun et al. [41] improved HER performance by introducing zinc without electro-catalytic activity into CoP. It was found that 8.3% zinc doped CoP only requires a 39mV over-potential to drive a current density of 10 mA/cm$^2$, and its HER activity is improved, which is superior to the original CoP (82 mV), Zn$_{0.04}$Co$_{0.96}$P (46 mV), and Zn$_{0.13}$Co$_{0.87}$P (66 mV). Xu et al. [42] reported an ultrafine CoMnP and N/P co-doped amorphous carbon self-supporting CoMnP-NPC/CC catalyst prepared on carbon cloth. The metal ligand was transformed in situ into an active phase carbon framework interconnect structure through a space constrained growth strategy. CoMnP crystals with a size less than 5 nm were completely surrounded by N/P co-doped amorphous carbon nano-domains, exposing the active sites completely. The doping of Mn and the strong coupling effect between CoMnP and carbon triggered the regulation of the electronic structure of the catalytic sites, while the hybridization with carbon space confinement improved conductivity and stability. The HER catalyst exhibits excellent catalytic activity and stability in alkaline electrolytes, requiring only 42 mV and 114 mV over-potentials to provide current densities of 10 mA.cm$^{-2}$ and 100 mA.cm$^{-2}$, respectively.

*(5) Two-dimensional transition metal phosphides*

In the process of electrolysis of water, protons or hydrated hydrogen ions obtain electrons at the cathode, undergo a reduction reaction, and produce hydrogen gas, which is called HER TMPs contain highly electronegative phosphorus atoms, which can capture positive charges during electro-chemical processes, accelerate charge transfer, and further accelerate the HER process. So far, researchers have prepared various two-dimensional TMPs for electrochemical HER research. For example, (Co, Ni) P@TiO Nano-sheet [43], ultra-thin bimetallic NiCoP nano-sheet [44], layered porous $NiCo_xP_y$-P/CC nano-sheet array [45], two-dimensional ultra-thin CoP nano-sheet aerogel [46], Zn-FeNi-P nano-sheet array [47], porous two-dimensional ultra-thin CoP nano-sheet [48], etc.

### 2.1.4  Transition metal sulfides

Metal sulfides are abundant and inexpensive resources, widely used in photoelectrochemistry and heavy metal removal, and have superior performance. Some sulfides have special properties due to their two-dimensional layered structure resembling graphene materials, which has attracted widespread attention from researchers. However, the HER performance of common sulfides is generally average, so designing sulfides with special structures is of great significance for developing low-cost and efficient HER catalysts. Transition metal sulfides exhibit excellent HER catalytic activity in alkaline media, but their electronic conductivity is low and durability is poor, requiring improvement through doping, mixing, heterostructures, and interface engineering [49].

Among numerous transition metal sulfides, $MoS_2$ has been widely studied due to its layered structure, high specific surface area, active edge sites, and low-energy bandgap. In theory, it has been speculated that the high activity of molybdenum disulfide comes from the sulfur edge. In 2005, Hinnemann et al. [50] discovered that the hydrogen bond free-energy of $MoS_2$ edge atoms is close to that of Pt. This discovery indicates that $MoS_2$ is a promising HER electro-catalyst. In order to further determine the actual active sites of $MoS_2$, Jaramillo et al. [51] prepared triangular $MoS_2$ single crystals of different sizes on Au (111) substrates and demonstrated a linear relationship between HER activity and the number of edge sites of $MoS_2$. The edge sites of $MoS_2$ exhibit high catalytic activity. Xie et al. [52] reported a strategy for designing defects on $MoS_2$ ultra-thin nano-sheets, which has been shown to significantly improve the electro-catalytic HER performance of $MoS_2$. The reason for its high activity is the additional edge active sites in the defect rich structure.

The structure and electronic arrangement of $WS_2$ are very similar to $MoS_2$, and it can also be used as an effective component of hydrogen evolution electro-catalysts. In 2012, Wu et al. [53] prepared a novel $WS_2$ nano-sheet catalytic material using $WO_3$ and S as starting materials through a simple ball milling sulfurization strategy Its unique lose stacking layer structure provides highly exposed edge active sites for hydrogen evolution reactions. Han et al. [54] achieved epitaxial growth of bent $WS_2$ nano-sheets in mesoporous graphene derived from nano-crystals by regulating the number of $WS_2$ layers, S vacancies, and abundant edge sites, in order to obtain high intrinsic hydrogen evolution activity $WS_2$@rGO catalyzer. Due to the synergistic effect of structure, strain, S vacancies, and edge sites $WS_2$@rGO The electrode exhibits Pt like properties in HER. In addition to layered sulfides ($MoS_2$ and $WS_2$), other transition metal sulfides (TMS) such as $FeS_2$, $NiS_2$, $CoS_2$, etc. have also attracted the attention of researchers due to their low cost, high conductivity, and fast charge transfer kinetics [55].

Among the numerous iron sulfide ($Fe_xS_y$) compounds composed of different stoichiometries, pyrite $FeS_2$ has multiple crystal forms and can achieve high hydrogen evolution activity through morphology engineering control [56-60]. It is often used as a HER electro-catalyst [56]. Jason et al. [57] synthesized $FeS_2$ with different morphologies (one-dimensional lines and two-dimensional disks) and studied their electro-catalytic performance. It was found that two-dimensional materials have more electro-catalytic active sites than one-dimensional materials, which accelerates the penetration rate of electrolyte and effectively utilizes the internal active sites of electrode materials to improve hydrogen evolution performance. Mesoporou's materials are three-dimensional materials that are easier to expose catalytic active sites than two-dimensional materials, making them star materials in the field of electro-catalysis. Miao et al. [58] obtained the precursor $Fe_2O_3$ through the sol gel method, and then successfully prepared $FeS_2$ nanoparticles with a large number of mesoporous structures through vulcanization. The unique mesoporous structure of this material allows $FeS_2$ to have a larger surface area and abundant active sites than commercial $FeS_2$, achieving a current density of 100 A.m$^{-2}$ at a low over-potential of -96 mV. Xu and Feng [60] prepared $Fe_7S_8/FeS_2$ heterostructure nano-sheets with interface structure and defect sites. By adding carbon powder to weaken the electron transfer barrier, $Fe_7S_8/FeS_2/C$ electro-catalysts were constructed. As expected, $Fe_7S_8/FeS_2/C$ requires an over-potential of 198 mV to achieve a hydrogen evolution reaction of 10 mA/cm$^2$. In addition, the overall water electrolysis in the two-electrode system maintained long-term stability for 24 hours. The excellent activity is related to the interface structure and surface defect sites, and the abundant active sites improve the charge transfer rate.

$Co_xS_y$ can exist in various stoichiometric forms, such as $Co_9S_8$, $CoS$, $Co_3S_4$, and $CoS_2$. It has rich redox properties and potential electrochemical performance and efficiency activity advantages, exhibiting excellent water electrolysis performance, superior to other transition metal chalcogenides. This endows them with diverse chemical structures and excellent hydrogen evolution performance. Faber et al. [61] successfully synthesized $CoS_2$ with different morphologies, such as thin films, nanowires, and nanowires, and systematically studied the relationship between their structure, activity, and stability. Among the three different morphologies, $CoS_2$ nanowires exhibit the highest HER catalytic performance and stability. This is because the nanowire electrode has a larger surface area, more hydrogen evolution active sites, and the structure of the electrode surface makes it easier for bubbles to be released, resulting in a faster gas release rate. Wang et al. [62] developed a sulfur rich $CoS_x$ catalyst with a hollow spherical silk cocoon structure using a simple hydrothermal method, which has a 3D conductive network structure. This unique 3D network structure promotes charge transfer during the HER process, enriches the active sites of the S-edge and bridging $S_2$-n, effectively reduces the activation barrier of the hydrogen evolution process, and makes the $CoS_x$ catalytic material rich in S stand out among HER catalysts. At a current density of 100 A.m$^{-2}$, it exhibits an over-potential of -42 mV for HER, with activity comparable to Pt/C materials. Meng and Xuan [63] prepared NF (foam nickel) electro-catalyst containing CoS nano-sheets by electrodeposition and annealed it. Due to its unique three-dimensional structure, this self-loaded catalyst has high intrinsic conductivity, abundant active sites, and excellent durability, making it a very durable and efficient HER catalyst.

The preparation of Co doped $VS_2$ nano-sheet materials also showed good HER activity during seawater electrolysis [64]. This is because Co doping can regulate the electronic structure of $VS_2$ nano-sheets, and sulfur elements covalently bind with other elements through electronic interactions. The $VS_2$ nano-sheets grown outward and vertically arranged can provide support to

Applications for Earth-Abundant Transition Metals        Materials Research Forum LLC
Materials Research Foundations 179 (2025) 259-319      https://doi.org/21741/9781644903711-10

avoid the aggregation of sulfur sites on the catalyst surface and $VS_2$ edges, promote the binding of adsorbed hydrogen atoms and the desorption of hydrogen molecules, thereby enhancing the catalytic effect of the modified material. By changing the structure of the catalyst, catalysis can also be activated through non-covalent activation modes.

$Ni_xS_y$ (NiS, $Ni_3S_2$, $Ni_7S_6$) has similar characteristics to cobalt sulfides, such as rich valence states, low cost, and environmental friendliness. Therefore, it has also received widespread attention from researchers [65]. Jiang et al. [66] prepared three crystal configurations of nickel sulfide (NiS, $NiS_2$, $Ni_3S_2$) of electrode materials using a microwave-assisted method, and found that $Ni_3S_2$ had the highest HER activity. Tong et al. [67] effectively controlled the in-situ epitaxial growth of $Ni_3S_2$ nano-sheets on nano-rods. The unique 3D homogeneous structure of this material provides a large number of catalytic active sites for hydrogen evolution reactions, demonstrating outstanding electro-catalytic hydrogen evolution activity. He et al. [68] synthesized nickel hydroxide nano-sheet arrays on foam nickel, and then carried out gas-phase fluorination on $Ni(OH)_2$/NF, followed by hydrothermal sulfuration to obtain F-$Ni_3S_2$/NF catalyst. It was found that the catalyst can achieve a current density of 10 mA.cm$^{-2}$ at an over-potential of 38 mV. The difference in electronegativity between fluorine and sulfur is significant. Introducing fluorine into transition metal sulfides can enhance the electronic interaction between transition metal cations and anions, adjust the electronic structure of $Ni_3S_2$ to shift the d-band electrons upward, which is beneficial for optimizing the adsorption energy of hydrogen and accelerating the dissociation of water molecules on the catalyst surface.

In recent years, paramagnetic nickel selenide nanomaterials, like nickel phosphide and nickel nitride, have become one of the highly promising catalysts due to their excellent d-electron configuration performance. Research has shown that during the HER process, the charge transfer from Se center to Ni center can improve the conductivity of the electrode and the upward shift of the d-band center, thereby enhancing the catalytic performance. Muhaiminl Islam [69] assembled a thin shell highly conductive titanium nitride nano-array (TiN@NiO-$NiSe_2$) on CC (carbon cloth) to synthesize a new catalyst, which promoted the number and types of electroactive sites in the system. Their research revealed that at a current density of 10 mA/cm$^2$, the hydrogen evolution over-potential of the catalyst only requires 115 mV. Wang and Zhou et al. [70] synthesized Co ($S_xSe_{1-x}$)$_2$ ternary alloy nano-wires (NWs) loaded with CFP (carbon-fiber paper). They found that Co ($S_{0.73}Se_{0.27}$)$_2$ nano-wires on CFP exhibited the highest electro-catalytic activity towards HER. The catalyst can maintain strong catalytic activity even after continuous electrolysis for 20 hours or 5000 cycles in an acidic electrolyte. Zhang and Du [71] synthesized a series of $NiSe_2$@$Ni_xS_y$ nano-arrays in situ on the surface of foam nickel. They found that the nano-array, as a three-dimensional electrode, can maximize the synergistic reaction between $NiSe_2$ and $Ni_xS_y$ and exhibit efficient hydrogen evolution performance. Among them, the $NiSe_2$@$Ni_xS_{y-0.3}$ material exhibits excellent HER performance, requiring only 148 mV over-potential at a current density of 50 mA/cm$^2$, which is one of the smallest values reported so far. In addition to nickel selenide, the Cui research group [72] synthesized the first transition metal selenides ($FeSe_2$, $CoSe_2$, $NiSe_2$) and found that $CoSe_2$ has the best electrochemical hydrogen evolution performance. The above metal selenides all have cubic pyrite type or orthorhombic pyrite type structures, in which the metal Fe atoms are located at the center of the octahedron composed of Se atoms. Chang et al. [73] prepared a layer of double doped nickel selenide nano-porous film containing Fe and P on the foam nickel substrate, and the film showed good HER activity in the process of seawater electrolysis. It is

Applications for Earth-Abundant Transition Metals                     Materials Research Forum LLC
Materials Research Foundations 179 (2025) 259-319          https://doi.org/21741/9781644903711-10

speculated that the good electrochemical hydrogen evolution activity may be related to the partially filled Eg band in $CoSe_2$.

## 2.1.5 Transition metal nitrides, oxides, and hydroxides

*(1) Transition metal nitride catalysts*

Transition metal nitride catalysts are interstitial compounds with a face centered cubic, close packed hexagonal, and simple cubic arrangements, exhibiting properties of covalent compounds, ionic compounds, and transition metals. The position of nitrogen atoms embedded in the gaps between metal atoms has unique electronic structure, high stability and catalytic activity, better conductivity, higher corrosion resistance, mechanical strength, and improved surface morphology. It is also the most promising HER catalyst for electrolytic hydrogen production [74]. In transition metal nitrides, nitrogen atoms have a smaller diameter and can be fully dissolved in the crystal structure of the transition metal, further enhancing the density of the metal crystal. At the same time, nitrogen atoms and metal atoms can produce a synergistic effect to adjust the d-band center of the transition metal and improve the catalytic activity of HER [75]. At present, some progress has been achieved in the research of transition metal nitrides. Miao et al. [76] doped nitrogen group elements of boron and nitrogen together into carbon nanotubes, and then used them as supporting materials to prepare hexagonal molybdenum nanotubes and catalyst materials. The doping of nitrogen element enhances the HER performance of the catalyst, and the increase of nitrogen element can also improve the catalytic performance of the catalyst. The different electronegativity and ionic radius between non-precious metal molybdenum and non-metallic elements can give the catalyst different hydrogen adsorption and desorption sites, thereby positively influencing the activity of the catalyst.

Wu et al. [77] prepared a novel Ni-MoN hydrogen evolution catalyst composed of dense crystalline Ni and MoN nano-particles on an amorphous MoN nano-rod matrix. The nano-rod structured $NiMoO_4.xH_2O$ precursor was obtained through a water bath reaction, followed by ammonia reduction to a hierarchical nano-rod nano-particle structure of Ni-MoN with high surface roughness. Its large surface area and multidimensional boundaries/defects can expose abundant active sites. The mesoporous structure (pore size of about 6.5 nm) and hydrophilic surfaces are conducive to rapid diffusion of electrolytes and rapid release of bubbles. This catalyst can achieve high current densities of 100 mA.cm$^{-2}$ and 1000 mA.cm$^{-2}$, respectively, in a 1 mol/L KOH solution at ultra-low over-potentials of 61 mV and 136 mV.

*(2) Transition metal oxide catalysts (TMOs)*

Transition metal oxide catalysts provide broad prospects for the development of hydrogen evolution catalysts with efficient electro-catalytic activity due to their low cost, high intrinsic activity, and strong stability [78,79]. Transition metal oxides composed of a single transition metal are the most common oxides in the metal oxide family, such as NiO, CoO, $TiO_2$, $MoO_3$, $WO_3$, $MnO_2$, and $CeO_2$. At present, researchers have achieved efficient activity of metal oxides as HER electro-catalysts through various methods [80]. Assembling controllable TMO nano-particles into a 3D mesh structure enriched with carbon substances on the surface TMO@C-QA It can accelerate the electronic transfer of materials and exhibit high conductivity and excellent electrochemical activity [81,82].

Bates et al. [83] prepared Ni-NiO-$Cr_2O_3$ catalyst with nickel and cobalt as metal substrates, which exhibited high hydrogen evolution activity and stability in seawater electrolysis for hydrogen production. It showed a low over-potential of 160 mV during seawater electrolysis and did not increase after long-term operation. Jin et al. [84] reported a catalyst S-Ni-$Fe_2O_4$, which utilizes oxygen vacancies in sulfur and spinel $NiFe_2O_4$. The introduction of S results in a highly electroactive surface in the crystal structure model, reducing the energy barrier for electron transferred from the site to the site, promoting the formation of intermediates in the sulfur hydrogen intermediate (S-H*), and a decreasing trend in adsorption energy of $H_2O$, thereby accelerating the HER reaction process. The S-Ni $Fe_2O_4$ catalyst has a HER over-potential of only 61 mV at a current density of 10 mA.$cm^{-2}$.

Due to its ease of synthesis, low cost, and adjustable composition, spinel oxides have aroused great interest among researchers in electrocatalysis. Zhu et al. [85] prepared a hollow $Co_3O_4$ microtube array ($Co_3O_4$ MTA) with graded porosity. This array has a hollow interior and a porous shell, exhibiting extremely high catalytic activity towards HER under alkaline conditions, with low initial over-potential and high-current density. Peng et al. [86] developed a series of unique necklace shaped spinel oxides ($NiCo_2O_4$, $CoMn_2O_4$, $NiMn_2O_4$). Compare to the original $NiCo_2O_4$, the reduced sample exhibits better HER activity, showing an over-potential of -135 mV with a Tafel's slope of 52 mV.$dec^{-1}$ in 1 mol.$L^{-1}$ KOH. In addition, the HER catalytic ability of spinel oxides can also be enhanced through traditional doping strategies [87].

(3) Metal (oxygen) hydroxide

Transition metal (oxygen) hydroxides are layered metal oxides with low cost, large specific surface area, good alkali stability, and unique electron distribution, which have great prospects for hydrogen evolution applications [88,89]. Recently, Zhang et al. [90] synthesized Fe substituted graded VOOH hollow sphere nanomaterials assembled from two-dimensional thin sheets. The optimized VOOH-3Fe exhibited a low over-potential of -93 mV in alkaline solutions at 100 A. $m^{-2}$. Li et al. [91] used anion ($S^{2-}$) doping to enhance the HER activity of CoOOH. In addition, the new component $CoO(OH)_{1-x}S_x$ generated in situ during the electrochemical process can further improve the HER performance of metal (oxygen) hydroxides.

## 2.1.6 Transition metal hybrid materials

N and P co-doped carbon shell coated $Cu_3P$ nanoparticles ($Cu_3P$@NPPC), bimetallic $Fe_xCo_{1-x}P$ nano-wires ($0 < x < 1$), $NiMoN_x$ nano-sheets, NiO-Ni carbon nano-tubes, copper coated nitrogen doped carbon nano-tubes (Cu@N-CNT). All of them showed excellent HER catalytic activity [92]. Ling et al. [93] prepared Ni and Zn double doped CoO nano-rod structures on conductive carbon-fiber paper (CFP), which exhibited excellent HER activity in alkaline solution. In a 1 mol.$L^{-1}$ KOH solution, at a current density of 10mA.$cm^{-2}$, the over-potential of Ni and Zn doped CoO nano-rods is 53mV. Panda et al. [94] uniformly embedded nickel platinum alloy $NiPt_3$ into the NiS amorphous matrix to form $NiPt_3$@NiS Heterostructure and deposited on a 3D porous conductive foam nickel (NF) substrate. The results indicate that it exhibits excellent HER catalytic activity in alkaline solutions. When the current density is10 mA.$cm^{-2}$ in a 1 mol. $L^{-1}$ KOH solution, $NiPt_3$@NiS. The over-potential of NF is only 12mV, and the Tafel's slope is 24 mV.$dec^{-1}$, which is lower than that of Pt and NiS synthesized by similar methods. Liu et al. [95] used a transition metal Co to replace half of the Mo atoms in the $Mo_2C$ lattice to construct CoMoC nanoparticles, which exhibited good HER activity under both acidic and alkaline conditions. In a 1 mol. $L^{-1}$ KOH solution with a current density of 10 mA.cm$^{-2}$, CoMoC exhibits an over-potential of 46 mV and a Tafel's slope of46mV.dec$^{-1}$, with a stability ofupto500 hours. Under the condition of 0.5 mol. L$^{-1}$ $H_2SO_4$, the over-potential of CoMoC is100 mV.

## 2.2    ORE Transition Metal Catalysts

Transition metal OER electro-catalysts have attracted widespread research interest due to their low cost and abundant crustal content. New strategies have been proposed to increase active sites, such as controlling morphology, element doping, and adjusting electronic structure and intermediate binding energy through element doping defect engineering to design efficient OER electro-catalysts [96-98].

### 2.2.1    Two-dimensional phosphides

TMPs have excellent HER and OER catalytic properties due to their high intrinsic catalytic activity, more active sites and strong electron transfer ability, and are feasible to replace noble metal catalysts. At present, researchers have prepared NiFeAl hybrid phosphide nano-sheet arrays [96] grown on NiFe foam, two-dimensional $CoP_3/Fe_2P$ ultra-thin nano-sheets[97] synthesized on foam nickel substrate, porous Mn CoP nano-sheets[98], $CeO_2/NiFeP/NF$ nano-sheets [99], and Fe CoP/CoO two-dimensional heterostructure nano-sheets [100] for OER research.

### 2.2.2    Metal oxides and hydroxides

*(1) Metal oxides*

Lv et al. [101] prepared a core-shell structure $IrO_2@Ti$ catalyzer. This nanoparticle helps to increase the active surface area of the catalyst, maintain good stability and metal conductivity, while reducing the amount of $IrO_2$ catalysts used. Xu et al. [102] reported a method of generating sufficient oxygen vacancies on $Co_3O_4$ nano-sheets through plasma etching strategy. At a current density of 10 mA/cm$^2$, its over-potential is only 153 eV, indicating excellent OER catalytic activity. In the field of seawater electrolysis for hydrogen production, many researchers have been conducting research on key materials. If manganese is used as the main material of the catalyst, a series of highly stable catalysts with high OER activity, such as manganese molybdenum oxide [103], can be prepared through metal or non-metal doping modification. The doping of metal elements such as molybdenum, tungsten, iron, and tin not only enhances the catalytic activity of OER, but also improves its stability.

*(2) Hydroxides*

Transition metal hydroxide catalysts are used as ORE catalysts. It is generally formed by surface reconstruction caused by electro oxidation of transition metal elemental catalysts during OER process, that is, transition metal oxides/hydroxides should be generated in situ on the catalyst surface, thus serving as the true catalytic active substance for oxygen production in OER reaction. Zhang et al. [104] prepared gel type FeCoW peroxides (W, Fe doped CoOOH). Under the conditions of 10 mA/cm$^2$ and a cycling stability of 500 h, the over-potential of FeCoW is the lowest at 191 mV. This performance is superior to Ni Fe based catalysts. DFT calculations indicate that the synergistic effect of Fe and W co doping with CoOOH optimizes the adsorption energy of the intermediate OH in the reaction. Chen et al. [105] reported a Ni CoOOH nano-sheet OER catalyst generated in situ through deep reconstruction of a pre catalyst and finally prepared a nickel cobalt based mixed pre catalyst (NiCo-H/NF). During the OER process, a vertical and interconnected Ni CoOOH nano-sheet structure was formed through deep reconstruction, and the unique nano-sheet array structure made the electrode exhibit super hydrophilicity and super hydrophobicity. The catalyst has excellent OER electro-catalytic performance, reaching a current density of 20 mA.cm$^{-2}$ at an ultra-low over-potential of 276 mV.

Zhou et al. [106] synthesized Fe doped Ni(OH)$_2$ nano-sheets using a simple and universal cation exchange method, which exhibited abundant defects and significantly enhanced OER catalytic activity. Compared with typical NiFe layered double hydroxides (LDH) nano-sheets, defect rich Ni$_{0.83}$Fe$_{0.17}$(OH)$_2$ nano-sheets have the lowest over-potential of 245 mV at a current density of 10 mA/cm$^2$. The excellent OER activity can be attributed to the abundant surface active sites and defects, and enhanced surface hydrophilicity. Transition metal carbonate hydroxides are also used as catalysts for ORE. It is a layered hydroxide salt composed of positively charged cations and intercalated anions in the interlayer region. Transition metal ions coordinate with two oxygen atoms and four hydroxide ions in the carbonate ion to form a twisted octahedron with (4+2) coordination. It has a typical layered structure, which not only facilitates mass transfer and ion migration between the electrode and electrolyte but also provides a large surface area to generate sufficient OER active sites [107]. Karmakar et al. [108] reported a nano-structured OER catalyst prepared by template free one-step reflux method for nickel substituted cobalt carbonate hydroxide hydrate (NCCHH). When the Ni substitution amount was 30% (molar fraction), the resulting NCCHH-30 catalyst exhibited an interconnected hollow ring layered nanostructure, allowing more active sites to contact the electrolyte and achieving an ultra-low OER over-potential of 141 mV@10 mA.cm$^{-2}$.

### 2.2.3 Transition metal hybrid materials

Transition metal hybrid materials exhibit excellent OER activity and stability in alkaline solutions due to their controllable electronic structure and low cost [109]. For example, Fe-CoP/CoO hybrid materials, FeCo$_x$S$_y$ nano-cluster carbon composites, CoNi$_2$Se$_4$ catalysts electro-deposited on the surface of carbon-fiber paper, Co$_3$O$_4$/Fe$_{0.33}$Co$_{0.66}$P nano-wires grown on foam, Co$_x$Fe$_{1-x}$N$_{0.5}$ nano-sheets prepared by ultrasonic treatment, nitrogen doped graphene wrapped Co$_3$ZnC/Co heterostructures, etc. have been studied as OER electro-catalyst materials and have shown good catalytic activity [110].

Luo et al. [111] synthesized Fe$_3$O$_4$/CoO interface nanostructures (Fe$_3$O$_4$/CoO CNTs) loaded on carbon nano-tubes (CNTs) using a two-step method of hydrothermal and calcination. The study showed that they exhibited high OER catalytic activity in alkaline solution. In a 1.0 mol.L$^{-1}$ KOH solution with a current density of 10 mA.cm$^{-2}$, the over-potential of Fe$_3$O$_4$/CoO CNTs is 270 mV and the Tafel's slope is 59 mV.dec$^{-1}$, exhibiting better OER performance than IrO$_2$. The wide pore structure and high specific surface area in Fe$_3$O$_4$/CoO CNTs facilitate the contact between catalytic active sites, and reaction intermediates, further enhancing the OER activity of the electro-catalyst.

Zhang et al. [112] generated transition metal nano-nets embedded with graphene films on commercial foam nickel, foam iron and foam nickel iron substrates(M-NM@G M=Ni, Fe, and NiFe), experiments results have shown excellent oxygen evolution activity under alkaline conditions. In a 1 mol.L$^{-1}$ KOH solution, the over-potential of NiFe-NM@G at a current density of 100 mA.cm$^{-2}$ is only 208 mV, and the charge transfer impedance value is the smallest, about 6.6 Ω, which is better than that of Ni-NM@G and Fe-NM@G. It exhibits extraordinary OER catalytic activity and efficient electron transfer ability.Huang et al. [113] reported a novel high-performance carbon fiber supported Co based OER catalyst (E-CoO$_x$/CF). The results showed good OER activity in 1 mol.L$^{-1}$ KOH, with over-potentials of 249 and 299 mV at current densities of 10 and 100 mA.cm$^{-2}$, respectively, which were superior to RuO$_2$/CF and exhibited good long-term stability. Electrochemical etched nickel oxide (E-NiO$_x$/CF) synthesized by the same method also demonstrated excellent OER catalytic performance. This study provides a simple and efficient strategy for developing vacancy rich defect metal oxides as efficient OER catalysts and paves the way for designing efficient electro-catalytic materials.

### 2.2.4  Other compounds

Nonprecious metal based catalysts not only include metal phosphides, oxides, and hydroxides, but also selenides, sulfides, nitrides, vanadate salts, and high entropy alloys [114-124]. Xu et al. [117] synthesized nano-structured nickel-iron diselenide (Ni$_x$Fe$_{1-x}$Se$_2$) as a template precursor for in-situ generation of a highly active NiFe oxide catalyst. The catalyst exhibits good OER activity, with an over-potential of only 195 mV at 10 mA/cm$^2$. Yu et al. [118] studied a transition metal nitride catalyst that can be used for alkaline seawater electrolysis NiMoN@NiFeN The three-dimensional core-shell structure catalyst doped with transition metals exhibited low over-potential during actual testing, with an over-potential of 398 mV at a current density of 1000 mA/cm$^2$, lower than the generation potential of hypochlorite ions (490 mV). In addition, the electrolytic cell operated continuously for 100 hours at a constant current of 500 mA/cm$^2$, with a voltage increase of less than 10%. However, these compounds lack stability in alkaline solutions with high oxidation potential. Therefore, it is necessary to pay attention to the understanding of the chemical properties of the real active sites of relevant OER catalysts.

Transition metal sulfides exhibit excellent OER performance due to their unique lattice structure and good electronic transport properties. However, there are still defects such as few active sites, poor conductivity and stability. Generally, transition metal sulfide OER catalysts such as single metal sulfides, bimetallic sulfides, multi metal sulfides, doping, and heterostructures are formed through surface modification, composition control, phase or structure-controlled design, etc. [119]. Sun et al. [120,] anchored the Zn S nano-spheres with rich Zn defects to the surface of NiCo$_2$S$_4$ nano-sheets, constructed a NiCo$_2$S$_4$/ZnS heterostructure OER catalyst, and adjusted its local electronic structure by integrating the heterogeneous interface and metal defects into one catalyst to enhance the electro-catalytic activity. The anchoring of defective ZnS nano-spheres suppresses the volume expansion of NiCo$_2$S$_4$ nano-sheets during cycling, ensuring the structural stability of the composite material. The prepared NiCo$_2$S$_4$/ZnS heterostructure hybrid exhibits excellent OER performance, with an ultra-low over-potential of 140 mV with 10 mA.cm$^{-2}$. Transition metal vanadate regulates the electronic structures of transition metal oxides and hydroxide oxides through the high valence metal vanadium, providing active sites for OER intermediates that are close to the optimal adsorption energy. The doping of V sites in transition metal oxides and hydroxides can endow OER intermediates with near optimal binding energy and exhibits lower over-potential compared to bare metal sites [121]. Shao et al. [122] synthesized NiFe vanadate, NiCo vanadate, and CoFe vanadate catalysts with different crystallinity. The NiFe-VO$_{x-0.5}$ catalyst with moderate crystallinity (mixed crystal and amorphous structure) not only has a nano-particle structure composed of small nano-sheets, but also has the highest oxygen vacancy content. The catalyst also exhibits a minimum OER over-potential of 267 mV at a current density of 10 mA.cm$^{-2}$. The combination of various transition metals and non-metallic elements can form high entropy amorphous alloy materials. The introduction of non-metallic elements in high entropy alloys cannot only turn them into amorphous alloy materials with a large number of unsaturated coordination atoms and defects but also optimize the electronic structure of the metal and promote electro-catalytic performance [123]. Wang et al. [124] prepared FeCoMoPB transition metal based amorphous high entropy nanomaterials using a simple chemical reduction method. Mo5 samples with amorphous nano-sheet structure exhibited OER over-potentials of 239, 281, and 331 mV at current densities of 10, 100, and 500 mA.cm$^{-2}$, respectively

## 2.3 HER/ORE multifunctional catalysts

### 2.3.1 Transition metal modified B, N co-doped carbon nanocomposites

*(1) Fe nanoparticles modified B, N co doped carbon nanocomposites*

Researchers have found that Fe atoms also have excellent ability to alter active sites and conductivity, causing changes in carbon materials. Therefore, Fe can be introduced into the co-doping of B and N to obtain abundant active sites, greatly improving the activity of the catalyst [125]. Lu et al. [125] successfully synthesized boron nitrogen co-doped/biomass carbon aerogel coated NiCoFe metal nanoparticles, and proved that it is an effective OER electro-catalyst. The best electro-catalysts have lower over-potentials and Tafel's slopes, even smaller than IrO$_2$ and RuO2. In addition, carbon aerogel as a matrix effectively slows down the aggregation and dissolution of metal nanoparticles, thus improving their stability. Hu et al. [126] utilized three-dimensional bark-like N-doped carbon (labeled as BNC) as a nano-reactor to prepare a series of ultra-thin metal borate (i.e., Co Bi, Ni Bi, and Fe Bi) nano-grids with abundant nano-pores, exhibiting excellent electro-catalytic activity. Liu et al. [127] successfully fabricated boron nitrogen co-doped carbon with Fe single atom anchoring sites using self-template method and high-temperature carbonization method, namely material Fe SA/BNC, which has a good HER activity.

*(2) Co nanoparticles modified B, N co-doped carbon nanocomposites*

Co atoms can also alter the structure of carbon atoms [128], form porous structures and increasing active sites, thereby enhancing the electro-catalytic activity of carbon. Therefore, Co can be introduced into the co-doping of B and N to obtain more pore structures and improve the activity of the catalyst. Guo et al. confirmed that B, N-co-doped carbon nano-sheets (Co-N, B CSs) exhibit excellent electro-catalytic activity for OER in zinc air batteries [129]. Yang et al. [130] doped B into glucose, and melamine derived carbon nitride (CN) in a convenient manner and introduced cobalt into boron doped carbon nitride (BNC) to form Co@BNC nanocomposite materials. At different Co contents Co@BNC in the series, the 12% Co@BNC exhibits an oxygen evolution reaction voltage of 1.67 V and exhibits good bifunctional electro-catalytic behavior. Sikdar et al. [128] proposed a dual-functional electro-catalyst consisting of boron nitrogen co-doped graphite two-dimensional carbon nanostructures impregnated with controlled amounts of transition metals (M-BCN; M=Co, Ni, Fe, Cu). The performance of Co-BCN is superior to that of Ni, Fe, and Cu based BCN catalysts, demonstrating strong dual-functional cycling stability. He et al. [131] used high-temperature pyrolysis synthesis to successfully prepare electro-catalytic materials with uniformly distributed cobalt nanoparticles in boron nitrogen co-doped nanotubes by introducing ZIF-67 into the precursor, exhibiting good catalytic activity and stability.

*(3) Ni nanoparticles modified B, N co-doped carbon nano-composites*

The addition of Ni can expose a large number of active sites [132], resulting in a large number of carbon defects, thereby improving the electro-catalytic activity of carbon. Therefore, Ni can be added to the co-doping of B and N to induce changes in the carbon structure, resulting in a larger specific surface area and improved catalytic activity of the catalyst. Guo et al. [133] used a simple and low-cost method to thermally decompose a three-dimensional nickel acetate [Ni(AC)$_2$.4H$_2$O] – hydroxyl phenyl boronic acid polyvinyl pyrrolidone precursor network prepared by electrospinning, and successfully synthesized a novel three-dimensional ultrafine Ni nano-particle embedded, B and N co-doped porous carbon nano-fiber (Ni@BNPCFsBNPCFs). Due to the advantages of structure and elemental composition, the best catalyst of Ni@BNPCFs-900BNPCFs-900 exhibits excellent HER electro-catalytic activity. Liu et al. [134] successfully fabricated NiCo/NBC, a boron nitrogen co-doped electro-catalyst with NiCo bimetallic atomic potential, using co-precipitation assisted chemical vapor deposition method. It has a special cage-like carbon nano-structure morphology, which may be formed due to the migration and aggregation of metal cobalt and nickel during pyrolysis, forming metal particles embedded in the surface and interior of the carbon substrate. The holes left after acid washing are favorable for oxygen transport in this morphology of the catalyst. Yu et al. [132] proposed a two-dimensional nano-sheet self-templating method for preparing bimetallic alloy nanoparticle decorated heteroatom doped carbon nano-sheet network electro-catalysts. The catalyst of NiCo@BNC-800 exhibits superior OER activity, but due to the special effect of B and N co-doping, a large number of active sites are exposed. Meanwhile, several layers of NiCo alloy coated with BNC shells were successfully synthesized, which exhibited high electro-catalytic activity and conductivity, further promoting the catalytic process.

*(4) Other transition metals (Cu, Mn, Mo, etc.) nano-particles modify B, N co-doped carbon nanocomposites*

Compared to commonly used and abundant transitional elements on Earth, other less commonly used transitional elements have also attracted the attention of some scholars and researchers. Liu et al. [135] successfully designed and synthesized porous carbon nanofibers modified with manganese oxide (Mn$_3$(BO$_3$)$_2$), embedded with iron carbide (Fe$_3$C), and co-doped with B and N (referred to as FexMny@BNPCFs, BNPCFs). The newly formed three-dimensional hybrid network exhibits excellent catalytic activity for electrolysis of water. Peng et al. [136] proposed a single atom Mn site anchored on a boron nitrogen co-doped carbon nanotube array (MnS A/BNC), which has abundant single atom active sites and strong synergistic effects between Mn and coordinating elements. It is an efficient electro-catalyst with activity and stability comparable to commercial Pt/C. Gao et al. [137] discovered a method of introducing environmentally friendly and low-cost biomass to assemble two-dimensional layered crystal surfaces, and successfully manufactured an ultra-thin boron nitrogen co-doped two-dimensional carbon composite structure catalyst embedded with nano-molybdenum carbide. This catalyst exhibits higher catalytic activity than platinum-like catalysts in HER reactions under alkaline conditions, with better electrochemical reaction kinetics and good reaction stability [138].

## 2.3.2   Transition metal hybrid materials

In recent years, researchers have made continuous efforts to prepare a variety of two-dimensional TMPs as electrolytic hydropower catalysts, such as ultra-thin heterogeneous Co$_2$P-Fe$_2$P nano-sheets[139], two-dimensional Ni (Co, Fe)PS [140], NiCoP/NF nano-sheets grown on nickel foam [141], CoP/Co$_2$P nano-sheets [142] and two-dimensional NiCoP ultra-thin nano-sheets [143]. Hua et al. [142] constructed a two-dimensional CoP nano-sheet array with interface enhancement using Co$_2$P nanoparticles as electro-catalysts. The CoP/Co$_2$P electro-catalyst exhibits excellent catalytic activity in a solution of 1 mol.L$^{-1}$ KOH. At a current density of 10 mA.cm$^{-2}$, the over-potential of HER is 68 mV and the Tafel's slope is 51 mV.dec$^{-1}$. The over-potential of OER is 256 mV, and the Tafel's slope is 40 mV.dec$^{-1}$, which is superior to most bi-functional cobalt based electro-catalysts for water electrolysis. Stability test results show that after continuous electrolysis for 58 hours, the current density of CoP/Co$_2$P electro-catalysts remains stable. Zhao et al. [143] reported a method for synthesizing two-dimensional Ni CoP ultra-thin nano-sheets assisted by boron. The optimized Ni CoP catalyst has a HER over-potential of 88 mV and a Tafel's slope of 41 mV.dec$^{-1}$ at a current density of 10 mA.cm$^{-2}$. The over-potential of OER is 290 mV, and the Tafel's slope is 66 mV.dec$^{-1}$. Ni atoms are dispersed into the CoP structure, forming unsaturated atom impurity NiCo$_{16-x}$P$_6$ sites. These sites effectively regulate the electronic structure of Co and P atoms, thereby improving the catalytic activity of HER and OER. Zhao et al. [144] successfully synthesized a novel two-dimensional nano-structure of NiCoP NiCoP exhibits excellent performance for both HER and OER in alkaline media When NiCoP catalyst is dissolved in 1 mol. L$^{-1}$ KOH solution and the current density is 10 mA.cm$^{-2}$, the over-potential of HER is 84 mV and the Tafel's slope is 67 mV.dec$^{-1}$; The over-potential of OER is 259 mV, and the Tafel's slope is 46 mV.dec$^{-1}$, which is better than most reported two-dimensional bimetallic phosphides.

In recent years, researchers have also made some progress in the field of transition metal hybrid materials for water electrolysis catalysts. For example: Fe doped Ni$_3$S$_2$ structure (Fe-Ni$_3$S$_2$/NF) loaded on foam nickel, Co doped MoS$_2$ nano-sheets, MoS$_2$/Ni$_3$S$_2$ heterostructures prepared on the surface of foam nickel, CoNi alloy encapsulated in ultra-thin graphene, bimetallic doped MoC$_2$ nano-particles wrapped in nitrogen doped graphene(NG-NiFe@MoC$_2$). Due to its excellent HER and OER activity, it has been studied as a fully decomposed hydroelectric catalyst [145-146]. Li et al. [147] developed a three-dimensional leaf shaped Co-Ni$_3$S$_2$ electro-catalyst on Ni mesh, which exhibited dual functional catalytic activity for hydrogen and oxygen evolution under alkaline conditions. The initial potential of OER for Co-Ni$_3$S$_2$ in a 1.0 mol.L$^{-1}$ NaOH solution is 1.3 V; When the current density is 10 mA.cm$^{-2}$, the over-potential of HER is 192 mV, and the Tafel's slope is 125 mV.dec$^{-1}$ Co-Ni$_3$S$_2$ exhibits excellent electro-catalytic activity in alkaline electrolytes. Co-Ni3S2 only requires 1.87 V to conduct a current density of 20 mA.cm$^{-2}$ in fully dissolved water in a two electrode system, and is considered a promising bifunctional electro-catalyst. Co-Ni$_3$S$_2$ has a unique leaf like structure, which can enhance the surface area of the material and expose more active sites; It can also accelerate the mass transfer between the electrolyte and the electro-catalyst, which is beneficial for improving the electrochemical performance of the catalyst. Lu et al. [148] proposed a method for constructing nitrogen doped graphene modified bimetallic Fe-Co-P nano-composites (FeCoP@NG) The experiment demonstrates that the material has good catalytic activity for hydrogen evolution, oxygen evolution, and complete water splitting in alkaline solutions. FeCoP@NG Under the condition of 1 mol.L$^{-1}$ KOH, the over-potential and Tafel's slope of HER are 189 mV and 66.7 mV.dec$^{-1}$, respectively, while the over-potential and Tafel's slope of OER are 269 mV and 49.2 mV.dec$^{-1}$, respectively. In the anodic and cathodic complete electrolysis test, a current density of 10 mA.cm$^{-2}$ can be provided at 1.63 V, and excellent stability over 30 hours is demonstrated. Liu et al. [149] synthesized a tri-functional electro-catalyst (Co$_2$P/Co NPC) with Co$_2$P/Co nano-particle embedded on nitrogen doped porous carbon.

Experiments have shown that it has excellent catalytic activity for oxygen evolution and complete water splitting under alkaline conditions. The OER catalytic performance of P and ZIF-67 with different ratios was verified in 1 mol · L$^{-1}$ KOH, and it was found that the best OER activity was achieved when the P/ZIF-67 mass ratio was 0.3. The over-potential at a current density of 10 mA.cm$^{-2}$ was 326 mV, and the Tafel's slope was 72.6 mV.dec$^{-1}$, which was superior to most reported transition metal phosphide catalysts.

Pang et al. [150] prepared an amorphous cobalt phosphate sulfide doped iron cobalt carbonate hydroxide (Co-S-P/Fe$_1$Co$_1$COH/NF) bifunctional catalyst. The HER over-potential of the catalyst is 73 mV@10 mA.cm$^{-2}$, and the OER over-potential is 160 mV @ 50 mA.cm$^{-2}$ In single-cell testing, only a battery voltage of 1.5 V is needed to achieve a current density of 10 mA.cm$^{-2}$ Liu et al. [151] reported a foam nickel supported bimetallic phosphide (Ni$_x$Co$_y$P) bifunctional catalyst. Achieved by treating with 0.005 mol/L CoCl$_2$ solution 0.005-NiCoP@NF The catalyst exhibits excellent activity, with a HER over-potential of 166 mV and an OER over-potential of 332 mV at 100 mA.cm$^{-2}$, and a single-cell voltage of 1.86 V at 100 mA.cm$^{-2}$. Zhu et al. [152] reported a phosphorus substituted cobalt selenide bifunctional catalyst for metal cobalt in HER and cobalt hydroxide oxide in OER. The CoSe$_{1.26}$P$_{1.42}$ sample exhibits the best HER and OER activity, reaching a current density of 10 mA.cm-2 with only 92 mV and 255 mV over-potentials. Li et al. [153] constructed a self-supporting heterogeneous bimetallic phosphide electrode C@CoP-FeP/FF bifunctional catalyst by liquid phase reaction on foam iron. The self-supporting C @ Co P-Fe P/FF electrode with porous structure can exhibit over-potentials of 154 mV (HER) and 297 mV (OER) at 100 mA.cm$^{-2}$ in alkaline seawater, and provide an overall alkaline water electrolysis current density of 100 mA.cm$^{-2}$ at a low voltage of 1.73 V. Zhang et al. [154] synthesized Co/Ni/Fe/Mn based amorphous high entropy phosphate oxide self-supporting electrode (CNFMPO), which exhibited excellent bi-functional electro-catalytic performance in alkaline water electrolysis. The HER and OER over-potentials were as low as 43 mV @ 10 mA.cm$^{-2}$ and 252 mV @ 10 mA.cm$^{-2}$, respectively, achieving a single-cell current density of 10 mA.cm$^{-2}$ at a low voltage of 1.54 V.

The electrochemical performance and long-term stability comparison of the above electrodes and other electrodes [155-] are listed in Table 1. The latest research on catalysts in the laboratory often outperforms ordinary industrial electrodes in terms of performance, but there are few new materials that can achieve stability over 100 hours in stability testing. The industrialization process of new materials still faces challenges and obstacles due to complex and difficult to scale processes, high research and production costs of new materials, and long-term stability and reliability that need to be verified.

Jothi et al. [160] electrodeposited NiCoP thin films on discarded copper wire (SCW), and the electro-catalytic performance of this bimetallic NiCoP/SCW were significantly better than that of single metal NiP/SCW and CoP/SCW. The over-potential at a current density of 10mA.cm$^{-2}$ was 178 mV (HER) and 220 mV (OER), respectively. In addition to bimetallic TMPs, tri-metal TMPs can also exhibit better electro-catalytic performance through multiple synergistic interactions between metal cations. For example, Xu's research group [161] developed a series of TMPs containing different metal components (M=Fe, Co, Ni) through chemical reduction and phosphating treatment. The OER activity of these catalysts is in the order of FeCoNiP > CoNiP > FeNiP > FeCoP > NiP > CoP > FeP. They found that when TMPs contain Co, the over-potential can be reduced in the low over-potential region; When TMPs contain Ni, it is helpful to increase the anode current in the high over-potential region Among these catalysts, bimetallic and ternary

TMPs have a significant promoting effect on the OER process, while ternary FeCoNiP has the best OER performance, with an over-potential of 200 mV at a current density of 10 mA/cm$^2$.

*Table 1 Transition metal-based bifunctional catalyst for water electrolysis*

| Electro-catalysts | OER $\eta$ 10/mV | HER $\eta$10/mV | Stability h@ mA.cm-$^2$ | Rref |
|---|---|---|---|---|
| Co-S-P/Fe$_1$Co$_1$COH/NF | 160($\eta$50) | 73 | 24 @38 for overall water splitting at 1.7V | [150] |
| 0.005-NiCoP@NF | 332($\eta$100) | 166($\eta$100) | OER:100 @100 HER:100 @100 | [151] |
| CoSe$_{1.26}$P$_{1.42}$ | 255 | 92 | OER:15 @255 over-potential HER:15 @10 | [152] |
| C@CoP-FeP/FF | 297($\eta$100) | 154 ($\eta$100) | OER:100 @100 HER:100 @100 | [153] |
| CNFMPO | 252 | 43 | OER:30 @50 HER:30 @50 | [154] |
| NiMoO$_x$/NiMoS | 186 | 38 | OER:50 @500 HER:60 @1000 | [155] |
| Ni-Fe-MoN-NT | 228 | 55 | OER:25 @244 mV over-potential HER:1000 CV cycles | [156] |
| FeCoNi-HNTA | 184 | 58 | OER:80 @200 HER:80 @200 | [157] |
| S-Fe-Ni | 200 | 25 | OER:50 @100 HER:50 @20 | [158] |
| Fe$_2$O$_3$/CuO (MH-TMO) | 220 | 70 | OER:100 @10 HER:100 @10 | [159] |

Hydrogen has extremely high-energy density and zero pollution emission capability, making it a clean and renewable-energy source that can replace fossil fuels. Electrolytic cells, as an efficient energy conversion and storage system, play a major role in the hydrogen production process. However, due to its high over-potential, the OER and HER reaction kinetics are slow, hindering the energy efficiency of the reaction. Currently, the production of hydrogen through electrolysis of water accounts for only 4% of the world's hydrogen production. In order to promote the practical application of electrolysis water in industry, designing efficient catalysts plays an important role in both OER and HER to minimize over-potential and improves catalytic efficiency. This article reviews some notable advances in nano-rare metal based and nano non-rare metal based electro-catalysts in the fields of HER and OER. At present, there are still some difficulties and challenges in commercializing hydrogen production through electrolysis of water, mainly manifested in: ①Developing non-precious metal OER electro-catalysts with high activity and long-term stability in acidic media is still a challenging research and development field; ② Limited understanding of the detailed catalytic mechanisms of transition metal based HER and OER electro-catalysts; ③ Due to the different mass loadings of catalysts on electrodes and substrate materials, different electrochemical measurement methods may affect the electron transfer rate. Therefore, it is difficult to directly compare various nano-structured catalyst materials based on performance descriptions. A more effective electro-catalyst screening strategy has to be established a standard evaluation scheme that can effectively compare the performance of catalysts from different research groups. In recent years, researchers' interest in catalyst nano-structures and lattice oxygen engineering has surged, and there is hope to make new progress in the design of OER and HER electro-catalysts with high activity and stability, and low cost, which is conducive to promoting the large-scale commercial application of hydrogen production by electrolysis of water.

## 3.  Proton exchange membrane electrolysis of water using Earth's abundant transition metal catalysts

### 3.1  Transition metal oxides

#### 3.1.1  Mn based oxides.

Inspired by the $Mn_4Ca$ clusters within the oxygen evolution centers (OECs) of biological photosynthesis, $MnO_x$ based materials are considered a potential OER catalyst. Mn based oxides have highly complex surface electrochemical states, with multiple oxidation states and crystal phases, and are prone to surface or near surface phase transitions at different pH and external voltages. Different valence states of Mn have different OER catalytic activities, among which $Mn^{3+}$ contributes the most to OER activity [162]. Although $Mn^{3+}$ undergoes disproportionation reaction under pH<9, resulting in low OER activity under acidic or neutral conditions, Mn based oxides have unique self-healing properties that can compensate for their dissolution in acidic media, providing conditions for achieving long-term stability of OER in acid. Nocera et al. [163] investigated the OER mechanism of Mn based oxides prepared by electrodeposition under alkaline, acidic, and neutral conditions. The study showed that $MnO_x$ prepared under pH=2-2.5 conditions exhibited significant self-healing properties, and $MnO_x$ thin films dissolved in the electrolyte could regrow on the electrode surface by electrodeposition. Take into account this phenomenon, researchers have developed numerous highly stable Mn based oxide acidic medium OER catalysts. Li et al. [164] synthesized $\gamma$ - $MnO_2$ on carbon paper and FTO (F-doped tin oxide conductive glass) using a simple pyrolysis method, achieving stability of over 8000 hours at pH=2 and current density of 10 $mA/cm^2$, and the performance is close to the actual production requirements. However, after running above 1.8 V for 120 hours, Mn dissolved in the form of $MnO_4^-$, and the over-potential increased by 600 mV. Based on the Pourbaix diagram of Mn [165], it can be seen that Mn based oxides in acidic media have a stable catalytic potential window between 1.6 and 1.75 V, but are sometimes difficult to maintain stability at high potentials. Therefore, it is necessary to further enhance the stability of Mn based oxides at high potentials. Coupling with more stable elements can further improve its stability, but it may sacrifice some activity. Frydendal et al. [166] used sputtering deposition to add a small amount of $TiO_2$ to $MnO_2$, which improved its stability (expected to reach 265 hours at 1.8 V), while the activity slightly decreased ($\sim$520) $mV@10$ $mA/cm^2$. Zhou et al. [167] provided stability by introducing Sb, which can effectively stabilize Mn3+octahedra in the Mn-Sb-O system. Pourbaix analysis confirmed that under the same operating conditions, Mn Sb rutile oxide alloy has better thermodynamic stability than $IrO_2$.

In response to the limited catalytic activity of Mn based oxide OER, researchers have also proposed a series of improvement plans[168-176]. Firstly, clarify the basic activity and stability characteristics of Mn based oxides with different crystal forms. Hayashi et al. [168] compared the acidic OER performance of four crystal forms, namely $\alpha$ - $MnO_2$, $\beta$-$MnO_2$, $\gamma$-$MnO_2$, and $\delta$-$MnO_2$, and found that the best performing $\alpha$-$MnO_2$ has an over-potential of 750 $mV@10$ $mA/cm^2$ and capable of stable operation at 1.9 V for 28 hours. Yang et al. [170] conducted a systematic study on the interlayer environment of $MnO_2$ nano-sheets by cation exchange with the first row of transition metal cations (including $Ni^{2+}$, $Co^{2+}$, $Cu^{2+}$, $Zn^{2+}$, and $Fe^{3+}$). The results showed that compared with the original $MnO_2$ nano-sheets, $MnO_2$ nano-sheets with Ni interlayer exhibited significantly enhanced OER activity and long-term stability. Secondly, the introduction of amorphous phase also serves to the improvement of catalytic activity. Huynh et al. [171] prepared water sodium manganese oxide phase manganese ($\delta$ - $MnO_2$) by electrodeposition, which transformed into a similar manganese oxide phase manganese ($\alpha$ - $Mn_3O_4$) during cyclic activation, while forming amorphous $\delta$ - $MnO_2$, exhibiting extremely high OER activity. At a pH of 2.5 and over-potential of 600 mV, the current density rose by two orders of magnitude, approaching the performance of Ru and Ir oxides.

In addition, doping or coupling with other elements can also enhance the catalytic activity of Mn based oxides. Prasad et al. [172] prepared F-doped $Cu_{1.5}Mn_{1.5}O_4$, which has a d-band close to $IrO_2$ and exhibits similar initial potential and stability to $IrO_2$. Van et al. [173] reported a $Ni_{0.5}Mn_{0.5}Sb_{1.7}O_y$ with a rutile structure, which achieved $(735\pm10)$ mV@10 mA/cm$^2$ in a 1 mol/L $H_2SO_4$ medium. In a recent report, Pan et al. [169] synthesized $Mn_8O_{10}Cl_3$, $Mn_{7.5}O_{10}Br_3$, and $\gamma$-$MnO_2$ on carbon cloth using a solvothermal method. $Mn_{7.5}O_{10}Br_3$ exhibited an OER over-potential of $(295 \pm 5)$ mV at a current density of 10 mA/cm$^2$ and maintained good stability for over 500 h in a PEM electrolysis cell, which is superior to state-of-the-art catalysts without precious metals and comparable to catalysts containing precious metals. Research suggests that $Mn_{7.5}O_{10}Br_3$ undergoes self-oxidation on its surface during the OER process, forming a tightly packed oxide layer. This optimizes the binding energy of intermediates and provides a dense protective layer to increase stability. In addition, the interaction between Mn and halogens is good for improving the electron transfer ability, thereby further enhancing the catalytic activity.

Finally, morphology engineering also has a positive effect on the catalytic performance of Mn based oxides. Ghadge et al. [174] synthesized $(Mn_{1-x}Nb_x)$ $O_2$:10F nano-rods formed by solid solution of F-doped $\alpha$ -$MnO_2$ and $NbO_2$, which were superior to pure $MnO_2$ catalyst (900 μm) mV@10 mA/cm$^2$) over-potential decreased to 680 mV@10 mA/cm$^2$, stable operation at 1.9 V for over 22 hours. In addition to the optimization of surface electronic structure by F doping, its one-dimensional structure provides high electrochemical activity specific surface area, more active sites, and lower charge transfer resistance, which are critical factors for improving catalytic activity. Zhao et al. [175] synthesized a single-phase manganese trioxide nano-plate by simple and rapid annealing of manganese chloride tetrahydrate in air. This two-dimensional material has an ultra-low over-potential of 210 mV at 10 mA/cm$^2$, with stability exceeding 20 hours and a Faraday efficiency of 99.1%. Theoretical calculations indicate that the theoretical over-potential of this catalyst is about 0.13 V lower than that of $RuO_2$. Wei et al. [176] obtained a PSC/$PbO_2$-$Mn_2O_3$ composite catalyst by loading Pb Sn Ca/$\alpha$-$PbO_2$/$\beta$-$PbO_2$ onto $Mn_2O_3$ nano-spheres. Thanks to the high electrochemical activity specific surface area and low charge transfer resistance provided by $Mn_2O_3$ nanospheres, the catalyst exhibited 621 mV@50 mA/cm$^2$ and 253 mV@ mA/cm$^2$ of the ultra-low over-potential and super-strong stability over 90 hours at 1.5 A/cm$^2$, respectively. Overall, Mn based oxides have the potential for acidic OER due to their self-healing properties, but their instability at higher temperatures (60-80 °C) and current densities urgently needs to be addressed. The detailed data of Mn based oxides are given in Table 2.

*Table 2 Performance of OER for Mn-based oxides*

| Catalysts | Medium | Performance mV@mA/cm$^2$ | Stability | Ref. |
|---|---|---|---|---|
| $\gamma$- $MnO_2$ | 1 mol/L $H_2SO_4$ | 489 @10 | 8000 h | [58] |
| Ti-$MnO_2$ | 0.05 mol/L $H_2SO_4$ | 520 @10 | 256 h | [60] |
| $\alpha$- $MnO_2$ | Na2SO4(pH=7.5) | 750 @10 | 28 h | [62] |
| $Mn_{7.5}O_{10}Br_3$ | 0.5 mol/L $H_2SO_4$ | 413 @10 | 500 h | [63] |
| $\delta$- $MnO_2$ | 0.5 mol/L $H_2SO_4$ | 540 @0.1 | 8 h | [65] |
| $Cu_{1.5}Mn_{1.5}O_4$:10F | 0.5 mol/L $H_2SO_4$ | 320 @10 | 24 h | [66] |
| $Ni_{0.5}Mn_{0.5}Sb_{1.7}O_y$ | 1 mol/L $H_2SO_4$ | (735±10) @10 | 168 h | [67] |
| $(Mn_{1-x}Nb_x)O_2$:10F | 1 mol/L $H_2SO_4$ | 680 @10 | 22 h | [68] |
| $R_{20}$-Mn | 0.5 mol/L $H_2SO_4$ | 210 @10 | 20 h | [69] |
| PSC/$PbO_2$-$Mn_2O_3$ | 3.06 g/L $H^+$ | 253 @10 | 90 h | [70] |

**3.1.2 Co based oxides.**

The d-electron theory of solid-state catalysis suggests that the excellent catalytic activity of precious metals is due to the high proportion of d-orbital electrons. The first transition metal element exhibits high catalytic activity due to its large number of d orbital electrons, but the dissolution of its oxide in acidic media is an urgent problem to be solved [177-178]. Fortunately, researchers have made some progress in improving the stability of Co based oxides. Leanne et al. [179] confirmed that metastable Co based films formed in situ on the surface in acidic media contribute to the stability of the OER process, which enables Co based oxides with superior catalytic activity to exhibit good acidic OER catalytic performance. Mondschein et al. [180] prepared $Co_3O_4$ nano-films on FTO substrates, and the robust $Co_3O_4$/FTO interface was found to be effective in 0.5 mol/L $H_2SO_4$ (570 μ g/L) mV@10 mA/cm$^2$ with a stable operation for over 12 hours. The study on the catalytic activity mechanism of Co based oxides found that the spinel structure $Co_3O_4$commonly used for OER catalysis contains octahedral coordinated $Co^{3+}$ and tetrahedral coordinated $Co^{2+}$ as active centers [181]. Yang et al. [182] confirmed by characterizing Ag doped $Co_3O_4$ with different atomic ratios of $Co^{2+}$/$Co^{3+}$ that $Co_3O_4$ rich in $Co^{2+}$ exhibits better acidic OER performance of 470 mV@10 mA/cm$^2$, and low-temperature calcination treatment is an effective method to promote the preferential growth of $Co^{2+}$ on the surface of $Co_3O_4$. In addition, Chen et al. [183] used DFT to study the interaction between water and the $Co_3O_4$ (110), confirming that the activity of the4f bridge site of Co is the strongest. Thanks to the support of these basic theories, the researchers further improved the OER activity and stability of Co based oxides. Yang et al. [184] synthesized high interfacial strength amorphous carbon coated $Co_3O_4$ nano-sheets on carbon paper using electrodeposition and two-step calcination method, which exhibited high OER activity of 370 mV@10 mA/cm$^2$ in 0.5 mol/L $H_2SO_4$ at a constant current density of 100 mA/cm$^2$, the lifetime can reach 86.8 hours. It can be seen that in-situ formation of a thin carbon coating can avoid surface failure of the inner catalyst, thereby improving the interfacial strength between the catalyst and the substrate, and helping to enhance the stability of OER in acidic media.

Secondly, researchers also used element doping to coordinate the catalytic activity and stability of Co based oxides. Sengeni et al. [185] introduced OER inert $Ti^{4+}$ into the $Co_3O_4$ spinel lattice ($Co_2TiO_4$) 513mV@10 mA/cm$^2$), it was found that the addition of $Ti^{4+}$ had a certain improvement in both activity and stability. Michael et al. [186] also pointed out that introducing elements such as Mn, Fe, and Pb into Co based oxide systems can enhance stability, while having a relatively minor impact on catalytic activity Among them, the Tafel's slopes between $CoO_x$, $CoMnO_x$, $CoPbO_x$, and $CoFePbO_x$ are close, which confirms that element doping can basically ensure the cooperative improvement of stability and catalytic activity of Co based oxides. In addition, Xiao et al. [187] embedded Mn into the lattice of $Co_3O_4$ spinel, which can extend the lifespan of the catalyst in acidic systems from several hours to hundreds or even thousands of hours. Without iR calibration, current densities exceeding 1000 mA/cm$^2$ can be obtained at potentials below 2 V. In addition, the stable operation time of the material at 200 mA/cm$^2$ exceeds 1500 hours, which is caused by the optimal binding energy between $Co_2MnO_4$ and OER intermediates. Its activation potential barrier is close to $IrO_2$, and the stable Mn-O bond inhibits the dissolution of O, making $Co_2MnO_4$ both highly active and stable. Huang et al. [188] found that the coupling of $CeO_2$ and $Co_3O_4$ can form a tightly bound interface, thereby expanding the electronic control effect at the two-phase interface, changing its local structural characteristics, and making $Co^{3+}$ on the surface of $Co_3O_4$ easily oxidized to OER active $Co^{4+}$ species, thereby improving the acidic OER catalytic activity (carbon paper substrate: 347mV@10 mA/cm$^2$; FTO substrate: 423 mV@10 mA/cm$^2$). Many scientists have considered the influence of catalyst morphology on the catalytic performance of Co based oxides. Yu et al. [189] reported a $Co_3O_4$ nanoparticle coated with N-doped C ($Co_3O_4$@C). The sheet-like nano-structure provides a high electrochemical activity specific surface area, and the N-doped C shell provides superior conductivity and protects the internal $Co_3O_4$ from corrosion. Therefore, the catalyst has a low over-potential of a 398 mV@10 mA/cm$^2$ and high stability over 40 hours. Fan et al. [190] prepared a traced amount of Mn doped cobalt spinel oxide nano-sheets using Co-MOF (Metal organic frameworks) precursor. The trace amount of Mn doping optimized the electronic structure of the active Co-sites, increased the covalency of Co-O, and significantly improved the service life of the electro-catalyst in the acidic OER process, allowing it to stably cycle for over24 hours at a voltage of 1.7 V. The sheet-like structure prepared by annealing the Co-MOF precursor provided a high specific surface area, with a double-layer capacity of up to 65.07 mF/cm$^2$, ensuring the dispersion of the active sites. The synergistic effect of the two resulted in a catalyst with an excellent performance of 431 mV@10 mA/cm$^2$. Based on Co-MOF precursor, Rajaet al. [191] in-situ grown Fe doped $Co_3O_4$ nano-composites derived from Co MOF on fluorine doped tin oxide (FTO) glass, with an over-potential as low as 396 mV@10 mA/cm$^2$, can run stably in 0.5 mol/L $H_2SO_4$ for more than 50 hours. Therefore, regulating the microstructure of catalysts can help improve the catalytic performance and stability of materials. In addition to exposing more active sites, the special microstructure can also enhance the stability of the catalyst by protecting the active material. Chen et al. [192] deposited approximately 4.4 nm thick amorphous $TiO_2$ on $Co_3O_4$, successfully increasing stability by three times while maintaining the original $Co_3O_4$ catalytic performance. Owing to the excellent catalytic activity provided by tetrahedral coordination of Co based oxides, their lattice can be stabilized through doping or morphology engineering to improve the stability of the catalyst. Therefore, the catalytic activity and stability of Co based oxides have been well synergistically regulated, as shown in Table 3 for detailed data.

*Table 3 Comparison of OER performance of Co-based oxides*

| Catalysts | Medium | Performance mV@mA/cm$^2$ | Stability | Ref. |
|---|---|---|---|---|
| $Co_3O_4$ films | 0.5 mol/L $H_2SO_4$ | 570 @10 | 12 h | [180] |
| Ag- $Co_3O_4$ | 0.5 mol/L $H_2SO_4$ | 470 @10 | 1000 cycles | [182] |
| $Co_3O_4$@C/CP | 0.5 mol/L $H_2SO_4$ | 370 @10 | 86.8 h | [184] |
| $Co_2TiO_4$ | 0.5 mol/L $H_2SO_4$ | 513 @10 | 10 h | [185] |
| $Co_2MnO_4$ | 0.5 mol/L $H_2SO_4$ | 720 @1000 | 1500h | [187] |
| $Co_3O_4CeO_2$ | 0.5 mol/L $H_2SO_4$ | Carbon paper: 47@10; FTO: 423@10 | 50 h | [188] |
| $Co_3O_4$@C | 1 mol/L $H_2SO_4$ | 398@10 | 40 h | [189] |
| $Mn_{0.08}$- $Co_3O_4$-400 | 0.5 mol/L $H_2SO_4$ | 431@10 | 20 h | [190] |
| Fe-$Co_3O_4$@C/FTO | 0.5 mol/L $H_2SO_4$ | 396@10 | 50 h | [191] |

### 3.1.3   Fe based oxides

Fe based oxides also have a certain potential in OER catalysis in acidic media. Messinger et al. [193] prepared a mixed polycrystalline film of magnetite ($\gamma$-$Fe_2O_3$) and hematite ($\alpha$-$Fe_2O_3$) with properties up to 650 mV@10 mA/cm$^2$. The Fe vacancies in highly conductive $\gamma$-$Fe_2O_3$ enhance the surface adsorption of water, while hematite acts as a noncatalytic support to stabilize the magnetic hematite catalyst from acid corrosion. The clever coupling of the two compounds results in OER stability of over 24 hours in 0.5 mol/L $H_2SO_4$. Its Tafel's slope is just 56 mV/dec, and the Faraday efficiency is close to 100%. Subsequently, they further investigated the effect of element doping on catalytic activity and stability, and found that 2% to 12% (atomic fraction) Co doping could improve the charge transfer performance of hematite thin films and effectively activate the prepared $Co_{0.05}Fe_{0.95}O_y$ thin film electrode (performance of 650 mV@10 mA/cm$^2$). When pH=0.3 and pH=2, the thin-film electrode can operate stably for over 50 hours and 85 hours respectively at a current density of 10 mA/cm$^2$ [194]. Zhao et al. [195] synthesized Fe doped titanium oxide nanowires (Fe/TiO$_x$-LNWs/Ti) on Ti foam. As an acid resistant OER catalyst, Fe/TiO$_x$-LNWs/Ti showed special OER electrocatalytic activity (126.2 mV/dec, 1.60 V@1.8 mA/cm$^2$) in 0.5 mol/L H2SO$_4$. Research and development have found that TiO$_2$ nano-wires can inhibit the dissolution of Fe ions, improve the stability of OER, and at the same time, the one-dimensional structure exposes more active sites, enhancing OER activity. In addition, coupling with PbO$_x$ [90] can further improve the catalytic stability of Fe based oxides. Based on the above discussion, Fe based oxides have excellent intrinsic catalytic activity, but their stability is severely lacking, especially relying on the coordination of other stable species. Research on their operation at high-current densities and high temperatures is relatively lacking. The performance data of Fe based oxides are described in detail in Table 4.

*Table 4 Comparison of OER performance of Fe-based oxides*

| Catalysts | Medium | Performance mV@mA/cm$^2$ | Stability | Ref. |
|---|---|---|---|---|
| c-$Fe_2O_3$ films | 0.5 mol/L $H_2SO_4$ | 650 @10 | 24 h | [193] |
| $Co_{0.05}Fe_{0.95}O_y$ | $H_2SO_4$ | 650 @10 | pH=0.3: 50h pH=2: 85 h | [194] |
| Fe-TiO$_x$-LNWs/Ti | 0.5 mol/L $H_2SO_4$ | 370 @1.8 | 20 h | [195] |
| CoFePbO$_x$ | $Na_2SO_4$ (pH=4) | 570 @1 | — | [196] |

## 3.2    Transition metal sulfides

Compared with transition metal oxides, transition metal sulfides have higher conductivity due to their generally lower band gap, and their intrinsic two-dimensional layered structure has become a research hotspot both domestically and internationally due to its advantages such as high specific surface area and high number of active sites [197]. In 2016, Konkena et al. [198] discovered that natural nickel chromium iron ore ($Fe_{4.5}Ni_{4.5}S_8$) can be directly used to catalyze hydrogen evolution reaction (HER) under acidic conditions and can maintain activity continuously for 170 hours at high current density ($\leq 650$ mA/cm$^2$). Given the excellent conductivity of nickel chromium iron ore and the outstanding OER performance of Ni and Fe, its OER performance has a great potential for development. Wang et al. [199] prepared sulfur doped NiS crystal particles (337 mV@10 mA/cm$^2$) on carbon nitrogen carriers, confirming that the strong hybridization of Ni and S not only provides favorable electronic structures for water adsorption and dissociation, but also enhances stability in acidic media. Hu et al. [200] used liquid polymer coated binary Prussian blue analog (PBA) nanoparticles as precursors to prepare ultrafine $Ni_4Fe_5S_8$ nanoparticles loaded on porous N, S-doped carbon networks. The pyridine-N, S-doped carbon network has strong water adsorption ability, and the Fe sites efficiently promote OER. The layered porous nanostructure promotes multiphase transport in the acidic OER process. The catalyst exhibits strong activity (200 mV@10 mA/cm$^2$) and stability in a 0.5 mol/L $H_2SO_4$ solution.

Based on the insights from the above research, cleverly designing two-dimensional layered structures with edge effects or constructing heterogeneous structures of multiple sulfide composites has become an effective means to optimize the OER activity of transition metal sulfide acidic media. Wu et al. [201] found that the stripped two-dimensional transition metal disulfides (1T-MoS$_2$, 2H-MoS$_2$, 1T-TaS$_2$, and 2H-TaS$_2$) exhibited Tafel's slopes of 260-360 mV/dec in acidic OER, with 1T-MoS$_2$ showing the best OER performance (420 mV@10 mA/cm$^2$), followed by 1T-TaS$_2$, 2H-MoS$_2$, and 2H-TaS$_2$. This suggests that the catalytic activity mainly comes from the edge sites rather than the surface. The edge of the 1T phase exhibits the lowest activation energy barrier, and the catalytic activity gradually increases during the 0-2000 cycles of CV cycling. Yang et al. [202] reported a hierarchical metal sulfide loaded on foam nickel. The MoS$_2$ and Co$_9$S$_8$ nano-sheets coordinated with each other were layered and fixed on the Ni$_3$S$_2$ nano-rod array to form a CoMoNiS-NF-xy three-dimensional composite with controllable morphology and composition (xy represents the mass ratio of Co and Mo substances, and the x : y range is 5:1~1:3). Among them, CoMoNiS-NF-31 showed the best electro-catalytic activity, showing ultra-low over-potential (HER of 113, 103 and 117 mV, OERs were 166, 228 and 405 mV respectively) under the current density of 10 mA/cm$^2$ in alkaline, acidic and neutral electrolytes, respectively, but their stability decreased significantly within 80 minutes. DFT calculations revealed that when Co$_9$S$_8$ is associated with MoS$_2$, the Fermi level decreases, which promotes electron transfer at the interface, changes the binding energy of adsorbed intermediates, and ultimately achieves a decrease in OER over-potential.

In addition, the introduction of Co is also a conventional means to coordinate the OER performance of transition metal sulfide acidic media. The hollow Co$_3$S$_4$@MoS$_2$ heterostructure synthesized in one pot using ZIF-67 by Guo et al. [203] exhibited excellent acidic mediator HER and OER dual functions. Xiong et al. [204] demonstrated excellent HER performance (60 mV@10 mA/cm$^2$) and certain OER performance of Co doped MoS$_2$ prepared by a simple hydrothermal method in 0.5 mol/L $H_2SO_4$. The above report has made significant breakthroughs in the OER performance of transition metal sulfide acidic media, but the issue of long-term stability remains a problem. Fully utilizing its edge active sites and stabilizing the main structure is expected to achieve a coordinated catalytic activity and stability. The performance data of transition metal sulfides are given in Table 5.

*Table 5 Comparison of OER performance of transition metal sulfides*

| Catalysts | Medium | Performance mV@mA/cm$^2$ | Stability | Ref. |
|---|---|---|---|---|
| NiS/SCN | 0.5 mol/L H$_2$SO$_4$ | 337@10 | 5 000cycles | [199] |
| P-NSC/Ni$_4$-Fe$_5$S$_8$-1000 | 0.5 mol/L H$_2$SO$_4$ | 200 @1 | 10 000 cycles | [200] |
| 1T-MoS$_2$ | 0.5 mol/L H$_2$SO$_4$ | 420 @10 | 2 000 cycles | [201] |
| CoMoNiS-NF-31 | 0.5 mol/L H$_2$SO$_4$ | 228 @10 | 80 min | [202] |
| Co-MoS$_2$ | 0.5 mol/L H$_2$SO$_4$ | 540 @10 | 5 000 s | [204] |

### 3.3 Transition metal phosphides

Transition metal phosphides (TMPs) are inexpensive and readily available, and their high specific surface area porous structure has abundant active sites, good conductivity, and structural stability. Transition metal phosphides (including M$_x$P$_y$ type phosphides and various forms of phosphates) exhibit excellent HER and OER performance. Parra et al. [205] reported a generalizable work in which they prepared various metal phosphides loaded on commercial carbon black powder using a simple pyrolysis method, and studied the HER, ORR, and OER catalytic performance and stability of carbon loaded Ni, Co, W, Cr, and Mo phosphides in acidic and alkaline environments one by one. Among them, carbon loaded Co$_2$P exhibits high kinetic activity towards HER, ORR, and OER in alkaline environments, but corrosion occurs in 0.1 mol/L HClO$_4$, resulting in the precipitation of single crystal Co$_3$O$_4$ nanoneedles as active sites. Although the true OER active site in the Co$_2$P system is not the phosphide itself, it indirectly confirms the potential of transition metal phosphides to catalyze OER in acidic media. Liu et al. [206] prepared Co$_2$P@NiCo$_2$O$_4$ bifunctional catalysts for HER and OER in acidic, neutral, and alkaline media, and observed the phenomenon of Co$_2$P transforming into cobalt oxide in OER. Although the performance of Co$_2$P@NiCo$_2$O$_4$ showed a significant decrease after 500 cycles of CV cycling, the comparison of the cycling stability between Co$_2$P@NiCo$_2$O$_4$ and NiCo$_2$O$_4$ confirmed that the electrochemical reactions occurring on the surface of Co$_2$P can alleviate the acidic corrosion of NiCo$_2$O$_4$. In addition, morphology also has a significant effect on the OER performance of transition metal phosphides in acidic media. Liu et al. [207] reported the synthesis of Mn doped porous FeP/Co$_3$(PO$_4$)$_2$ nano-sheets (PMFCP) supported on conductive carbon cloth. This efficient 3D self-supporting adhesive free integrated catalyst exhibits excellent catalytic performance over a wide pH range. When used for acidic OER in 0.5 mol/L H$_2$SO$_4$, it only requires a voltage of 1.75 V to achieve a current density of 10 mA/cm$^2$, which is similar to commercial Ir based catalysts. They believe that the catalytic performance comes from the doping of Mn and the porous structure on the catalyst surface. Guan et al. [208] reported a nano array formed by embedding hollow CoP nanoparticles into N-doped carbon nanotubes (NC-CNT/CoP). Irregular hollow CoP nano-spheres were synthesized through the nanoscale Kirkendall effect, forming a synergistic effect with N-doped carbon nanotubes. The NC array served as the conductive support and protection for CoP. Carbon nanotubes can further increase the electrode surface area and promote electron transfer. Irregular hollow CoP nano-spheres embedded in carbon matrix can provide abundant active sites. This catalyst based on NC-

CNT/CoP only requires a voltage of 1.66 V to achieve a current density of 10 mA/cm$^2$ in $H_2SO_4$, and has stability for over 20 hours. The Co/CoP nanoparticles (Mott Schottky electrocatalysts) developed by Janus et al. [209] and the NiAlP nano-wall array containing metal vacancies [210] have also shown good catalytic activity in acidic OER, but their stability still faces great challenges.

In addition to crystal structure, amorphous structure also has a positive influence on OER under acidic conditions. Hu et al. [211] explored a class of bulk amorphous NiFeP materials with superior macroscopic conductivity and active sites that can be directly used as device electrodes. In a 0.05 mol/L $H_2SO_4$ medium, the bulk amorphous $Ni_{40}Fe_{40}P_{20}$ (NFP$_{40}$) material exhibited high OER activity (540 mV@10 mA/cm$^2$) and stability (able to execute stably for more than 30 hours at a potential close to 2 V vs. RHE). They believe that the high activity of NFP$_{40}$ comes from the highly active sites formed by the synergistic action of coordination unsaturated Ni, Fe, and P. In the acidic OER process, Ni dissolves more from the surface than Fe, and the dissolution of Ni leaves a layer of iron phosphate on the surface, which can generate active sites and significantly promote the OER reaction kinetics. The large amount of retained Fe further enhances the OER performance. To investigate the source of stability of NFP$_{40}$, they compared the surface morphology of NFP$_{40}$ after 5 minutes of etching with $HNO_3$ and its long-term stability in 0.05 mol/L $H_2SO_4$. They also compared their P K-edge spectra after 5 minutes of OER in acidic media. It was noted that cracks appeared on the surface of NFP$_{40}$ after $HNO_3$ etching, while non etched NFP$_{40}$ tended to produce more phosphate during OER. Combining the OER stability of both acidic media (NFP$_{40}$ stability exceeds 30 hours, while the performance of NFP$_{40}$ after $HNO_3$ etching begins to decline after 20 hours), it can be considered that the dense surface and the production of phosphate contribute to the improvement of stability. In summary, transition metal phosphides have excellent intrinsic catalytic activity, and inducing their spontaneous conversion to phosphates in acidic media can effectively improve catalytic stability. The performance data of transition metal phosphides are given in Table 6.

*Table 6 Comparison of OER performance of transition metal phosphides*

| Catalysts | Medium | Performance mV@mA/cm$^2$ | Stability | Ref. |
|---|---|---|---|---|
| $Co_2P@NiCo_2O_4$ | 0.5 mol/L $H_2SO_4$ | 390@10 | 500 cycles | [206] |
| Mn-FeP/$Co_3(PO_4)_2$ | 0.5 mol/L $H_2SO_4$ | 520@10 | 30 000 s | [207] |
| NC-CNT/CoP | 0.5 mol/L $H_2SO_4$ | 430 @10 | 20 h | [208] |
| Co/CoP | 0.5 mol/L $H_2SO_4$ | 660 @1 | 12 h | [209] |
| NiAlP | 0.5 mol/L $H_2SO_4$ | 340 @10 | 10 000 s | [210] |
| NFP$_{40}$ | 0.05 mol/L $H_2SO_4$ | 540 @10 | 30 h | [211] |

### 3.4 Polyoxometalates

Polyoxometalates (POMs) are a class of multi metal oxide cluster compounds constituted by the connection of transition metal ions through oxygen. They are considered to be intermediate substances that combine the characteristics of molecular catalysts and solid oxides, and have excellent properties such as electron rich, strong redox ability, and oxidation stability in strong oxidizing environments [212-213]. Sara et al. [108] designed a central nine cobalt core connected by hydroxyl and phosphate hydrogen bonds, supported by three notched Keggin type polyphosphate tungstate ligands, consisting of $[Co_9(H_2O)_6(OH)_3(HPO_4)_2(PW_9O_{34})_3]_{16}$ polyanions. It exhibits high and stable activity as a homogeneous OER catalyst at neutral pH and can catalyze OER continuously for several days in a sodium hypochlorite medium. This achievement has inspired the application of polyoxometalates in the field of OER catalysis. Marta et al. [216] designed insoluble salts with cesium or barium resistance to cations on cobalt phosphotungstate polyanions, where the barium salt OER performance of cobalt phosphotungstate polyanions is even better than $IrO_2$ (189 mV@1 mA/cm² at pH<1). In this study, they also found that carbon paste conductive carriers with hydrocarbon binders can improve the stability of metal oxide catalysts in acidic media by offering a hydrophobic environment. Han et al. [216] found that in the polyoxometalate cluster $[\{Co_4(OH)_3(PO_4)\}_4(SiW_9O_{34})_4]_3^{2-}$ (insoluble barium salt), the catalytic activity was significantly enhanced by replacing Co with Fe (0.5 mol/L $H_2SO_4$ 385 mV@10 mA/cm², a decrease of 66 mV). These polyoxometalates also exhibited superior stability and showed no signs of decomposition after 24 hours of operation. Polyoxometalates have shown great potential in the catalytic activity and stability of OER in acidic media, as shown in Table 7.

*Table 7 Comparison of OER performance of POMs*

| Catalysts | Medium | Performance mV@mA/cm² | Stability | Ref. |
|---|---|---|---|---|
| Ba[Co-POM]/CP | $H_2SO_4$ (pH<1) | 189 @1 | 24 h | [215] |
| $[\{Co_4(OH)_3(PO_4)\}_4^{4-}(SiW_9O_{34})_4]_3^{2-}$ | 0.5 mol/L $H_2SO_4$ | 385@10 | 24 h | [216] |

**3.5    Other catalysts**

In addition to the acidic medium OER catalysts introduced above, there have also been reports proposing other diverse forms of potential OER catalysts (Table 8), such as intermetallic compounds with relatively simple configurations. Jared et al. [217] synthesized an electrode composed of the intermetallic compound $Ni_2Ta$ using high-temperature arc melting and powder metallurgy reaction. The $Ni_2Ta$ electrode melted by the arc releases oxygen for more than 66 hours at a current density of 980 mV@10 mA/cm²), the corrosion rate is two orders of magnitude lower than pure nickel. In addition, this material design method can be extended to other elements of the first transition metal, including intermetallic compounds $Fe_2Ta$ and $Co_2Ta$ systems, and alloying 3d transition metals with tantalum can significantly improve acid stability at oxidation potential. Maxine et al. [218] used $TiB_2$ particles as OER electro-catalysts in 1.0 mol/L $HClO_4$ (560 ± 20 µ g/L) mV@10 mA/cm², Faraday efficiency greater than 96%), exhibiting an ultra-low dissolution rate of 0.24 µ g/cm²/h. Lei et al. [219] designed a molecular iron nitrogen coordinated carbon nanofiber ($FeN_4$/NF/EG, 294 mV@10 mA/cm²) with graphene as the carrier for electrochemical exfoliation. The DFT study found that the $FeN_4$ structure improves the electro-catalytic OER performance by reducing the potential barrier and optimizing the reaction pathway. Shen et al. [220] synthesized a porous Si based intermetallic compound $Fe_5Si_3$. During the OER process, Si in the material dissolves and forms a layer of amorphous $SiO_2$ covering the surface of the material, which protects the catalyst from corrosion and improves its stability.

Multivariate hybrids also exhibit excellent OER catalytic performance in acidic media. Leyla et al. [221] used $H_2O_2$ assisted liquid-phase exfoliation (LPE) combined with chemical etching to prepare HER and OER bifunctional electro-catalysts loaded on single-walled carbon nanotubes (SWCNTs) with porous sheet-like $MoSe_2$ and spherical $Mo_2C$ hybridization. The hybridization of $MoSe_2$ and $Mo_2C$ crystals promoted the Volmer reaction, achieving satisfactory catalytic activity in both acidic and alkaline media (0.5 mol/L $H_2SO_4$, HER: 490 mV@10 mA/cm$^2$, OER: 197 mV@10 mA/cm$^2$; 1 mol/L KOH, HER: 89 mV@10 mA/cm$^2$, OER: 241 mV@10 mA/cm$^2$. Han et al. [222] designed a highly gas repellent N-doped WC nanoarray (N-WC) as a dual functional electro-catalyst for HER and OER. N-doping increased the intrinsic activity at the atomic level, and the unique micro nano structure provided high specific surface area and superhydrophilic and superhydrophobic properties, achieving high HER and OER activity in acidic media (0.5 mol/L $H_2SO_4$) (HER: 310 mV@200 mA/cm$^2$, OER: 470 mV@60 mA/cm$^2$). In the fully dissolved water experiment, N-WC provided a low initial potential of 1.4 V vs RHE as the anode and cathode, respectively. Before 1.7 V vs. RHE, the current density could reach 30 mA/cm$^2$. The OER performance comparison of the above catalysts is given in Table 8.

*Table 8 Comparison of OER performance of other catalysts*

| Catalysts | Medium | Performance mV@mA/cm$^2$ | Stability | Ref. |
|---|---|---|---|---|
| $Ni_2Ta$ | 0.5 mol/L $H_2SO_4$ | 980 @10 | 66 h | [217] |
| $TiB_2$ | 1 mol/L $HClO_4$ | 560±20 @10 | 144 h | [218] |
| $FeN_4/NF/EG$ | 0.5 mol/L $H_2SO_4$ | 294 @10 | 24 h | [219] |
| SWCNTs/MoSe$_2$- 2: $Mo_2C$ | 0.5 mol/L $H_2SO_4$ | 197 @10 | 30 min | [221] |
| N-WC | 0.5 mol/L $H_2SO_4$ | 470 @60 | 60 h | [221] |

Other acid stable multicomponent nonprecious metal-based oxides have also been reported. Based on the Pourbaix diagram in aqueous solution, oxide catalysts in acidic media were designed. Norskov et al. [117] conducted high-throughput screening calculations on over 11000 two-dimensional materials from different databases and found that nearly 35 materials theoretically have the ability to exist stably under acidic ORR or OER conditions. Agnes et al. [224] confirmed that $BiO_x$ thin films exhibit a certain over-potential (780-810) mV@10 mA/cm$^2$, it can operate stably for more than 110 hours at pH 1.0-2.25. Anjli et al. [225] developed a new Pourbaix algorithm (which has been included in the pymatgen code and the Materials Project website), achieving Pourbaix stability calculation for complex multi-component systems and developing Pourbaix diagrams for binary or ternary compounds, providing a theoretical basis for the study of OER stability in acidic media of multicomponent oxides. Aniketa et al. [226] screened the entire $(MnCoTaSb)O_x$ composition space and observed that components with high initial activity, including cobalt and manganese oxides, can provide excellent stability after introducing antimony and tantalum components.

In addition, understanding the reconstruction of catalysts under acidic conditions is beneficial for designing efficient and acid stable OER catalysts. In fact, as early as 1990, none in situ XPS studies already confirmed the surface oxidation of iridium oxide films during the oxygen evolution process [218]. During the OER process, catalyst cation/anion leaching, cation/anion vacancy, and multi-component co permeation can all cause reconstructions. Non precious metal based catalysts with reversible reconstruction phenomena during OER under acidic conditions have also been reported, such as cobalt containing polyoxometalates [215] and strontium titanate [218]. Taking amorphous Ir nano-sheets saturated with 0.1 mol/L $HClO_4$ for $O_2$ as an example [229], in-situ EXAFS measurements confirmed that a decrease in the bond length of Ir-C/O was observed when the applied potential increased from 1.16 V to 1.48 V, due to the formation of intermediate species or $O_2$ on the catalyst surface. However, when returning to the initial potential, the bond length of Ir-

C/O and the valence state of Ir can be restored to their original state. These results indicate that amorphous Ir catalysts have high component stability. It is worth mentioning that introducing traces amounts of precious metals on top of some non-precious metal-based catalyst is also a way to improve the OER activity and stability in acidic media. The clever introduction of trace precious metals, such as single atom catalysts [229-230] or extremely low doping perovskites, pyrochlore catalysts [232-233], is expected to optimize the charge distribution and orbital state of non-precious metal catalysts, thereby regulating their OER performance.

In this section, we reviewed the research progress of non-precious metal OER catalysts for PEM electrolysis water systems at home and abroad, mainly revealing the evolution of catalysts in acidic medium OER process from multiple aspects such as morphology, composition, crystal structure, self-healing, and reconstruction, providing theoretical reference for the development of efficient and stable new acidic medium OER catalysts in the future In recent years, PEM electrolysis hydrogen production systems have gradually shifted from laboratory research to industrial applications. Nonprecious metal based OER catalysts are limited in their large-scale application due to issues with activity and stability, performance, and production costs, and need to be strengthened in the future.

## 4. Earth rich transition metal catalysts for photocatalytic water splitting to produce hydrogen

The development and utilization of clean and renewable green energy is becoming more and more urgent. Solar energy has the advantages of abundant reserves, wide distribution, and no pollution, and is regarded as an ideal form of renewable energy [234-236]. However, solar energy has problems such as low energy flow density and discontinuous spatiotemporal distribution, which require a reasonable energy conversion method for its effective storage and utilization. Compared with other hydrogen production methods, photocatalytic water splitting for hydrogen production is milder and more effective and has received widespread attention. The photocatalytic water splitting process can be divided into three basic steps: light absorption, charge separation and migration, and surface redox reaction. Catalysts can effectively improve charge separation efficiency, provide reactive sites, and inhibit the occurrence of photocatalytic corrosion of catalysts, thereby enhancing water splitting efficiency. Catalysts can also enhance surface redox kinetics by activating water molecules, thereby improving the overall solar energy conversion efficiency of photocatalytic reactions. The core of photocatalytic water splitting for hydrogen production is the catalyst, among which the most widely used is the precious metal platinum (Pt) based catalyst [234-236]. Precious metal Pt based catalysts are prepared by loading Pt particles into semiconductor materials, where the semiconductor material absorbs light and forms electron hole pairs, and the Pt particles provide catalytic active centers to decompose water into hydrogen gas. Although precious metal Pt based catalysts have high photocatalytic efficiency for water splitting and hydrogen production, they are expensive and have scarce resources, making them unsuitable for large-scale commercial applications [236]. In recent years, various transition metal catalysts have been developed and some progress has been made [237-268]. We will introduce as follows.

Applications for Earth-Abundant Transition Metals                    Materials Research Forum LLC
Materials Research Foundations 179 (2025) 259-319          https://doi.org/21741/9781644903711-10

## 4.1  TiO₂ photocatalysts

### 4.1.1  TiO₂ single-phase photocatalysts

TiO$_2$ has stable performance, is environmentally friendly, non-toxic and harmless, and is a typical photocatalysts. When light is irradiated on TiO$_2$ samples and the energy of photons is greater than their bandgap width (3.0-3.2 eV), the energy conditions for water splitting can be achieved [237]. However, its high bandgap and fast recombination rate of photo generated charge carriers seriously affect the hydrogen evolution efficiency of the catalyst. Scientists have improved this catalyst using three methods: surface structure adjustment, metal/non-metal ion doping, and precious metal precipitation. Kumaravel et al. [238] prepared porous reduced TiO$_2$ with oxygen vacancy gradient based on catalyst surface structure tuning method. The porous structure provides a larger surface area and active sites, which is good for increasing the photocurrent density. Due to the higher porosity, which can improve the photoelectric conversion efficiency, partially decreased porous TiO$_2$ oriented structures are expected to become electrode materials for future photocatalysis. Metal/non-metal ion doping and precious metal precipitation are important methods for improving TiO$_2$ photocatalysts. Wang et al. [239] prepared tin doped titanium dioxide photocatalysts and found that Sn$^{4+}$ doping transformed mixed phase TiO$_2$ into single rutile phase TiO$_2$, improving water splitting efficiency. Chen et al. [240] prepared titanium dioxide nano-sheets doped with main group metal ions, which significantly improved the photocatalytic activity of the catalyst. This may be ascribed to the enhanced UV visible-light absorption ability and accelerated catalyst charge transfer/separation. Almomani et al. [241] found that adding Co to the TiO$_2$ lattice changed its lattice size and surface morphology, enhancing its catalytic activity. Research has shown that appropriate doping conditions can significantly affect the chemical properties of the final material and improve the photocatalytic hydrogen evolution effect of the catalyst.

### 4.1.2  TiO₂ composite photocatalysts

Combining TiO$_2$ with suitable semiconductors through heterojunction method can improve the photocatalytic performance of the catalyst [242]. Huang et al. [243] prepared Cu$_3$Mo$_2$O$_9$/TiO$_2$ hybrid, and the p-n heterojunction formed between Cu$_3$Mo$_2$O$_9$ and TiO$_2$ facilitates charge separation. The hydrogen production rate of this hybrid is 14.5times thatof the original catalyst. Wang et al. [244] prepared SnO$_2$/TiO$_2$ heterostructures, in which abundant oxygen vacancies can serve as active sites for catalytic reactions. The oxide interface of the modified catalyst has uniform and tight contact, which is conducive to the separation and migration of photogenerated carriers. Yu et al. [245] prepared NiO nanoparticles doped TiO$_2$ nanotube p-n heterojunction photocatalysts with a larger specific surface area, which showed a hydrogen production rate four times higher than TiO$_2$ without any co catalysts. This effect is due to the enhanced photo responsiveness of the catalyst to visible light by NiO nanoparticles, while the formation of p-n heterojunctions induces an intrinsic electric field, promoting the transfer of photo generated electrons and increasing the migration and separation rates of charge carriers. Yue et al. [246] prepared TiO$_2$/SiO$_2$ composite gel, and the porous core-shell material obtained has good specific surface area and high photocatalytic efficiency. The hydrogen evolution effects of several TiO$_2$ composite photocatalysts are given in Table 9.

*Table 9 Hydrogen evolution effect of TiO$_2$ composite photocatalysts*

| Medium | Catalysts | Rate of H$_2$ Production /[$\mu$mol·(g·h)$^{-1}$] | Condition | Ref. |
|---|---|---|---|---|
| Deionized water | Cu$_3$Mo$_2$O$_9$/TiO$_2$ | 3 401.90 | 300 W Xenon lamp | [243] |
| Deionized water | NiO/TiO$_2$ | 228.00 | 300 W Xenon lamp | [245] |
| Deionized water | TiO$_2$@SiO$_2$ | 410.61 | Ultraviolet radiation (at room temperature) | [246] |
| Deionized water | Cu$_2$O/TiO$_2$ | 11 000.00 | 100 MW/cm$^2$ Xenon lamp | [247] |
| NaCl solution | H-CoS/CdS | 143.10 | ($\lambda$>420 nm) | [248] |
| Deionized water | Cu$_2$S/CdZnS | 5 904.00 | 300 W Xenon lamp ($\lambda$≥420 nm) | [249] |
| Deionized water | Zn$_{0.3}$Cd$_{0.7}$S/15%NiS$_{1.97}$ | 22637.00 | Simulated Sunline | [250] |
| Deionized water | MoS$_2$/Zn$_{0.5}$Cd$_{0.5}$S | 388.20 | 300 W Xenon lamp | [251] |
| Deionized water | Ni$_2$P@UiO-66-NH$_2$/Zn$_{0.5}$Cd$_{0.5}$S | 40910.00 | 300 W Xenon lamp ($\lambda$>400 nm) | [252] |

Cu$_2$O has a narrow bandgap (2.2 eV) and abundant reserves, with good visible-light absorption ability, and is considered one of the most promising materials for practical use [247]. Lü et al. [247] combined Cu$_2$O nanoparticles with mesoporous TiO$_2$ to obtain Cu$_2$O/TiO$_2$ composite photocatalysts. The mesoporous structure of Cu$_2$O makes the material more stable, and the narrow bandgap of Cu$_2$O enhances the visible-light absorption ability of the catalyst and increases the migration rate of charge carriers, thereby improving the hydrogen evolution rate of the catalyst. Chen et al. [253] prepared Cu$_2$O/TiO$_2$ heterogeneous materials and tested that the catalyst exhibited excellent photocatalytic activity when the addition of TiO$_2$ was 70%. The hydrogen production within 6 hours was 416 times and 49 times higher than that of Cu$_2$O and TiO$_2$, respectively. The reason for this phenomenon is the appropriate bandgap structure between Cu$_2$O and TiO$_2$.

Red phosphorus is an emerging non-metallic photocatalyst with excellent light absorption ability, but it is still constrained by the problem of fast recombination of photo generated charge carriers. Huang et al. [254] designed and fabricated RP modified titanium dioxide spheres, with an optimal hydrogen production rate of 215.5 $\mu$ mol/(g.h). The hollow sphere structure and RP light absorber can enhance the light absorption capacity of the catalyst, while the heterojunction induced interface charge migration promotes photo induced charge separation and improves the photocatalytic hydrogen production performance.

### 4.2 Transition metal sulfides

Transition metal sulfide catalysts have the advantages of narrow bandgap, strong light collecting ability, low cost, adjustable structure, and a wide variety of types, and are considered the best choice to replace precious metal Pt based catalysts. Transition metal sulfides are constituted by the combination of transition metal elements such as nickel (Ni) and molybdenum (Mo) with sulfur

(S) element [255-264]. Transition metal sulfide catalysts can absorb light to form photo generated electron hole pairs, separate photo generated electron-hole pairs to form free charges and provide catalytic active centers to decompose water molecules into hydrogen gas. They exhibit good catalytic performance and stability in the process of photocatalytic water splitting for hydrogen production summarize the research progress of transition metal sulfide catalysts in photocatalytic water splitting for hydrogen production in recent years, discuss the key factors affecting the catalytic efficiency of transition metal sulfide catalysts in photocatalytic water splitting for hydrogen production, analyze the problems and defects of transition metal sulfide catalysts in photocatalytic water splitting for hydrogen production, and future research trends.

The commonly used transition metal sulfide catalysts can be divided into two categories: binary ($A_xS_y$) and multicomponent ($A_xB_yS_z$) [253-284]. Binary transition metal sulfide catalysts are composed of two elements: transition metal (A) and sulfur. Multivariate transition metal sulfides are composed of two or more transition metal elements combined with sulfur element [255]. For a single binary transition metal sulfide catalyst, light exposure can easily cause oxidation reactions between the surface transition metal elements and sulfur elements (i.e. photo corrosion phenomenon), resulting in the loss of surface sulfur elements and the formation of a layer of transition metal oxides with lower catalytic activity on its surface, thereby reducing the catalytic activity and stability of the catalyst [256]. In addition, a single binary transition metal sulfide catalyst has poor ability to separate and transport photo generated electron-hole pairs, which also reduces the photocatalytic activity and stability of a single binary transition metal sulfide catalyst for water splitting and hydrogen production [257-262]. Hao et al. [263] modified single atom Pt on the surface of CdS nanoparticles, and the synergistic metal semiconductor interaction promoted the migration of surface photogenerated carriers in CdS, resulting in high photo fuel conversion efficiency. Dong et al. [264] studied the crystal structure, electronic properties, light absorption properties, and photocatalytic hydrogen production activity of copper doped zinc sulfide. The substitution of Cu and S vacancies reduced the bandgap width of ZnS.

The improved catalyst has a higher effective mass ratio of photo generated holes to electrons, which suppresses the recombination of photo generated carriers and enhances the photocatalytic activity of the catalyst. Research has found that composite catalysts are formed by combining two or more binary transition metal sulfides or by loading transition metal sulfides onto other materials, such as $MoS_2/CdS$ [266] $MoS_2@TiO_2$ [266], CdS/ $Mo_2C@C$ [267], CdS/WC [268], CuS/CdS [269], $CdS/Co_3S_4$ [270], $CdS/FeS_2$ [271], CdInS [272], and CdS/NiS [273] exhibit superior resistance to photo corrosion compared to single binary transition metal sulfide catalysts,as showed in Table 10. The good interface formed between different components on the composite catalyst effectively improves the separation efficiency of photo generated electron-hole pairs and suppresses the recombination of photo generated electrons and holes. On this basis, the composite catalyst exhibits much higher catalytic activity and stability than the single binary transition metal sulfide catalyst.

*Table 10 Comparison of hydrogen production from photocatalytic water splitting by transition metal sulfide catalysts*

|  | Catalysts | Rate of H$_2$ Production | Condition Xenon lamp ($\lambda \geq 420$ nm) | Ref. |
|---|---|---|---|---|
| Binary system | MoS$_2$/CdS | 71.24 mmol/(g·h) | 300 W | [265] |
|  | MoS$_2$@TiO$_2$ | 8.43 μmol/(cm$^2$·h) | 300 W | [267] |
|  | CdS/WC | 1331.7 μmol/h | 300 W | [268] |
|  | CuS/CdS | 561.7 μmol/h | 300 W | [269] |
|  | CdS/Co$_3$S$_4$ | 1083.9 μmol/h | 300 W | [270] |
|  | CdS/FeS$_2$ | 107.56 mL/g | 300 W | [271] |
|  | Cd—In—S | 660.5 μmol/h | 300 W | [272] |
|  | CdS/NiS | 39.68 mmol/(g·h) | 300 W | [273] |
| Multicomponent system | MoS$_2$-CdS/GDY | 17.99 mmol/(g·h) | 5 W Diode | [276] |
|  | CdS/Ni$_3$S$_2$@C | 1164.7 μmol/h | 300 W | [277] |
|  | Mn$_x$Cd$_{1-x}$S | 185.95 μmol/h | 300 W | [278] |
|  | Mn$_{0.2}$Cd$_{0.8}$S/α-MnS | 335 μmol/h | 300 W | [280] |
|  | CoS$_{1.097}$@ZIS | 2632.33 μmol/(g·h) | 300 W | [281] |
|  | CuS@ZnIn$_2$S$_4$ | 7910 μmol/(g·h) | 300 W | [282] |

**4.3    g-C$_3$N$_4$ photocatalyst**

The g-C$_3$N$_4$ has the advantages of suitable band structure and environmental friendliness, and has great potential in hydrogen evolution [285-287]. In order to overcome its limited specific surface area and fast recombination rate of photo generated carriers, researchers have made many improvements to g-C$_3$N$_4$ photocatalysts, such as morphological optimization, metal/non-metal doping, and heterojunction construction [288]. Rao et al. [289] performed morphological control on g-C$_3$N$_4$ and obtained a photocatalytic material of g-C$_3$N$_4$ with pore structure and carbon vacancies. The pore structure endows new photocatalytic materials with higher specific surface area and new active surfaces. The introduction of carbon vacancies increases the mobility of photo generated carriers and improves the visible-light response capability of the catalyst. Wang et al. [290] synthesized a P-doped g-C$_3$N$_4$ photo-catalyst with a porous structure morphology based on the idea of non-metallic doping. Compared with unmodified g-C$_3$N$_4$, the modified photo-catalyst has a better visible-light response, and better photo generated electron-hole separation efficiency. Experimental results have demonstrated that the modified photo-catalyst has better hydrogen evolution effect and stability. Liu et al. [291] prepared mesoporous g-C$_3$N$_4$ Co modified with sulfur doping and nitrogen defects. The introduction of mesoporous increased the surface area to 153.9 m$^2$/g and provided more active sites. S doping and nitrogen defects jointly optimize the band structure, reduce bandgap energy, expand light absorption capacity, and generate more photos generated electrons. The separation rate and migration rate of photo generated carriers in the catalyst are strengthened, resulting in a 48.4-fold increase in hydrogen volatilization rate compared to before. The nitride composite photo-catalyst based on the construction of heterostructures is also a research focus of photocatalytic hydrogen production catalysts. Combining semiconductor catalysts with carbon materials to form composite photo-catalysts for studying nanostructures has become a new research field. Carbon spheres, as a common carbon material, have the advantages of good conductivity and high thermal chemical stability [292]. Sun et al. [293] prepared g-$C_3N_4$/$WO_3$ carbon microsphere composite photo-catalysts, and the newly prepared photocatalysts showed improved photocatalytic hydrogen evolution ability and chemical stability, as well as good conductivity, promoting the transfer of photo generated electrons in g-$C_3N_4$ nano-sheets. The g-$C_3N_4$ and $WO_3$ formed a Z-shaped heterojunction on the surface of g-$C_3N_4$/$WO_3$ carbon microsphere composite photo-catalyst. The addition of carbon microspheres also increases the specific surface area of the composite photo-catalyst. Under sunlight irradiation, the hydrogen production efficiency of the catalyst for photocatalytic water splitting is 70.54 times that of pure g-$C_3N_4$. The combination of sulfides with g-$C_3N_4$ catalyst is a common way for researchers to improve the catalytic performance of g-$C_3N_4$ catalyst. The hydrogen evolution efficiency of several g-$C_3N_4$ composite photo-catalysts is given in Table 11.

*Table 11 Hydrogen evolution effect of g-$C_3N_4$ composite photocatalysts*

| Medium | Catalysts | Rate of $H_2$ Production | Condition Xenon lamp $(\lambda \geq$ 420 nm$)$ | Ref. |
|---|---|---|---|---|
| Deionized water | $MoS_2$/g-$C_3N_4$ | 12 000.00 | 300 W | [294] |
| Deionized water | ZnS-$NiS_2$/g-$C_3N_4$ | 366.90 | 300 W | [295] |
| Deionized water | 2D(10-SnS)/g-$C_3N_4$ | 818.93 | 300 W | [296] |
| Deionized water | $Cu_3P$/g-$C_3N_4$ | 343.00 | 300 W | [297] |

The addition of some phosphides and oxides can also enhance the catalytic ability of g-$C_3N_4$ catalyst. Zhou et al. [297] synthesized graphite nitride carbon-based catalysts loaded with copper phosphide. UV visible spectroscopy indicates that after loading copper phosphorus, the light absorption ability of the catalyst is enhanced, and the separation and transfer ability of photoexcited charge carriers is greatly improved. In addition, copper phosphide/g-$C_3N_4$ photocatalyst exhibits a relatively high hydrogen evolution rate, while copper phosphide itself has no photocatalytic activity. Therefore, copper phosphide acts as a Co catalyst in the photocatalytic hydrogen production process. Liu et al. [298] designed 2D crystalline g-$C_3N_4$ (referred to as CCN) and 2D $TiO_2$ (referred to as TO) nano-sheets using electrostatic self-assembly technology. The crystalline g-$C_3N_4$ photocatalyst introduced $TiO_2$ through interface effects, which facilitated the reduction of $H^+$ to $H_2$ and improved photocatalytic activity.

The photocatalytic water splitting process for hydrogen production utilizes abundant and environmentally friendly solar energy, and the entire process has no $CO_2$ emissions, fully meeting the needs of green and sustainable development. Transition metal sulfide catalysts have advantages such as abundant resources and low prices, making them highly suitable for large-scale commercial applications as photocatalytic catalysts for water splitting and hydrogen production. Overall, $TiO_2$ photocatalysts have the advantages of high chemical stability, corrosion resistance, and strong oxidation-reduction ability of photo generated electron-hole pairs generated by excitation. However, the bandgap of $TiO_2$ photocatalyst is wide. The efficiency of solar-energy utilization is low, and the recombination rate of charge carriers is too fast, which limits the hydrogen production rate of photocatalyst. Transition metal sulfide photocatalysts have the advantages of narrow bandgaps, strong light collection ability, and low work function, but they are still constrained by the fast recombination of electron-hole pairs. Graphitic carbon nitride (g-$C_3N_4$) has been widely studied due to its suitable bandgap width, good chemical stability, and ability to maintain stability under high-temperature conditions. It also has the advantages of being non-toxic and easy to prepare. However, in practical applications, factors such as rapid recombination of photo generated electron-hole pairs, poor visible-light absorption ability, and low specific surface area greatly limit its photocatalytic hydrogen evolution efficiency. $TiO_2$ photocatalyst and graphite phase nitrogen carbide (g-$C_3N_4$) catalyst have the advantage of low cost Due to factors such as cost, efficiency, and catalyst types of photocatalytic hydrogen production. The method of photocatalytic water splitting for hydrogen production is still difficult to achieve large-scale use. Further research is needed to find catalysts with suitable prices, design more convenient modified photocatalyst methods, and develop photocatalysts with better hydrogen evolution capabilities.

**Conclusion and Prospective**

The development of highly active, low-cost, and durable earth rich metal oxidants is the key to achieving electro-catalytic and photocatalytic water splitting for hydrogen production. For this purpose, hydrogen evolution catalysts such as transition metal elements, transition metal carbides, phosphides, sulfides, selenides, nitrides, oxides and hydroxides, and transition metal hybrid materials have been developed and achieved good results. Two-dimensional phosphides, metal oxides and hydroxides, transition metal hybrid materials, sulfides, nitrides, vanadate salts, and high entropy alloys have been developed as oxygen evolution catalysts, achieving the expected design goals. We have developed Fe, Co, Ni, and nano particle modified B, N co-doped carbon nanocomposites, which are mostly HER/ORE multifunctional catalytic materials such as transition metal layered nanomaterials and transition metal hybrid materials. At present, the conductivity and intrinsic activity of catalysts are mainly optimized through electronic structure control, morphology control, and surface/interface control. On this basis, the electro-catalytic oxygen evolution performance of transition metals has been significantly improved, but there is still a certain distance from practical application. In terms of proton exchange membrane water electrolysis catalysts, people have developed Mn, Co, Fe based transition metal oxides, two-dimensional layered structures or heterogeneous structures composed of various sulfides, supported Ni, Co, W, Cr and Mo phosphides, multi metal oxides and intermetallic compounds, and other compounds for water electrolysis. The evolution of the catalyst in the OER process has been revealed in multiple aspects such as morphology, composition, crystal structure, self-healing, and reconstruction, and has achieved certain results. This provides a theoretical reference for the subsequent development of efficient and stable new acidic medium OER catalysts. However, the large-scale application of such OER catalysts is limited due to issues with activity and stability, performance, and production costs, and further research is needed. The core of photocatalytic water splitting for hydrogen production is the catalyst, and precious metals are commonly used catalysts. Although they have high efficiency in photocatalytic water splitting for hydrogen production, they are expensive and have scarce resources, making them unsuitable for large-scale commercial applications. Therefore, people have developed $TiO_2$ single-phase and suitable photocatalysts, transition metal sulfides, and transition metal sulfide catalysts, which have significantly improved hydrogen evolution performance. However, there is still a certain distance from practical applications. In order to achieve industrialization and industrial application as soon as possible. In the future, further in-depth research should be conducted in the following areas:

(1) Identifying the catalytic activity and stability sources of catalysts is crucial for guiding the design and synthesis of efficient and stable catalysts. Therefore, further clarification is needed on the catalytic sites and dynamic catalytic mechanisms of the catalyst. During the reaction process, the catalyst may undergo morphological or compositional evolution on the surface and even undergo dynamic changes in composition of the electrolyte. Traditional reaction mechanisms are difficult to explain this complex process of change. With the development of high-end precision characterization technology, the dynamic changes in HER/OER and photo catalysis can be captured through advanced in-situ techniques. For example, in situ XAS characterization of O or H bond lengths at different potentials can help clarify the effect of lattice strain on the adsorption energy of HER/OER intermediates; The characterization of catalyst composition changes under different potentials using in-situ Raman spectroscopy and in-situ infrared spectroscopy helps to clarify the species that truly participate in catalytic reactions in multi-component catalytic systems and their changes during the catalytic process, and is also beneficial for identifying their dynamic surface reconstruction processes; The combination of isotope labeling experiments and in-situ spectroscopy techniques to characterize the pathways of intermediate changes in catalytic reaction processes helps to clarify and quantify the surface-active sites of catalysts; Combining the changes in the composition and morphology of actual catalytic sites revealed by in-situ TEM, in order to improve the catalytic process and reveal the dynamic catalytic mechanism. Therefore, the combination of in-situ characterization technology and dynamic characterization technology with theoretical computational modeling can effectively explore reaction mechanisms and guide experiments, while greatly reducing the trial-and-error costs of a large number of experiments, thus effectively solving the problems and challenges faced by catalysts.

(2) Multi-element catalysts, especially composite catalysts, have complex components and diverse structures, and the catalytic reaction mechanism that occurs on their surfaces is not yet clear. Exploring the electro-catalytic reaction mechanism and pathway can provide guidance for material design and preparation. Therefore, in-situ characterization technology and dynamic characterization technology combined with theoretical calculation modeling can effectively explore the reaction mechanism and guide experiments, while greatly reducing the trial-and-error costs of a large number of experiments, thus effectively solving the problems and challenges faced by catalysts. Theoretical calculation simulation research is helpful for exploring reaction mechanisms and pathways, especially for the mechanism analysis of transition metals and their modified catalytic materials with complex structures. Theoretical calculations can clarify the synergistic mechanism between transition metals, heteroatoms, and composite hybrid materials, and gain a more comprehensive and in-depth understanding of the influencing factors and reaction mechanisms of transition metal modified or hybrid catalyst performance, in order to guide experiments to design efficient catalysts more reasonably. Although theoretical calculation methods have been widely used in the exploration and catalytic performance analysis of catalysts, most of the models currently used are highly idealized and do not consider the reaction conditions that change over time in the real reaction process, such as lattice strain, solvation effect, electrolyte pH value, and oxygen concentration. Therefore, in the process of theoretical research, Researchers should strive to make the theoretical calculation simulation process as close as possible to real reaction conditions. At the same time, by combining machine learning methods such as deep

learning and neural networks, high-performance catalysts can be more accurately predicted and quickly screened to guide experimental synthesis.

(3) Carry out high-throughput calculations based on material databases to assist in the screening of HER/OER and photocatalytic catalysts with stable media. A large amount of reliable data obtained from existing experiments and simulations will drive the screening, prediction, and evaluation of catalytic materials. Based on materials genetic engineering such as the Inorganic Crystal Structure Database (ICSD), Cambridge Structural Database, and Open Quantum Materials Database, combined with traditional computational materials science (such as DFT) and artificial intelligence (such as machine learning ML) methods, theoretical analysis methods for screening and optimizing HER/OER electro-catalysts and photocatalysts are developed, laying the foundation for the widespread application of electrolysis and photocatalytic water technology.

(4) Industrial-grade water electrolysis or photocatalytic hydrogen production requires ensuring the quality of large and different batches of catalysts to meet commercial performance. Therefore, researchers should optimize catalyst performance while simplifying the preparation process as much as possible to improve the dispersion and intrinsic activity of active sites in the material, while achieving precise control.

**Reference**

[1] S. Jiao, X. Fu, S. Wang, et al. Perfecting electro-catalysts via imperfections: Towards the large-scale deployment of water electrolysis technology. Energy & Environmental Science, 14 (2021) 1722-1770. https://doi.org/10.1039/D0EE03635H

[2] W.Y. Jiang, C.Y. Han, Y. Liu, et al. Research progress of hydrogen purification technology under the background of carbon neutrality. Chemical Engineering of Oil & Gas, 52 (2023) 38-49.

[3] M.A. Ashraf, O. Sanz, C. Italiano, et al. Analysis of Ru/La-Al$_2$O$_3$ catalyst loading on alumina monoliths and controlling regimes in methane steam reforming. Chemical Engineering Journal, 334 (2018) 1792-1807. https://doi.org/10.1016/j.cej.2017.11.154

[4] M. Gur, E.D. Canbaz. Analysis of syngas production and reaction zones in hydrogen oriented underground coal gasification. Fuel, 269 (2020) 11733. https://doi.org/10.1016/j.fuel.2020.117331

[5] X. Huang, B.Y. Zhao, B.G. Lougou, et al. Research progress on methane steam reforming for hydrogen production. Chemical Engineering of Oil & Gas, 51 (2022) 53-61.

[6] X.X. Zou, Y. Zhang. Noble metal-free hydrogen evolution catalysts for water splitting. Chemical Society Reviews 44 (2015) 5148-5180 https://doi.org/10.1039/C4CS00448E

[7] V.R. Stamenkovic, D. Strmcnik, P.P. Lopes, et al. Energy and fuels from electrochemical interfaces. Nature Materials 16 (2017) 57-69 https://doi.org/10.1038/nmat4738

[8] J. Turner, G. Sverdrup, M.K. Mann, et al. Renewable hydrogen production. International Journal of Energy Research 32 (2008) 379-407 https://doi.org/10.1002/er.1372

[9] H.Y. Qu, X.W. He, Y.B. Wang, et al. Electrocatalysis for the oxygen evolution reaction in acidic media: progress and challenges. Applied Sciences 11 (2021) 4320 https://doi.org/10.3390/app11104320

[10] G.Q. Zhao, K. Rui, S.X. Dou, et al. Heterostructures for electrochemical hydrogen evolution reaction: a review. Advanced Functional Materials 28 (2018) 1803291 https://doi.org/10.1002/adfm.201803291

[11] S.C. Shang, X.Y. Zhang, X.L. Liu, et al. Design and construction of cocatalysts for photocatalytic overall water splitting. Acta Physico-Chimica Sinica 36 (2020) 15-25

[12] X.H. Lu, L. Tian. Research progress on key materials for anode of hydrogen production by electrocatalytic water splitting. Anhui Chemical Industry (2022)

[13] X.L. An, B. Wang, Q.G. Cao, et al. Research progress of transition metal-based hydrogen evolution catalysts. Chemical Engineering (China)/Huaxue Gongcheng 51(7) (2023)

[14] J.H. Xu, Y.C. Wang, Y.T. Yin, et al. Research progress of industrial electrolytic seawater hydrogen production technology and electrode materials[J/OL]. Low-Carbon Chemistry and Chemical Industry 1-10 (2024) [2024-09-28]

[15] G.Z. Li, W. Fu, G.C. Wang, et al. Research progress of self-supporting transition metal-based electrode materials in the field of hydrogen production by electrolyzed water. Shandong Chemical Industry 53 (2024) 117-119

[16] Q. Wang, X. Xiang, C.M. Yang. Application progress of transition metal sulfur/oxides in alkaline electrolyzed water hydrogen production. Journal of Natural Science of Hunan Normal University 45 (2022) 124-135

[17] J.G. Ma, M.Y. Yang, G.L. Zhao, et al. Ni electrodes with 3D-ordered surface structures for boosting bubble releasing toward high current density alkaline water splitting. Ultrason Sonochem 96 (2023) 106398 https://doi.org/10.1016/j.ultsonch.2023.106398

[18] R.B. Levy, M. Boudart. Platinum-like behavior of tungsten carbide in surface catalysis. Science 181 (1973) 547-549 https://doi.org/10.1126/science.181.4099.547

[19] J.R. Kitchin, J.K. Nørskov, M.A. Barteau, et al. Trends in the chemical properties of early transition metal carbide surfaces: a density functional study. Catalysis Today 105 (2005) 66-73 https://doi.org/10.1016/j.cattod.2005.04.008

[20] H. Vrubel, X.L. Hu. Molybdenum boride and carbide catalyze hydrogen evolution in both acidic and basic solutions. Angewandte Chemie International Edition 51 (2012) 12703-12706 https://doi.org/10.1002/anie.201207111

[21] P. Xiao, Y. Yan, X.M. Ge, et al. Investigation of molybdenum carbide nano-rod as an efficient and durable electro-catalyst for hydrogen evolution in acidic and alkaline media. Applied Catalysis B: Environmental 154/155 (2014) 232-237 https://doi.org/10.1016/j.apcatb.2014.02.020

[22] C.F. Yang, R. Zhao, H. Xiang, et al. Ni-activated transition metal carbides for efficient hydrogen evolution in acidic and alkaline solutions. Advanced Energy Materials 10 (2020) 2002260 https://doi.org/10.1002/aenm.202002260

[23] P. Liu, J.A. Rodriguez. Catalysts for hydrogen evolution from the [NiFe] hydrogenase to the Ni2P(001) surface: the importance of ensemble effect. Journal of the American Chemical Society 127 (2005) 14871-14878 https://doi.org/10.1021/ja0540019

[24] E.J. Popczun, J.R. McKone, C.G. Read, et al. Nano-structured nickel phosphide as an electro-catalyst for the hydrogen evolution reaction. Journal of the American Chemical Society 135 (2013) 9267-9270 https://doi.org/10.1021/ja403440e

[25] R. Zhang, X.X. Wang, S.J. Yu, et al. Ternary NiCO2Px nanowires as pH-universal electro-catalysts for highly efficient hydrogen evolution reaction. Advanced Materials 29 (2017) 1605502 https://doi.org/10.1002/adma.201605502

[26] Y. Lin, K. Sun, S. Liu, et al. Construction of CoP/NiCoP nanotadpoles heterojunction interface for wide pH hydrogen evolution electrocatalysis and supercapacitor. Advanced Energy Materials 9 (2019) 1901213 https://doi.org/10.1002/aenm.201901213

[27] M.T. Chen, J.J. Duan, J.J. Feng, et al. Iron, rhodium-codoped Ni2P nano-sheets arrays supported on nickel foam as an efficient bifunctional electro-catalyst for overall water splitting. Journal of Colloid and Interface Science 605 (2022) 888-896 https://doi.org/10.1016/j.jcis.2021.07.101

[28] H. Zhang, W.Q. Li, X. Feng, et al. A chainmail effect of ultrathin N-doped carbon shell on Ni2P nano-rod arrays for efficient hydrogen evolution reaction catalysis. Journal of Colloid and Interface Science 607 (2022) 281-289 https://doi.org/10.1016/j.jcis.2021.08.169

[29] Z. Li, X. Dou, Y. Zhao, et al. Enhanced oxygen evolution reaction of metallic nickel phosphide nano-sheets by surface modification Inorganic Chemistry Frontiers 3 (2016) 1021-1027 https://doi.org/10.1039/C6QI00078A

[30] M. Liu, J. Li. Cobalt Phosphide Hollow Polyhedron as Efficient Bifunctional Electro-catalysts for the Evolution Reaction of Hydrogen and Oxygen. ACS Applied Materials & Interfaces 8 (2016) 2158-2165 https://doi.org/10.1021/acsami.5b10727

[31] Y. Pan, Y. Lin, Y. Chen, et al. Cobalt phosphide-based electro-catalysts: synthesis and phase catalytic activity comparison for hydrogen evolution. Journal of Materials Chemistry A 4 (2016) 4745-4754 https://doi.org/10.1039/C6TA00575F

[32] M.Y. Lu, R.Z. Yang, J.H. Tian. Research progress on preparation, characterization and application of transition metal phosphides in electrolyzed water. Chinese Battery Industry 24 (2020) 84-93

[33] Y. Zhang, Y.W. Liu, M. Ma, et al. A Mn-doped Ni2P nano-sheet array: An efficient and durable hydrogen evolution reaction electro-catalyst in alkaline media. Chemical Communications 53 (2017) 11048-11051 https://doi.org/10.1039/C7CC06278H

[34] L.L. Wen, J. Yu, C.C. Xing, et al. Flexible vanadium-doped Ni2P nano-sheet arrays grown on carbon cloth for an efficient hydrogen evolution reaction. Nanoscale 11 (2019) 4198-4203 https://doi.org/10.1039/C8NR10167A

[35] Y.Q. Sun, L.F. Hang, Q. Shen, et al. Mo doped Ni2P nanowire arrays: An efficient electro-catalyst for the hydrogen evolution reaction with enhanced activity at all pH values. Nanoscale 9 (2017) 16674-16679 https://doi.org/10.1039/C7NR03515B

[36] Y.Q. Sun, T. Zhang, X.Y. Li, et al. Bifunctional hybrid Ni/Ni2P nano-particles encapsulated by graphitic carbon supported with N,S modified 3D carbon framework for highly efficient overall water splitting. Advanced Materials Interfaces 5 (2018) 1800473 https://doi.org/10.1002/admi.201800473

[37] Y.Q. Sun, K. Xu, Z.H. Zhao, et al. Strongly coupled dual zerovalent nonmetal doped nickel phosphide nano-particles/N,B-graphene hybrid for pH-universal hydrogen evolution catalysis. Applied Catalysis B: Environmental 278 (2020) 119284 https://doi.org/10.1016/j.apcatb.2020.119284

[38] Z.J. Yang, Y.X. Tuo, Q. Lu, et al. Hierarchical Cu3P-based nanoarrays on nickel foam as efficient electro-catalysts for overall water splitting. Green Energy & Environment 7 (2020) 236-245 https://doi.org/10.1016/j.gee.2020.09.002

[39] N. Wang, X. He, Y.J. Chen, et al. S and Co co-doped Cu3P nanowires self-supported on Cu foam as an efficient hydrogen evolution electro-catalyst in artificial seawater. Journal of Porous Materials 28 (2021) 763-771 https://doi.org/10.1007/s10934-021-01032-0

[40] Y. Shi, B. Zhang. Recent advances in transition metal phosphide nanomaterials: synthesis and applications in hydrogen evolution reaction. Chemical Society Reviews 45 (2016) 1529-1541 https://doi.org/10.1039/C5CS00434A

[41] T. Liu, D. Liu, F. Qu, et al. Enhanced Electrocatalysis for Energy-Efficient Hydrogen Production over CoP Catalyst with Nonelectroactive Zn as a Promoter. Advanced Energy Materials 7 (2017) 1700020 https://doi.org/10.1002/aenm.201700020

[42] Y.C. Xu, S.T. Wei, L.F. Gan, et al. Amorphous carbon interconnected ultrafine CoMnP with enhanced Co electron delocalization yields Pt-like activity for alkaline water electrolysis. Advanced Functional Materials 32 (2022) 2112623 https://doi.org/10.1002/adfm.202112623

[43] X.F. Liu, H. Qi, B. Zhu, et al. Boosting electrochemical hydrogen evolution of porous metal phosphides nano-sheets by coating defective TiO2 overlayers. Small 14 (2018) 1802755 https://doi.org/10.1002/smll.201802755

[44] X.D. Lyu, X.T. Li, C. Yang, et al. Large-Size, porous, ultrathin NiCoP nano-sheets for efficient electro/photocatalytic water splitting. Advanced Functional Materials 30 (2020) 1910830 https://doi.org/10.1002/adfm.201910830

[45] S.Y. Song, M.J. Guo, S.S. Zhang, et al. Plasma-assisted synthesis of hierarchical NiCoxPy nano-sheets as robust and stable electro-catalyst for hydrogen evolution reaction in both acidic and alkaline media Electrochimica Acta, 331 (2020) 135431. https://doi.org/10.1016/j.electacta.2019.135431

[46] H. Li, X.L. Zhao, H.L. Liu, et al. Sub-1.5 nm ultrathin CoP nano-sheet aerogel: Efficient electro-catalyst for hydrogen evolution reaction at all pH values Small, 14 (2018) e1802824. https://doi.org/10.1002/smll.201802824

[47] L. Sun, Y. Dang, A.P. Wu, et al. Synchronous regulation of morphology and electronic structure of FeNi-P nano-sheet arrays by Zn implantation for robust overall water splitting Nano Research, 16 (2023) 5733-5742. https://doi.org/10.1007/s12274-022-5245-y

[48] Z.T. Li, P. Zhou, Y.X. Zhou, et al. Ultrathin and porous CoP nano-sheets as an efficient electro-catalyst for boosting hydrogen evolution behavior at a broad range of pH International Journal of Hydrogen Energy, 51 (2024) 1279-1286. https://doi.org/10.1016/j.ijhydene.2023.09.181

[49] X.Y. Wu, X.P. Han, X.Y. Ma, et al. Morphology-controllable synthesis of Zn-Co-mixed sulfide nano-structures on carbon fiber paper toward efficient rechargeable zinc-air batteries

Materials Research Forum LLC
https://doi.org/21741/9781644903711-10

and water electrolysis ACS Appl Mater Interfaces, 9 (2017) 12574-12583. https://doi.org/10.1021/acsami.6b16602

[50] Hinnemann, B., Moses, P.G., Bonde, J., et al. Biomimetic hydrogen evolution: $MoS_2$ nanoparticles as catalyst for hydrogen evolution Journal of the American Chemical Society, 127 (2005) 5308-5309. https://doi.org/10.1021/ja0504690

[51] T.F. Jaramillo, K.P. Jørgensen, J. Bonde, et al. Identification of active edge sites for electrochemical H2 evolution from $MoS_2$ nanocatalysts Science, 317 (2007) 100-102. https://doi.org/10.1126/science.1141483

[52] J.F. Xie, H. Zhang, S. Li, et al. Defect-rich $MoS_2$ ultrathin nano-sheets with additional active edge sites for enhanced electro-catalytic hydrogen evolution Advanced Materials, 25 (2013) 5807-5813. https://doi.org/10.1002/adma.201302685

[53] Z.Z. Wu, B. Fang, A. Bonakdarpour, et al. $WS_2$ nano-sheets as a highly efficient electro-catalyst for hydrogen evolution reaction Appl Catal B Environ, 125 (2012) 59-66. https://doi.org/10.1016/j.apcatb.2012.05.013

[54] W.Q. Han, Z.H. Liu, Y.B. Pan, et al. Designing champion nano-structures of tungsten dichalcogenides for electro-catalytic hydrogen evolution Adv Mater, 32 (2020) e2002584. https://doi.org/10.1002/adma.202002584

[55] Y.P. Ye, J.S. Rong, J. Cao, et al. Restructuring electronic structure via W doped 1T $MoS_2$ for enhancing hydrogen evolution reaction Applied Surface Science, 579 (2021) 152216. https://doi.org/10.1016/j.apsusc.2021.152216

[56] G.C. Qi, Y. Zhang, K. Pan. Controllable preparation of $FeS_2$/rGO and its electrocatalytic hydrogen evolution characteristics Journal of Engineering of Heilongjiang University, 10(1) (2019) 40-44, 52.

[57] D. Jasion, J.M. Barforoosh, Q. Qiao, et al. Low-dimensional hyperthin $FeS_2$ nano-structures for efficient and stable hydrogen evolution electrocatalysis ACS Catal, 5 (2015) 6653-6657. https://doi.org/10.1021/acscatal.5b01637

[58] R. Miao, B. Dutta, S. Sahoo, et al. Mesoporous iron sulfide for highly efficient electro-catalytic hydrogen evolution J Am Chem Soc, 139 (2017) 13604-13607. https://doi.org/10.1021/jacs.7b07044

[59] Q. Wang, X. Xiang, C.M. Yang. Application progress of transition metal sulfur/oxides in alkaline electrolyzed water hydrogen production Journal of Natural Science of Hunan Normal University, 45 (2022) 124-135.

[60] Y. Xu, T.T. Feng, Z.J. Cui, et al. Fe7S8/$FeS_2$/C as an efficient catalyst for electro-catalytic water splitting International Journal of Hydrogen Energy, 46 (2021) 39216-39225. https://doi.org/10.1016/j.ijhydene.2021.09.159

[61] M.S. Faber, R. Dziiedzic, M.A. Lukowski, et al. High-performance electro-catalysis using metallic cobalt pyrite ($CoS_2$) micro- and nano-structures Journal of the American Chemical Society, 136 (2014) 10053-10061. https://doi.org/10.1021/ja504099w

[62] C. Wang, T.Y. Wang, J.J. Liu, et al. Facile synthesis of silk-cocoon S-rich cobalt polysulfide as an efficient catalyst for the hydrogen evolution reaction. Energy Environ Sci, 11 (2018) 2467-2475. https://doi.org/10.1039/C8EE00948A

[63] L.X. Meng, H.C. Xuan, G.H. Zhang, et al. Rational construction of uniform CoS/NiFe$_2$O$_4$ heterostructure as efficient bifunctional electro-catalysts for hydrogen evolution and oxygen evolution reactions Electrochimica Acta, 404 (2022) 139596. https://doi.org/10.1016/j.electacta.2021.139596

[64] M. Zhao, M. Yang, W. Huang, et al. Synergism on electronic structures and active edges of metallic vanadium disulfide nano-sheets via Co doping for efficient hydrogen evolution reaction in seawater ChemCatChem, 13 (2021) 2138-2144. https://doi.org/10.1002/cctc.202100007

[65] C. Song. Construction of active sites of nickel-based transition metal electrocatalysts and research on hydrogen evolution performance under alkaline conditions Changchun: Jilin University, 2021.

[66] N. Jiang, Q. Tang, M.L. Sheng, et al. Nickel sulfides for electro-catalytic hydrogen evolution under alkaline conditions: a case study of crystalline NiS, NiS$_2$, and Ni$_3$S$_2$ nano-particles Catal Sci Technol, 6 (2016) 1077-1084. https://doi.org/10.1039/C5CY01111F

[67] M.M. Tong, L. Wang, P. Yu, et al. Ni3S2 nano-sheets in situ epitaxially grown on nano-rods as high active and stable homojunction electro-catalyst for hydrogen evolution reaction ACS Sustain Chem Eng, 6 (2018) 2474-2481. https://doi.org/10.1021/acssuschemeng.7b03915

[68] W.J. He, L.L. Han, Q.Y. Hao, et al. Fluorine-anion-modulated electron structure of nickel sulfide nano-sheet arrays for alkaline hydrogen evolution. ACS Energy Lett, 4 (2019) 2905-2912. https://doi.org/10.1021/acsenergylett.9b02316

[69] A. Mi, A. Dtt, A. Thn, et al. Efficient synergism of NiO-NiSe2 nano-sheet-based heterostructures shelled titanium nitride array for robust overall water splitting. Journal of Colloid and Interface Science, 612 (2021) 121-131. https://doi.org/10.1016/j.jcis.2021.12.137

[70] K. Wang, C.J. Zhou, D. Xi, et al. Component-control-lable synthesis of Co(SxSe1-x)2 nanowires supported by carbon fiber paper as high-performance electrode for hydrogen evolution reaction. Nano Energy, 18 (2015) 1-11. https://doi.org/10.1016/j.nanoen.2015.10.001

[71] C.Y. Zhang, X.Q. Du, Y.H. Wang, et al. NiSe$_2$@Ni$_x$S$_y$ nano-rod on nickel foam as efficient bifunctional electrocat-alyst for overall water splitting. International Journal of Hydrogen Energy, 46 (2021) 34713-34726. https://doi.org/10.1016/j.ijhydene.2021.08.046

[72] D. Kong, J.J. Cha, H.T. Wang, et al. First-row transition metal dichalcogenide catalysts for hydrogen evolution reaction. Energy&Environmental Science, 6 (2013) 3553-3558. https://doi.org/10.1039/c3ee42413h

[73] J. Chang, G. Wang, Z. Yang, et al. Dual-doping and synergism toward high-performance seawater electrolysis. Advanced Materials, 33 (2021) 2101425. https://doi.org/10.1002/adma.202101425

[74] Z. Liu, D. Liu, L. Zhao, et al. Efficient overall water splitting catalyzed by robust FeNi$_3$N nano-particles with hollow interiors. Journal of Materials Chemistry A, 9 (2021) 7750-7758. https://doi.org/10.1039/D1TA01014J

[75] H. Wang, J. Li, K. Li, et al. Transition metal nitrides for electrochemical energy applications. Chemical Society Reviews, 50 (2021) 1354-1390. https://doi.org/10.1039/D0CS00415D

[76] J. Miao, Z. Lang, X. Zhang, et al. Polyoxometalate derived hexagonal molybdenum nitrides(MXenes) supported by boron, nitrogen Co doped carbon nanotubes for efficient electrochemical hydrogen evolution from seawater. Advanced Functional Materials, 29 (2019) 1805893. https://doi.org/10.1002/adfm.201805893

[77] L.B. Wu, F.H. Zhang, S.W. Song, et al. Efficient alkaline water/seawater hydrogen evolution by a nano-rod-nano-particle-structured Ni-MoN catalyst with fast water-dissociation kinetics. Adv Mater, 34 (2022) e2201774. https://doi.org/10.1002/adma.202201774

[78] Z. Dou. In-situ growth of carbon dots and cobalt tetroxide on carbon cloth for electrochemical hydrogen evolution and oxygen evolution research. Shenyang: Liaoning University, 2020.

[79] C. Shu, S. Kang, Y. Jin, et al. Bifunctional porous nonprecious metal WO$_2$ hexahedral networks as an electro-catalyst for full water splitting. Journal of Materials Chemistry A, 5(2017) 9655-9660. https://doi.org/10.1039/C7TA01527E

[80] H.L. Yue. Research on hydrogen evolution reaction activity of NiO-based thin film materials under alkaline conditions. Xiamen: Xiamen University, 2019.

[81] H.L. Guo, J. Zhou, Q.Q. Li, et al. Emerging dual-channel transition-metal-oxide quasiaerogels by self-embedded templating. Adv Funct Mater, 30 (2020) 2000024. https://doi.org/10.1002/adfm.202000024

[82] L.W. Lin, S.Q. Piao, Y. Choi, et al. Nano-structured transition metal nitrides as emerging electro-catalysts for water electrolysis: Status and challenges. EnergyChem, 4 (2022) 100072. https://doi.org/10.1016/j.enchem.2022.100072

[83] M.K. Bates, Q. Jia, N. Ramaswamy, et al. Composite Ni/NiO-Cr$_2$O$_3$ catalyst for alkaline hydrogen evolution reaction. The Journal of Physical Chemistry C, 119 (2015) 5467-5477. https://doi.org/10.1021/jp512311c

[84] J. Jin, J. Yin, H.B. Liu, et al. Atomic sulfur filling oxygen vacancies optimizes H absorption and boosts the hydrogen evolution reaction in alkaline media. Angew Chem Int Ed Engl, 60 (2021) 14117-14123. https://doi.org/10.1002/anie.202104055

[85] Y.P. Zhu, T.Y. Ma, M. Jaroniec, et al. Self-templating synthesis of hollow Co$_3$O$_4$ microtube arrays for highly efficient water electrolysis. Angew Chem Int Ed Engl, 56 (2017) 1324-1328. https://doi.org/10.1002/anie.201610413

[86] S. Peng, F. Gong, L. Li, et al. Necklace-like multishelled hollow spinel oxides with oxygen vacancies for efficient water Electrolysis. J Am Chem Soc, 140 (2018) 13644-13653. https://doi.org/10.1021/jacs.8b05134

[87] L.L. Huang, D.W. Chen, G. Luo, et al. Zirconium-regulation-induced bifunctionality in 3D cobalt-iron oxide nano-sheets for overall water splitting. Adv Mater, 31 (2019) e1901439. https://doi.org/10.1002/adma.201901439

[88] S.K. Diao, X. Zhao, Z.T. Yu, et al. Research progress of key materials for alkaline water electrolyzers. Journal of the Chinese Ceramic Society, 52 (2024) 1841-1860.

[89] Z.J. Chen, X.G. Duan, W. Wei, et al. Recent advances in transition metal-based electro-catalysts for alkaline hydrogen evolution. J Mater Chem A, 7 (2019) 14971-15005. https://doi.org/10.1039/C9TA03220G

[90] J. Zhang, R.J. Cui, C.C. Gao, et al. Cation-modulated HER and OER activities of hierarchical VOOH hollow architectures for high-efficiency and stable overall water splitting. Small, (2019) 15(47) e1904688. https://doi.org/10.1002/smll.201904688

[91] H.L. Li, W. Yuan, Q. Wang, et al. Two-dimensional cobalt oxy-hydrate sulfide nano-sheets with modified t2g orbital state of CoO6-x octahedron for efficient overall water splitting. ACS Sustain Chem Eng, 7 (2019) 17325-17334. https://doi.org/10.1021/acssuschemeng.9b04256

[92] P. Yu, F.M. Wang, T.A. Shifa, et al. Earth abundant materials beyond transition metal dichalcogenides: a focus on electrocatalyzing hydrogen evolution reaction. Nano Energy, 58 (2019) 244-276. https://doi.org/10.1016/j.nanoen.2019.01.017

[93] T. Ling, T. Zhang, B.H. Ge, et al. Well-dispersed nickel- and zinc-tailored electronic structure of a transition metal oxide for highly active alkaline hydrogen evolution reaction. Advanced Materials, 31 (2019) e1807771. https://doi.org/10.1002/adma.201807771

[94] C. Panda, P.W. Menezes, S.L. Yao, et al. Boosting electro-catalytic hydrogen evolution activity with a NiPt3@NiS heteronano-structure evolved from a molecular nickel-platinum precursor. Journal of the American Chemical Society, 141 (2019) 13306-13310. https://doi.org/10.1021/jacs.9b06530

[95] G.J. Liu, H.P. Bai, Y.J. Ji, et al. A highly efficient alkaline HER Co-Mo bimetallic carbide catalyst with an optimized Mo d-orbital electronic state. Journal of Materials Chemistry A, 7 (2019)12434-12439. https://doi.org/10.1039/C9TA02886B

[96] Y.Y. Wu, Z.J. Xie, Y. Li, et al. In-situ self-reconstruction of Ni-Fe-Al hybrid phosphides nano-sheet arrays enables efficient oxygen evolution in alkaline. International Journal of Hydrogen Energy, 46 (2021) 25070-25080. https://doi.org/10.1016/j.ijhydene.2021.05.060

[97] Y.Y. Yang, J.Y. Yang, C. Kong, et al. Heterogeneous cobalt-iron phosphide nano-sheets formed by in situ phosphating of hydroxide for efficient overall water splitting. Journal of Alloys and Compounds, 926 (2022)166930. https://doi.org/10.1016/j.jallcom.2022.166930

[98] Y.H. Liu, N. Ran, R.Y. Ge, et al. Porous Mn-doped cobalt phosphide nano-sheets as highly active electro-catalysts for oxygen evolution reaction. Chemical Engineering Journal, 425 (2021) 131642. https://doi.org/10.1016/j.cej.2021.131642

[99] L. Du, X.Y. Gao, G.Y. Wang, et al. $CeO_2$ nano-particles decorated on porous Ni-Fe bimetallic phosphide nano-sheets for high-efficient overall water splitting. Materials Research Letters, 11 (2023) 159-167. https://doi.org/10.1080/21663831.2022.2131372

[100] X.M. Hu, S.L. Zhang, J.W. Sun, et al. 2D Fe-containing cobalt phosphide/cobalt oxide lateral heterostructure with enhanced activity for oxygen evolution reaction. Nano Energy, 56 (2019) 109-117. https://doi.org/10.1016/j.nanoen.2018.11.047

[101] H. Lü, Y.K. Song, C.P. Hao. Research on optimization of membrane electrode of solid polymer electrolytic cell. Journal of Central South University (Science and Technology), 46 (2015) 4671-4678.

[102] L. Xu, Q.Q. Jiang, Z.H. Xiao, et al. Plasma-engraved $Co_3O_4$ nano-sheets with oxygen vacancies and high surface area for the oxygen evolution reaction. Angewandte Chemie International Edition, 55 (2016) 5277-5281. https://doi.org/10.1002/anie.201600687

[103] K. Fujimura, T. Matsui, H. Habazaki, et al. The durability of manganese-molybdenum oxide anodes for oxygen evolution in seawater electrolysis. Electrochimica Acta, 45(14):2297-2303 (2000). https://doi.org/10.1016/S0013-4686(00)00316-9

[104] B. Zhang, X.L. Zheng, O. Voznnyy, et al. Homogeneously dispersed multimetal oxygen-evolving catalysts. Science, 352 (2016) 333-337. https://doi.org/10.1126/science.aaf1525

[105] M.P. Chen, D. Liu, J.X. Feng, et al. In-situ generation of Ni-CoOOH through deep reconstruction for durable alkaline water electrolysis. Chem Eng J, 443 (2022) 136432. https://doi.org/10.1016/j.cej.2022.136432

[106] Q. Zhou, Y.P. Chen, G.Q. Zhao, et al. Active-site-enriched iron-doped nickel/cobalt hydroxide nano-sheets for enhanced oxygen evolution reaction. ACS Catalysis, 8 (2018) 5382-5390. https://doi.org/10.1021/acscatal.8b01332

[107] B.R. Guo, Y.N. Ding, H.H. Huo, et al. Recent advances of transition metal basic salts for electro-catalytic oxygen evolution reaction and overall water electrolysis. Nanomicro Lett, 15 (2023) 57. https://doi.org/10.1007/s40820-023-01038-0

[108] A. Karmakar, S.K. Srivastava. Hierarchically hollow interconnected rings of nickel substituted cobalt carbonate hydroxide hydrate as promising oxygen evolution electro-catalyst. Int J Hydrog Energy, 47 (2022) 22430-22441. https://doi.org/10.1016/j.ijhydene.2022.05.062

[109] X.R. Kou, C.Y. Li, Y.Y. Qi, et al. Research progress of transition metal hybrid materials in electrolyzed water. Chemical Research, 33 (2022) 290-298. DOI:10.14002/j.hxya.2022.04.002.

[111] Y.T. Luo, H.D. Yang, P. Ma, et al. $Fe_3O_4$/CoO interfacial nano-structure supported on carbon nanotubes as a highly efficient electro-catalyst for oxygen evolution reaction . ACS Sustainable Chemistry & Engineering, 8 (2020) 3336-3346. https://doi.org/10.1021/acssuschemeng.9b07292

[112] J. Zhang, W.J. Jiang, S. Niu, et al. Organic small molecule activates transition metal foam for efficient oxygen evolution reaction . Advanced Materials, 32 (2020) e1906015. https://doi.org/10.1002/adma.201906015

[113] H.W. Huang, C. Yu, H.L. Huang, et al. Activation of transition metal oxides by in-situ electro-regulated structure-reconstruction for ultra-efficient oxygen evolution . Nano Energy, 58 (2019) 778-785. https://doi.org/10.1016/j.nanoen.2019.01.094

[114] Y. Pei, Y. Cheng, J.Y. Chen, et al. Recent developments of transition metal phosphides as catalysts in the energy conversion field. Journal of Materials Chemistry A, 6 (2018) 23220-23243. https://doi.org/10.1039/C8TA09454C

[115] W.C. Ke, Y. Zhang, A.L. Imbault, et al. Metal-organic framework derived iron-nickel sulfide nano-rods for oxygen evolution reaction. International Journal of Hydrogen Energy, 46 (2021) 20941-20949. https://doi.org/10.1016/j.ijhydene.2021.03.207

[116] T.N. Zhou, J. Bai, Y.H. Gao, et al. Selenide-based 3D folded polymetallic nano-sheets for a highly efficient oxygen evolution reaction. Journal of Colloid and Interface Science, 615 (2022) 256-264. https://doi.org/10.1016/j.jcis.2022.01.139

[117] X. Xu, F. Song, X.L. Hu. A nickel iron diselenide-derived efficient oxygen-evolution catalyst. Nature Communications, 7 (2016) 12324. https://doi.org/10.1038/ncomms12324

[118] L. Yu, Q. Zhu, S. Song, et al. Non-noble metal-nitride based electro-catalysts for high-performance alkaline seawater electrolysis. Nature Communications, 10 (2019) 5106. https://doi.org/10.1038/s41467-019-13092-7

[119] R.Z. He, X.Y. Huang, L.G. Feng. Recent progress in transition-metal sulfide catalyst regulation for improved oxygen evolution reaction. Energy Fuels, 36 (2022):6675-6694. https://doi.org/10.1021/acs.energyfuels.2c01429

[120] J. Sun, H. Xue, N.K. Guo, et al. Synergetic metal defect and surface chemical reconstruction into NiCo2S4/ZnS heterojunction to achieve outstanding oxygen evolution performance. Angew Chem Int Ed Engl, 60 (2021) 19435-19441. https://doi.org/10.1002/anie.202107731

[121] J. Jiang, F.F. Sun, S. Zhou, et al. Atomic-level insight into super-efficient electro-catalytic oxygen evolution on iron and vanadium Co-doped nickel(oxy)hydroxide. Nat Commun, 9 (2018) 2885. https://doi.org/10.1038/s41467-018-05341-y

[122] W.J. Shao, M.J. Xiao, C.D. Yang, et al. Assembling and regulating of transition metal-based heterophase vanadates as efficient oxygen evolution catalysts. Small, 18 (2022) e2105763. https://doi.org/10.1002/smll.202105763

[123] T.X. Nguyen, Y.C. Liao, C.C. Lin, et al. Advanced high entropy perovskite oxide electro-catalyst for oxygen evolution reaction. Adv Funct Materials, 31 (2021) 2101632. https://doi.org/10.1002/adfm.202101632

[124] Q.Q. Wang, Z. Jia, J.Q. Li, et al. Attractive electron delocalization behavior of FeCoMoPB amorphous nanoplates for highly efficient alkaline water oxidation. Small, 18 (2022) e2204135. https://doi.org/10.1002/smll.202204135

[125] Q. Runqing, S. Danielkobina, W. Wenbo, G. Shanhe, L. Jun, D. Arulappan, L. Mengxian, L. Xiaomeng. Boron, nitrogen co-doped biomass-derived carbon aerogel embedded nickel-cobalt-iron nano-particles as a promising electro-catalyst for oxygen evolution reaction. Journal of Colloid and Interface Science, 613 (2022) 126-135. https://doi.org/10.1016/j.jcis.2022.01.029

[126] H. Qi, M. Guomin, H. Zhen, W. Ziyu, H. Xiaowan, C. Xiaoyan, Z. Qianling, L. Jianhong, H. Chuanxin. General Synthesis of Ultrathin Metal Borate Nanomeshes Enabled by 3D Bark-

Like N-Doped Carbon for Electrocatalysis. Adv. Energy Mater, 9 (2019) 1901130. https://doi.org/10.1002/aenm.201970109

[127] S. Liu. Preparation of non-metal doped iron coordination catalyst and its application in zinc-carbon dioxide battery. Tianjin: Tianjin University of Technology, 2022.

[128] N. Nivedita, S. Philipp, M. Danea, D. Stefan, Q. Thomas, B. Anncathrin, C. Steffen, M. Martin, M. Justus, S. Wolfgang. Trace Metal Loading of B-N-Co-doped Graphitic Carbon for Active and Stable Bifunctional Oxygen Reduction and Oxygen Evolution Electro-catalysts. ChemElectroChem 8 (2021) 1685-1693. https://doi.org/10.1002/celc.202100374

[129] Y.Y. Yingying, Y. Pengfei, Z. Jianan, H. Yongfeng, A. Ibrahimsana, W. Xin, Z. Jigang, X. Huicong, S. Zhibo, X. Qun, M. Shichun. Carbon nano-sheets containing discrete Co-$N_x$-$B_y$-C active sites for efficient oxygen electrocatalysis and rechargeable Zn-Air batteries. ACS Nano, 12 (2018) 1894-1901. https://doi.org/10.1021/acsnano.7b08721

[130] Y. Zhou, J.J. Jingjing, N. Hongwei, H. Jia, C. Hanbing, X. Meng. Cobalt on boron-doped carbon nitride for a novel bifunctional electro-catalyst in zinc-air batteries. Energy Storage, 4 (2022) 313. https://doi.org/10.1002/est2.313

[131] D.H. He. Preparation and electrochemical performance research of bifunctional electrocatalyst based on ZIF-67. Hangzhou: Zhejiang University, 2018.

[132] Y.L. Chunlin, H. Shaoyun, L. Lecheng, Z. Xingwang. Synthesis of NiCo Alloy Nano-particle-Decorated B,N-Doped Carbon Nano-sheet Networks via a Self-Template Strategy for Bifunctional Oxygen-Involving Reactions. ACS Sustainable Chem. Eng, 7 (2019) 14394-14399. https://doi.org/10.1021/acssuschemeng.9b04164

[133] G. Fei, L. Zhuo, Z. Yiyong, X. Jie, Z. Xiaoyuan, Z. Chengxu, D. Peng, L. Tingting, Z. Yingjie, L. Mian. Tiny Ni Nano-particles Embedded in Boron- and Nitrogen-Codoped Porous Carbon Nanowires for High-Efficiency Water Splitting. ACS Appl. Mater. Interfaces, 14 (2022) 24447-24461. https://doi.org/10.1021/acsami.2c04956

[134] D.X. Liu. Preparation and electrocatalytic performance research of transition metal/non-metal heteroatom co-doped carbon nanomaterials. Qinhuangdao: Yanshan University, 2019.

[135] L. Zhuo, G. Fei, C. Lei, B. Xiangjie, L. Tingting, L. Mian. Fabrication of manganese borate/iron carbide encapsulated in nitrogen and boron co-doped carbon nanowires as the accelerated alkaline full water splitting bi-functional electro-catalysts. Journal of Colloid and Interface Science, 629 (2023) 179-192. https://doi.org/10.1016/j.jcis.2022.09.068

[136] P. Xianyun, H. Junrong, M. Yuying, S. Jiaqiang, Q. Gaocan, Q. Yongji, Z. Shusheng, Q. Yuan, L. Jun, L. Xijun. Bifunctional single-atomic Mn sites for energy-efficient hydrogen production. Nanoscale, 13 (2021) 4767. https://doi.org/10.1039/D0NR09104A

[137] Z.M. Gao, C.H. Wu. Preparation and electrochemical hydrogen evolution reaction performance research of nano-molybdenum carbide/boron and nitrogen co-doped two-dimensional carbon composite structure catalyst. Journal of Chinese Electron Microscopy Society, 37 (2018) 38-44.

[138] F. Guo, Y.B. Wang, L. Xiao, et al. Research progress of hydrogen production by electrolyzed water of transition metal functionalized boron and nitrogen doped carbon nanocomposites. Nonferrous Metals Equipment, 37 (2023) 26-30..

[139] J.Z. Zhang, Z.C. Zhang, H.B. Zhang, et al. Prussian-Blue-Analogue-Derived ultrathin $Co_2P$-$Fe_2P$ nano-sheets for universal-pH overall water splitting. Nano Letters, 23 (2023) 8331-8338. https://doi.org/10.1021/acs.nanolett.3c02706

[140] Y. Jeung, H. Roh, K.J. Yong. Co-anion exchange prepared 2D structure Ni(Co,Fe)PS for efficient overall water electrolysis. Applied Surface Science, 576 (2022) 151720. https://doi.org/10.1016/j.apsusc.2021.151720

[141] D.D. Wang, Y.W. Zhang, T. Fei, et al. NiCoP/NF 1D/2D biomimetic architecture for markedly enhanced overall water splitting. ChemElectroChem, 8 (2021) 3064-3072. https://doi.org/10.1002/celc.202100487

[142] Y.P. Hua, Q.C. Xu, Y.J. Hu, et al. Interface-strengthened CoP nano-sheet array with $Co_2P$ nano-particles as efficient electro-catalysts for overall water splitting. Journal of Energy Chemistry, 37 (2019) 1-6. https://doi.org/10.1016/j.jechem.2018.11.010

[143] Y.F. Zhao, J.Q. Zhang, Y.H. Xie, et al. Constructing atomic heterometallic sites in ultrathin nickel-incorporated cobalt phosphide nano-sheets via a boron-assisted strategy for highly efficient water splitting. Nano Letters, 21 (2021) 823-832. https://doi.org/10.1021/acs.nanolett.0c04569

[144] L. Zhao, M. Wen, Y.K. Tian, et al. A novel structure of quasi-monolayered NiCo-bimetal-phosphide for superior electrochemical performance. Journal of Energy Chemistry, 74 (2022) 203-211. https://doi.org/10.1016/j.jechem.2022.07.017

[145] M. Wang, L. Zhang, Y.J. He, et al. Recent advances in transition-metal-sulfide-based bifunctional electro-catalysts for overall water splitting. Journal of Materials Chemistry A, 9 (2021) 5320-5363. https://doi.org/10.1039/D0TA12152E

[150] L.Q. Pang, Q. Ma, C.D. Zhu. Multifunctional amorphous co phosphosulfide-coated Fe-co carbonate hydroxide for highly efficient overall water splitting. J Electron Mater, 52 (2023) 1808-1818. https://doi.org/10.1007/s11664-022-10194-9

[151] H.B. Liu, J.C. Li, Y.Q. Zhang, et al. Boosted water electrolysis capability of $Ni_xCo_yP$ via charge redistribution and surface activation. Chem Eng J, 473 (2023) 145397. https://doi.org/10.1016/j.cej.2023.145397

[152] Y.P. Zhu, H.C. Chen, C.S. Hsu, et al. Operando unraveling of the structural and chemical stability of P-substituted $CoSe_2$ electro-catalysts toward hydrogen and oxygen evolution reactions in alkaline electrolyte. ACS Energy Lett, 4 (2019) 987-994. https://doi.org/10.1021/acsenergylett.9b00382

[153] J.W. Li, Y.Z. Hu, X. Huang, et al. Bimetallic phosphide heterostructure coupled with ultrathin carbon layer boosting overall alkaline water and seawater splitting. Small, 19 (2023) e2206533. https://doi.org/10.1002/smll.202206533

[154] H.M. Zhang, L.H. Zuo, Y.H. Gao, et al. Amorphous high-entropy phosphoxides for efficient overall alkaline water/seawater splitting. J Mater Sci Technol, 173 (2024) 1-10. https://doi.org/10.1016/j.jmst.2023.08.003

[155] P.L. Zhai, Y.X. Zhang, Y.Z. Wu, et al. Engineering active sites on hierarchical transition bimetal oxides/sulfides heterostructure array enabling robust overall water splitting. Nat Commun, 11 (2020) 5462. https://doi.org/10.1038/s41467-020-19214-w

[156] C.L. Zhu, Z.X. Yin, W.H. Lai, et al. Fe-Ni-Mo nitride porous nanotubes for full water splitting and Zn-air batteries. Adv Energy Mater, 8 (2018) 1802327. https://doi.org/10.1002/aenm.201802327

[157] H.Y. Li, S.M. Chen, Y. Zhang, et al. Systematic design of superaerophobic nanotube-array electrode comprised of transition-metal sulfides for overall water splitting. Nat Commun, 9 (2018) 2452. https://doi.org/10.1038/s41467-018-04888-0

[158] Z.H. Zang, Q.J. Guo, X. Li, et al. Construction of a S and Fe co-regulated metal Ni electro-catalyst for efficient alkaline overall water splitting. J Mater Chem A, 11 (2023) 4661-4671. https://doi.org/10.1039/D2TA09802D

[159] F. Hu, D.S. Yu, M. Ye, et al. Lattice-matching formed mesoporous transition metal oxide heterostructures advance water splitting by active Fe-O-Cu bridges. Adv Energy Mater, 12 (2022) 2200067. https://doi.org/10.1002/aenm.202200067

[160] V.R. Jothi, R. Bose, H. Rajan, et al. Harvesting Electronic Waste for the Development of Highly Efficient EcoDesign Electrodes for Electro-catalytic Water Splitting. Advanced Energy Materials, 8(2018) 1-11. https://doi.org/10.1002/aenm.201802615

[161] J. Xu, J. Li, D. Xiong, et al. Trends in activity for the oxygen evolution reaction on transition metal(M=Fe,Co,Ni)phosphide pre-catalysts. Chemical Science, 9 (2018) 3470. https://doi.org/10.1039/C7SC05033J

[162] T. Takashima, K. Hashimoto, R. Nakamura. Development of a new catalytic system for the oxidation of alcohols. Journal of the American Chemical Society, 134 (2012) 1519. https://doi.org/10.1021/ja206511w

[163] M. Huynh, D.K. Bediako, D.G. Nocera. Electrocatalytic water oxidation on nickel-based electrodes. Journal of the American Chemical Society, 136 (2014) 6002. https://doi.org/10.1021/ja413147e

[164] A. Li, H. Ooka, N. Bonnet, et al. Sustainable hydrogen production via photoelectrochemical water splitting. Angewandte Chemie, International Edition, 58 (2019) 5054. https://doi.org/10.1002/anie.201813361

[165] H.Y. Su, Y. Gorlin, I.C. Man, et al. Investigation of catalysts for $CO_2$ reduction. Physical Chemistry Chemical Physics, 14 (2012) 14010. https://doi.org/10.1039/c2cp40841d

[166] R. Frydendal, E.A. Paoli, I. Chorkendorff, et al. High-performance catalysts for energy conversion. Advanced Energy Materials, 5 (2015) 1500991. https://doi.org/10.1002/aenm.201500991

[167] L. Zhou, A. Shinde, J.H. Montoya, et al. Catalytic activity of transition metal oxides for oxygen evolution. ACS Catalysis, 8 (2018) 10938.

[168] T. Hayashi, N. Bonnet-Mercier, A. Yamaguchi, et al. Exploring new pathways for energy storage. Royal Society Open Science, 6 (2019) 190122. https://doi.org/10.1098/rsos.190122

[169] S.J. Pan, H. Li, D. Liu, et al. Mechanisms of electrocatalytic $CO_2$ reduction. Nature Communications, 13 (2022) 2294.

[170] Y. Yang, X. Su, L. Zhang, et al. Innovative catalysts for fuel cell applications. ChemCatChem, 11 (2019) 1689. https://doi.org/10.1002/cctc.201802019

Materials Research Forum LLC
https://doi.org/21741/9781644903711-10

[171] M. Huynh, C. Shi, S.J. Billinge, et al. Structure-activity relationship in electrocatalysts. Journal of the American Chemical Society, 137 (2015) 14887. https://doi.org/10.1021/jacs.5b06382

[172] P.P. Patel, M.K. Datta, O.I. Velikokhatnyi, et al. Nanostructured materials for energy applications. Scientific Reports, 6 (2016) 28367. https://doi.org/10.1038/srep28367

[173] I.A. Moreno-Hernandez, C.A. Macfarland, C.G. Read, et al. Advances in energy and environmental science. Energy & Environmental Science, 10 (2017) 2103. https://doi.org/10.1039/C7EE01486D

[174] S.D. Ghadge, O.I. Velikokhatnyi, M.K. Datta, et al. Performance evaluation of solar energy technologies. ACS Applied Energy Materials, 3 (2020) 541. https://doi.org/10.1021/acsaem.9b01796

[175] Z. Zhao, B. Zhang, D. Fan, et al. Catalytic properties of novel materials for industrial applications. Journal of Catalysis, 405 (2022) 265. https://doi.org/10.1016/j.jcat.2021.12.009

[176] J. Wei, J. Wang, X. Wang, et al. Electrochemical processes for energy storage. Electrochimica Acta, 432 (2022) 141221. https://doi.org/10.1016/j.electacta.2022.141221

[177] N. Blanchard, V. Bizet. Chemical processes for sustainable development. Angewandte Chemie, International Edition, 58 (2019) 6814. https://doi.org/10.1002/anie.201900591

[178] C.J. Chang, Y.P. Zhu, J.L. Wang, et al. Nanomaterials for energy storage and conversion. Journal of Materials Chemistry A: Materials for Energy and Sustainability, 8 (2020) 19079.

[179] L.G. Bloor, P.I. Molina, M.D. Symes, et al. New pathways for catalytic synthesis. Journal of the American Chemical Society, 136 (2014) 3304. https://doi.org/10.1021/ja5003197

[180] J.S. Mondschein, J.F. Callejas, C.G. Read, et al. Material design for energy applications. Chemistry of Materials, 29 (2017) 950. https://doi.org/10.1021/acs.chemmater.6b02879

[181] H.Y. Wang, S.F. Hung, H.Y. Chen, et al. Advances in electrocatalysis for hydrogen production. Journal of the American Chemical Society, 138 (2016) 36. https://doi.org/10.1021/jacs.5b10525

[182] K.L. Yan, J.Q. Chi, J.Y. Xie, et al. Innovative renewable energy solutions. Renewable Energy, 119 (2018) 54. https://doi.org/10.1016/j.renene.2017.12.003

[183] J. Chen, A. Selloni. Theoretical insights into catalytic processes. The Journal of Physical Chemistry Letters, 3 (2012) 2808. https://doi.org/10.1021/jz300994e

[184] X.L. Yang, H. Li, A.Y. Lu, et al. Nanostructured materials for energy applications. Nano Energy, 25 (2016) 42. https://doi.org/10.1016/j.nanoen.2016.04.035

[185] S. Anantharaj, K. Karthick, S. Kundu. Metal-organic frameworks in energy applications. Inorganic Chemistry, 58 (2019) 8570. https://doi.org/10.1021/acs.inorgchem.9b00868

[186] M. Huynh, T. Ozel, C. Liu, et al. Advances in energy conversion technologies. Chemical Science, 8 (2017) 4779. https://doi.org/10.1039/C7SC01239J

[187] A.L. Li, S. Kong, C.X. Guo, et al. Emerging catalysts for sustainable energy. Nature Catalysis, 5 (2022) 109. https://doi.org/10.1038/s41929-021-00732-9

[188] J.Z. Huang, H.Y. Sheng, R.D. Ross, et al. New methodologies for energy storage. Nature Communications, 12 (2021) 3036. https://doi.org/10.1038/s41467-021-23390-8

[189] J. Yu, F.A. Garcés-Pineda, J. González-Cobos, et al. Innovations in environmental sustainability. Nature Communications, 13(2022) 4341.

[190] R.Y. Fan, H.Y. Zhao, Y.N. Zhen, et al. Advances in fuel technology. Fuel, 333 (2023) 126361. https://doi.org/10.1016/j.fuel.2022.126361

[191] D.S. Raja, P-Y. Cheng, C-C. Cheng, et al. Application of Catalytic Processes for the Removal of Environmental Contaminants, Applied Catalysis, B: Environmental, 303 (2022) 120899. https://doi.org/10.1016/j.apcatb.2021.120899

[192] T. Tran-Phu, H. Chen, R. Daiyan, et al. High-Performance Supercapacitors with Enhanced Cycling Stability, ACS Applied Materials & Interfaces, 14 (2022) 33130. https://doi.org/10.1021/acsami.2c05849

[193] W.L. Kwong, C.C. Lee, A. Shchukarev, et al. Metal-Organic Frameworks as Catalysts for Chemical Reactions, Journal of Catalysis, 365 (2018) 29. https://doi.org/10.1016/j.jcat.2018.06.018

[194] W.L. Kwong, C.C. Lee, A. Shchukarev, et al. Advances in Catalysis Using Nanostructured Materials, Chemical Communications, 55(2019) 5017. https://doi.org/10.1039/C9CC01369E

[195] L.L. Zhao, Q. Cao, A.L. Wang, et al. Nanomaterials for Energy Conversion and Storage, Nano Energy, 45(2018) 118.

[196] S.A. Bonke, K.L. Abel, D.A. Hoogeveen, et al. Sustainable Catalysis in Organic Synthesis, ChemPlusChem, 83 (2018) 704. https://doi.org/10.1002/cplu.201800020

[197] J. Lin, P. Wang, H. Wang, et al. Materials for High-Efficiency Solar Cells, Advanced Science, 6 (2019) 1900246.

[198] B. Konkena, K. Junge Puring, I. Sinev, et al. Understanding Catalytic Mechanisms in Green Chemistry, Nature Communications, 7 (2016) 12269. https://doi.org/10.1038/ncomms12269

[199] L. Wang, L.L. Cao, X.K. Liu, et al. Nanostructured Catalysts for Energy Applications, Journal of Physical Chemistry C, 124 (2020) 2756. https://doi.org/10.1021/acs.jpcc.9b09796

[200] Q. Hu, G.D. Li, X.F. Liu, et al. Materials for Energy Storage and Conversion, Journal of Materials Chemistry A: Materials for Energy and Sustainability, 7 (2019) 461.

[201] J.J. Wu, M.J. Liu, K. Chatterjee, et al. Interface Engineering in Energy Storage Materials, Advanced Materials Interfaces, 3 (2016) 1500669.

[202] Y. Yang, H.Q. Yao, Z.H. Yu, et al. Synthesis of High-Efficiency Photocatalysts, Journal of the American Chemical Society, 141 (2019) 10417.

[203] Y.N. Guo, J. Tang, H.Y. Qian, et al. Investigating the Properties of Novel Catalytic Materials, Chemistry of Materials, 29 (2017) 5566. https://doi.org/10.1021/acs.chemmater.7b00867

[204] Q. Xiong, X. Zhang, H. Wang, et al. Design and Characterization of Catalysts for Sustainable Processes, Chemical Communications, 54 (2018) 3859. https://doi.org/10.1039/C8CC00766G

[205] A. Parra-Puerto, K.L. Ng, K. Fahy, et al. Developing Catalytic Systems for Green Chemistry, ACS Catalysis, 9 (2019) 11515. https://doi.org/10.1021/acscatal.9b03359

[206] Y. Liu, F. Yang, W. Qin, et al. Advances in Colloidal and Interfacial Science, Journal of Colloid and Interface Science, 534 (2019) 55. https://doi.org/10.1016/j.jcis.2018.09.017

[207] H. Liu, X. Peng, X. Liu, et al. Sustainable Chemical Processes Using Renewable Resources, ChemSusChem, 12 (2019) 1334. https://doi.org/10.1002/cssc.201802437

[208] C. Guan, H. Wu, W. Ren, et al. Innovations in Materials for Energy Sustainability, Journal of Materials Chemistry A: Materials for Energy and Sustainability, 6 (2018) 9009.

[209] Z.H. Xue, H. Su, Q.Y. Yu, et al. Advanced Energy Materials for Sustainable Development, Advanced Energy Materials, 7 (2017) 1602355. https://doi.org/10.1002/aenm.201602355

[210] W.R. Cheng, H. Zhang, X. Zhao, et al. Emerging Materials for Energy Conversion and Storage, Journal of Materials Chemistry A: Materials for Energy and Sustainability, 6 (2018) 9420.

[211] F. Hu, S. Zhu, S. Chen, et al. Nanostructured Materials for Energy Applications, Advanced Materials, 29 (2017) 1606570. https://doi.org/10.1002/adma.201601925

[212] N.I. Gumerova, A. Rompel. Recent Advances in Catalysis Research, Nature Reviews Chemistry, 2 (2018) 0112. https://doi.org/10.1038/s41570-018-0112

[213] M. Stuckart, K.Y. Monakhov. Novel Catalysts for Sustainable Energy Applications, Journal of Materials Chemistry A: Materials for Energy and Sustainability, 6 (2018) 17849. https://doi.org/10.1039/C8TA06213G

[214] S. Goberna-Ferrón, L. Vigara, J. Soriano-López, et al. Inorganic Catalysts in Modern Chemistry, Inorganic Chemistry, 51 (2012) 11707. https://doi.org/10.1021/ic301618h

[215] M. Blasco-Ahicart, J. Soriano-López, J.J. Carbó, et al. The Role of Catalysis in Green Chemistry, Nature Chemistry, 10 (2018) 24. https://doi.org/10.1038/nchem.2874

[216] X.B. Han, D.X. Wang, E. Gracia-Espino, et al. Catalytic Reactions for Energy Conversion, Chinese Journal of Catalysis, 41(2020) 853. https://doi.org/10.1016/S1872-2067(20)63538-0

[217] J.S. Mondschein, K. Kumar, C.F. Holder, et al. Advances in Inorganic Chemistry for Energy Applications, Inorganic Chemistry, 57 (2018) 6010. https://doi.org/10.1021/acs.inorgchem.8b00503

[218] M.J. Kirshenbaum, M.H. Richter, M. Dasog. Recent Developments in Catalysis and Its Applications, ChemCatChem, 11 (2019) 3877. https://doi.org/10.1002/cctc.201801736

[219] C.J. Lei, H.Q. Chen, J.H. Cao, et al. Innovative Materials for Energy Conversion, Advanced Energy Materials, 8 (2018) 1870119. https://doi.org/10.1002/aenm.201801587

[220] B. Shen, Y. He, Z. He, et al. Sustainable Approaches to Colloidal Chemistry, Journal of Colloid and Interface Science, 605 (2022) 637. https://doi.org/10.1016/j.jcis.2021.07.127

[221] L. Najafi, S. Bellani, R. Oropesa-Nuñez, et al. Nanoengineering for Energy Applications, ACS Nano, 13(2019) 3162. https://doi.org/10.1021/acsnano.8b08670

[222] N.N. Han, K.R. Yang, Z.Y. Lu, et al. Emerging Trends in Green Catalysis, Nature Communications, 9 (2018) 924.

[223] A. Jain, Z. Wang, J.K. Nørskov. Exploring Catalytic Mechanisms in Energy Applications, ACS Energy Letters, 4 (2019) 1410. https://doi.org/10.1021/acsenergylett.9b00876

[224] A.E. Thorarinsdottir, C. Costentin, S.S. Veroneau, et al. Materials for Renewable Energy Technologies, Chemistry of Materials, 34 (2022) 826. https://doi.org/10.1021/acs.chemmater.1c03801

[225] A.M. Patel, J.K. Nørskov, K.A. Persson, et al. Understanding the Catalytic Properties of Advanced Materials, Physical Chemistry Chemical Physics, 21 (2019) 25323. https://doi.org/10.1039/C9CP04799A

[226] A. Shinde, J.R.J. Jones, D. Guevarra, et al. Electrocatalysis for Renewable Energy Conversion, Electrocatalysis, 6 (2015) 229. https://doi.org/10.1007/s12678-014-0237-7

[227] L.C. Seitz, C.F. Dickens, K. Nishio, et al. High-Efficiency Energy Conversion Materials, Science, 353 (2016) 1011. https://doi.org/10.1126/science.aaf5050

[228] X. Liang, L. Shi, Y. Liu, et al. Advances in Angewandte Chemistry for Energy Applications, Angewandte Chemie, International Edition, 58 (2019) 7631. https://doi.org/10.1002/anie.201900796

[229] G. Wu, X.S. Zheng, P.X. Cui, et al. Nature-Inspired Approaches to Catalysis, Nature Communications, 10 (2019) 4855.

[230] F. Luo, H. Hu, X. Zhao, et al. Nano-Enabled Energy Conversion Technologies, Nano Letters, 20 (2020) 2120. https://doi.org/10.1021/acs.nanolett.0c00127

[231] J.Y. Xu, J.J. Li, Z. Lian, et al. Catalysis for Sustainable Energy Solutions, ACS Catalysis, 11 (2021) 3402. https://doi.org/10.1021/acscatal.0c04117

[232] J. Kim, P.C. Shih, K.C. Tsao, et al. Materials Innovations for Advanced Chemical Processes, Journal of the American Chemical Society, 139 (2017) 12076. https://doi.org/10.1021/jacs.7b06808

[233] H.J. Song, H. Yoon, B. Ju, et al. Future Perspectives on Advanced Energy Materials, Advanced Energy Materials, 11 (2021) 2002428. https://doi.org/10.1002/aenm.202002428

[234] T.W. Wang, C.S. Duanmu, X.Y. Meng, et al. Research progress on performance optimization of transition metal sulfide catalysts and photocatalytic water splitting for hydrogen production[J/OL]. Low Carbon Chemistry and Chemical Industry, 1-10[2024-09-19]. http://kns.cnki.net/kcms/detail/51.1807.TQ.20240528.1830.008.html.

[235] S.C. Shang, X.Y. Zhang, X.L. Liu, et al. Design and construction of cocatalysts for photocatalytic overall water splitting. Acta Phys. -Chim. Sin., 36 (2020) 1905007. https://doi.org/10.3866/PKU.WHXB201905083

[236] D. Wang, F.Q. Li, W.G. Pan, Q.Z. Zhu, J. Wu, Z.Z. Guan. Research progress of photocatalytic hydrogen production catalysts. Applied Chemical Industry, 53 (2024) 223-237.

[237] J. Zhang, L.J. Hao, T.T. Zhang, et al. Research progress on application of Bi-loaded TiO2 catalyst in photocatalysis field. Applied Chemical Industry, 51(2022) 3597-3603.

[238] V. Kumaravel, S. Mathew, J. Bartlett, et al. Photocatalytic hydrogen production using metal doped TiO2: A review of recent advances. Applied Catalysis B: Environmental, 2019, 244:1021-1064. https://doi.org/10.1016/j.apcatb.2018.11.080

[239] H. Wang, L. Song, L. Yu, et al. Charge transfer between $Ti^{4+}$, $Mn^{2+}$ and Pt in the tin doped $TiO_2$ photocatalyst for elevating the hydrogen production efficiency. Applied Surface Science, 2022, 581 (2022) 152202. https://doi.org/10.1016/j.apsusc.2021.152202

[240] D. Chen, H. Gao, Y. Yao, et al. Pd Loading, Mn+(n = 1, 2, 3) metal ions doped $TiO_2$ nanosheets for enhanced photocatalytic $H_2$ production and reaction mechanism. International Journal of Hydrogen Energy, 47 (2022) 10250-10260. https://doi.org/10.1016/j.ijhydene.2022.01.112

[241] F. Almomani, K. Aljaml, R. Bhosale. Solar photo-catalytic production of hydrogen by irradiation of cobalt co-doped $TiO_2$. International Journal of Hydrogen Energy, 46 (2021) 12068-12081. https://doi.org/10.1016/j.ijhydene.2020.07.164

[242] P. Zhang, L. Yu, X. Lou. Construction of heterostructured $Fe_2O_3$-$TiO_2$ microdumbbells for photoelectrochemical water oxidation. Angewandte Chemie-International Edition, 57 (2018) 15076-15080. https://doi.org/10.1002/anie.201808104

[243] W. Huang, Z. Fu, X. Hu, et al. Efficient photocatalytic hydrogen evolution over Cu3Mo2O9/TiO2 p-n heterojunction. Journal of Alloys and Compounds, 904 (2022) 164089. https://doi.org/10.1016/j.jallcom.2022.164089

[244] H. Wang, J. Liu, X. Hu, et al. Engineering of $SnO_2$/$TiO_2$ heterojunction compact interface with efficient charge transfer pathway for photocatalytic hydrogen evolution. Chinese Chemical Letters, 37 (2022) 101123. https://doi.org/10.1016/j.cclet.2022.01.018

[245] C. Yu, M. Li, D. Yang, et al. NiO Nano-particles dotted $TiO_2$ nano-sheets assembled nanotubes p-n heterojunctions for efficient interface charge separation and photocatalytic hydrogen evolution. Applied Surface Science, 2021, 568 (2021) 150981. https://doi.org/10.1016/j.apsusc.2021.150981

[246] X. Yue, H. Li, Y. Qiu, et al. A facile synthesis method of $TiO_2$@$SiO_2$ porous core shell structure for photocatalytic hydrogen evolution. Journal of Solid State Chemistry, 300(2021) 122250. https://doi.org/10.1016/j.jssc.2021.122250

[247] S. Lü, Y. Wang, Y. Zhou, et al. Oxygen vacancy stimulated direct Z-scheme of mesoporous $Cu_2O$/$TiO_2$ for enhanced photocatalytic hydrogen production from water and seawater. Journal of Alloys and Compounds, 868 (2021) 159144. https://doi.org/10.1016/j.jallcom.2021.159144

[248] S. Liu, Y. Ma, D. Chi, et al. Hollow heterostructure CoS/CdS photocatalysts with enhanced charge transfer for photocatalytic hydrogen production from seawater. International Journal of Hydrogen Energy, 47 (2022) 9220-9229. https://doi.org/10.1016/j.ijhydene.2021.12.259

[249] G. Wang, Y. Quan, K. Yang, et al. EDA-assisted synthesis of multifunctional snowflake-Cu2S/CdZnS S-scheme heterojunction for improved the photocatalytic hydrogen evolution.

Journal of Materials Science & Technology, 121 (2022) 28-39.
https://doi.org/10.1016/j.jmst.2021.11.073

[250] L. Lu, Y. Ma, H. Liu, et al. Controlled preparation of hollow $Zn_{0.3}Cd_{0.7}S$ nanospheres modified by $NiS_{1.97}$ nano-sheets for superior photocatalytic hydrogen production. Journal of Colloid and Interface Science, 606 (2022) 1-9. https://doi.org/10.1016/j.jcis.2021.08.006

[251] Y. Zhang, D. Lu, H. Li, et al. Enhanced visible Light-Driven photocatalytic hydrogen evolution and stability for noble metal-free $MoS_2/Zn_{0.5}Cd_{0.5}S$ heterostructures with W/Z phase junctions. Applied Surface Science, 586 (2022) 152770. https://doi.org/10.1016/j.apsusc.2022.152770

[252] A. Wang, L. Zhang, X. Li, et al. Synthesis of ternary $Ni_2P@UiO-66-NH_2/Zn_{0.5}Cd_{0.5}S$ composite materials with significantly improved photocatalytic $H_2$ production performance. Chinese Journal of Catalysis, 43(2022) 1295-1305. https://doi.org/10.1016/S1872-2067(21)63912-8

[253] J. Chen, M. Liu, S. Xie, et al. $Cu_2O$-Loaded $TiO_2$ heterojunction composites for enhanced photocatalytic $H_2$ production. Journal of Molecular Structure, 1247 (2022) 131294. https://doi.org/10.1016/j.molstruc.2021.131294

[254] G. Huang, W. Ye, C. Lü, et al. Hierarchical red phosphorus incorporated $TiO_2$ hollow sphere heterojunctions toward superior photocatalytic hydrogen production. Journal of Materials Science & Technology, 108 (2022) 18-25. https://doi.org/10.1016/j.jmst.2021.09.026

[255] L. Qi, K. Dong, T. Zeng, et al. Three-dimensional red phosphorus: A promising photocatalyst with excellent adsorption and reduction performance. Catalysis Today, 314 (2018) 42-51. https://doi.org/10.1016/j.cattod.2018.01.002

[256] X.Y. Meng, J.J. Li, P. Liu, et al. Long-term stable hydrogen production from water and lactic acid via visible-light-driven photocatalysis in a porous microreactor . Angewandte Chemie International Edition, 62 (2023) e202307490. https://doi.org/10.1002/anie.202307490

[257] B.R. Zhao, X.J. Wang, P. Liu, et al. Visible-light-riven photocatalytic $H_2$ production from H2O boosted by hydroxyl groups on alumina . Industrial & Engineering Chemistry Research, 61 (2022) 6845-6858. https://doi.org/10.1021/acs.iecr.2c00767

[258] Z.H. Lv, P. Liu, Y.Y. Zhao, et al. Visible-light-driven photocatalytic $H_2$ production from $H_2O$ boosted by anchoring Pt and CdS nano-particles on a NaY zeolite. Chemical Engineering Science, 255 (2022) 117658. https://doi.org/10.1016/j.ces.2022.117658

[259] Y.H. Li, Y.Q. Li, J.W. Shang, et al. Application of metal sulfides in energy conversion and storage. Chinese Chemical Letters, 34 (2023) 107928. https://doi.org/10.1016/j.cclet.2022.107928

[260] D. Nayak, R. Thangavel. Strain modulated electronic and photocatalytic properties of MoS2/WS2 heterostructure: A DFT study. ACS Applied Electronic Materials, 5 (2023) 302-316. https://doi.org/10.1021/acsaelm.2c01344

[261] Q.H. Zhu, Q. Xu, M.M. Du, et al. Recent progress of metal sulfide photocatalysts for solar energy conversion. Advanced Materials, 34 (2022) 2202929. https://doi.org/10.1002/adma.202202929

315

[262] C. Prasad, N. Madkhali, J.S. Won, et al. CdS based heterojunction for water splitting: A review. Materials Science and Engineering: B, 292 (2023) 116413. https://doi.org/10.1016/j.mseb.2023.116413

[263] X.L. Hao, X.S. Chu, X.Y. Liu, et al. Synergetic metal-semiconductor interaction: Single-atomic Pt decorated CdS nano-photocatalyst for highly water-to-hydrogen conversion. Journal of Colloid and Interface Science, 621 (2022) 160-168. https://doi.org/10.1016/j.jcis.2022.04.053

[264] M. Dong, P. Zhou, C. Jiang, et al. First-principles investigation of Cu-doped ZnS with enhanced photocatalytic hydrogen production activity. Chemical Physics Letters, 668 (2017) 1-6. https://doi.org/10.1016/j.cplett.2016.12.008

[265] H.M. Zhao, H.T. Fu, X.H. Yang, et al. $MoS_2$/CdS rod-like nanocomposites as high-performance visible light photocatalyst for water splitting photocatalytic hydrogen production . International Journal of Hydrogen Energy, 47(2022) 8247-8260. https://doi.org/10.1016/j.ijhydene.2021.12.171

[266] Y.P. Liu, Y.H. Li, F. Peng, et al. 2H- and 1T- mixed phase few-layer $MoS_2$ as a superior to Pt co-catalyst coated on $TiO_2$ nano-rod arrays for photocatalytic hydrogen evolution. Applied Catalysis B: Environmental, 241 (2019) 236-245. https://doi.org/10.1016/j.apcatb.2018.09.040

[267] Y.X. Pan, J.B. Peng, S. Xin, et al. Enhanced visible-light-driven photocatalytic $H_2$ evolution from water on noble-metal-free CdS-nano-particle-dispersed $Mo_2C@C$ nanospheres. ACS Sustainable Chemistry & Engineering, 5 (2017) 5449-5456. https://doi.org/10.1021/acssuschemeng.7b00787

[268] Y.X. Pan, H.Q. Zhuang, H. Ma, et al. Tungsten carbide hollow spheres flexible for charge separation and transfer for enhanced visible-light-driven photocatalysis . Chemical Engineering Science, 194 (2019) 71-77. https://doi.org/10.1016/j.ces.2018.01.022

[269] F. Zhang, H.Q. Zhuang, W.M. Zhang, et al. Noble-metal-free CuS/CdS photocatalyst for efficient visible-light-driven photocatalytic $H_2$ production from water. Catalysis Today, 330 (2019) 203-208. https://doi.org/10.1016/j.cattod.2018.03.060

[270] F. Zhang, H.Q. Zhuang, J. Song, et al. Coupling cobalt sulfide nano-sheets with cadmium sulfide nano-particles for highly efficient visible-light-driven photocatalysis. Applied Catalysis B: Environmental, 226 (2018) 103-110. https://doi.org/10.1016/j.apcatb.2017.12.046

[271] M.S. Goh, H. Moon, H. Park, et al. Sustainable and stable hydrogen production over petal-shaped $CdS/FeS_2$ S-scheme heterojunction by photocatalytic water splitting. International Journal of Hydrogen Energy, 47 (2022) 27911-27929. https://doi.org/10.1016/j.ijhydene.2022.06.118

[272] Y. Cao, Y.Y. Zhao, P. Liu, et al. Boosted visible-light-driven photocatalytic $H_2$ production from water on a noble-metal-free Cd-In-S photocatalyst . Industrial & Engineering Chemistry Research, 63 (2024) 1834-1842. https://doi.org/10.1021/acs.iecr.3c03875

[273] W.B. Zhang, Q.Y. Xu, X.Q. Tang, et al. Construction of a transition-metal sulfide heterojunction photocatalyst driven by a built-in electric field for efficient hydrogen evolution

under visible light . Journal of Colloid and Interface Science, 649 (2023) 325-333. https://doi.org/10.1016/j.jcis.2023.06.080

[274] M.M. Jiang, J. Xu, P. Munroe, et al. 1D/2D CdS/WS$_2$ heterojunction photocatalyst: First-principles insights for hydrogen production . Materials Today Communications, 35 (2023) 105991. https://doi.org/10.1016/j.mtcomm.2023.105991

[275] F. Xue, H.C. Wu, Y.T. Liu, et al. CuS nano-sheet-induced local hot spots on g-C$_3$N$_4$ boost photocatalytic hydrogen evolution . International Journal of Hydrogen Energy, 48 (2023): 6346-6357. https://doi.org/10.1016/j.ijhydene.2022.05.087

[276] L.J. Zhang, Y.L. Wu, J.K. Li, et al. Amorphous/crystalline heterojunction interface driving the spatial separation of charge carriers for efficient photocatalytic hydrogen evolution . Materials Today Physics, 27 (2022) 100767. https://doi.org/10.1016/j.mtphys.2022.100767

[277] Y. Cui, Y.X. Pan, H.L. Qin, et al. A noble-metal-free CdS/Ni$_3$S$_2$@C nanocomposite for efficient visible-light-driven photocatalysis . Small Methods, 2 (2018) 1800029. https://doi.org/10.1002/smtd.201800029

[278] Y.X. Sun, Y. Li, J.J. He, et al. Controlled synthesis of MnxCd1-x for enhanced visible-light driven photocatalytic hydrogen evolution. Chinese Journal of Structural Chemistry, 42 (2023) 100145. https://doi.org/10.1016/j.cjsc.2023.100145

[279] J.M. Wang, J. Luo, D. Liu, et al. One-pot solvothermal synthesis of MoS$_2$-modified Mn$_{0.2}$Cd$_{0.8}$S/MnS heterojunction photocatalysts for highly efficient visible-light-driven H$_2$ production. Applied Catalysis B: Environmental, 241 (2019) 130-140. https://doi.org/10.1016/j.apcatb.2018.09.033

[280] X.J. Feng, H.S. Shang, J.L. Zhou, et al. Heterostructured core-shell CoS$_{1.097}$@ZnIn$_2$S$_4$ nano-sheets for enhanced photocatalytic hydrogen evolution under visible light . Chemical Engineering Journal, 457 (2023) 141192. https://doi.org/10.1016/j.cej.2022.141192

[281] H.T. Fan, Z. Wu, K.C. Liu, et al. Fabrication of 3D CuS@ZnIn$_2$S$_4$ hierarchical nanocages with 2D/2D nano-sheet subunits p-n heterojunctions for improved photocatalytic hydrogen evolution . Chemical Engineering Journal, 433(2022) 134474. https://doi.org/10.1016/j.cej.2021.134474

[282] L.L. Lu, Y.J. Ma, H.Q. Liu, et al. Controlled preparation of hollow Zn$_{0.3}$Cd$_{0.7}$S nanospheres modified by NiS$_{1.97}$ nano-sheets for superior photocatalytic hydrogen production . Journal of Colloid and Interface Science, 606 (2022) 1-9. https://doi.org/10.1016/j.jcis.2021.08.006

[283] S.J. Liu, Y. Ma, D.J. Chi, et al. Hollow heterostructure CoS/CdS photocatalysts with enhanced charge transfer for photocatalytic hydrogen production from seawater . International Journal of Hydrogen Energy, 47 (2022) 9220-9229. https://doi.org/10.1016/j.ijhydene.2021.12.259

[284] S.Y. Chang, H.J. Gu, H.H. Zhang, et al. Facile construction of a robust CuS@NaNbO$_3$ nano-rod composite: A unique p-n heterojunction structure with superior performance in photocatalytic hydrogen evolution . Journal of Colloid and Interface Science, 644 (2023) 304-314. https://doi.org/10.1016/j.jcis.2023.04.111

[285] Y. Zhou, L. Zhang, J. Liu, et al. Brand new P-doped g-$C_3N_4$: Enhanced photocatalytic activity for $H_2$ evolution and Rhodamine B degradation under visible light. Journal of Materials Chemistry A, 3 (2015) 3862-3867. https://doi.org/10.1039/C4TA05292G

[286] Y. Wang, S. Zhao, Y. Zhang, et al. Facile synthesis of self-assembled g-$C_3N_4$ with abundant nitrogen defects for photocatalytic hydrogen evolution. ACS Sustainable Chemistry & Engineer, 6 (2018) 10200-10210. https://doi.org/10.1021/acssuschemeng.8b01499

[287] X. Wang, K. Maeda, A. Thomas, et al. A metal-free polymeric photocatalyst for hydrogen production from water under visible light. Nature Materials, 8 (2009) 76-80. https://doi.org/10.1038/nmat2317

[288] C. Yang, J. Qin, Z. Xue, et al. Rational design of carbon-doped $TiO_2$ modified g-$C_3N_4$ via in-situ heat treatment for drastically improved photocatalytic hydrogen with excellent photostability. Nano Energy, 41 (2017) 1-9. https://doi.org/10.1016/j.nanoen.2017.09.012

[289] F. Rao, J. Zhong, J. Li. Improved visible light responsive photocatalytic hydrogen production over g-$C_3N_4$ with rich carbon vacancies. Ceramics International, 48 (2022), 1439-1445. https://doi.org/10.1016/j.ceramint.2021.09.130

[290] C. Wang, C. Yang, J. Qin, et al. A facile template synthesis of phosphorus-doped graphitic carbon nitride hollow structures with high photocatalytic hydrogen production activity. Materials Chemistry and Physics, 275 (2022) 125299. https://doi.org/10.1016/j.matchemphys.2021.125299

[291] Y. Liu, Y. Zhang, L. Shi. One-step synthesis of S-doped and nitrogen-defects co-modified mesoporous g-$C_3N_4$ with excellent photocatalytic hydrogen production efficiency and degradation ability. Colloids and Surfaces A: Physicochemical and Engineering Aspects, 641 (2022) 128577. https://doi.org/10.1016/j.colsurfa.2022.128577

[292] H. Tang, W. Cheng, Y. Yi, et al. Nano zero valent iron encapsulated in graphene oxide for reducing uranium. Chemosphere, 278 (2021) 130229. https://doi.org/10.1016/j.chemosphere.2021.130229

[293] M. Sun, Y. Zhou, T. Yu, et al. Synthesis of g-$C_3N_4$/$WO_3$-carbon microsphere composites for photocatalytic hydrogen production. International Journal of Hydrogen Energy, 47 (2022) 10261-10276. https://doi.org/10.1016/j.ijhydene.2022.01.103

[294] H. Yuan, F. Fang, J. Dong, et al. Enhanced photocatalytic hydrogen production based on laminated $MoS_2$/g-$C_3N_4$ photocatalysts. Colloids and Surfaces A: Physicochemical and Engineering Aspects, 2022, 641 (2022) 128575. https://doi.org/10.1016/j.colsurfa.2022.128575

[295] D. Wei, Y. Liu, X. Shao, et al. Cooperative effects of zinc-nickel sulfides as a dual cocatalyst for the enhanced photocatalytic hydrogen evolution activity of g-$C_3N_4$. Journal of Environmental Chemical Engineering, 10 (2022) 107216. https://doi.org/10.1016/j.jece.2022.107216

[296] W. Lei, F. Wang, X. Pan, et al. Z-Scheme $MoO_3$-2D SnS nano-sheets heterojunction assisted g-$C_3N_4$ composite for enhanced photocatalytic hydrogen evolutions. International Journal of Hydrogen Energy, 47 (2022) 10877-10890. https://doi.org/10.1016/j.ijhydene.2022.01.139

Materials Research Forum LLC

https://doi.org/21741/9781644903711-10

[297] H. Zhou, R. Chen, C. Han, et al. Copper phosphide decorated g-$C_3N_4$ catalysts for highly efficient photocatalytic $H_2$ evolution. Journal of Colloid and Interface Science, 610 (2022) 126-135. https://doi.org/10.1016/j.jcis.2021.12.058

[298] J. Liu, H. Zhou, J. Fan, et al. In situ oxidation of ultrathin $Ti_3C_2T_x$ MXene modified with crystalline g-$C_3N_4$ nano-sheets for photocatalytic $H_2$ evolution. International Journal of Hydrogen Energy, 47 (2022) 4546-4558. https://doi.org/10.1016/j.ijhydene.2021.11.059

Applications for Earth-Abundant Transition Metals        Materials Research Forum LLC
Materials Research Foundations 179 (2025) 320-338     https://doi.org/21741/9781644903711-11

# Chapter 11

# Earth-Abundant Metals-Based Catalysts for Hydrogen Evolution Reaction

K. Keerthi[1], E.A. Lohith[1], Sada Venkateswarlu[2], K. Siva Kumar[3]*, N.V.V. Jyothi[1]*

[1]Department of Chemistry, Sri Venkateswara University, Tirupati-517502, A.P, India

[2]Nanotechnology Centre, Centre for Energy and Environmental Technologies, VSB-Technical University of Ostrava, 17. Listopadu 2172/15, 70800 Ostrava-Poruba, Czech Republic

[3]Department of Chemistry, S.V. Arts College, TTD, Tirupati 517502, A.P, India

nvvjyothi67@gmail.com, Sivakumarkasi64@gmail.com

**Abstract**

A crucial step in the search for sustainable energy is the hydrogen evolution reaction (HER), especially when it comes to water splitting for the production of hydrogen. Because of their exceptional stability and efficiency, platinum-group metals (PGMs) have long been the standard catalyst for HER. However, the hunt for substitute catalysts made from earthly plentiful metals has been fueled by the rarity and high cost of PGMs. This chapter provides a thorough overview of recent developments in the creation of metal-based catalysts for HER that are abundant on earth. We explore the structural design approaches, synthesis procedures, and catalytic properties of different metal-based catalysts, such as those made from transition metals, their compounds, and alloys. Particular focus is placed on surface engineering techniques, electronic structural alterations, and mechanistic discoveries that have significantly improved their catalytic efficiency. This chapter also addresses the difficulties in maintaining these catalysts' activity and durability under operating settings and provides insights into potential future research avenues for the goal of producing hydrogen in a way that is both affordable and sustainable.

**Keywords**

Earth-Abundant Metals, Catalysts, Water Splitting, HER

## Contents

## 1.    Introduction

Environmental and energy-related concerns are two of the most important ones for the sustainable growth of human society. Over 85% of the energy used worldwide comes from fossil fuels including coal, oil, and natural gas. Fossil fuels are valuable natural resources used in several chemical industries [1]. However, their use as fuel may lead to rapid depletion. Fossil fuel combustion releases harmful greenhouse gases such as $CO_2$, $NO_x$, and $SO_x$, causing significant environmental issues. Replacing fossil fuels with renewable and carbon-neutral energy sources is desirable [2].

The hydrogen evolution reaction (HER) is an important electrochemical process for producing hydrogen gas, a clean and renewable energy source [3]. As global energy demands change toward sustainable alternatives, HER plays an important role in technologies like water electrolysis, which produces hydrogen gas from water. This reaction is crucial to the development of hydrogen fuel cells, which provide a zero-emission energy source [4,5]. Furthermore, HER is useful in energy storage systems since hydrogen may be stored and used as fuel. The efficiency and cost-effectiveness of HER have a considerable impact on hydrogen's feasibility as a large-scale energy carrier, resulting in research into catalysts and systems that improve its performance [6,7]. As a result, HER has the potential to transform sustainable energy technology and contribute to global decarbonization efforts. Effective catalysts must be used to lower the energy barriers and improve the reaction rate to make the HER more efficient [8-10].

Although the noble metal platinum is a very effective HER catalyst, its exorbitant cost and limited supply prevent its widespread use. Scientists have been trying to create stable, affordable, and effective materials to take the role of platinum [11-13]. This has prompted many studies into Earth-abundant metals-based catalysts, which are more readily available, reasonably priced, and ecologically friendly materials [14]. Earth-abundant metals are important in the HER because of their ability to provide cost-effective and sustainable alternatives to scarce noble metals [15]. As global energy demand rises, effective hydrogen production methods, particularly electrochemical water splitting, become increasingly important [16]. Earth-abundant metals including copper, manganese, cobalt, nickel, and molybdenum have emerged as attractive possibilities for HER catalysts, with both economic and environmental benefits [17].

This chapter examines the synthesis, performance, and design of Earth-abundant metal catalysts for HER, with a particular emphasis on the strategies used to increase their efficiency, including surface changes, alloying, and nano structuring. We also highlight the recently developed catalysts based on Metal-organic frameworks (MOFs), MXenes, and materials derived from them such as metal nanoparticles, metal oxide nanoparticles [18 – 23], etc. A viable strategy for improving HER is a combination of MOFs with MXenes. When coupled, the special qualities of both materials can greatly increase catalytic performance.

## 1.1 Signification of earth-abundant metals

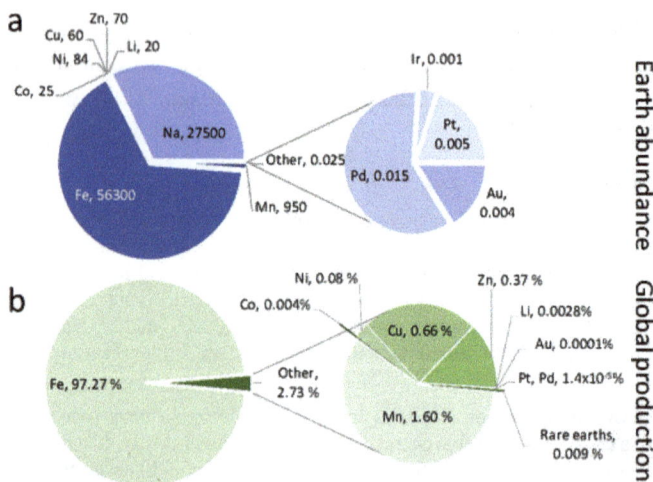

*Fig 1. a) The abundance of certain metals in the world's crust (in ppm) b) the percentage of worldwide output in selected metals in 2020. Reprinted and took permission from ACS [26].*

Earth-abundant metals (EAMs) have significance as they have the potential to revolutionize many industrial processes, especially those related to energy storage, catalysis, and environmental sustainability [24]. In many applications, they are appealing substitutes for precious metals due to their greater availability and less expensive price. EAMs like iron, nickel, and cobalt are being used more frequently in catalytic processes, especially in the pharmaceutical industry, where they are replacing expensive palladium to create sustainable and affordable procedures [25]. In

Materials Research Forum LLC
https://doi.org/21741/9781644903711-11

electrochemical energy storage and conversion technologies such as fuel cells and batteries, EAMs are essential catalysts that facilitate processes like hydrogen evolution and oxygen reduction [26]. The abundance of certain metals in earth's curst and percentage was shown in Fig. 1. Even though EAMs have many benefits, there are still issues with maximizing their catalytic efficiency and completely comprehending their environmental effects. To fully realize their promise in sustainable technology, more research is necessary [27].

## 1.2 Role of hydrogen energy in the fuel sector

In the fuel industry, hydrogen energy is becoming progressively more important. It provides a clean substitute for fossil fuels and aids in the decarbonization of numerous applications [28]. It is a flexible energy carrier with potential applications in power generation, transportation, and long-duration energy storage. The combustion of hydrogen releases no pollutants and only water vapor as a byproduct, making it a more environmentally friendly fuel than fossil fuels [29]. To combat climate change and reduce greenhouse gas emissions, the shift to a hydrogen economy is necessary because hydrogen can replace carbon-intensive fuels in industries that are difficult to decarbonize [30]. For hydrogen to be integrated into other areas, where it is now used minimally, a complete infrastructure is necessary [31]. Although hydrogen offers a promising way forward for sustainable energy, to fully realize its potential in the fuel sector, production and infrastructure development difficulties need to be addressed.

## 2. Catalyst synthesis methods

### 2.1 Top-down

The process of creating nanomaterial catalysts through top-down synthesis transforms bulk materials into structures at the nanoscale, providing distinct benefits for catalysis.

Top-down synthesis of nanoparticles entails breaking down bulk materials into nanoscale particles by mechanical, chemical, or physical processes. Milling and laser ablation are all regularly used techniques for reducing material size. Although this approach enables large-scale production, it may have limits in terms of homogeneity and surface control. The following are the different techniques used for Top-down synthesis.

### 2.1.1 Chemical etching

Chemical etching is a top-down technique for fabricating and modifying nanoparticles. It involves selectively removing material from the surface of bulk materials or pre-synthesized structures via chemical processes. In this procedure, an etchant is a chemical solution that dissolves or reacts with the material to reduce its size or form it into nanoparticles. Feng Wu et al., discussed by using FePc and CNT as precursors, straightforward self-assembly techniques were used to create the CNT enameled with ultrasmall $Fe_2O_3$ nanoparticles ($Fe_2O_3$/CNT). To prepare a suspension, 50 mg of acid-oxidized CNT was ultrasonically distributed in 10 mL of $H_2O$. After mixing 100 mg of FePc fine particles with 50 mL of ammonia water (25%), the suspension was ultrasonically treated at 40°C for 24 hours. The purplish-blue residues were recovered by high-speed centrifugation after assembly, and they were heat-treated for one hour at 400°C in air after being fully leached with water and ethanol to remove the FePc/CNT precursor. They used a FePc/CNT as a precursor various proportions of FeP/CNT as the final catalyst were prepared [32].

### 2.1.2 Mechanical ball milling

Mechanical ball milling is a popular process for creating nanomaterials, alloys, and composites through high-energy impacts between balls and powders in a spinning container. This approach reduces particle size and promotes phase changes, making it excellent for creating nanocrystalline materials with superior characteristics. A. Ambrosi et al., explained the process of $MoS_2$ using an industrial ball mill. In the steel drum roller on the throne, 500g of $MoS_2$ was combined with 5 Kg of 10 mm and 2 Kg of 20 mm stainless steel balls [33]. L. Zhang et al. reported their work involved hydrogenating 6g Mg powder at 380°C for 2 hours, milling it at 450 rpm for 5 hours, and preparing $MgH_2Fe$ composites via mechanical ball milling. The combinations were combined with $MgH_2$ and Fe nanosheets before being ball-milled in an argon environment of 0.1 MPa. Commercial Fe powders were also processed with $MgH_2$ under identical circumstances [34].

### 2.1.3 Laser ablation

Laser ablation is a precision technology that uses high-intensity laser beams to remove material from the surface through vaporization or fragmentation. It is valued for its precision, speed, and non-contact nature. The procedure involves heating and ejecting material from the surface using a laser pulse, with precision settings that may be adjusted. F. Davodi et al., prepared the Nanoparticles of Nickel-Iron alloy encased in an ultra-thin graphene shell (NiFe@UTG) were produced by laser ablation of a target picosecond laser (ps-laser) of the $Ni_{80}Fe_{20}$ alloy in acetone [35].

### 2.1.4 Sputtering

Sputtering is a physical vapor deposition process that uses ion bombardment to expel material from a target, resulting in thin layers. These expelled atoms then settle onto a substrate in a vacuum. Its precision and ability to create homogenous films make it popular in the electronics, optics, and semiconductor industries. Binary Ni-Mo alloy nanorods were produced by co-depositing pure nickel and molybdenum in a multi-source sputtering method, as described by L. Zhang et al. Before sputtering, the vacuum chamber was inflated to a residual pressure of less than $1 \times 10^{-3}$ Pa. During the deposition process, the sputtering pressure was kept at 4.0 Pa and the working gas was high-purity argon at 40 sccm. Nickel and Molybdenum cathode powers were maintained at 80W and 180W, correspondingly. High-purity argon, a vacuum chamber, and a 5-minute pre-sputtering procedure were all required [36]. B. Goa et al. stated by using DC/RF reactive magnetron sputtering in a clean argon environment at room temperature, the 3D flower-like $MoS_2$ was formed on the candle soot substrate. A 99.99% $MoS_2$ target was used to deposit the $MoS_2$ layer, which was then uniformly rotated at 6 rmp after being pre-sputtered for 20 minutes. The depositions were conducted using an 80W $MoS_2$ target power and an optimized fixed flow rate of 30 sccm argon [37].

### 2.2 Bottom up

The bottom-up strategy in nanoparticle synthesis is creating nanoparticles at the atomic or molecular level, assembling smaller units into larger structures. Unlike the top-down strategy, which involves breaking down bulk materials into smaller particles, the bottom-up method provides more exact control over nanoparticle size, shape, and composition. The following are the different techniques used for Bottom-up synthesis, which provide advantages in terms of cost-effectiveness, scalability, and the capacity to generate extremely uniform particles with desired attributes.

### 2.2.1 Hydro/solvothermal

Solvothermal synthesis is a process for producing materials, particularly nanomaterials, that involves dissolving a precursor in a solvent and heating it at high temperatures and pressure. This technique accelerates crystal formation while providing fine control over material attributes such as size and shape. It is extensively used in catalysis and energy storage, allowing for the development of sophisticated materials such as nanoparticles, zeolites, and metal-organic frameworks. Ni(acac)2 was combined with ethanol and oleylamine (OAm) in a Teflon-lined reactor for solvothermal synthesis, according to S.A. Abbas et al. After 20 hours of the reaction at 180°C, NiEt-OAm1, NiEt-OAm4, and NiEt-OAm6 were produced, correspondingly. The products were centrifuged multiple times [38]. Jebaslinhepzybai. B.T. et al. study focuses on producing cobalt nitrate hexahydrate ($Co_2P$) from ethylene glycol, red phosphorus, Nafion, and super P carbons. The synthesis entails dissolving $Co(NO_3)_2.6H_2O$ in a mixture of EG and distilled water, adding red phosphorous, and treating at 200°C for 24 hours. The product is next rinsed and oven-dried. The growth of $Co_2P$ NP under solvothermal conditions, solvent effects, and cobalt acetate as a substitute precursor [39].

### 2.2.2 Coprecipitation

Co-precipitation is a simple chemical synthesis procedure that involves precipitating ions from a solution at the same time to make homogeneous materials, particularly nanoparticles. It enables fine control over materials composition and is commonly used in sectors such as catalysis and environmental remediation. By altering parameters such as pH and temperature, the qualities of the resultant materials can be tuned to specific uses. D. Chand et al. created the $NiFe_2O_4$ via a conventional co-precipitation process using deionized water, $Ni(NO_3)_2.6H_2O$, and $FeCl_3.6H_2O$. The mixture was stirred for 15 minutes, and the NaOH was added. A crimson precipitate was formed, which was then dried and calcined at 475°C [40]. K. Uma et al.'s research focuses on the creation of nanorod composites with analytical grade reagents and chemicals. Zinc acetate, ferric nitrate, sodium hydroxide, and DI water were combined with NaOH, ferric nitrate, zinc acetate, and DI water, and the mixture was heated in an autoclave lined with Teflon, after a 24-hour hydrothermal treatment at 200°C and annealed for four hours at 450°C. ZnO nanorods were produced with $\alpha$-$Fe_2O_3$ nanoparticles using a co-precipitation technique [41].

### 2.2.3 Solgel

The sol-gel synthesis method is a versatile way to create nanoscale materials, particularly ceramics and thin films. Chemical reactions involving metal alkoxides or salts are used to convert a colloidal solution ("sol") into a solid network ("gel"). This approach enables fine control over material composition and structure, making it appropriate for use in optics, electronics, catalysis, and biomedicine. Its low processing temperatures and capacity to include additives increase its utility in the production of composite and hybrid materials. Ran Miao et al. created the mesoporous $FeS_2$ nanoparticles, amorphous $Fe_2O_3$ must be generated using the inverse micelle sol-gel method, heated at 150°C, and chemically converted using sulfur sources such as sulfur powder and $H_2S$ gas [42].

Materials Research Forum LLC
https://doi.org/21741/9781644903711-11

### 3. Hydrogen evolution reaction

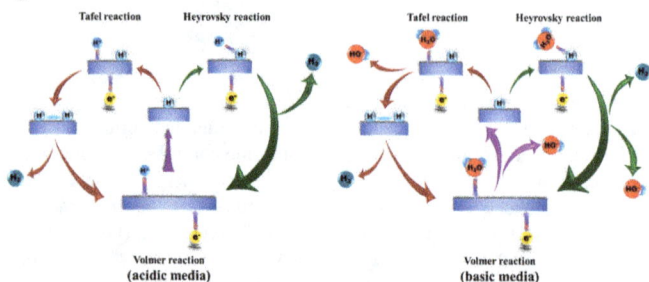

*Fig 2. HER electrocatalysis mechanism in two different media. Reprinted and took permission from Elsevier [43]*

Hydrogen gas ($H_2$) is produced electrochemically from water or an aqueous solution using the hydrogen evolution reaction (HER). This reaction holds significance in a multitude of domains, such as electrochemistry, energy storage, and renewable energy technologies. The mechanism of HER was shown in Fig. 2. At an electrochemical cell's cathode, water molecules undergo a hydrogen ion separation process (HER) to produce electrons ($e^-$) and hydrogen ions ($H^+$). The reaction between electrons and hydrogen ions produces hydrogen gas [43].

**Acidic**

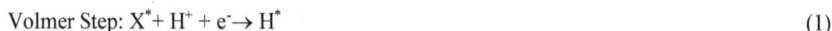
Volmer Step: $X^* + H^+ + e^- \rightarrow H^*$ (1)

Heyrovsky step: $H^* + H^+ + e^- \rightarrow H_2 + {}^*$ (2)

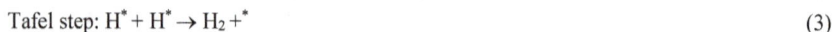
Tafel step: $H^* + H^* \rightarrow H_2 + {}^*$ (3)

$2(H^+ + e^-) \leftrightarrow H_2$ (4)

**Alkaline**

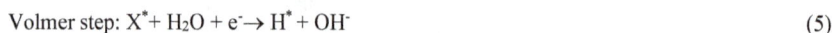
Volmer step: $X^* + H_2O + e^- \rightarrow H^* + OH^-$ (5)

Heyrovsky step: $H^* + H_2O + e^- \rightarrow H_2 + OH^- + {}^*$ (6)

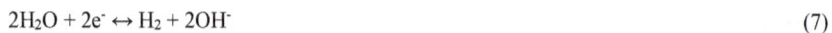
$2H_2O + 2e^- \leftrightarrow H_2 + 2OH^-$ (7)

$X^*$- catalyst with active sites, * active site

The HER usually takes place at an electrode coated with a catalyst (palladium, platinum, or other transition metals), lowering the reaction's energy barrier and leading to more efficient hydrogen production. The HER rate is dependent on several variables, including the electrode material, temperature, pH of the solution, and the presence of any catalyst or additives. The reaction rate generally increases at higher temperatures and lower pH (more acidic conditions). But using Pt, and Pd are very cost-effective and cannot be easily affordable. Hence there is a need for alternate metals replacing the traditional catalysts. Earth-abundant metals however can be a good choice for their use as electrocatalysts in HER reaction.

## 3.1 Electrochemical

An economical and sustainable approach for producing hydrogen at low cost is a water-splitting reaction. However, to produce hydrogen, an affordable, effective, and stable electrocatalyst is needed to reduce overpotential. For the HER process, transition metal sulfides, phosphides, carbides, and selenides have demonstrated strong activity at room temperature and pressure. In his study, Y.Du reports an easy method for utilizing transition metal oxides as precursors to create a variety of CoP-MnP, CoP, and CoP-FeP, nanostructures and synthesis in Fig 3. Fe and Mn doping can modify the electrical structure of P and Co centers in varying degrees, influencing HER nanostructures and electrocatalytic activity. CoP's morphology has correspondingly altered from cylinder-like to microflower-like, due to doping with Fe or Mn atoms. Fe salts stimulate the evolution of FeP, forming a structure resembling a microflower with increased catalytic regions. Because of their characteristic nanostructure and optimized electronic structure, CoP-FeP and CoP-MnP exhibit higher electrochemical activity for HER in 0.5 M $H_2SO_4$, 1.0 M PBS, and 1.0 M KOH when compared to pure CoP. It has been described how to improve HER performance on CoP-FeP and CoP-MnP, which is helpful when designing and manufacturing stable, high-performance transition metal phosphide-supported electrocatalysts over all pH HER in a controlled manner [44].

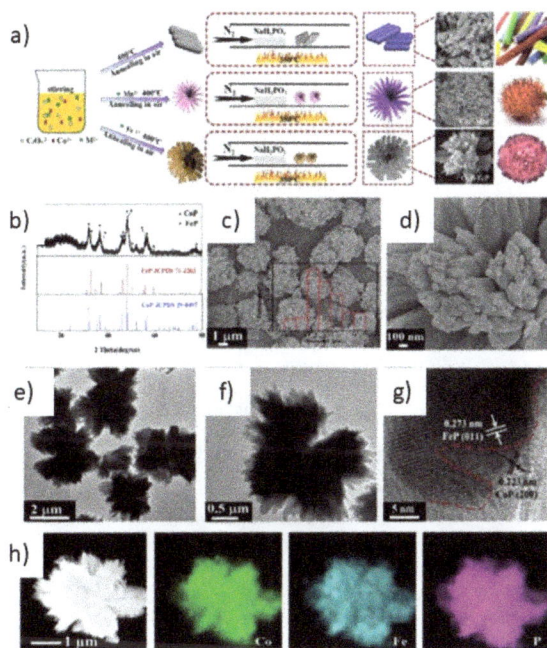

Fig 3. a) The synthesis mechanism showing various nanostructures b) XRD pattern, c-d) SEM images e-f) TEM images. g) HR-TEM image and h) Elemental mapping of the CoP-FeP. Reprinted and took permission from Elsevier [44].

Wu et al. demonstrated the synthesis of a porous molybdenum carbide nano-octahedron using a Cu-supported MOF [HKUST-1; $Cu_3(BTC)_2(H_2O)_3$]. The MOF's porous network contained guest polyoxometalate units ($H_3PMo_{12}O_{40}$) [45]. The designation for this MOF precursor is NENU-5 with the formula of $[Cu_2(BTC)_{4/3}(H_2O)_2]_6[H_3PMo_{12}O_{40}]$. After NENU-5 is directly pyrolyzed in $N_2$ gas flow, molybdenum carbide ($MoC_x$) and elemental Cu are formed and embedded in an amorphous carbon matrix. Cu is eliminated using a $FeCl_3$ solution, leaving $MoC_x$ nano-octahedrons for electrocatalytic HER in Fig 4. $MoC_x$'s electrocatalytic efficacy in HER is evaluated in electrolytes that are both acidic and alkaline. A 142 mV overpotential and a corresponding Tafel slope of 53 mV dec$^{-1}$ are needed to achieve a current density of 10 mA cm-2 in 0.5 M $H_2SO_4$. These values are 151 mV and 59 mV dec$^{-1}$ in 1M KOH. Consequently, the catalyst exhibits exceptional stability in both alkaline and acidic electrolytes.

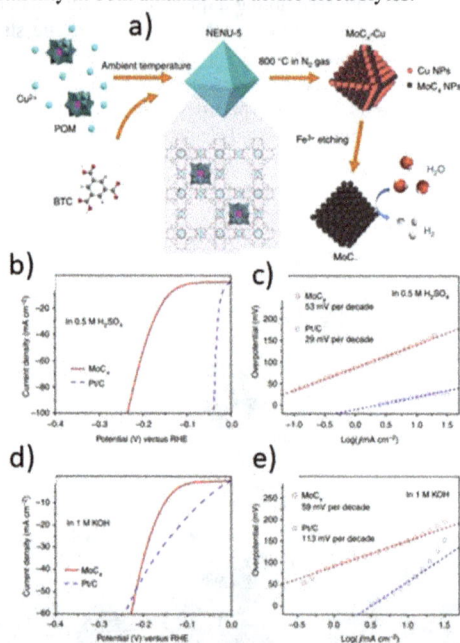

*Fig 4. a) The synthesis of NENU-5 nano-octahedrons with Mo-based POMs in the pores of the HKUST-1 host. Formation of $MoC_x$-Cu nano-octahedrons after 800°C annealing. Using $Fe^{+3}$ etching to remove metallic Cu nanoparticles and create porous $MoC_x$ nano-octahedrons for electrocatalytic hydrogen generation. b) Polarization curve at 2 mVs$^{-1}$ and c) Tafel plots in 0.5M $H_2SO_4$, d) polarization curve at 2 mVs$^{-1}$, e) Tafel plots in 1M KOH. Reprinted and took permission from Springer [45]. Zhang et al. introduced catalytically activated POMs ($PMo_{12}$) into MOF (Co-BTC) using a solvent-thermal method as shown in Fig 5., They produced $CoMoO_4$ hollow tubes ($CoMoO_4$ HTs) as an electrocatalyst, outperforming most cobalt catalysts in HER and OER activity and stability in alkaline electrolytes [46].*

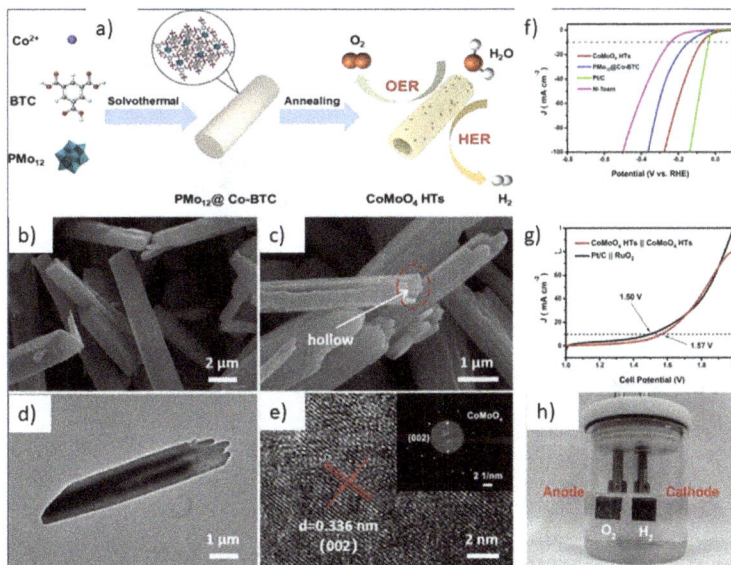

*Fig 5. a) Diagrammatic representation of the CoMoO₄ HTs synthesis process, b,c) SEM images of PMo 12@Co-BTC and CoMoO₄ HTs, d) TEM image, e) HRTEM and SAED pattern (inset). f) LSV curves for Ni foam in 1.0 M KOH, Pt/C, PMo 12@Co-BTC, and CoMoO4 HTs for the HER. g) The catalysts LSV curves for total water splitting in 1.0 M KOH, and h) An electronic image of the electrolytic cell apparatus. Reprinted and took permission from Elsevier [46].*

Breaking the HER's technological barrier requires designing and developing high-performance, inexpensive Pt-free electrode materials supported on transition metals that are abundant on Earth. B. Shen et al. reported a stable and practical bottom-up approach to the co-assembly process-based controlled synthesis of 3D interweaving $Ti_3C_2T_x$ MXene-reduced graphene oxide network-confined Ni-Fe layered double hydroxide nanosheets (LDH/MX-RGO). Fig.6a shows the synthesis of LDH/MX-RGO. Numerous advantageous textural qualities, including great accessible surface areas, copious porosity, plenty of bare active centers, optimum electronic structure, and sufficient electron-conducting pathways, are provided by this newly designed configuration [47]. Thus, compared to bare LDHs, $Ti_3C_2T_x$, RGO, as well binary LDH/MX, and LDH/RGO catalysts, the as-derived LDH/MXRGO catalyst with an adequate LDH concentration (60% wt.) demonstrates substantially higher HER electrocatalytic capabilities.

Materials Research Forum LLC
https://doi.org/21741/9781644903711-11

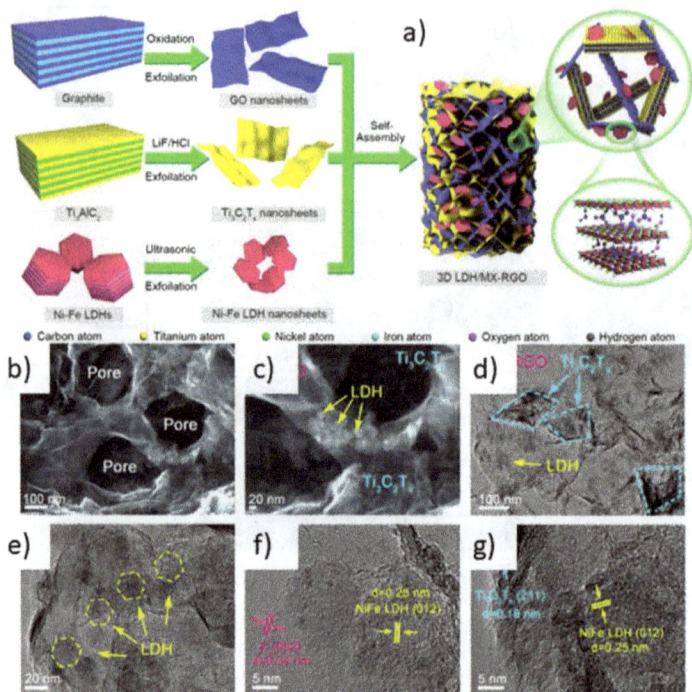

*Fig 6. a) Diagram illustrating the 3D LDH/MX-RGO nanoarchitecture's stereo assembly, Images of LDH/MX-RGO nanoarchitecture are shown in b-c) FE-SEM, d-e) TEM and f-g) HR-TEM. Reprinted and took permission from Elsevier [47]*

### 3.1.1 Acidic medium

The development of electrocatalysts based on earth-abundant transition metals for HER and general water splitting has recently received much attention. In acidic media, the HER is vital for electrochemical reactions like fuel cells and water electrolysis. Because there are more protons ($H^+$) in this medium than in alkaline media, the proton transfer process is simpler and kinetics are faster. Pt is widely employed in proton exchange membrane (PEM) fuel cells and electrolyzers because it is an extremely effective catalyst in acidic environments. Sustainable hydrogen production requires HER in acidic conditions, yet, obstacles like electrode corrosion and the expensive price of precious metals like Pt continue to be major hindrances to broader use.

### 3.1.2 Alkali medium

Though it has slower kinetics than in acidic media because water serves as the proton source, the HER in an alkaline medium is essential for electrochemical hydrogen production. The reaction becomes more complex as a result of the extra energy needed for water dissociation. Because platinum loses its effectiveness in alkaline environments, Ni, Mo, and Co are being investigated as potential substitutes. Effective catalyst design is therefore essential. Applications like alkaline water electrolysis and fuel cells, which are necessary for the sustained production of hydrogen, depend on HER in alkaline media. It is further highlighted by the impact that $OH^-$ ion adsorption has on charge distribution and reaction kinetics on electrode surfaces. This procedure is essential for developing affordable, abundant materials for large-scale hydrogen generation and promoting sustainable energy solutions. Ni nanoparticles with various morphologies were produced by S. A. Abbas et al. for HER in an alkaline solution. Here, simple alcohols and oleylamine (OAm) underwent solvothermal reactions to transform $Ni(acac)_2$ into Ni metal nanoparticles. The shape of the resultant Ni nanoparticles was mostly influenced by the alcohol solvent's OAm/Ni molar ratio. On top of the multi-pod Ni particles created by $Ni(OH)_2$/NiO aggregates, echinoid Ni nanoparticles were made through the addition of an atom. Under the same reaction conditions, weblike feet particles ($NiIPA$-$OAm_4$) with plate borders were also seen in isopropanol. In an alkaline solution, the produced Ni nanoparticle's catalytic activity for HER was assessed. Compared to the $NiIPA$-$OAm_4$ the $NiET$-$OAm_4$ with urchin-like morphology was significantly more active. An increase in the OAm/Ni molar ratio resulted in a corresponding rise in the exchange current density of Ni catalysts. The $NiEt$-$OAm_4$ outperformed commercial Pt catalysts. In electrocatalytic performance during a stability test lasting 100 kiloseconds at 1.5 V (vs Hg/HgO) in 1.0 M NaOH due to its high stability [38]. According to R. Wang et al. study the preparation of Co-NC composites with a distinct hierarchical structure at various temperatures was achieved by using the 2D leaf-like Co-MOFs as the precursor Fig. 7c. To guarantee a large surface area, the stereoscopic holes were first evenly placed over the entire carbon sheet. Second, the cobalt nanoparticles were firmly fixed on the N-carbon and the resulting composites maintained the distinct structure of the original 2D Co-based MOFs. The catalytic active sites are formed by the combination of cobalt species with N-carbon, and the resulting Co-NC materials exhibit increased HER activity in both acidic and alkaline conditions [48].

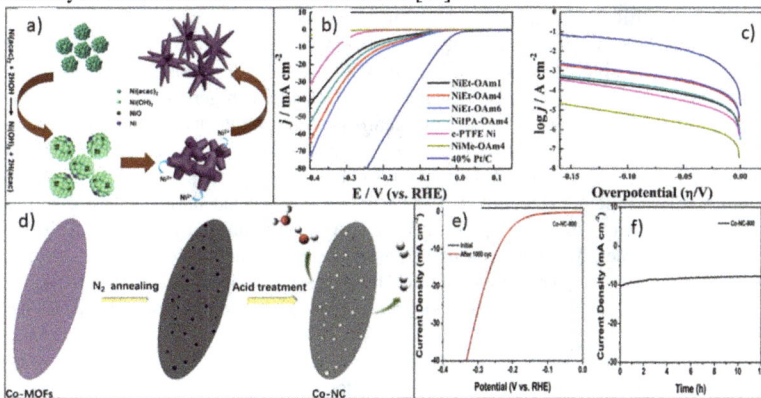

Fig 7. a) Schematic illustration of Ni nanoparticles, b) Linear sweep voltammograms of Ni nanoparticles for the HER demonstrated catalytic activity at a scan rate of 50 mV s-1, and c) Tafel slopes linear sweep at a scan rate of 1 mV s⁻¹. Reprinted and took permission from Elsevier [38, 48].

### 3.2 Photochemical

The development of solar-to-hydrogen technology relies heavily on photochemical hydrogen evolution, which provides a sustainable method of producing hydrogen fuel for use in fuel cells, industrial operations, and transportation. There are two types of processes in photochemical reactions based on their source of energy either electric or solar power.

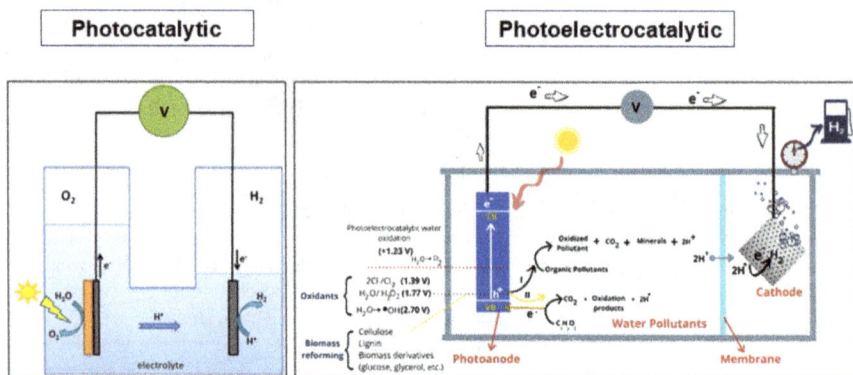

*Fig 8. Graphical representation of photocatalytic and photoelectrocatalytic. Reprinted with permission from MDPI [49, 50].*

### 3.2.1 Photocatalytic

In photocatalytic HER, light energy (usually solar energy) is used to produce hydrogen gas ($H_2$) from water in the presence of a photocatalyst. Remarkably, nickel-based catalysts, such as NIFeP, $Ni_2P$, NiFe-layered double hydroxide, and Ni/NiO/CNT are more desirable for water splitting [51]. This approach is critical to creating sustainable and clean hydrogen fuel generation. Finding effective and stable photocatalysts that can efficiently absorb sunlight and enhance the HER, which happens when water splits into hydrogen and oxygen, is a common topic of study in this field. K. Uma et al. successfully created ZnO nanorods/$\alpha$-$Fe_2O_3$ composites using the hydrothermal technique. With a limited evolution of hydrogen, the synthesized photocatalyst exhibits high photocatalytic efficiency. Comparing the ZnO/$\alpha$-$Fe_2O_3$ materials, they provide the best HER results. With the successful separation and transfer of photogenerated electrons as well as the avoidance of electron-hole pair recombination. The electrolyte's penetration and diffusion can be aided by the three-dimensional networked structure. ZnO/$\alpha$-$Fe_2O_3$ on Ni foam electrodes perform significantly better for HER with a low onset overpotential (-125 mV), according to electrochemical tests. The obtained results will enable the production of inexpensive materials-based ZnO-based hybrid devices [41]. X Liu et al. successfully synthesized a metallic $Ni_5P_4$ with a carnation like hierarchical morphology and a porous structure by employing $Ni(OH)_2$ that resembled flowers as the precursor. The synthesized $Ni_5P_4$ was then employed as a very effective cocatalyst to increase the performance of the $H_2$ production process over HCN and the photocatalytic water-splitting reaction. Exceptional photocatalytic $H_2$ generation performance of 1157.5 $\mu molg^{-1}h^{-1}$ and high EQE of 8.4% at 420 nm were demonstrated by the HCN/1%$Ni_5P_4$

hybrid. The hybrid including $Ni_2P$ nanoflowers and $NiS_2$ aggregates as cocatalysts (161.9 and 593.1 $\mu molg^{-1}h^{-1}$, correspondingly) was produced from the similar $Ni(OH)_2$ precursor as the $HCN/1\%Ni_5P_4$ hybrid, which had a substantially greater photocatalytic $H_2$ generation activity [52].

### 3.2.2 Photoelectrocatalytic

The goal of photoelectrocatalytic (PEC) processes is to improve chemical reactions by applying an electric field and light. PEC procedures integrate the concepts of electrocatalysis with photocatalysis. In photo electrocatalysis, a semiconductor substance absorbs light to produce electron-hole pairs. Subsequently, these charge carriers drive redox reactions at the electrode surfaces, achieving the intended chemical reaction. By utilizing renewable solar energy, this technique provides a clean hydrogen fuel source. PEC provides green hydrogen and renewable energy but, prices associated with materials, durability, and efficiency are challenges.

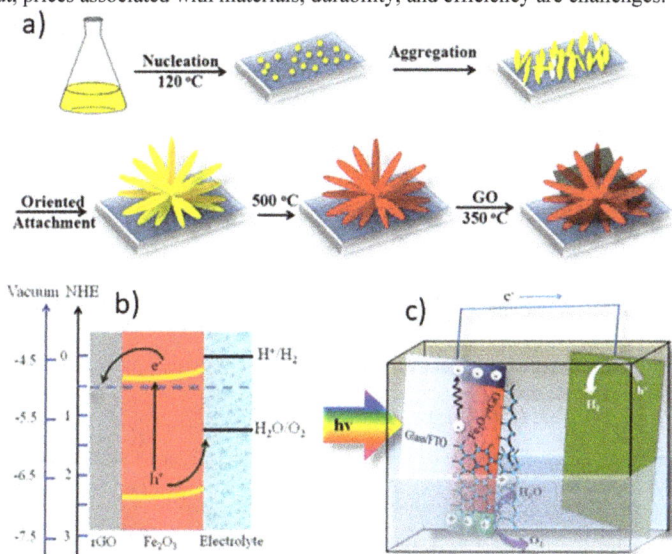

Fig 9. a) Diagram showing the general synthesis pathway for the Fe2O3-rGO hybrid photoanode in 3D, resemblilng an urchin, b) energy band matching and c) photoelectrochemical cell using the Fe2O3-rGO photoanode. Reprinted and took permission from Elsevier [53].

A.G. Tamirat et al. [53] used a simple synthetic method to merge an ultra-thin rGO sheet with a 3D urchin-like nanostructured $Fe_2O_3$ to create an effective water-splitting photoanode. The synthesis of Fe2O3-rGO ws shown in Fig. 9a. When present in ideal quantity, the ultra-thin rGO sheet functions as a surface passivation layer to speed up the sluggish water oxidation reaction and as a conducting scaffold to absorb photogenerated electrons from the host $Fe_2O_3$ photocatalyst and so reduce electron-hole recombination. They conducted PEC experiments in the existence of a 0.5 M $H_2O_2$ hole scavenger to examine the duple effect of rGO on the PEC activity of $Fe_2O_3$, which provides direct proof of the charge separation and injection capabilities.

Materials Research Forum LLC
https://doi.org/21741/9781644903711-11

## Conclusion

The importance of sustainable and affordable catalysts in the HER is examined in the chapter. The significance of EAMs as substitutes for precious metal catalysts is first highlighted, especially in light of the relevance of hydrogen energy for the fuel industry's future. The chapter then delves into various synthesis techniques, techniques for green synthesis that prioritize environmental sustainability are also covered. The chapter also focuses on HER. It covers electrochemical reactions in alkaline and acidic media, also photochemical methods like photocatalytic and photo electrocatalytic reactions. Finally, because EAM based catalysts are accessible, affordable, and have catalytic activity comparable to that of precious metals, they offer a potential option for large-scale hydrogen production. To optimize these catalysts and expand the hydrogen energy landscape, sophisticated synthesis techniques and a deeper comprehension of HER processes in various settings are needed.

## References

[1] U. Shankar, A. Basu, Energy and Sustainability, In Oxford University Press eBooks 2024 pp. 767-782. https://doi.org/10.1093/oxfordhb/9780198884682.013.43

[2] F. Martins, C. Felgueiras, M. Smitkova, N. Caetano, Analysis of Fossil Fuel Energy Consumption and Environmental Impacts in European Countries, Energies, 2019, 12(6), pp. 964. https://doi.org/10.3390/en12060964

[3] G. Shi, T. Tano, D. A. Tryk, A. Iiyama, M. Uchida, K. Terao, H. Osada, M. Yamaguchi, K. Tamoto, K. Kakinuma, Nanostructured Pt-NiFe Oxide Catalyst for Hydrogen Evolution Reaction in Alkaline Electrolyte Membrane Water Electrolyzers, ACS Catalysis, 2024, 14(12), pp. 9460-9468. https://doi.org/10.1021/acscatal.4c01685

[4] S. Xie, H. Dong, X. Peng, P.K. Chu, Non-precious Electrocatalysts for the Hydrogen Evolution Reaction, Innovation discovery, 2024, 1(2), pp. 11. https://doi.org/10.53964/id.2024011

[5] G.K. Gebremariam, A.Z. Jovanović, I.A. Pašti, The Effect of Electrolytes on the Kinetics of the Hydrogen Evolution Reaction. Hydrogen, 2023, 4(4), pp. 776-806. https://doi.org/10.3390/hydrogen4040049

[6] Z. Hou, Y. Guo, F. Chen, X. Chi, J. Jiao, W. Zhang, S. Zhou, Z. Hu, Performance analysis of hydrogen storage systems with oxygen recuperation for intermittent renewable energy generation system, Journal of Energy Storage, 2024, 86, 111150. https://doi.org/10.1016/j.est.2024.111150

[7] Z. Xie, Q. Jin, G. Su, W. Lu, A Review of Hydrogen Storage and Transportation: Progresses and Challenges. Energies, 2024 17(16), 4070. https://doi.org/10.3390/en17164070

[8] M. Younas, M. Yar, H. AlMohamadi, T. Mahmood, K. Ayub, A.L. Khan, M. Yasin, M.A. Gilani, A rational design of covalent organic framework supported single atom catalysts for hydrogen evolution reaction: A DFT study. International Journal of Hydrogen Energy, 2023, 51, pp. 758-773. https://doi.org/10.1016/j.ijhydene.2023.07.062

[9] Y. Wang, T. Wang, H. Arandiyan, G. Song, H. Sun, Y. Sabri, C. Zhao, Z. Shao, S. Kawi, Advancing Catalysts by Stacking Fault Defects for Enhanced Hydrogen Production: A Review. Advanced Materials, 2024 https://doi.org/10.1002/adma.202313378

[10] A. Bazan, G. García, A.M. Baena-Moncada, E. Pastor, Ni Foam-Supported NiMo Catalysts for the HER, Electrochemical Society, MA2022-0, 2022, 1(34), 1390. https://doi.org/10.1149/MA2022-01341390mtgabs

[11] X. Gao, S. Dai, Y. Teng, Q. Wang, Z. Zhang, Z. Yang, M. Park, H. Wang, Z. Jia, Y. Wang, Y. Yang, Ultra-Efficient and Cost-Effective Platinum Nanomembrane Electrocatalyst for Sustainable Hydrogen Production. Nano-Micro Letters, 2024, 16(1). https://doi.org/10.1007/s40820-024-01324-5

[12] H. Niu, Q. Wang, C. Huang, M. Zhang, Y. Yan, T. Liu, W. Zhou, Noble Metal-Based Heterogeneous Catalysts for Electrochemical Hydrogen Evolution Reaction, Applied Sciences, 2023, 13(4), 2177 https://doi.org/10.3390/app13042177

[13] K. Cheng, L.C.J. Smulders, L.L. Van Der Wal, J. Oenema, J.D. Meeldijk, N.L. Visser, G. Sunley, T. Roberts, Z. Xu, E. Doskocil, H. Yoshida, Y. Zheng, J. Zečević, P. De Jongh, K.P. De Jong, Maximizing noble metal utilization in solid catalysts by control of nanoparticle location. Science, 2022, 377(6602), 204-208. https://doi.org/10.1126/science.abn8289

[14] J. Yu, F.G. Pineda, J.R. Galan-Mascaros, Earth-Abundant Metal Oxides as Oxygen Evolution Electrocatalysts in 1 M H2SO4, Meeting Abstracts (Electrochemical Society. CD-ROM), 2024, MA2024-01(34), 1782. https://doi.org/10.1149/MA2024-01341782mtgabs

[15] G. Xu, C. Cai, W. Zhao, Y. Liu, T. Wang, Rational design of catalysts with earth-abundant elements. Wiley Interdisciplinary Reviews Computational Molecular Science, 2022, 13(4). https://doi.org/10.1002/wcms.1654

[16] P. Wang, J. Zheng, X. Xu, Y. Zhang, Q. Shi, Y. Wan, S. Ramakrishna, J. Zhang, L. Zhu, T. Yokoshima, Y. Yamauchi, Y. Long, Unlocking Efficient Hydrogen Production: Nucleophilic Oxidation Reactions Coupled with Water Splitting. Advanced Materials. 2024, 36, 2404806. https://doi.org/10.1002/adma.202404806

[17] N. Giddaerappa, N.N. Kousar, N.S. Kumar, M. Hojamberdiev, L.K. Sannegowda, Substrate-Driven Electrocatalysis of Natural and Earth-Abundant Pyrite Towards Oxygen Evolution Reaction. Electrochimica Acta, 2023, 475, 143575. https://doi.org/10.1016/j.electacta.2023.143575

[18] H. Nemamcha, N. Vu, D.S. Tran, C. Boisvert, D.D. Nguyen, P. Nguyen-Tri, Recent progression in MXene-based catalysts for emerging photocatalytic applications of CO2 reduction and H2 production: A review, The Science of the Total Environment, 2024, 172816. https://doi.org/10.1016/j.scitotenv.2024.172816

[19] Y. Lan, H. He, C. Liu, J. Qin, L. Luo, F. Zhu, Y. Zhao, J. Zhang, L. Yang, H. Huang, Ultrasmall Pd nanocrystals confined into Co-based metal organic framework-decorated MXene nanoarchitectures for efficient methanol electrooxidation, Journal of Power Sources, 2024, 603, 234438. https://doi.org/10.1016/j.jpowsour.2024.234438

[20] P. Rodrigues, H. Bangali, E.A.M. Saleh, S.R. Hamza, B.S. Mirzaev, F. Ghali, B.M. Hussien, S.B. Hussein, R.T. Habash, Y.F. Mustafa, Metal-organic framework/MXene

nanohybrid composites as an emerging electrochemical sensing platform for food safety and biomedical monitoring: From synthesis to application, Electrochimica Acta, 2024, 494, 144424. https://doi.org/10.1016/j.electacta.2024.144424

[21] N. Kitchamsetti, J.S. Cho, A roadmap of recent advances in MXene@MOF hybrids, its derived composites: Synthesis, properties, and their utilization as an electrode for supercapacitors, rechargeable batteries and electrocatalysis, Journal of Energy Storage, 2024, 80, 110293. https://doi.org/10.1016/j.est.2023.110293

[22] Z. Sun, R. Wang, V.E. Matulis, K. Vladimir, K. Structure, Synthesis, and Catalytic Performance of Emerging MXene-Based Catalysts, Molecules, 2024, 29(6), 1286. https://doi.org/10.3390/molecules29061286

[23] G. Maduraiveeran, M. Sasidharan, W. Jin, Earth-abundant transition metal and metal oxide nanomaterials: Synthesis and electrochemical applications, Progress in Materials Science, 2019, 106, 100574. https://doi.org/10.1016/j.pmatsci.2019.100574

[24] M.A. Khan, S. Khan, S. Sengupta, B.N. Mongal, S. Naskar, Earth abundant transition metal complexes as molecular water oxidation catalysts, Coordination Chemistry Reviews, 2024, 504, 215679. https://doi.org/10.1016/j.ccr.2024.215679

[25] P.J. Chirik, K.M. Engle, E.M. Simmons, S.R. Wisniewski, Collaboration as a Key to Advance Capabilities for Earth-Abundant Metal Catalysis. Organic Process Research & Development, 2023, 27(7), pp.1160-1184. https://doi.org/10.1021/acs.oprd.3c00025

[26] S. Kment, A. Bakandritsos, I. Tantis, H. Kmentová, Y. Zuo, O. Henrotte, A. Naldoni, M. Otyepka, R.S. Varma, R. Zbořil, Single Atom Catalysts Based on Earth-Abundant Metals for Energy-Related Applications. Chemical Reviews, 2024. https://doi.org/10.1021/acs.chemrev.4c00155

[27] A.M. Sadeq, R.Z. Homod, A.K. Hussein, H. Togun, A. Mahmoodi, H.F. Isleem, A.R. Patil, A.H. Moghaddam, Hydrogen energy systems: Technologies, trends, and future prospects. The Science of the Total Environment, 2024, 939, 173622. https://doi.org/10.1016/j.scitotenv.2024.173622

[28] M.A. Habib, G.A. Abdulrahman, A.B. Alquaity, N.A. Qasem, Hydrogen combustion, production, and applications: A review. Alexandria Engineering Journal, 2024, 100, 182-207. https://doi.org/10.1016/j.aej.2024.05.030

[29] S.E. Hosseini, Hydrogen fuel, a game changer for the world's energy scenario. International Journal of Green Energy, 2023, 21(6), pp.1366-1382. https://doi.org/10.1080/15435075.2023.2244050

[30] N. Victor, C. Nichols, Future of Hydrogen in the U.S. Energy Sector: MARKAL Modeling Results, Applications in Energy and Combustion Science, 2024, 100259. https://doi.org/10.1016/j.jaecs.2024.100259

[31] R.M. Bullock, J.G. Chen, L. Gagliardi, P.J. Chirik, O.K. Farha, C.H. Hendon, C.W. Jones, J.A. Keith, J. Klosin, S.D. Minteer, R.H. Morris, A.T. Radosevich, T.B. Rauchfuss, N.A. Strotman, A. Vojvodic, T.R. Ward, J.Y. Yang, Y. Surendranath, Using nature's blueprint to expand catalysis with Earth-abundant metals. Science, 2020, 369(6505). https://doi.org/10.1126/science.abc3183

[32] F. Wu, Z. Chen, H. Wu, F. Xiao, S. Du, C. He, Y. Wu, Z. Ren, In Situ Catalytic Etching Strategy Promoted Synthesis of Carbon Nanotube Inlaid with Ultrasmall FeP Nanoparticles as Efficient Electrocatalyst for Hydrogen Evolution, ACS Sustainable Chemistry & Engineering, 2019, 7(15), pp.12741-12749. https://doi.org/10.1021/acssuschemeng.9b00998

[33] A. Ambrosi, X. Chia, Z. Sofer, M. Pumera, Enhancement of electrochemical and catalytic properties of MoS2 through ball-milling, Electrochemistry Communications, 2015, 54, pp.36-40. https://doi.org/10.1016/j.elecom.2015.02.017

[34] L. Zhang, L. Ji, Z. Yao, N. Yan, Z. Sun, X. Yang, X. Zhu, S. Hu, L. Chen, Facile synthesized Fe nanosheets as superior active catalyst for hydrogen storage in MgH2, International Journal of Hydrogen Energy, 2019, 44(39), pp.21955-21964. https://doi.org/10.1016/j.ijhydene.2019.06.065

[35] F. Davodi, E. Mühlhausen, D. Settipani, E. Rautama, A. Honkanen, S. Huotari, G. Marzun, P. Taskinen, T. Kallio, T. Comprehensive study to design advanced metal-carbide@garaphene and metal-carbide@iron oxide nanoparticles with tunable structure by the laser ablation in liquid, Journal of Colloid and Interface Science, 2019, 556, pp.180-192. https://doi.org/10.1016/j.jcis.2019.08.056

[36] L. Zhang, K. Xiong, Y. Nie, X. Wang, J. Liao, Z. Wei, Sputtering nickel-molybdenum nanorods as an excellent hydrogen evolution reaction catalyst, Journal of Power Sources, 2015, 297, pp.413-418. https://doi.org/10.1016/j.jpowsour.2015.08.004

[37] B. Gao, X. Du, Y. Ma, Y. Li, Y. Li, S. Ding, Z. Song, C. Xiao, 3D flower-like defected MoS2 magnetron-sputtered on candle soot for enhanced hydrogen evolution reaction. Applied Catalysis B Environment and Energy, 2019, 263, 117750. https://doi.org/10.1016/j.apcatb.2019.117750

[38] S.A. Abbas, M.I. Iqbal, S. Kim, K. Jung, K. Catalytic Activity of Urchin-like Ni nanoparticles Prepared by Solvothermal Method for Hydrogen Evolution Reaction in Alkaline Solution, Electrochimica Acta, 2017, 227, pp. 382-390. https://doi.org/10.1016/j.electacta.2017.01.039

[39] B.T. Jebaslinhepzybai, T. Partheeban, D.S. Gavali, R. Thapa, M. Sasidharan, One-pot solvothermal synthesis of Co2P nanoparticles: An efficient HER and OER electrocatalysts, International Journal of Hydrogen Energy, 2021, 46(42), pp. 21924-21938. https://doi.org/10.1016/j.ijhydene.2021.04.022

[40] D. Chanda, J. Hnát, M. M. Paidar, J. Schauer, K. Bouzek, Synthesis and characterization of NiFe2O4 electrocatalyst for the hydrogen evolution reaction in alkaline water electrolysis using different polymer binders, Journal of Power Sources, 2015, 285, pp. 217-226. https://doi.org/10.1016/j.jpowsour.2015.03.067

[41] K. Uma, E. Muniranthinam, S. Chong, T.C. Yang, J. Lin, Fabrication of Hybrid Catalyst ZnO Nanorod/α-Fe2O3 Composites for Hydrogen Evolution Reaction, Crystals, 2020, 10(5), 356. https://doi.org/10.3390/cryst10050356

[42] R. Miao, B. Dutta, S. Sahoo, J. He, W. Zhong, S.A. Cetegen, T. Jiang, S.P. Alpay, S.L. Suib, Mesoporous Iron Sulfide for Highly Efficient Electrocatalytic Hydrogen Evolution. Journal of the American Chemical Society, 2017, 139(39), pp. 13604-13607. https://doi.org/10.1021/jacs.7b07044

[43] P. Yu, F. Wang, T.A. Shifa, X. Zhan, X. Lou, F. Xia, J. He, Earth abundant materials beyond transition metal dichalcogenides: A focus on electrocatalyzing hydrogen evolution reaction. Nano Energy, 2019, 58, pp. 244-276. https://doi.org/10.1016/j.nanoen.2019.01.017

[44] Y. Du, Z. Wang, H. Li, Y. Han, Y. Liu, Y. Yang, Y. Liu, L. Wang, Controllable synthesized CoP-MP (M=Fe, Mn) as efficient and stable electrocatalyst for hydrogen evolution reaction at all pH values, International Journal of Hydrogen Energy, 2019, 44(36), pp. 19978-19985. https://doi.org/10.1016/j.ijhydene.2019.06.036

[45] H.B. Wu, B.Y. Xia, L. Yu, X. Yu, X.W. Lou, Porous molybdenum carbide nano-octahedrons synthesized via confined carburization in metal-organic frameworks for efficient hydrogen production. Nature Communications, 2015, 6(1). https://doi.org/10.1038/ncomms7512

[46] Z. Zhang, J. Ran, E. Fan, S. Zhou, D. Chai, W. Zhang, M. Zhao, G. Dong, Mesoporous CoMoO4 hollow tubes derived from POMOFs as efficient electrocatalyst for overall water splitting, Journal of Alloys and Compounds, 2023, 968, 172169. https://doi.org/10.1016/j.jallcom.2023.172169

[47] B. Shen, H. Huang, Y. Jiang, Y. Xue, H. He, 3D interweaving MXene-graphene network-confined Ni-Fe layered double hydroxide nanosheets for enhanced hydrogen evolution, Electrochimica Acta, 2022, 407, 139913. https://doi.org/10.1016/j.electacta.2022.139913

[48] R. Wang, P. Sun, Q. Yuan, R. Nie, X. Wang, MOF-derived cobalt-embedded nitrogen-doped mesoporous carbon leaf for efficient hydrogen evolution reaction in both acidic and alkaline media. International Journal of Hydrogen Energy, 2019, 44(23), pp.11838-11847. https://doi.org/10.1016/j.ijhydene.2019.03.029

[49] C. Liao, C. Huang, J.C.S. Wu, Hydrogen Production from Semiconductor-based Photocatalysis via Water Splitting. Catalysts, 2012, 2(4), pp. 490-516. https://doi.org/10.3390/catal2040490

[50] S. Pitchaimuthu, K. Sridharan, S. Nagarajan, S. Ananthraj, P. Robertson, M.F. Kuehnel, A. Irabien, M. Maroto-Valer, Solar Hydrogen Fuel Generation from Wastewater-Beyond Photoelectrochemical Water Splitting: A Perspective, Energies, 2022, 15(19), 7399. https://doi.org/10.3390/en15197399

[51] B.M. Hunter, W. Hieringer, J.R. Winkler, H.B. Gray, A.M. Müller, Effect of interlayer anions on [NiFe]-LDH nanosheet water oxidation activity. Energy & Environmental Science, 2016, 9(5), pp. 1734-1743. https://doi.org/10.1039/C6EE00377J

[52] X. Liu, Y. Zhao, X. Yang, Q. Liu, X. Yu, X. Li, H. Tang, T. Zhang, Porous Ni5P4 as a promising cocatalyst for boosting the photocatalytic hydrogen evolution reaction performance, Applied Catalysis B Environment and Energy, 2020, 275, 119144. https://doi.org/10.1016/j.apcatb.2020.119144

[53] A.G. Tamirat, W. Su, A.A. Dubale, C. Pan, H. Chen, D.W. Ayele, J. Lee, B. Hwang, (2015). Efficient photoelectrochemical water splitting using three dimensional urchin-like hematite nanostructure modified with reduced graphene oxide. Journal of Power Sources, 2015, 287, pp.119-128. https://doi.org/10.1016/j.jpowsour.2015.04.042

Applications for Earth-Abundant Transition Metals
Materials Research Foundations 179 (2025) 339-355

Materials Research Forum LLC
https://doi.org/21741/9781644903711-12

# Chapter 12

# Earth-Abundant Metals in Chemical Industries

Puspendu Choudhury, Mriganka Sekhar Manna, Soma Nag*

Department of Chemical Engineering, National Institute of Technology Agartala, Tripura-
799046, India

*somanag.nita@gmail.com

## Abstract

Earth-abundant metals, *viz.*, iron, cobalt, nickel, manganese, copper, zinc, *etc.* have provided great industrial and economic growth for prolonged time. The substantial availability of these metals in the Earth's crust and the lower costs of extraction trigger widespread applications of them in civil engineering, chemical and manufacturing industries, and the energy sector. The pivotal role in various applications, in manufacturing materials, chemicals, fertilizers, pesticides, biological activities, catalysis, and energy storage, has become evident through significant modifications and chemical transformations of these metals. The recent advances in catalysis, synthesis of new chemicals, applications as biosensors and cofactors of crucial biomolecules, green chemistry, and energy storage have been illustrated in this chapter.

**Keywords**

Extraction, Ores, Catalysis, Transition Metal, Phytoremediation, Alloy

## Contents

## 1.      Introduction

Earth-abundant metals exist in significant amounts on the Earth's crust to be available for, viable and sustainable for industrial developments. The extraction of these metals is easier due to their abundance in the respective ores. On the other hand, the rare and precious metals are often costly due to difficulty in extraction. The abundant metals include iron, cobalt, nickel, manganese, copper, sodium, potassium, zinc, etc. The widespread abundance leads to their inexpensiveness as well as minimizes the ecological footprint connected with their several processing steps such as mining, extraction, and refining. The metals have found their applications in various fields such as structural materials, civil engineering, manufacturing industries, etc. The increasing uses of many abundant metals in catalysis and energy storage have brought about the industrial revolution in chemical engineering, manufacturing industries and energy sector. The one of the most abundant metals irons is mainly used in various structures in the field of civil engineering. But, it has a very crucial role as the catalyst in the manufacturing of ammonia which is again the one of the essential raw materials for urea production. Another common abundant metal is nickel, and it has the role in the transformation of iron for different mechanical properties of it. However, nickel has a crucial role in the hydrogenation reactions for the transformation of oils for better use. Nickel is also used in the cross-coupling processes. On the other hand, cobalt effectively plays a role in hydro-functionalization while manganese promotes oxidation reactions.

The earth-abundant metals have versatile applications including the catalysis for facilitating complex transformation through heterogenic catalytic chemical reactions. One of the sub-classifications of earth-abundant metals is the group of transitional metals. They are the chemical elements in the $d$-block of the periodic table (Group 3-12). However, the elements of groups 3-12 are often not considered. Moreover, inner transition metals like actinide or lanthanide (the $f$-block) may be considered as transition metals. The twenty-four transition elements are metals. These metals are glossy and have good thermal and electrical conductivities. Most of them are hard and strong except for the copper family. They are strong refractory and electro positive and have high density. The transition elements have high melting and boiling points (except groups 11 and 12). Several of them have boiling point above 2000 °C and three (Ta, W and Re) of them have boiling point nearly 3000 °C. They have more than one oxidation state to be bound to several ligands to form colored coordination complexes. Due to the presence of valence electrons, they can take part in bond formation in two shells instead of one. Various alloys are formed to be used as catalysts either in various forms such as elements, oxides, and complexes. Most of them are strongly paramagnetic due to the presence of unpaired $d$-electrons. These metals (iron, cobalt and nickel, gadolinium) are ferromagnetic near room temperature. Metals such as titanium, iron, nickel, and copper are used structurally and in electrical technology. The transition metals form many useful alloys with one another and with other metallic elements. These elements mostly dissolve in mineral acids, except the noble metals, such as platinum, silver, and gold, which are unaffected by acids.

Apart from the manufacturing and catalysis earth-abundant metals have applications in energy sectors such as in fuel cells and batteries. The applications of some of these metals are involved in electrochemical reactions associated with the technologies where they serve as catalysts in oxygen reduction as well as splitting of water for the production of hydrogen. Their use at catalysts increases the rate of reaction otherwise necessitate precious metals. The alternative chemical processes with the use of these metals as catalysts empower the progress of more efficient methods that are economically more sustainable for the production of energy and better chemical feedstocks as well. The more uses of earth-abundant metals promotes the goal of global sustainability, addresses the concerns about the gradual and continuous environmental hazards as well as the scarcity of resources. With the exploration of improve properties and application thereby, these metals are assured for significant impact on advancement in the field of green chemistry. Moreover, these enhanced properties of earth-abundant metals lead to the more sustainable and environmentally benign manufacturing practices, chemical technologies that result in a more resilient and eco-friendly economy.

The heterogeneous catalysis is an essential aspect in efficient synthesis of molecules and materials. More than 90% of chemical conversion relies on catalytic reactions. The new advancements in the area of synthesis of nano-catalyst and their use have been proved to be potential by the easier availability of earth-abundant metals. The metals like iron, cobalt, nickel, manganese, and copper, are more readily available and less expensive. As a result, they not only provide a sustainable alternative to precious metals but also align with the principle of green chemistry with a solution to address resource scarcity. Nano-size earth-abundant metals in the size range of 1-100 nm possess very large surface area and are advantageous for the catalytic application for the transformation reactions. This property of nanomaterials provides enhanced reactivity for the catalysis. Moreover, the unique electronic properties of nano-size metals also facilitate catalytic reactivity. The large surface area yields more active sites for reactions and facilitates catalytic reactions such as hydrogenation, electro-catalysis, conversion of biomass, etc. The iron nanoparticles have a widespread use in heterogeneous hydrogenation reactions owing to their inexpensive and non-toxic nature.

The use of earth-abundant metals for the replacement of scary and precious rare metals is potentially evident in the future. Chemists, scientist, and researchers are engaged in discovering approaches for the reduction of the use of rare metals in bimetallic systems in connection with their scarcity. The earth-abundant metals are combined with noble metals to achieve the required mechanical, chemical, and thermal stability of bimetallic. The success of this exploration may trigger improved catalytic activity. Moreover, this innovation may lead to minimization of the environmental as well as economic impacts due to the use of earth-abundant metals in place of expensive rare metals. Moreover, scientific investigation is in progress on earth-abundant metals for their activities in plasmon-enhanced catalysis. This catalytic activity is based on the property of the metals to absorb light. The absorbed light energy facilitates the increased rate of chemical reactions more efficiently. Therefore, the plasmon-enhanced catalytic activities of earth-abundant metals deliver energy in chemical processes in more sustainable and efficient way.

Transition metals are the important group of metals that fall under earth-abundant metals. The catalytic nature of transition metals has enormously transformed chemical synthesis. The synthesized chemicals have significant impacts on chemical industries and economic growth as well. The research and innovation on metal complexation has led to numerous Nobel Prizes for recognition of contribution of these complexes to various new reactions, metathesis, and cross-coupling processes between precious metals like palladium and rhodium. However, the scarcity of the resource of palladium and rhodium and the environmental hazards of using them put up limitation in the cross-coupling. But, this limitation has prompted attention in the use of earth-abundant element (EAE) catalysis in their replacement. This emerging area suggests sustainable alternatives for the enabling of reactions which are not accessible with precious metals i.e. cross-coupling between sp2−sp3 and sp3−sp3 orbitals, and different transformation processes such as photocatalytic and electro-catalytic reactions.

Recent advancements in the topic "Advances and Applications in Catalysis with Earth-Abundant Metals" focus on the potential of metals like cobalt, manganese, iron, etc. These metals are researched to be applied in formation of novel bonds and in various hydro-functionalization reactions. The scope of the reactions, the limitations of the processes and mechanistic pathways are to be emphasized to assist the implementation of these new catalysts in industrial scenario. However, the utilization of EAE catalysts requires the focus on sustainability of the processes with these catalysts. Screening technologies with substantial efficiency are important for the optimization of these new reactions to ensure the scalability and commercial viability of the processes. Therefore, the partnership between research community and industry personnel is crucial to modify these catalysts and create trustworthy supply chains. Finally, motivation and innovation may be applied to address the global challenges in the field of material science and chemistry. This new catalysis may align with sustainable development and rightful resource utilization [1].

## 2. Natural sources of earth-abundant metals

The cyclic natural processes are involved for the existence of metals in the environment i.e. in the earth crust since formation of the earth. However, the human activities, mostly the urbanization and rapid industrial growth, have dictated the distribution of metals between various compartments of the environment. This distribution of metals across land has been controlled by the localized industrial pollution, rock geochemistry, and unique geological conditions. Different minerals in the rocks decide the geochemical variation of the rocks and their types. The characteristics of soils and sediments formed by the breaking of the rocks along with their mineralogy and chemistry decide whether the metals halt near the source, move away over the distances or transport in water. The volcanic activity associated with the released of gas and fluid from the earth's crust introduces metals into surface environments, either in the atmosphere or in land and ocean floor. A geochemically varied earth's surface with varied composition of metals is created thereby. This led to the localized areas with varied high or low levels of a particular metal. This variation impacts biological health, either through toxicity or deficiency based on biodegradation of metals. The outermost layer of the earth's crust constitutes about 1% of its total mass and is composed of solid rocks and minerals. They have been forming over geological time scales, millions of years. This layer is crucial for the availability of earth-abundant metals which are the building blocks of the modern civilization and industrial propensity. The crust is rich in a variety of elements, metals, non-metals and metalloids which are essential to geological and biological systems. The metals available in abundance are termed as earth-abundant metals that have a crucial role in the industrial revolution. The properties of the nine earth-abundant metals have been depicted in Table 1. A brief description of the natural sources of earth-abundant metals is given in following sections.

*Table 1: The properties of abundant metals*

| Sl. No. | Metals | Atomic number | Mol. weight | Density at NTP (kg/m³) | Abundance on earth crust (%) |
|---------|--------|---------------|-------------|------------------------|------------------------------|
| 1 | Oxygen (O) | 8 | 15.999 u | 1.429 | 46.1 |
| 2 | Silicon (Si) | 14 | 28.085 u | 2330 | 28.2 |
| 3 | Aluminium (Al) | 13 | 26.981 u | 2710 | 8.23 |
| 4 | Iron (Fe) | 26 | 55.845 u | 7800 | 5.63 |
| 5 | Calcium (Ca) | 20 | 40.078 u | 1540 | 4.15 |
| 6 | Sodium (Na) | 11 | 22.989 u | 4669.2 | 2.36 |
| 7 | Potassium (K) | 19 | 39.098 u | 860 | 2.33 |
| 8 | Magnesium (Mg) | 12 | 24.305 u | 1429-3200 | 2.09 |
| 9 | Titanium (Ti) | 22 | 47.867 u | 4540 | 0.565 |

## 2.1 Aluminium (Al)

Aluminium is the most abundant metallic element that make up about 8.2% of the earth's crust. It is available in hard rocks and ores in the form of various compounds such as aluminium sulphate, aluminium carbonate, and aluminium oxide rather than in its pure elemental form. The aluminium is widely used in daily life due to its lightweight and malleable properties. This silvery-white metal can be easily transformed into the desired shape and size compatible to its target applications. The above-mentioned characteristics along with excellent corrosion resistance trigger its widespread use in utensils, others household container. The malleability and the lightweight are used in its application as structural material for aeroplane [2].

## 2.2 Iron (Fe)

The iron has the second most availability and use as earth-abundant metals. Iron constitutes about 5.6% of the Earth's crust which can be available for extraction and use. However, it is available in more amounts in the earth's core. It is a shiny, greyish metal which has been essentially used for the progress of human civilization for centuries. Iron is very strong, but flexible with versatility.

Hence it has been an essential raw material of structures in civil engineering and manufacturing industries. Approximately 1.3 billion tonnes of crude steel are manufactured from iron per annum. The various forms of steel are the building blocks of the modern industrialization and economic growth is an alloy of iron and carbon [3].

## 2.3 Calcium (Ca)

The next abundant metal is calcium which makes up about 4.2% of the earth's crust. Calcium is extracted from various rocks such as limestone, chalk, marble, pearls, and shells. It is a soft, silvery-white metal and plays a very important role in biology and physiology apart from its greater use in chemical industries. The calcium compounds like limestone comprising calcium carbonate ($CaCO_3$) and others are used in construction activities and in manufacturing of cement. Moreover, calcium carbonate has the application in steel production for the removal of silica impurities and enhancement of quality of the steels. It is biologically important as calcium is an essential element for maintaining human health. The strength and structure of human bones and teeth are maintained by the combination of calcium with other minerals to form hard crystals. The bone density along with overall skeletal health are maintained by consumption of calcium [4].

## 2.4 Sodium (Na)

Sodium constitutes about 2.4% of the earth's crust. It is a soft, highly reactive material. Therefore, it is found in its compound forms. It reacts with air and water very vigorously. The most available and usable form of sodium is sodium chloride known as common salt. Sodium chloride has wide applications as kitchen staple as it is widely used for enhancement of food flavour. Sodium compounds such as sodium hydroxide (NaOH) and sodium bicarbonate ($NaHCO_3$) have wide uses in the processes such as soap making, glass production, and pH regulation. Sodium is biologically important too as it is essential for all living organisms. It plays a key role in physiology for the maintenance of fluid balance, muscle contraction, and nerve function [5].

## 2.5 Magnesium (Mg)

Magnesium makes up about 2.3% of the earth's crust and extracted largely from the ores like magnesite and dolomite. The lightest element potassium is 75% lighter than steel, 50% lighter than titanium, and 33% lighter than aluminium. Due to very low lightweight, it is used in products like laptop, mobiles, luggage bags, car seats, etc. It burns brightly when in flares, fireworks, and sparklers are applied to it. The biological importance of potassium lies in the fact that it is an essential element for the production of chlorophylls in plants. The photosynthesis is carried out by the help of chlorophylls. The photosynthesis provides the energy and $O_2$ required for the existence of lives on the earth [6].

## 2.6 Potassium (K)

Potassium constitutes about 2.1% of the earth's crust. It is highly reactive in the presence of both oxygen and hydrogen. So, it is available primarily in the form of compounds rather than in its elemental state. Potassium chloride (KCl) is the more prevalent compound and is extensively used in fertilizers. Potassium is one of the essential macro nutrients of plants to enhance plant growth. Another significant compound of potassium is potassium carbonate ($K_2CO_3$). It plays a key role in production processes for glass, soap, detergents, etc. The biological function of potassium lies in its use for maintaining proper electrolyte levels and fluid balance in the body. The muscle contraction, nerve function, and overall cellular activity are controlled by the appropriate regulation of sodium and potassium in the blood [7].

## 2.7 Titanium (Ti)

Titanium makes up about 0.6% of the earth's crust. It is extracted from ores such as rutile and ilmenite. The combination of exceptional properties like high strength and low weight makes it so valuable for various applications. The alloys of titanium are used in spacecraft, aircraft, spacecraft, and missiles because of their lightness and durability at the same time. The pipes made of titanium alloys are used in the condenser of power plant. The biocompatibility of titanium enables medical applications such as bone repairing, surgical implants, and dental prosthetics.

## 3. Extraction and purification

The earth-abundant metals are available on the rocks and ores of respective metals. Very often, more than one metal in their respective compounds remain in complex combinations in the ores. The extraction and purification of metals from the ores are very difficult. It needs application of multiple unit operations and reaction schemes in series for the economic extraction of earth-abundant metals. The extraction processes of three important metals are elaborated in the following sections.

## 3.1 Aluminium

The extraction of aluminium involves several complex stages.

**Bayer Process:** Bauxite ore is first crushed, washed, and dried. It is then mixed with sodium hydroxide (NaOH) under high pressure and temperature to form a solution of aluminium oxides. Later oxides are precipitated and recovered.

**Hall-Heroult Process:** Alumina is dissolved in molten cryolite and subjected to electrolysis. A strong electrical current reduces alumina to pure aluminium metal at the cathode while oxygen is formed at the anode. The following is the electrochemical reaction at a high temperature (950 °C) for this process.

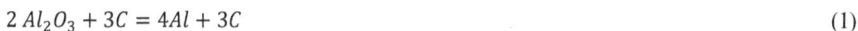

$$2\,Al_2O_3 + 3C = 4Al + 3C \tag{1}$$

These processes are highly energy-intensive and smelters for carrying out the reaction are arranged near energy sources. Energy efficiency and environmental sustainability are important aspects to look after for aluminium extraction. High-purity aluminium (over 99%) is produced through two main processes. Initially, aluminium oxide is extracted from bauxite ore using the Bayer process. In the second step, the Hall–Heroult process performs electrolytic reduction to obtain pure aluminium. Most of the processing plants are installed close to the site of bauxite mining to minimize transportation costs of bauxite ores. The production of aluminium oxide starts with washing and crushing of bauxite ore. Then the crushed bauxite mixed with calcium oxide and the mixture is finely ground once again. The resulting powder is dissolved in high-purity sodium hydroxide. The mixture is processed in high-pressure autoclaves to produce sodium aluminate. This solution is then decomposed to form the crystals of aluminium hydroxide. The crystals are filtered and washed before separation from the caustic soda solution which is recycled. Finally, the aluminium hydroxide crystals are heated to over 1000°C in a process called calcination. The calcination yields pure aluminium oxide ($Al_2O_3$) powder. The metallic aluminium is produced from $Al_2O_3$ in the electrolytic reduction process known as Hall–Heroult electrolysis process. Here, aluminium oxide ($Al_2O_3$) from the Bayer process is melted with cryolite and subjected to electrolysis at temperatures below 900 °C. The resulting liquid aluminium is separated from the electrolyte by using vacuum siphon. It is then transferred to foundry furnaces for refining which is accomplished in two different methods. The passing of chlorine through molten aluminium forms chlorides as impurities to be removed. The second method is by using electrolytic reduction of aluminium-copper alloy. Both of the method ensure the production of high quality aluminium [8].

## 3.2  Iron

The iron ore is mixed with limestone and coke and fed into a blast furnace for the extraction of iron from concentrated ores. The blast furnace operates by blowing hot air from the bottom, where coke burns and reaches temperatures up to 2200 K. This high temperature causes coke to react with oxygen to produce carbon monoxide (CO). The CO and heat rise through the furnace and react with the ore, which is introduced from the top. In the upper parts of the furnace, which are cooler than the bottom, iron ore minerals like hematite ($Fe_2O_3$) and magnetite ($Fe_3O_4$) are reduced to ferrous oxide (FeO). The reactions that occur in the Blast furnace at high temperatures (900-1500 K) are as follows:

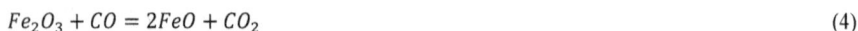

$$3Fe_2O_3 + CO = 2Fe_3O_4 + CO_2 \qquad (2)$$

$$Fe_3O_4 + 4CO = 3Fe + 4CO_2 \qquad (3)$$

$$Fe_2O_3 + CO = 2FeO + CO_2 \qquad (4)$$

On the other hand the reactions that occur in the Blast furnace at lower temperature (500-800 K) are given below:

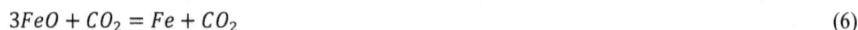

$$C + CO_2 = 2CO \qquad (5)$$

$$3FeO + CO_2 = Fe + CO_2 \qquad (6)$$

The limestone decomposes to form calcium oxide (CaO) in the Blast furnace. The produced CaO reacts with impurities like silicates to produce slag, which can be easily separated from the molten iron. The iron produced by this way is called as pig iron which contains 3–4% carbon. Carbon along with other impurities such as sulphur and silicon helps making the iron hard and brittle. This pig iron is re-melted with iron scraps and coke. The melt is exposed to a blast of hot air to reduce the carbon content and remove the impurities. The iron so produced is known as cast iron, which has a lower carbon content (2–3%) and is even harder than pig iron [9].

## 3.3   Calcium

Calcium comprises about 3% of the earth's crust and is fifth most abundant element and the third most abundant metal after aluminium and iron. It is predominantly extracted from sedimentary deposits such as limestone, dolomite, marble, chalk, and aragonite. The key calcium ores include gypsum, anhydrite, fluorite, and apatite. The methods of extraction of calcium involve two primary methods viz. Davy's electrolysis and aluminium reduction. The first method is practiced by Russia and China. Here, calcium is extracted by electrolyzing molten calcium chloride. On the other hand, the U.S. and Canada produce calcium by reducing lime with aluminium at high temperatures. China, Russia, and the United States are the leading producers, with annual outputs of approximately 10,000-12,000 tonnes, 6,000- 8,000 tonnes, and 2,000-4,000 tonnes, respectively. On the other hand, Canada and France are minor producers of calcium. The global production of calcium reached about 24,000 tonnes in 2005, with the U.S. consuming nearly half of this amount.

## 4.   Modification to enhance the properties

The pure form of the most of the metals may not be used for they have one or more weak properties in terms of strength, softness, ductility or brittleness etc. Very often, metals are mixed with other metals in varied proportion to achieve the desired properties for their target applications. The mixed metals are known as alloys. Different alloys of a base metal yields varied properties for different applications. Two very important earth-abundant metals (iron and aluminium) that are modified by mixing other metals in varied proportion to them to achieve an array of useful properties and application are demonstrated below.

## 4.1   Modification of iron

The metals iron is a fundamental metal known for its versatility, but pure iron (wrought iron) is soft and not very strong. With increasing carbon content iron becomes stronger but less malleable. Cast iron, with about 5% carbon, is very hard but brittle. Steel is produced by purifying cast iron, removing impurities, and adjusting the carbon content to between 0.1% and 1.5%. Older methods like the open hearth or Bessemer processes use hot air to remove impurities, while modern techniques like the Kaldo or L.D. processes use pure oxygen for faster, cleaner results. Apart from the varied proportions of carbon, chromium or nickel are added to steel to enhance its properties. The stainless steel is produced by mixing 18% chromium and 8% nickel [10]. Different steels are produced with carbon as the main allowing element as illustrated below.

*Plain carbon steels*

Plain carbon steels are classified based on their carbon content into four main types.

**Mild steel (Low carbon steel):** Mild steel contains 0.05-0.3% carbon. It is soft, tough, and ductile, making it suitable for products like wires, nails, chains, and structural components such as tubes and girders.

**Structural steel (Medium carbon steel):** Medium carbon steel contains 0.3 to 0.6% carbon. This steel has high tensile strength. It is shock-resistant, and performs well under heat. It is used in different heavy structures such as railway components, machine parts, springs, and other engineering applications.

**High carbon steel:** High carbon steel contains 0.6-0.9% carbon. Known for its hardness, toughness, and resistance to wear, It is used for tools, drills, knives, and other high-strength applications.

**Tool Steel (Higher carbon steel):** Higher carbon steel contains 0.9-1.5% carbon. These are high-quality steels that can be hardened to produce sharp cutting edges, making them ideal for cutting tools, lathe tools, and other precision instruments.

*Alloy steels*

The steels containing other elements in addition to carbon to enhance their properties are known as alloy steels. Various transition metals like nickel, cobalt, chromium, vanadium, molybdenum, and tungsten are added in varied proportions with the steel. These elements give alloy steels a wide range of unique properties, making them suitable for various applications. Alloy steels are classified based on the purpose of the modifications as:

**By chemical composition:** Different alloying elements are added to the steel to achieve specific chemical properties.

**By structure:** The arrangement of elements affects the steel's microstructure and its resulting properties.

**By purpose:** Alloy steels are tailored for particular uses, such as improving strength, hardness, or resistance to wear.

**Brittle steels**

Brittle steels are prone to breaking or shattering under stress. Excessive phosphorus (more than 0.06%) and sulphur in steel or iron contribute to brittleness. Sulphur causes "hot shortness," making the metal brittle at high temperatures, while phosphorus leads to "cold shortness," causing brittleness at low temperatures.

**4.2     Modification of aluminium**

The modification of aluminium is also very important for its different applications owing to the modified properties. The key methods for modification involve tailoring its properties and shaping for specific applications. The properties like strength and resistance to corrosion are achieved by the introduction of alloying metals. On the other hand, heat treatment processes such as solution heat treating and aging improve mechanical properties. Surface treatments like anodizing or powder coating enhance durability and appearance. Machining techniques such as milling and drilling allow for precise shaping of aluminium components. Similarly, casting or extrusion methods create desired forms by pouring molten aluminium into moulds or forcing it through dies. Overall modification has a crucial role in optimizing the desired properties of aluminium fit for various uses in industries and households [11].

**5.     Applications of earth abundant metals**

Earth-abundant metals such as iron, cobalt, nickel, manganese, copper, and zinc offer sustainable alternatives to rare and precious metals. The substantial availability of these metals in the earth's crust and lower costs of extraction provide this opportunity. A pivotal role is played with the modification and transformation of these metals in various applications, particularly in catalysis and energy storage. For instance, iron is widely used in the production of ammonia and steel, while cobalt and nickel are essential in hydrogenation reactions and battery technologies. This versatility aligns with global sustainability goals, as the shift towards these metals helps mitigate concerns about resource scarcity and environmental degradation. The natural cycling of metals is influenced by geological processes and human activities, resulting in a varied geochemical landscape. For example, aluminium is the most abundant metallic element in the earth's crust, found primarily in compound forms like bauxite, and is prized for its lightweight, malleable, and corrosion-resistant properties. Iron constitutes about 5.6% of the crust and is fundamental to civil engineering and manufacturing. Calcium, another abundant metal, plays crucial roles in both industry and biology, while sodium and potassium are vital for various industrial processes and biological functions. The extraction and purification of these metals involve complex methods. The Bayer process and Hall-Heroult electrolysis are key techniques for producing high-purity aluminium, although these processes are highly energy-intensive. Iron extraction typically occurs in blast furnaces, where iron ore is reduced with carbon, yielding pig iron, which is further refined into cast iron. Calcium extraction employs methods like Davy's electrolysis and aluminium reduction. The applications of different earth-abundant metals in various sectors are discussed in the following sections.

## 5.1 Applications as chemicals and consumables

**a. Common uses**: The aluminum is used as the structural metal for making of aircrafts in aerospace industry. The low density of aluminium helps reduce the weight of the aircrafts. The lightweight of aircraft improves fuel efficiency and performance. Overall, aluminium's combination of lightness, The high strength  and resistance to corrosion makes it an essential material in many industries and everyday products [12]. For the same reason of lightweight and non-reactive properties aluminium foil is used for packaging, cooking, and food preservation. Many kitchen utensils such as trays, pans and pots are often made from aluminium as it is easy to clean and resistant to rust. The stainless steel an alloy of iron is also very strong and corrosion resistant. The steel is very durable and police surface is available. Therefore, kitchen appliances and medical instruments are made of steel. On the other hand, cast iron contains 3-5 % carbon and more brittle as compared to steel to be used for manufacturing many structural objects such as pipes, valves, and pumps. The excellent cast-ability and machinability make cast iron so valuable for heavy-duty applications. Elemental iron is an integral part of haemoglobin that helps transport oxygen in animal body to sustain life [13]. Calcium carbonate ($CaCO_3$) known as limestone is applied in construction for buildings, and production of cement. It is also used in steel manufacturing. The silica impurities are removed to improve the quality of the steel. Sodium chloride known as table salt is widely used to enhance the flavour of food, balance sweetness, and reduce bitterness.

**b. Manufacturing fertilizers:** Many earth-abundant metals are the building blocks of fertilizers as the metals have the biological activities for maintenance of plant growth.  So, they have an important role in manufacturing fertilizers and  in maintaining soil health They often act as the essential nutrients for plant growth [14]. Calcium (Ca) is a major component of lime (calcium carbonate), which is used to improve soil structure and pH. Magnesium sulphate (Epsom salts) is used as a fertilizer to address magnesium deficiencies in soil.

### c. Indirect activity as pesticides

Calcium, potassium, and magnesium are primarily used in agriculture to improve soil health and plant nutrition. But these elements also help in pest and disease management and control indirectly rather than acting as direct pesticides. The good health of the plants resists the attack of pests and indirectly control the pest [15]. The following are the chemicals that help in pest control of the plants. Calcium Chloride ($CaCl_2$) is not a pesticide, it is sometimes used in agriculture to help manage soil conditions and improve plant health. This can make plants more resilient to diseases and pests. Potassium ulphate ($K_2SO_4$) and otassium nitrate ($KNO_3$) are used as fertilizers to enhance plant growth and stress resistance. Healthy plants with adequate potassium are more robust and better able to fend off pests and diseases. Magnesium sulphate ($MgSO_4$) commonly known as epsom salts is used to correct magnesium deficiencies in plants, maintain the chlorophyll production for overall plant health. Again, the good health of the plant improves resistance to pests and diseases. In a nutshell, calcium, potassium, and magnesium are not pesticides, but proper supply of these as nutrient certainly improves plant health and resilience. The healthy plants resist, and reduce the impact of pests and diseases.

### 5.2    Application as catalysts

Many commercial syntheses of inorganic and organic compounds are accomplished by the use of transitional metals such as Fe, Ni, Cr, Ti, Cu, etc. Ammonia synthesis by Haber process uses iron and molybdenum catalysts. Manganese dioxide is used for catalytic decomposition of $H_2O_2$, whereas vanadium pentoxide catalyses the transformation of $SO_2$ to $SO_3$ in the sulphuric acid production by contact process. Other metals such as platinum, chromium, etc. also have enormous applications in the synthesis of inorganic and organic compounds.

Transition-metal catalysis has fundamentally transformed chemical synthesis, earning recognition through multiple Nobel Prizes for advancements involving precious metals like poalladium and rhodium. These metals play a crucial role in various industrial applications, enabling key reactions such as cross-coupling and hydrogenation. However, their widespread use raises significant concerns about sustainability, cost, and limited reaction scope due to their low natural abundance and the environmental impact associated with their extraction. To address these issues, there is a growing focus on earth-abundant element (EAE) catalysis, which utilizes more environmentally friendly and readily available metals like cobalt, manganese, and iron. This shift not only mitigates sustainability concerns but also introduces new reactivity and transformation capabilities that precious metals often cannot achieve efficiently. For instance, EAE catalysis facilitates novel bond formations and reaction types that enhance synthetic pathways in both academic and industrial settings.

The advancement of high-throughput screening technologies further accelerates this transition by enabling rapid testing of diverse reaction conditions, thereby facilitating the discovery of new compounds and processes. These technologies allow chemists to explore a broader range of chemical space, identify previously unexamined structures, and optimize reaction conditions more effectively, fostering sustainable practices in the pharmaceutical and chemical industries. As EAE catalysis continues to evolve, collaboration between academic researchers and industry practitioners will be crucial. Such partnerships can help bridge the gap between fundamental research and practical application, ensuring that innovative discoveries in EAE catalysis are translated into scalable, efficient technologies that meet the changing demands of the chemical landscape. Ultimately, this synergy has the potential to reshape chemical synthesis for a more sustainable future, reducing reliance on precious metals while expanding the toolkit available to chemists [16].

### 5.3 Applications in nanotechnology

Nanotechnology is a subject for designing of nano-size materials, their synthesis, and application in various fields of science and engineering such as sensors, catalysis, production and storage of energy, separation processes, and wastewater treatment. The investigation and exploration of nanostructured materials have surged in every field of technology. But, it has been accomplished thoroughly in the area of catalysis in the recent past. The transition metals are the main group of catalytic species that are transformed to nanomaterials. The nanoscale catalysts are appreciated for their unique properties such as high surface area, selectivity, and reusability compared to bulk catalysts. The rates of adsorption and reaction increase several fold making them advantageous for many electrochemical applications other than catalysis. The electrochemical technology involves many engineering and scientific areas like medicine, sensors, biosensors, and environmental science. The high sensitivity and specificity are available for a range of applications with these sensors and biosensors made of novel transition metals and by the application of nanotechnology. The earth-abundant transition metals and their oxides (such as Fe, Co, Ni, Cu, and Mn) as nanomaterials are emerging as promising candidates for the production of sustainable energy too. The cost-effectiveness, biocompatibility, and high catalytic capabilities are advantageous, making them suitable to be applicable for various energy conversion and storage technologies. The demand for high-performance nanomaterials made of earth-abundant transition metals and their oxides is likely to grow as research continues with success in this field. The laboratory innovations in this regard may be transformed to practical and scalable technologies in near future [17].

### 5.4 Application of transitional metals with biological importance

The importance of earth-abundant metals in biology and physiology are discussed here as the synthetic compounds of these metals are used as supplements and is synthesized in chemical reactions. The synthesis, modification, separation, and purification are very much dependent on the principle of chemical engineering and practices. The earth-abundant metals, specifically several transition metals such as iron, cobalt, copper, etc. constitute the biologically important molecules and take part in the biochemical processes of living systems including humans. Sodium is essential for all living organisms. It plays a critical role in maintaining fluid balance within cells and throughout the body. It involves in nerve function and muscle contraction, making it vital for many physiological processes. On the other hand, iron is the most widespread and crucial transition metal. It has enormous functions in living systems. The iron containing protein (haemoglobin) transports oxygen and takes part in electron transfer in oxidation–reduction reactions. The biochemical action of cobalt, a trace element has a role in synthesis of vitamin $B_{12}$ and related coenzymes. Also, many enzymes use copper as their cofactors and their enzymatic activities are enabled by the presence of copper both in plants and animal. Nitrogen-fixing bacteria utilize enzymes that contain iron. This enzyme helps in catalysis the reduction of nitrogen gas to nitrogenous compounds. The functions of transition metals in biological systems are studied in bioinorganic chemistry.

## 6.   Challenges and remediation

The mining, extraction, purification and transportation of the metals are energy intensive and detrimental to the environmental sustainability. The uneven distribution of the ores in different regions, the varied proportions of composition of target metals in their respective are a bottleneck for the extraction of the metals with economic viability. Different challenges in this regard have been discussed in this section. Aluminium is a vital material with widespread application primarily extracted from bauxite. Bauxite is abundant in world scenario, but concentrated in specific regions like Guinea and Brazil. The extraction process of aluminium comprises bauxite mining, alumina refining, and molten salt electrolysis. Ninety percent of these processes emit $CO_2$, are energy-intensive, particularly smelting, often powered by coal, especially in China. The production process creates pollutants like particulate matter and sulfur dioxide, impacting both the environment and worker health. The inert anode process is available to replace the convention smelting process, but it is still not economically viable and hence not widely used. Transitioning to cleaner energy sources and adopting advanced technologies, such as inert anode process are essential for enhancing environmental sustainability in the aluminium industry. Iron mining is also a considerable extent of negative impact on environmental. Specifically, the iron ore tailings-waste materials contain harmful heavy metals such as cadmium, chromium, and lead leading to water pollution with carcinogenic heavy metals. The iron ores are converted into tailings of approximately 1.4 billion tons each year. The harmful compounds of the tailings disrupt receiving waterbody and pollute it with dissolved metal. This in turn, leads to acid mine drainage to severely affect aquatic ecosystems. In addition, mining operations generate air pollutants with toxic gases and particulate matters. Therefore, an effective management strategies and comprehensive environmental assessments are very important for the sake of surrounding ecosystems and the health of local populations. The pollution problem associated with the mining, extraction and purification of earth-abundant metals may be controlled in the upgradation of production process, use of cleaner energy, use of green technology. But, despite all these measures implementation, pollution of waterbody cannot be resisted. The residual pollution may be reduced by very recent techniques of phytoremediation that are discussed here.

It is a favourable strategy for the removal of contaminants from and water. The plantation and growth of hyper accumulator plants are accomplished. These hyper accumulator plants have the unique ability to selectively absorb and concentrate the specific earth-abundant metals from soil and water. The phytoremediation works via two mechanisms viz. phyto-extraction phyto-stabilization. The pollutants (metals) are absorbed via the roots and translocated to the above-ground parts of the plants. On the other hand, phyto-stabilization mechanism immobilizes metals in the soil and thereby resists their movement and stops their flow into the food chain. The selection of the right plant species is the key to the success of these methods as the plants are specific to the metal pollutants in varying capacities. Various factors such as soil pH, organic content, nutrient levels and the presence of chelating agent like ethylenediaminetetraacetic acid (EDTA) influence the uptake of pollutants.

Alternatively, multiple plant species are cultivated together to improve the overall accumulation of multiple earth-abundant metals. Various intercropping strategies such as mixed cropping, row intercropping, relay cropping, etc. are applied for better result. The intercropped between maize and chickpeas has improved cadmium absorption showcasing the benefit of biodiversity for the phyto-remediation of heavy metals. Organic farming, i.e. replacement of synthetic pesticides by organic ones may play crucial role in remediation of earth-abundant metals. Promotion of biodiversity and organic farming enhances soil health and adopts natural pest control. The convention water treatment methods with these novel phytoremediation can be amalgamated for better result. Again, a dietary regulation may help relief in cleansing the body once exposure of pollutants occurs. The foods such as cilantro, garlic, and chlorella have the chelating properties. The presence of these chelating agents in the body by food intake may help in binding and facilitating the excretion of earth-abundant metals from the body. Therefore, a multi-faceted approach comprising phytoremediation, intercropping, organic farming, and dietary strategies, in combination, may help manage the remediation of the earth-abundant metals. The effect of pollution on human health from the mining, extraction and application of earth-abundant metals may be controlled by this combine effort [18].

## 7.    Recent developments

As the chemical industry increasingly embraces earth-abundant metals, it not only promotes sustainability, but also fosters innovative approaches to chemical synthesis. This transition supports responsible resource utilization and environmental stewardship, ultimately contributing to a more resilient and eco-friendly chemical economy. By harnessing the properties of these metals, researchers and industries can drive advancements in green chemistry and develop efficient processes for energy and materials production. The recent advancement in the catalysis by earth-abundant transition metals is noteworthy, specifically for the dihydrosilylation of alkynes. The alkynes are typically favoured to undergo monohydrosilylation owing to their distinctive electronic and steric properties. But, in the recent development of the reaction scheme, the employment of dual catalyst systems or the improved single catalyst can be able to perform the monohydrosilylation of various alkynes. Additionally, the notable effectiveness of the transition metals such as iron (Fe), cobalt (Co), and nickel (Ni) have been achieved by research communities. In a co-based system i.e. use of other agents with the transition metals (Fe), selectivity of aryl alkynes can be improved. The Fe converts aliphatic terminal alkynes into 1,1-disilylalkanes selectively with the help of phenylsilane and ethyl magnesium bromide. On the other hand, the newly developed Ni catalyst facilitates direct synthesis reaction at atmospheric condition in absence of strong reductant. However, this emerging development in the catalysis is not complete yet. Further exploration of the development of the catalysis, specifically in the selective addition reaction of alkynes for sterically hindered substrates is necessary. Though, the insertion of alkynes into metal bonds has been proposed, an isolation of key intermediates may provide better insights in the addition reaction mechanism and enhance understanding of these reactions [19].

## Conclusions

The earth-abundant metals are embraced by chemical industries for sustainable economic growth with innovative approaches. The synthesis of new chemicals and transformation of metals play a central role in various applications as structural materials, chemicals, fertilizers, pesticides, and for energy storage. The new and novel applications have become evident through significant modifications and chemical transformations of these metals. The creation of alloyed forms of these metals and advancements in metal nano-catalysis increase their usefulness in a variety of applications, such as battery technology and hydrogenation. Earth-abundant metals play a key role in the production of effective fertilizers that promote robust plant growth and, in turn, help increase plant resistance to pests and enhance the sustainability of agriculture as a whole. As the importance

Materials Research Forum LLC
https://doi.org/21741/9781644903711-12

of sustainability in chemical processes rises, these metals will be crucial in promoting a greener, more resilient economy while tackling global concerns in environmental stewardship and resource management.

## References

[1] Advances and Applications in Catalysis with Earth-Abundant Metals, pp. 1677-1679, 2023. https://doi.org/10.1021/acs.organomet.3c00292

[2] E. S. Gorlanov, L. I. Leontev, Directions in the technological development of aluminium pots, J. Min. Inst., 266 (2024) 46-259, 2024.

[3] M. Podgórska, M. Jóźwiak, Heavy metals contamination of post-mining mounds of former iron-ore mining activity, Int. J. Environ. Sci. Technol. 21 (4) (2024) 4645-4652. https://doi.org/10.1007/s13762-023-05206-y

[4] Z. Wang, Y. Zeng, Z. Ahmed, H. Qin, I. A. Bhatti, and H. Cao, Calcium-dependent antimicrobials: Nature-inspired materials and designs, Exploration, no. October 2023, https://doi.org/10.1002/EXP.20230099

[5] B. Sayahpour, Quantitative analysis of sodium metal deposition and interphase in Na metal batteries, Energy Environ. Sci. 17(3) (2024) 1216-1228 https://doi.org/10.1039/D3EE03141A

[6] G. O, M. A, P. T, Potassium Heterogeneous Distribution in Soil and Its Uptake by Zea mays, vol. 27, no. 5, pp. 10-21, 2024. https://doi.org/10.9734/jalsi/2024/v27i5656

[7] A. K. Sharma, Magnesium Alloys, Galvanotechnik, vol. 115, no. 5, pp. 567-573, 2024.

[8] G. Liu et al., Recent advances and future trend of aluminum alloy melt purification: A review, J. Mater. Res. Technol. 28 (2024) 647-4662. https://doi.org/10.1016/j.jmrt.2024.01.024

[9] J. Le Coze, Purification of iron and steels a continuous effort from 2000 BC to AD 2000, Materials Transactions, 41 (2000) 219-232. https://doi.org/10.2320/matertrans1989.41.219

[10] M. de Bouw, I. Wouters, J. Vereecken, and L. Lauriks, Iron and steel varieties in building industry between 1860 and 1914 - A complex and confusing situation resolved, Constr. Build. Mater. 23 (2009), 2775-2787. https://doi.org/10.1016/j.conbuildmat.2009.03.009

[11] A. Kurt, I. Uygur, and E. Cete, Surface modification of aluminium by friction stir processing, J. Mater. Process. Technol. 211 (2011) 313-317. https://doi.org/10.1016/j.jmatprotec.2010.09.020

[12] J. Hirsch, Recent development in aluminium for automotive applications, Trans. Nonferrous Met. Soc. China, 24 (2014) 1995-2002. https://doi.org/10.1016/S1003-6326(14)63305-7

[13] A. Oarga-Mulec, U. Luin, and M. Valant, Back to the future with emerging iron technologies, RSC Adv. 14 (2024), 20765-20779. https://doi.org/10.1039/D4RA03565H

[14] A. L. Srivastav et al., Sustainable options for fertilizer management in agriculture to prevent water contamination: a review, Environ. Dev. Sustain. 26 (2024) 8303-8327. https://doi.org/10.1007/s10668-023-03117-z

[15] B. Mansfield et al., A new critical social science research agenda on pesticides, Agric. Human Values, 41 (2024) 395-412. https://doi.org/10.1007/s10460-023-10492-w

[16] Š. Kment et al., Single Atom Catalysts Based on Earth-Abundant Metals for Energy-Related Applications, Chem. Rev. 2024. https://doi.org/10.1021/acs.chemrev.4c00155

[17] G. Maduraiveeran, M. Sasidharan, and W. Jin, Progress in Materials Science Earth-abundant transition metal and metal oxide nanomaterials : Synthesis and electrochemical applications, Prog. Mater. Sci. 106 (2019) 100574 https://doi.org/10.1016/j.pmatsci.2019.100574

[18] J. Briffa, E. Sinagra, and R. Blundell, Heavy metal pollution in the environment and their toxicological effects on humans, Heliyon, 6 (2020) 1024-1032. https://doi.org/10.1016/j.heliyon.2020.e04691

[19] P.-F. L. Dai Qiang-Qiang; Qu, Jian-Ping; Kang, Yan-Biao, Recent Developments on the Earth-Abundant-Metal-Catalyzed α,β-Dehydrogenation of Carbonyl Compounds, Synthesis (Stuttg). 56 (2024) 2213-2222. https://doi.org/10.1055/a-2232-8882

Applications for Earth-Abundant Transition Metals
Materials Research Foundations 179 (2025) 356-399

Materials Research Forum LLC
https://doi.org/21741/9781644903711-13

# Chapter 13

# Transition-Metal Based Catalyst for Petrochemistry

Rupam Chatterjee[1], Subhajit Kundu[1], Debarati Mitra[1*]

[1]Department of Chemical Technology, University of Calcutta, Kolkata - 700009, India.

*debarati.che@gmail.com

## Abstract

Petrochemistry is a study focused on the conversion of crude petroleum into useful raw materials for various industries as well as their transformation into valuable end products namely fuels and petrochemicals. These transformations are achieved by applying various physical and chemical processes. The myriad of chemical changes, which transpires in these processes, often require the aid of various catalysts for the purpose of hastening up the chemical reaction as well as to improve the conversion and yield of desired products. In petrochemistry, earth abundant transition metals such as iron, cobalt, nickel, molybdenum, aluminium, tungsten and zirconium find applications. Some of the catalytic processes utilizing transition metal catalyst in petroleum refining industry include hydrotreatment, reforming, and catalytic cracking, while in the petrochemical industry, they catalyze mainly polymer and organic solvent synthesis. The transition metal catalysts used in the well-established processes in petrochemistry along with emerging technologies utilizing the same category of catalysts are discussed in this book chapter. The nature of a catalyst is not only determined by the transition metal used but also on the characteristics of the support materials which is also addressed in this chapter.

## Keywords

Petroleum Refining, Petrochemical Processes, Catalysis, Transition Metal Catalysts, Catalyst Support

## Contents

## 1.   Introduction

Catalysis is a process where a substance known as catalyst is used to provide a new lower activation energy pathway in a chemical reaction, thereby accelerating the reaction. The term catalysis was coined by Berzalius (1). Catalytic processes are an integral part of various industries, since this can not only reduce the production cost by decreasing time and energy required in a chemical reaction but also can increase the selectivity and yield of targeted product. However, catalysis does not enhance or mitigate the thermodynamics and the equilibrium composition of a reaction mixture. When a catalyst is added to a reaction mixture it does not undergo chemical transformation therefore it can be separated out from the mixture and reused time and again. Thus, catalysis can play a crucial role pertaining to the sustainability and economy of the industrial sector (2).

Depending on the nature of the reaction mixture's and catalyst's phases, it can be classified as homogenous catalysis (both are in the same phase) and heterogeneous catalysis (both are in different phases). In a homogenous catalytic reaction, the catalytic activity is higher compared to heterogeneous catalysis since the reaction rate is not diffusion controlled. The selectivity of this type of catalyst is also high, however these catalysts are difficult to separate out from the reaction media and the product/s might require a further purification step (3).

Whereas, in heterogeneous catalysis, the catalyst is generally a solid which adsorbs the reactants, thereby increasing reactant concentration near its surface and providing suitable reaction condition. After the reaction, the product gets desorbed from the surface of the catalyst. This type of catalyst also has good selectivity and can be easily separated from the reaction media (3). Two categories of metal and metal compounds, which are prominently used as catalysts, are noble metals and transition metals. Their catalytic activity stems from the electronic configuration of these metals and compounds. The limited availability coupled with their high value restrains the use of noble metals and entails the need for alternative catalysts. Transition metals and their respective compounds can bestow substitute catalysts in some reactions, which implement noble metals. The resistance of transition metal based catalyst to poisoning is also a boon in this regard. The transition metal oxide catalyst and mixed oxide catalysts possess tunable surface properties and hetero-structures which enhance product selectivity. The surface area and surface properties of catalyst are also influenced by the support material therefore; proper selection of the support material also plays a key role in the catalytic activity (2).

Petroleum sector is a key industry dealing with purification and transformation of complex mixture of hydrocarbons which necessitates the incorporation of catalytic reactions in various processes including hydrodesulfurization, hydrocracking, hydrogenation, catalytic cracking, catalytic reforming etc. Therefore, it is necessary to determine suitable catalysts and catalyst modification which can expedite these processes (1, 3).

The petrochemical sector is another industry which is as equally important. The feedstock for this industry mainly comes from crude petroleum. The conversion of these feedstocks into useful products often involves catalytic reactions which mandate probing into catalytic technology (3).

## 2.    Conventional catalyst synthesis processes:

Catalysts can be categorized as supported and bulk catalyst based on the distribution of active agent. In bulk catalyst, the active agent is distributed throughout the mass of the catalyst whereas, for supported catalyst the active agent is distributed on the surface of an inert substance known as a support.Bulk catalysts are mainly synthesized by three routes: precipitation, crystallization and gelation/hydrothermal processes (4).

In the precipitation method, a solution of required concentration containing precursor salt/s is made which is then subjected to precipitation either by addition of a precipitating agent or by varying physicochemical parameters to induce nucleation, followed by aggregation. This precipitate is then washed to remove remaining precursor impurities, dried and calcined at an appropriate temperature (4).

Crystallization approach for the synthesis of bulk catalyst is similar to the precipitation method except this process is more time intensive because of the slower nucleation process required for the formation of crystal structures. The advantage of this process is the larger crystals are more uniform and easily separated from solution, however catalysts synthesized by this method exhibit lower porosity and surface area which are important factors for catalytic activity (4, 5).

Bulk catalyst synthesis by sol-gel process is performed by the following steps: formation of sol by the hydrolysis of metal salts (having suspended particle size ranging from 1 to 1000nm), condensation reaction of the moieties in the sol state leading to gelation by the formation of macromolecules, gel ageing for structural hardening, washing of the gel and finally drying and calcinations (5).

The hydrothermal process is used predominantly for the synthesis of zeolite catalysts which are basically composed of sodium alumino silicate structures. The process involves preparation of a solution of precursors which is allowed to age in order to form a gel. This gel is hydrothermally reacted to form crystalline structures which on further calcination results in the formation of a crystalline lattice structure known as zeolite. The precursor salts and the molar composition in solution, pH of the gel formed, gel ageing time, hydrothermal treating time along with the temperature and also the calcination time and temperature determine the nature and activity of the zeolite catalyst. The gel structure thus formed after the hydrothermal treatment can be shaped before calcination to produce shaped catalysts (4).

Some methods for the synthesis of supported catalysts are impregnation, deposition precipitation, reductive deposition, colloidal synthesis, adsorption, melt infiltration, atomic layer deposition. A brief description of each of these methods is given in fig.1.

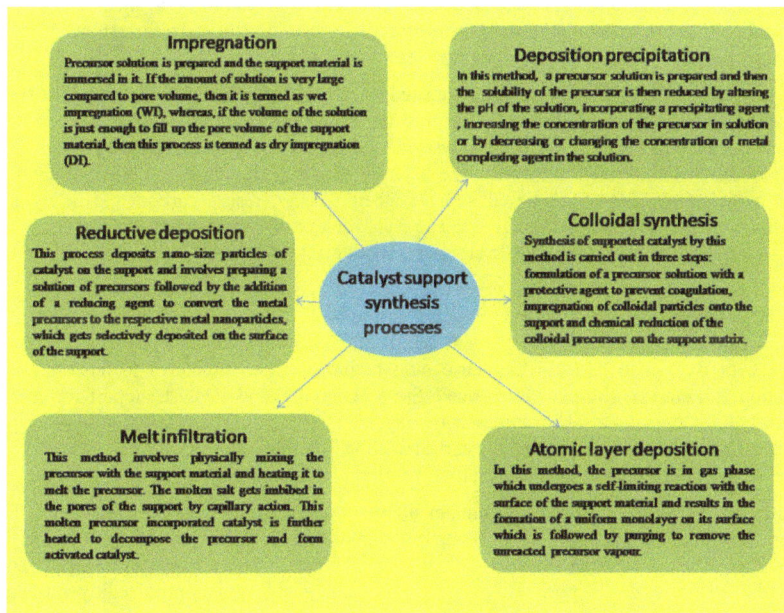

*1 Figure: Some industrial methods for synthesis of supported catalysts (6, 7)*

## 3. Function of catalyst support

A catalyst support though an inert material, serves the following important purposes:

- enhances the mechanical and thermal properties of the catalyst,
- provides better distribution of the active catalyst substance,
- the porosity of the support can be manipulated during its synthesis which increases the surface area of the catalyst, reduces the cost of catalyst due to the need for lower amount of catalytic material (8, 9, 10).

These materials can also provide other functions including: resistance to catalyst sintering, poisoning and carbon deposition (asin the case of cracking catalysts) (11).

Based on the type and nature of materials used, catalyst support can be broadly classified as: silica based, carbon based, pereovskite based, zeolite based, hydrotalcite based, mayenite based and ceramic foam type (12).

Silica based catalyst support exhibit high surface area, high sintering resistance and structural stability at elevated temperatures. The large specific surface area is due to the layered nature of silica structure. This property also facilitates in better bonding of metals to the support which also prevents agglomeration and provides better distribution in case of multi metal (alloy) based catalyst. This characteristic lamellar structure of silica also hinders coke formation, in case of cracking operation, which can cause catalyst deactivation (11).

Biochar, activated carbon and coal coke are used as carbon based catalyst supports. This type of catalyst supports are generally used for tar cracking catalysts. These materials are highly porous which improves the distribution of catalyst and its natural adsorption tendency helps in the transport of reactant through the pores. The inherent alkalinity, due to the presence of alkali and alkali earth metals, of carbon based support albeit weakly, resists carbon and heavy metal deposition which can deactivate catalysts. The disadvantage of carbon support is that it can react if used in reforming operations thus leading to slow catalyst depletion (11).

Perovskites are a newly emerging class of materials which can be used as catalyst support. These have an octahedral structure with a structural formula of $ABO_3$. The 'A' position contains an alkaline earth or lanthanide group element and the 'B' position contains a transition metal. The presence of oxygen in non-stoichiometric ratio facilitates these structures to possess oxygen migration and storage abilities which make these materials good catalyst support for cracking and reforming operations. The presence of oxygen in the perovskite structure also helps to oxidize the coke thus resisting catalyst deactivation by coke deposition. This structure also enhances the dispersion of metal in its lattice, resulting in better catalytic activity (11).

Zeolites are basically a mixture of sodium alumino silicates in various compositions. These materials can be used as catalysts or as catalyst supports based on the processes in which these are applied. They have large specific surface area, inherent surface acidity, high thermal ability and specific pore size. The pore size of zeolites is tunable and can be optimized by varying the composition and synthesis condition parameters (6, 11).

Hydrotalcite type materials have a general structural formula $[M(II)_{1-x}M(III)_x(OH)_2]^{x+}(A^{n-}_{x/n})_m \cdot H_2O$ where M(II) and M(III) represents divalent and trivalent metal ions and 'A" represents the interlayer counter ion. This type of supported catalysts is synthesized by reduction or calcinations of the hydrocalcite precursor. The merits of hydrocalcite are: very good dispersion of the catalyst on the support, good thermal stability, large specific surface area, alkaline characteristics of the surface and resistance to coke formation due to the nanocomposite structures. These materials also exhibit high catalytic activity in the steam reforming process along with resistance to sintering and sulfur poisoning. If calcium oxide based hydrocalcite is used as reforming catalyst, the carbon dioxide gets adsorbed and hydrogen is produced by water gas shift reaction (11).

Mayenite is a crystalline material containing nano-sized pores. The molecular composition is $Ca_{12}Al_{14}O_{33}$ with free oxygen in the lattice structure. Nickel based mayenite supported catalyst is used for steam reforming which has the following advantages: reduced coke forming tendency, provides resistance to sulfur poisoning and surface dispersion of catalyst metal is improved (6,7,11).

Ceramic foam as the name suggests is a highly porous ceramic material. This type of catalyst support have good heat resistance, good mechanical strength are chemically inert and is used to increase the bed porosity of a packed bed which results in less pressure drop, enhances turbulence and better heat and mass transfer. For reformers, this provides advantage of less coke formation due to mitigation of dead spaces which occurs in dense beds. Three conventional methods for synthesizing this type of catalyst support are: polymeric sponge replica method, direct foaming method, and pore-forming method. The high amount of interconnected mesopores provide high surface to volume ratio which implies that these type of supportedcatalysts can be used in reactors with lower holdup volume and smaller size catalyst particles can be utilized efficiently. The larger tortuous pores also amplify the turbulence in the interior of the catalyst such that there is better contact of the reactant with the catalyst surface (12,13).

## 4. Role of transition metal catalysts in petroleum refining:

### 4.1 Hydrocracking

The quality of crude has deteriorated over the years and the majority of the crude available contains large amount of heavy fractions. The demand for these heavy fractions is less and combined with the fact that these are lower value products, compared to the commercial fuel fractions (diesel, kerosene, jet fuel, gasoline), it is necessary to upgrade these heavier fractions into high demand,lighter fuel fractions. In refineries, two types of catalytic cracking operations are carried out (14,15):

- cracking in presence of hydrogen (hydrocracking) and
- cracking without hydrogen (catalytic cracking)

The former process involves simultaneous cracking of heavier hydrocarbons to form alkene or carbocation intermediates followed by hydrogenating the products. Therefore, the catalyst utilized in this process is a bi-functional catalyst consisting of an acid site (for bond breaking) and a hydrogenation site (for the production of saturates after bond cleavage). The constituents of feed for hydrocracking process are: long chain n-alkane, naphthenes, aromatics and polynuclear aromatics. Supported transition metal catalyst containing a particular combination of nickel, cobalt, iron, tungsten and/or molybdenum impregnated on an alumina-silica or zeolite support is utilized in this process (16,17). The support (zeolite or aluminosilicate) have inherent acidic sites, which carries out the functions of cracking and isomerisation. The enhancement of acidity of the catalyst support can be achieved by the incorporation of halogens, which improves the cracking ability. Molybdenum and tungsten (oxides or sulphides) are used as promoter to enhance the catalytic activity in hydrogenation-dehydrogenation and heterolytic reactions. In order to increase the selectivity of hydrocracking, tri-fulctional and tetra-functional catalysts are necessary, in which rare earth and noble metals are incorporated as catalytic promoters. However, these catalysts are more expensive than the conventional transition metal based catalysts. In Co/Mo and Ni/Mo catalysts though the molybdenum is incorporated as a promoter, but each component of these catalysts promote each other. The intense processing conditions are 80–200 bar and at temperatures in the range of 300–450°C, which necessitate the need for robust catalyst support (18). The characteristics of some commercial hydrocracking catalyst forms (manufactured by TL:TRILOBE®) are depicted in table 1.

*Table 1: Characteristics of hydrocracking catalysts (TL:TRILOBE®) (19)*

|  | RM-430 | RN-410 | RN-400 |
|---|---|---|---|
| Active metals | Group VII and/or VI metals | | |
| Size (mm) | 1.6, 2.5 | 1.3, 1.6, 2.5 | 1.3 |
| Bulk density (gm/cm3) | 0.54 | 0.65 | 0.65 |
| Crush strength (N/mm) | 15.5 | 21.3 | 22.2 |
| Pore volume (cm$^3$/g) | 0.87 | 0.67 | 0.67 |
| Surface area (m$^2$/g) | 150 | 155 | 220 |

## 4.2    Catalytic cracking

The catalytic cracking operation uses heavy hydrocarbon feedstocks viz. vacuum gas oil and atmospheric residue to produce lighter more utilitarian hydrocarbons. This process also uses light naphtha feedstock to produce gasoline range hydrocarbons. Cracking of heavy (long chain) hydrocarbons can be achieved by thermal treatment; however this process is prone to coking due to thermal degradation. Mitigation of coking and production of useful products at lower temperature can be achieved by the use of catalyst, which lowers the activation energy required for the C-C bond cleavage (20). The catalysts utilized in this process are either ZSM-5 zeolite catalysts (used conventionally in fluidized catalytic cracking operation) or transition metal oxide based catalysts, which might also have noble metal oxides incorporated to enhance the catalytic activity and selectivity. The catalysts are selected based on the desired product and the type of feed. Transition metal oxide based catalysts can either contain a single metal oxide (oxides of chromium, cerium, zirconium, titanium and aluminium) or complex metal oxides, which consist of a combination of the aforementioned metals. In the complex metal oxide based catalysts, aluminium oxide plays a crucial role of stabilizing the metal complex at high temperatures. The incorporation of iron, manganese, zinc, molybdenum and tungsten oxides as promoters are integrated into these catalysts to regulate the surface acidity which in turn dictates the cracking activity of the catalysts and the final products generated. The molybdenum oxide is of significance in this regard as it enhances the acid strength which mitigates coking (21, 22). A schematic of catalytic cracking of n-hexane is depicted in fig. 2.

n-hexane $\xrightarrow{\text{ZnO}}$ hexene $\xrightarrow{\text{H-ZSM-5}}$ lower olefins $\xleftarrow{\text{H-ZSM-5}}$ Oligomers $\xrightarrow{\text{ZnO}}$ aromatic hydrocarbons

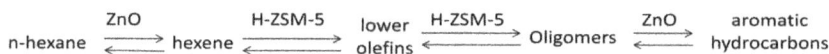

*Figure2: Catalytic cracking of n-hexane (19)*

## 4.3    Hydrotreating

Hydrotreating of distillates are done for hydrodenitrogenation (HDN), hydrodesulfurization(HDS) and hydrodearomatization (HDA) functions. The HDN and HDS processes are crucial to mitigate air pollution, due to the $SO_x$ and $NO_x$ (sulfur and nitrogen oxides respectively) emissions and to abide by the environmental regulations. This process also improves the fuel quality due to saturation of hydrocarbons (23). Heavy residues are also hydrotreated for hydrodemetallization (HDM) and hydrogenation (HYD). This process is carried out in packed bed reactors (PBRs) operating at a hydrogen pressure in the range on 0.7 to 5 MPa and temperature of 350°C to 450°C. For hydrotreating of vacuum residue, high hydrogen pressure is necessitated due to the presence of contaminants (heterocyclic and metallic compounds) (23, 24). The catalysts used in this process mainly serves three functions viz. bond cleavage of long chain hydrocarbons followed by hydrogenation, saturation of unsaturates and saturation and dearomatization of aromatic rings. The conventional hydrotreating catalysts consist of molybdenum or tungsten sulfide active sites with nickel or cobalt as promoter supported on γ-alumina. The hydrodesulfurization process conventionally utilizes a catalyst consisting of oxides of nickel, cobalt and molybdenum. The nickel oxide along with molybdenum oxide as promoter assists in hydrogenation of heterocyclic sulfur compounds to form saturates, which makes the sulfur atom easily accessible for desulfurization. The cobalt oxide along with molybdenum as the promoter facilitates in the sulfur bond cleavage resulting in desulfurization of the organic molecules. The characteristics of a standard hydrotreating catalyst is shown in table 2. The sulfur is removed in the form of hydrogen sulfide ($H_2S$) which is further converted to elemental sulfur in the downstream Claus process. The HDS catalyst inevitably gets deactivated after prolonged usage and needs to be replaced (23). Therefore, it is necessary to monitor the catalyst performance to understand when the catalyst needs to be replaced. This catalyst activity is determined by equation (1) (25)

$$D_s = \tfrac{1}{4} D_{s0} e^{At} \qquad\qquad (1)$$

where,

Ds = desulfurization activity
$D_{s0}$ = initial desulfurization activity

A = deactivation rate, °F/bpp (barrel of feed per pound of catalyst)

t = catalyst life, bpp (barrel of feed per pound of catalyst)

*Table 2: Characteristics of a standard Ni Mohydrotreating catalyst of AZKO NOBEL (19)*

| Properties of NiMo-alumina (INTR1R) catalyst | |
|---|---|
| Pellet size(cm) | 0.127 |
| Surface area (m$^2$/gm) | 140 |
| EDR (gm/cm$^3$) | 4.6 |
| Bulk density (gm/cm$^3$) | 0.6 |
| Pellet length (mm) | 4.5 |
| Bulk crush strength (kg/cm$^2$) | 7.8 |

The hydrotreating catalysts are manufactured by co-mulling which is a type of incipient impregnation method followed by drying and calcinations. The transitions metals exist as oxides after the calcinations process which needs to be converted to sulfides as per the requirement of modern refineries. This is achieved by sulfiding which can either be achieved in in-situ or ex-situ. The in-situ sulfiding is accomplished by loading the reactor with the oxide state of the catalyst followed by charging with $H_2S/H_2$ mixture to convert the oxides to sulfides. However, this process is not environmentally friendly due to the usage of $H_2S$, also in this process, the catalytic reactor needs to be pretreated before the hydrotreating reaction commences. Therefore, ex-situ sulfiding is preferred by modern refineries. In this process, the transition metal oxide catalyst is treated with an organo sulfur compound (dimethyldisulfide or dimethylsulfide) or exposed to sulfur containing fuel feedstock. The sulfur compounds gets impregnated in the catalyst lattice and on thermal treatment, the oxides in the catalyst are transformed to sulfides (23, 25). Characteristics of conventional ex-situ sulfide catalysts are reported in table 3.

*Table 3: Characteristics of presulfidedhydrotreating catalysts (19)*

| Characterization parameters | Ni-Mo/Al$_2$O$_3$ | Co-Mo/Al$_2$O$_3$ |
|---|---|---|
| Chemical composition (wt. %) (dry basis) | | |
| MoO$_3$ | 13.20 | 13.71 |
| NiO | 4.03 | - |
| CoO | - | 3.4 |
| Physical nature | | |
| Pore dia. (mm) | 0.96 | 1.01 |
| Surface area (m$^2$/g) | 312.0 | 270.4 |
| Pore volume (H$_2$O adsorption ) (cm$^3$/gm) | 0.70 | 0.69 |

## 4.4 Catalytic reforming

Catalytic reforming is a crucial process in petroleum refineries, aimed at upgrading naphtha into high-octane reformate, which is essential for gasoline production and serves as a precursor for petrochemical feedstocks (26). Traditional catalytic reforming processes use precious metals like platinum and rhodium, which are costly and can impact the overall economics of the refinery. Consequently, there has been significant interest in developing low-cost transition metal-based catalysts that can provide comparable performance at a fraction of the cost (27, 28).

Applications for Earth-Abundant Transition Metals                    Materials Research Forum LLC
Materials Research Foundations 179 (2025) 356-399          https://doi.org/21741/9781644903711-13

Catalytic reforming involves the conversion of naphtha, a complex mixture of hydrocarbons, into high-octane components through a series of reactions, including dehydrogenation, isomerization, and cyclization. These reactions are conducted at a temperature range of 450°C to 550°C, and under pressures of 10 to 30 atm. The primary reactions include (29, 30):

- Dehydrogenation

- Isomerization

- Cyclization

These reactions are facilitated by catalysts that promote the desired chemical transformations while suppressing undesired side reactions. Some of the earth abundant transition metals which are utilized as catalysts for catalytic reforming are: iron, nickel and cobalt. These metals are used in their oxide state supported on alumina and are synthesized mainly by the co-precipitation route. In case of the iron based catalysts, the active sites can activate C-H bonds in hydrocarbons, enabling transformations viz. dehydrogenation of paraffins to olefins, which are subsequently converted into aromatics (26, 29, 30).

Nickel catalyzes reforming reactions by facilitating the dehydrogenation of paraffins to form olefins, which can then be further transformed into aromatics. Nickel active sites play the role of activation of the hydrocarbons, which is crucial for effective reforming (31, 32).

The cobalt based catalysts promotes dehydrogenation and cyclization reactions. It helps in breaking C-H bonds and forming intermediates that lead to the production of high-octane reformate (33, 34).

## 5.    Role of transition metal catalysts in petrochemical industry:

### 5.1    Methanol from syngas

The synthesis of methanol from synthesis gas (syngas) is a vital process in the petrochemical industry, forming the basis for producing a wide range of chemicals and fuels. Syngas, primarily composed of carbon monoxide (CO) and hydrogen ($H_2$), is converted into methanol ($CH_3OH$) using transition metal catalysts. The $CO:H_2$ ratio is generally maintained as 1:2. Steam is also utilized in the reaction to mitigate the formation of undesirable byproducts. This process is highly significant since methanol is the feedstock for various products, including formaldehyde, acetic acid, and also used as a base material for many plastics and fuels (biodiesel). The use of transition metal catalysts in this synthesis is crucial for enhancing efficiency, selectivity, and overall process economics (35).

The primary reaction for methanol synthesis from syngas is given as follows:

$$CO + 2H_2 \longrightarrow CH_3OH \qquad (i)$$

This reaction, known as the methanol synthesis reaction, typically occurs in the presence of a catalyst under specific conditions of temperature and pressure. The process also involves secondary reactions such as water-gas shift (35):

$$CO + H_2O \longrightarrow CO_2 + H_2 \quad (ii)$$

The transition metal based catalysts which are utilized in this process include: copper, iron and cobalt. Copper oxide based catalysts supported on alumina is conventionally utilized for methanol synthesis from syngas. Zinc oxide is used as a promoter for copper based catalysts. These catalysts are highly effective for the hydrogenation of CO. The copper metal sites facilitate the adsorption and activation of CO and $H_2$, enabling the formation of methanol. The presence of zinc oxide in the catalyst helps to enhance the stability and performance of the copper sites. The synthesis typically occurs at temperatures ranging from 200°C to 300°C and pressures between 50 to 100 atm. These conditions help to achieve a high rate of reaction while maintaining good selectivity for methanol (36).

Iron catalysts, although less common than copper-based catalysts due to lower activity, are used in methanol synthesis, often in the form of iron-based mixed-metal catalysts (containing copper) or supported on materials like alumina or silica. These catalysts facilitate the reaction by activating CO and $H_2$ and promoting their reaction to form methanol. Iron catalysts are typically used at higher temperatures, ranging from 300°C to 400°C, and moderate pressures. The higher temperatures can lead to a higher rate of reaction but may also lead to coke formation (36, 37, 38).

Cobalt is another transition metal used in methanol synthesis, often in the form of cobalt oxide on a support material such as alumina silica. These catalysts facilitate methanol synthesis by enabling the activation and transformation of CO and $H_2$ into methanol. Cobalt based catalysts are also less utilized industrially due to their lower activity compared to copper. The reaction conditions for these catalysts are similar to that of iron based catalysts (35).

## 5.2 Catalytic production of ethylene:

Ethylene ($C_2H_4$) is a key raw material in the production of various chemicals and polymers, including polyethylene, ethylene glycol, styrene etc. The traditional method for ethylene production is steam cracking, where hydrocarbons like naphtha or ethane are subjected to high temperatures (around 750°C to 900°C) in the presence of steam, resulting in the cleavage of long-chain hydrocarbons into smaller molecules, including ethylene (39). This process is however energy and cost intensive, which has led to the development of low-cost transition metal catalysts that offer a cheaper alternative for ethylene production. These catalysts can help reduce operational costs while maintaining high efficiency and selectivity. The three transition metals which are utilized as catalysts in this process are: iron, nickel and cobalt. The alumina foam supported iron catalysts have good activity and low cost. These catalysts facilitate the activation of hydrocarbons and the formation of ethylene through mechanisms like oxidative dehydrogenation and oxidative coupling processes. The temperature typically ranges from 300°C to 500°C, and the process is carried out under moderate pressures. The presence of oxygen or steam is often necessary to drive the reaction (39, 40).

The alumina supported nickel catalyst is another alternative catalyst utilized in the synthesis of ethylene. The main role of nickel based catalyst is the catalytic dehydrogenation of the precursor hydrocarbon (methane or ethane) through the cleavage of C-H bonds. The advantage of this catalyst is that, the reaction can be carried out at atmospheric pressure though, the temperature requirement is higher (500°C and 800°C) than the iron catalyzed process (39).

Although not commonly utilized, silica supported cobalt oxide catalyst also finds application in the catalytic synthesis of ethylene. The feedstock for ethylene production process utilizing this type of catalyst is either ethane or higher paraffins. Cobalt catalyzes the C-C and C-H bond cleavage reactions resulting in the formation of ethylene and other short chain unsaturated hydrocarbons.This process requires the same temperature and pressure parameters similar to the aforementioned nickel based catalyst process (41).

### 5.3    Catalytic naphtha cracking

Naphtha cracking is a crucial process in the petrochemical industry used to nickel-based olefins like ethylene, propylene, and butylene from naphtha (42). Traditionally, this process has relied on high-cost catalysts, often containing noble metals or complex materials. However, research into low-cost transition metal catalysts for naphtha cracking has gained significant interest due to their potential to reduce production costs while maintaining efficiency and selectivity. Naphtha cracking involves breaking down the long-chain hydrocarbons in naphtha (a complex mixture of hydrocarbons derived from crude oil) into smaller, more valuable molecules through high-temperature reactions. The catalytic cracking process proceeds under a temperature range of 500°C to 800°C, and at atmospheric to moderate pressures (42, 43). The key reaction is represented as follows:

$$C_6H_{12} \longrightarrow C_2H_4 + C_4H_8 (iii)$$

The transition metals which are employed in this process are: iron, nickel and cobalt. The transition metals are used in their oxide state supported on γ-alumina in case of iron and nickel whereas in case of cobalt, zeolite (ZSM-5) is used as the catalyst support (43, 44).

Iron catalysts facilitate the cracking process by providing active sites for the cleavage of C-C bonds in hydrocarbons. The iron catalyzes the formation of smaller hydrocarbon fragments through mechanisms like oxidative dehydrogenation or catalytic cracking (45).

In the naphtha cracking process utilizing nickel-based catalyst, nickel catalyzes the rupture of C-C bonds in naphtha, forming smaller olefins from larger hydrocarbons promotinghydrogenation and dehydrogenation reactions which are crucial for efficient cracking (46).

The cobalt catalysts form olefins through cracking of naphtha, providing active sites for the cleavage of hydrocarbons (46).

### 5.4    Synthesis of vinyl chloride monomer by direct chlorination of ethylene:

The transition metal-catalyzed synthesis of vinyl chloride monomer (VCM) from ethylene is a crucial industrial process, due to its role as a key precursor in the production of polyvinyl chloride (PVC), a widely used plastic. The direct chlorination of ethylene ($C_2H_4$) involves its reaction with chlorine ($Cl_2$) to produce vinyl chloride ($C_2H_3Cl$). Transition metal catalysts are pivotal in this process, providing the necessary activation and selectivity for the reaction (47).

The direct chlorination of ethylene to produce vinyl chloride is represented by the following reaction:

$$C_2H_4 + Cl_2 \longrightarrow C_2H_3Cl + HCl \qquad (iv)$$

In this reaction, ethylene reacts with chlorine gas to form vinyl chloride and hydrochloric acid (HCl). This reaction is exothermic and requires careful control to avoid over-chlorination and to maximize the yield of vinyl chloride. The most common transition metal catalysts used in this process are iron and aluminium based. Metallic catalysts have very good catalytic activity, however these are prone to coking and deactivation by chloride deposition. Therefore alumina supported chloride salts are preferred as catalysts in this process (48, 49).

The iron based catalyst facilitates the formation of chlorine radicals, which then react with ethylene. The iron catalyst helps in the generation of these radicals through the dissociation of chlorine molecule into chloride radicals. These radicals are highly reactive and can initiate the chlorination of ethylene to form vinyl chloride. This catalytic process is typically conducted at temperatures ranging from 100°C to 200°C and under atmospheric to moderate pressures (49, 50).

If aluminum based catalyst is used for vinyl chloride synthesis, it performs the role of a Lewis acid and forms a complex with chlorine. This complex then facilitates the formation of chlorine radicals or electrophiles that react with ethylene to produce vinyl chloride. The aluminum chloride helps in the selective chlorination by stabilizing intermediates. The aluminium catalyzed reaction typically occurs at temperatures between 50°C to 150°C and moderate pressure conditions (50).

## 5.5    Vinyl acetate monomer synthesis

The synthesis of vinyl acetate monomer (VAM) is a significant industrial process, as VAM is a key intermediate used in the production of various polymers, adhesives, and coatings. The conventional synthesis of VAM involves the catalytic reaction of ethylene ($C_2H_4$), acetic acid ($CH_3COOH$), and oxygen ($O_2$). Ethylene reacts with acetic acid and oxygen to form vinyl acetate and water.While traditional catalysts like palladium and silver are highly effective, they are expensive. Consequently, research into more economical transition metal catalysts for this process is crucial for reducing production costs and improving economic feasibility (51, 52).

The reaction for synthesizing vinyl acetate is (53):

$$C_2H_4 + CH_3COOH + O_2 \longrightarrow CH_2CH(OCOCH_3) + H_2O \qquad (v)$$

The reaction is facilitated by transition metal-based catalysts copper, iron and cobalt, which activate the reactants and promote the oxidative coupling needed to produce vinyl acetate.

Copper is implemented as catalyst either in its metallic form or as an oxide supported on alumina matrix. This type of catalyst facilitates the activation of ethylene and acetic acid. The copper sites on the catalyst help form an intermediate complex that reacts with oxygen to produce vinyl acetate. Copper-catalyzed reactions typically operate at temperatures between 200°C to350°C and under a pressure range of 1 to10 atm. The oxygen partial pressure is crucial to mitigate the formation of byproducts due to over or under oxidation.Cobalt based catalysts are used as metal supported on alumina and the catalytic mechanism is similar to that of copper (54, 55).

Iron is a lesser used catalyst in this process due to poor efficiency. The catalytic activity of iron-based catalyst is increased by the incorporation of catalytic promoters like molybdenum or copper. Iron-catalyzed processes are typically conducted at higher temperatures, ranging from 300°C to 500°C. The reaction conditions need to be optimized to balance cost and performance (56).

## 5.6    Acrylonitrile synthesis

The transition metal-catalyzed synthesis of acrylonitrile is a pivotal process in the chemical industry due to its importance as a monomer in the production of synthetic fibers, resins, and elastomers. Acrylonitrile ($C_3H_3N$) is synthesized primarily through the **Propylene Ammoxidation Process**, which involves the direct oxidation of propylene ($C_3H_3$) in the presence of ammonia ($NH_3$) and oxygen ($O_2$) (57, 58).

The general reaction for the synthesis of acrylonitrile is:

$$C_3H_6 + NH_3 + 1.5O_2 \longrightarrow C_3H_3N + 3H_2O \qquad \text{(vi)}$$

In this reaction, propylene reacts with ammonia and oxygen to produce acrylonitrile and water. The reaction is highly selective and requires efficient catalysts to ensure high yields and minimize the formation of byproducts. This process incorporates bimetallic oxide catalysts viz. iron-molybdenum (FeMo) and vanadium –antimony (VSb) supported on alumina or silica (59, 60).

In case of the FeMo catalyst, the presence of both metals is crucial for stabilizing transition states and enhancing overall catalytic activity. Iron typically catalyzes the initial oxidation of propylene, forming intermediate species, while molybdenum helps in the final conversion of these intermediates to acrylonitrile. The reaction typically occurs at temperatures ranging from 300°C to 500°C and pressures 1 to 5 atmospheres (59, 61, 62).

For the acrylonitrile synthesis process utilizing the VSb catalyst, the vanadium acts as the active site for the oxidation of propylene, while antimony improves selectivity towards acrylonitrile by stabilizing the intermediate oxidation states. This combination helps in achieving a high yield of acrylonitrile (63, 64).

Catalysts used in acrylonitrile synthesis can suffer from deactivation over time due to coking or poisoning by by-products. Regular catalyst regeneration is necessary to maintain performance. Regeneration typically involves oxidation treatments to remove accumulated carbon deposits and restore catalyst activity (64).

## 5.7    Acetaldehyde synthesis from acetylene and ammonia

Acetaldehyde from acetylene and ammonia is a notable reaction in industrial organic chemistry. This process involves the hydroaminomethylation of acetylene, a method which uses transition metal catalysts to convert acetylene ($C_2H_2$) and ammonia ($NH_3$) into acetaldehyde ($CH_3CHO$). This reaction is significant for producing acetaldehyde, a key intermediate in the synthesis of various chemicals, including acetic acid, acrylic acid, and pharmaceuticals (65,66).The overall reaction can be summarized as follows:

$$C_2H_2 + NH_3 \rightarrow CH_3CHO + NH_2R \qquad \text{(vii)}$$

$NH_2R$ represents an ammonium by-product that depends on the specific conditions and catalyst used.

In this reaction, acetylene reacts with ammonia and oxygen to produce acetaldehyde and water. The process typically involves a series of steps facilitated by a transition metal catalyst, which includes the activation of acetylene, insertion of ammonia, and subsequent oxidation to yield acetaldehyde. Apart from the conventionally used palladium-based catalyst, nickel and copper-based catalysts are also utilized, although their efficiency is lower, but these are more economically viable (67, 68).

The alumina supported nickel catalyst works by coordinating with acetylene and facilitating the insertion of ammonia. The catalytic reactions typically operate at temperatures of 200°C to 350°C. Higher partial pressure of acetylene and ammonia ensures better conversion and yield (67, 69, 70).

Copper catalysts are used either in the form of copper(I) chloride or copper(II) oxide bulk catalyst. The reaction mechanism is similar to that of nickel-based catalysts. The copper-catalyzed reactions generally take place at moderate temperatures (150°C to 250°C) and require careful control of oxygen levels (71).

Transition metal catalysts used in this process can suffer from deactivation due to factors such as coking or poisoning. Therefore, regular regeneration or replacement is necessary to maintain catalyst activity. One of the primary regeneration methods is the oxidative treatment which eliminates the coke and impurity deposits from the active sites on the catalysts (71).

### 5.8 Catalytic dehydrogenation of propane and higher paraffins

Dehydrogenation of propane and higher paraffins is a significant industrial process used to produce valuable olefins, such as propylene and butenes, which are important feedstocks for the petrochemical industry. Dehydrogenation involves the removal of hydrogen atoms from alkanes to form alkenes. Transition metal catalysts are crucial for facilitating this reaction due to their ability to activate and dehydrogenate hydrocarbons effectively (72, 73, 74).

The dehydrogenation reactions are depicted as follows:

$$C_3H_8 \longrightarrow C_3H_6 + 2H_2 \qquad \text{(viii)}$$

$$C_4H_{10} \longrightarrow C_4H_8 + 2H_2 \qquad \text{(ix)}$$

The conventional catalysts utilized in this process are either platinum or palladium supported on alumina. These catalysts are highly selective and provides large yield of target products. New investigations have developed cheaper transition metal-based catalyst alternatives consisting of iron and chromium oxides (FeCr) (75, 76). This catalyst also has good selectivity but lower yield compared to the noble metal counterparts. The FeCr catalyzes the dehydrogenation reaction by the activation of the C-H bonds and promote the removal of hydrogen. The reaction with this catalyst is typically conducted at high temperatures (500°C to 800°C) and moderate pressure is maintained to mitigate coke formation. The drawback of this process is that lower pressures can help drive the reaction forward by removing hydrogen gas, which shifts the equilibrium towards olefin formation. Lower pressure based processes are only feasible with noble metal-based catalysts since the reaction temperature is also lower (200°C to 300°C) (77, 78).

## 5.9    Direct oxidation process for methacrylate synthesis

The catalytic production of methyl methacrylate (MMA) by direct oxidation is a vital industrial process for producing this important monomer used in the manufacture of acrylic resins, coatings, and plastics. MMA is synthesized through the direct oxidation of isobutylene (2-methylpropene) in the presence of methanol and an oxidizing agent (79, 80). The reaction is depicted as follows:

$$C_4H_8 + CH_3OH + O_2 \longrightarrow C_5H_6O_2 + H_2O \qquad (x)$$

The reaction is typically carried out using transition metal catalysts namely palladium, silver and vanadium to enhance the efficiency and selectivity of the oxidation process.Among the three metals which are used as catalysts in this process, vanadium is the most widely available and therefore a cost effective alternative(80, 81).

Vanadium based catalyst is used in the form of vanadium pentoxide ($V_2O_5$). The conversion is achieved by the catalytic activation of oxidizing species, which reacts with methanol and isobutylene to form methyl methacrylate.The use of vanadium catalysts typically requires high temperatures (200°C to 400°C) compared to the other catalysts and controlled oxygen levels to ensure efficient oxidation and selectivity. The pressure is maintained in a range of 1 to 10 bardepending on the reaction temperature and the concentration of reactants. Higher pressures can increase the reaction rate but must be carefully managed to avoid undesirable side reactions (82).

## 5.10    Propylene glycol form propylene oxide

Propylene glycol, a versatile chemical widely used in pharmaceuticals, cosmetics, and as an industrial solvent. The synthesis process involves the hydrolysis of propylene oxide, an epoxide compound, to yield propylene glycol. Transition metal catalysts are pivotal in facilitating this reaction due to their ability to effectively activate and hydrolyze the epoxide ring (83, 84).

The hydrolysis of propylene oxide ($C_3H_6O$) to propylene glycol ($C_3H_8O_2$) is represented by the following reaction:

$$C_3H_6O + H_2O \longrightarrow C_3H_8O_2 \qquad (xi)$$

The hydrolysis of propylene oxide is typically carried out at moderate temperatures, usually between 80°C and 150°C. Higher temperatures help in increasing the reaction rate and ensuring complete conversion of propylene oxide.The reaction is generally performed under atmospheric pressure or slightly elevated pressures. This helps maintain the liquid phase and ensure efficient contact between the reactants and the catalyst. A sufficient amount of water must be present to drive the reaction towards the formation of propylene glycol. Typically, a stoichiometric or excess amount of water is used (85).

The transition metal-based catalyst systems utilized in this reaction can be categorized as metal acid catalyst (titanium dioxide catalyst and zirconium oxide) and Lewis acid catalyst (aluminium chloride) (86).

Materials Research Forum LLC
https://doi.org/21741/9781644903711-13

Titanium dioxide, particularly in its anatase form, is often used as a catalyst for the hydrolysis of propylene oxide. It can be modified with acidic sites to enhance its catalytic activity. Titanium dioxide provides acidic sites that facilitate the ring-opening of the propylene oxide epoxide. Once the epoxide ring is opened, water can react with the intermediate to form propylene glycol (86, 87).

Zirconium dioxide, sometimes doped with other metals (molybdenum) or acidic groups, can also be used to catalyze this reaction. Similar to titanium dioxide, zirconium dioxide offers acidic sites that help in breaking the epoxide ring and promoting the reaction with water (88, 89).

Aluminum chloride is a Lewis acid that can be used as a catalyst for the hydrolysis of propylene oxide. This catalyst interacts with the epoxide ring to make it more susceptible to nucleophilic attack by water, thereby facilitating the ring-opening and hydrolysis process (90, 91, 92).

### 5.11 Isobutanol and n-butanol to n-butylene

The transition metal-catalyzed conversion of isobutanol and n-butanol to n-butylene is an important reaction in organic synthesis and petrochemical processing. This process typically involves the conversion of alcohols to alkenes generally involves a dehydration reaction where water is removed from the alcohol to form an alkene (93, 94). For isobutanol (2-methyl-1-butanol) and n-butanol (1-butanol), the target product is n-butylene (also known as 1-butene) (95, 96).

The dehydration of alcohols to form alkenes can be represented as follows:

For **n-butanol**:

$$C_4H_9OH \longrightarrow C_4H_8O + H_2O \qquad \text{(xii)}$$

For **isobutanol**:

$$C_4H_{10}O \longrightarrow C_4H_8 + H_2O \qquad \text{(xiii)}$$

The dehydration reactions are typically conducted at elevated temperatures, ranging from 150°C to 300°C. The reaction is generally performed under atmospheric pressure, but conditions may vary based on the specific setup and catalyst (96).

Transition metal catalysts are essential in facilitating the dehydration of alcohol. The process typically involves acid-catalyzed dehydration, and while traditional acid catalysts (like sulfuric acid) are commonly used, transition metals can also play a crucial role, particularly when supported on acidic materials (93, 96).

Zeolites, particularly those modified with metals like platinum or gallium, can be used as solid acid catalysts. They provide the acidic sites necessary for dehydration reactions while also offering high surface area for catalytic interactions.Zeolites facilitate the removal of a water molecule from the alcohol, leading to the formation of the alkene. The transition metal component can enhance catalytic activity by stabilizing reaction intermediates (97, 98, 99).

Alumina supported with transition metals like chromium or zirconium can act as effective dehydration catalysts. The metal oxides on alumina provide acidic sites necessary for the dehydration reaction. They help in protonating the alcohol, leading to the formation of a carbocation intermediate, which then loses a water molecule to form the alkene (100).

Dehydration catalysts also include transition metal oxides liketungsten oxide ($WO_3$) and molybdenum oxide ($MoO_3$).These oxides provide acidic sites that promote the formation and stabilization of carbocation intermediates, facilitating the loss of a water molecule and the formation of the alkene (101, 102).

Some transition metal sulfides can also be used in dehydration reactions, providing an alternative to metal oxides. Transition metal sulfides offer acidic sites that assist in the dehydration process by promoting the formation of carbocation intermediates (103).

### 5.12   Oxo route for n-butanol synthesis

The transition metal-catalyzed synthesis of n-butanol via the oxo route, also known as the hydroformylation or oxo-synthesis is an important industrial process. This method involves the conversion of alkenes into aldehydes, which are then further reduced to alcohols. In the case of n-butanol, the process starts with the hydroformylation of butene to form butanal, which is subsequently reduced to n-butanol. This synthesis utilizes transition metal catalysts to efficiently facilitate these reactions (104, 105).

The oxo process consists of two main steps:

- Hydroformylation of butene to butanal

- Reduction of butanal to n-butanol

The first step in the synthesis of n-butanol is the hydroformylation of butene ($C_4H_8$) to produce butanal ($C_4H_7CHO$). This reaction involves the addition of a formyl group (–CHO) to the alkene, which is catalyzed by a transition metal complex consisting of either rhodium or cobalt (106, 107).

The second step involves the reduction of butanal ($C_4H_9CHO$) to n-butanol ($C_4H_{10}OH$). This step can be accomplished using various reducing agents (sodium borohydride) or catalysts (palladium) (104).

The cobalt based catalyst used in hydroformylation includes cobalt carbonate and cobalt acetate. They are less expensive than rhodium and are employed in larger-scale operations.The reaction involves the coordination of the alkene to the cobalt center, followed by oxidative addition of carbon monoxide (CO) and hydrogen ($H_2$). The cobalt complex facilitates the insertion of CO and $H_2$ into the alkyl-rhodium bond to form the aldehyde. Hydroformylation with cobalt catalysts generally requires a temperature range of 100°C to 200°C and pressure in the range of 30 to 70 bar. Cobalt catalysts are often used in conjunction with specific ligands to improve selectivity and activity (107, 108, 109).

In the reduction step, palladium catalyst is used either as palladium deposited on carbon or as palladium chloride. These catalysts facilitate the hydrogenation of butanal by providing active sites for the adsorption and hydrogenation of the carbonyl group, converting it to an alcohol.

These reactions are typically performed at moderate temperatures (50°C to 100°C) and under hydrogen pressure (1 to 10 bar) (110, 111).

**Sodium borohydride** (NaBH₄) is a widely used reducing agent for the reduction of aldehydes to alcohols. This donates hydride ions (H⁻) to the carbonyl group of butanal, resulting in the formation of n-butanol. The reduction with NaBH₄ is usually conducted at room temperature and atmospheric pressure (112, 113).

### 5.13 Methacrylic acid from isobutylene

The conversion of isobutylene (or isobutene) to methacrylic acid is an important reaction in industrial organic synthesis, particularly for producing methacrylic acid, a valuable monomer used in the manufacture of acrylic polymers and resins. This process involves several steps, including the oxidative cleavage and subsequent oxidation of isobutylene (114, 115). The conversion of isobutylene ($C_4H_8$) to methacrylic acid ($C_4H_4O_2$) typically involves two main steps (114):

- Oxidative cleavage of isobutylene to 2-methyl-2-butenal

- Oxidation of 2-methyl-2-butenal to methacrylic acid

The first step is facilitated by transition metal catalysts (molybdenum and tungsten) that can effectively promote the cleavage of carbon-carbon double bonds. These transition metals are used in oxide form supported on sintered alumina.Molybdenum-based catalysts provide the necessary active sites to facilitate the cleavage of the double bond in isobutylene and also catalyze the formation of peroxide intermediates that react with isobutylene, leading to the formation of 2-methyl-2-butenal (116).

The tungsten oxides are used due to their strong oxidative properties which promote oxidative cleavage by facilitating the formation of reactive oxygen species that attack the double bonds in isobutylene.The reaction is typically carried out at elevated temperatures, ranging from 200°C to 400°C, to ensure the oxidative cleavage proceeds efficiently and under atmospheric or slightly elevated pressures, depending on the specific catalyst and reaction conditions (117, 118).

The second step involves further oxidation of 2-methyl-2-butenal to methacrylic acid. This reaction converts the aldehyde group of 2-methyl-2-butenal into a carboxylic acid group, yielding methacrylic acid. This process is catalyzed by noble metal catalysts viz. platinum and palladium. Carbon is used as a catalytic support in this case to mitigate the coke formation. These catalysts assist in the oxidation of 2-methyl-2-butenal by providing a suitable environment for the oxidation reaction to occur. This process is typically performed at temperatures ranging from 100°C to 200°C and under atmospheric pressure (119, 120, 121).

### 5.14 MTBE from isobutene and methanol

Methyl tert-butyl ether (MTBE) is an important industrial reagent used to produce a high-octane additive for gasoline. The process involves the acid-catalyzed reaction of isobutene with methanol. The conversion of isobutene ($C_4H_8$) and methanol ($CH_3OH$) to MTBE ($C_5H_{12}O$) is represented by the following reaction (122):

$$C_4H_8 + CH_3OH \longrightarrow C_5H_{12}O \qquad \text{(xiv)}$$

In this reaction, isobutene reacts with methanol to form MTBE and water.Typically, excess methanol is used to drive the reaction towards the formation of MTBE and to ensure high conversion rates. The reaction is catalyzed either by acid catalysts or noble metal catalysts (122, 123, 124).

Zeolite-Y or a mixture of tungsten and molybdenum oxide is used as solid acid catalysts. These solid acid catalysts provide acidic sites necessary for the formation of the carbocationintermediate, which is crucial for the etherification process. The zeolite catalyst effectively facilitates the reaction between isobutene and methanol to produce MTBE (123). Whereas thetungsten molybdenum mixed oxide catalyst provides the necessary acidic sites and can enhance the selectivity towards MTBE formation.The reaction is typically carried out at moderate temperatures, ranging from 100°C to 200°C (125). This temperature range is sufficient to ensure that the reaction proceeds at a reasonable rate without causing excessive side reactions or deactivating the catalyst. A near atmospheric or slightly elevated pressure is maintained which sustains the proper phase and concentration of reactants (126).

Transition metal catalysts can suffer deactivation due to factors like coking or fouling. Regular regeneration is necessary to maintain catalyst performance which involves burning offthe carbon deposits (127).

## 5.15 Butadiene from butane

Butadiene is an important compound in the chemical industry, particularly for the production of synthetic rubbers and other valuable chemicals. Butadiene, a conjugated diene, is primarily synthesized through the dehydrogenation of butane ($C_4H_{10}$). This process relies on specific transition metal catalysts that facilitate the removal of hydrogen atoms from butane, resulting in the formation of butadiene ($C_4H_6$). The conversion of butane to butadiene involves the dehydrogenation of butane (128). The general reaction can be summarized as:

$$C_4H_{10} \longrightarrow C_4H_6 + 2H_2 \qquad \text{(xv)}$$

In this reaction, butane loses hydrogen to form butadiene, a conjugated diene with two double bonds (128, 129). The transition metal catalysts used in this process include: platinum, palladium and nickel. From these three catalyst alternatives nickel, especially in the form of Raney nickel or alumina supported nickel oxide catalysts, are cheaper and resistant to deactivation by poisoning due to the presence of impurities in the feed mixture. These catalystsaccelerate the dehydrogenation by providing active sites for the breaking of C-H bonds (130,131). While nickel is less expensive than platinum or palladium, it generally operates at higher temperatures typically around 500°C to 700°C which leads to coke forming tendencies (132).

## 5.16 Maleic anhydride synthesis

Maleic anhydride ($C_4H_2O_3$), a key intermediate in the production of various chemicals, including resins, lubricants, and pharmaceuticals, is primarily synthesized through the oxidative dehydrogenation of n-butane or benzene (133, 134). The reactions are as follows:

### Oxidative dehydrogenation of n-butane

$$2C_4H_{10} + 5O_2 \longrightarrow 2C_4H_2O_3 + 4H_2O \qquad \text{(xvi)}$$

## Oxidative dehydrogenation of benzene

$$C_6H_6 + 3O_2 \longrightarrow C_4H_2O_3 + 2CO_2 + 2H_2O \qquad (xvii)$$

Vanadium and molybdenum-based catalysts are used for both the feedstocks.

Vanadium-based catalysts are the most commonly used in the form of vanadium pentoxide ($V_2O_5$), often supported on γ-alumina to enhance the surface area of the catalyst, which provide the necessary oxidation state changes to facilitate the reaction, converting n-butane or benzene into maleic anhydride with high selectivity.The reaction utilizing n-butane is typically conducted at a temperature range 300°C - 500°C (133) which ensures that the oxidation and dehydrogenation processes are sufficiently activated. Whereas, for benzene, a higher temperature range (400°C to 600°C) is necessary to provide sufficient activation energy for the reaction and achieve high conversion rates (134, 135).

Molybdenum oxide ($MoO_3$) is also used as a catalyst for this reaction and is usually combined with vanadium to enhance performance and increase yield. The reaction conditions required for this catalyst is similar to the $V_2O_5$ based process (135, 137).

### 5.17   Isoprene from isopentane

Isoprene (2-methyl-1,3-butadiene) is an important monomer used in the manufacture of synthetic rubber, such as polyisoprene. The conversion from isopentane (2-methylbutane) to isoprene involves dehydrogenation and dehydrocyclization processes which is facilitated by transition metal catalysts (138). The conversion proceeds by the following steps:

- Dehydrogenation of isopentane to isopentenes

- Dehydrocyclization of isopentenes to isoprene

The dehydrogenation of isopentane to form isopentenes, an intermediate compound, requires a suitable transition metal catalyst that can facilitate the removal of hydrogen atoms from isopentane (139). This process utilizes platinum, palladium and nickel-based catalysts. The nickel-based catalysts either supported (on silica or alumina) or in the form of Raney nickel are used for the bulk commercial synthesis process to keep the production cost low.Raney nickel, a form of nickel with a high surface area, is used in industrial settings for its cost-effectiveness and activity.This catalytic dehydrogenation process is typically carried out at elevated temperatures, ranging from 300°C to 500°C. Higher temperatures are required to overcome the activation energy barrier for dehydrogenation.The reaction is usually conducted under atmospheric or reduced pressure to promote the removal of hydrogen gas and drive the reaction forward (140).

The second step, the dehydrocyclization of isopentenes to form isoprene, uses transition metal catalyst (zirconium, chromium or molybdenum) to quicken the formation of a conjugated diene structure, resulting in isoprene (141).

Zirconium-based catalyst, are used in the form of metallocenes which facilitates the formation of conjugated diene structure needed for isoprene synthesis (141). Chromium catalysts are used in the form of chromium oxide ($CrO_3$) which provides the necessary activation for cyclization and conjugation (142, 143).Molybdenum-based catalysts (in the form of molybdenum oxide ($MoO_3$)

can be used in the dehydrocyclization process. Molybdenum catalysts are known for their ability to promote multiple-step reactions involving cyclization and diene formation. The dehydrocyclization reaction generally occurs at high temperatures, typically around 400°C to 600°C. Similar to dehydrogenation, dehydrocyclization is often conducted under reduced pressure for the removal of by-products and enhances the reaction efficiency (144).

### 5.18    Polypropylene synthesis

Polypropylene (PP), is a versatile thermoplastic polymer which is commonly used in packaging, automotive parts, textiles, and many other applications. The synthesis of polypropylene typically involves the polymerization of propylene ($C_3H_6$), a three-carbon olefin (145, 146). The polymerization process can be conducted using various types of catalysts, but the most common and commercially important ones are Ziegler-Natta catalystsandmetallocene catalysts (146, 147).

**Ziegler-Natta catalysts** are a class of transition metal catalysts that are widely used for the polymerization of olefins, including propylene. These catalysts were first developed by Karl Ziegler and Giulio Natta in 1963 (147). This class 6ofcatalyst, typically consist of a combination of a transition metal compounds, such as titanium tetrachloride ($TiCl_4$) or titanium trichloride ($TiCl_3$), and an organoaluminum compound, such as triethylaluminum (TEA) or trimethylaluminum (TMA) (148, 149). The transition metal compound is responsible for the catalytic activity, while the organoaluminum compound acts as a co-catalyst to activate the transition metal. The polymerization mechanism involves the insertion of propylene monomers into the growing polymer chain. The transition metal center in the catalyst forms a complex with the propylene monomer, aiding the addition of the monomer to the polymer chain. This process continues, resulting in the formation of polypropylene. The Ziegler-Natta catalysts can produce isotactic polypropylene, where the methyl groups are aligned in a regular pattern along the polymer chain. The polymerization is typically conducted at temperatures ranging from 50°C to 80°C. This range ensures optimal activity of the Ziegler-Natta catalyst while controlling the polymerization rate and molecular weight of the polypropylene. The pressure range is generally between 1 to 10 atm which ensures efficient monomer conversion (147, 148, 149).

**Metallocene catalysts** are another class of transition metal catalysts used for polypropylene synthesis. These catalysts are based on metallocenes, which are organometallic compounds featuring a transition metal sandwiched between two cyclopentadienyl rings (150). A typical metallocene catalyst used in polypropylene synthesis is cyclopentadienyl-based, such as dimethylsilyl-bis(cyclopentadienyl)titanium ($Cp_2TiMe_2$). The well-defined structure of metallocene catalysts allows for precise control over the polymerization process, resulting in polypropylene with specific properties. This catalyst facilitates the polymerization of propylene by coordinating the monomer to the transition metal center and promoting the insertion of the monomer into the growing polymer chain. These catalysts can produce polypropylene with a narrow molecular weight distribution and controlled stereochemistry, such as isotactic or syndiotactic polypropylenemetallocene-catalyzed polymerization is performed at controlled temperatures and pressures similar to Ziegler-Natta catalysts, often at lower pressures compared to traditional catalysts. This allows for the production of high-quality polypropylene with consistent properties (151, 152).

## 5.19    Aniline from nitrobenzene

Aniline, or aminobenzene, is an important industrial compound used in dyes, pharmaceuticals, and polymers. The transformation from phenol and nitrobenzene to aniline typically involves the Beckmann rearrangement andhydrogenation (153). The reaction schemes are given as follows:

**Reduction of nitrobenzene to aniline:**

$$C_6H_5NO_2 + 3H_2 \longrightarrow C_6H_5NH_2 + 2H_2O \qquad (xviii)$$

The reduction of nitrobenzene to aniline is catalyzed by platinum (Pt), palladium (Pd) or Raney nickel (Ni) catalysts. The catalytic selectivity varies as Pt>Pd>Ni, however the former two are prone to catalytic poisoning by impurities (heteroatoms) in feed (154, 155). Therefore, the Raney nickel (a highly porous form of nickel) is the preferred catalyst for bulk manufacturing of aniline from nitrobenzene. The reaction is generally carried out at moderate temperatures, usually between 50°C and 150°C, depending on the catalyst and desired reaction rate and the hydrogen pressure is maintained in the range of 30 to 50 atm (156, 157, 158).

Solvents viz. ethanol or water can also be incorporated in the reaction mixture to manage catalyst dispersion.After the hydrogenation reaction, the product mixture contains aniline, unreacted nitrobenzene, and water. Aniline is typically isolated through distillation or extraction methods. Additional purification steps may include recrystallization to obtain high-purity aniline (159).

## 5.20    Styrene from ethyl benzene

Styrene is a crucial monomer for the manufacture of polystyrene and other polymers. This transformation primarily involves the dehydrogenation of ethylbenzene ($C_6H_5C_2H_5$) to styrene ($C_6H_5CH=CH_2$). Ethylbenzene is an aromatic hydrocarbon with an ethyl group attached to a benzene ring, while styrene is a vinyl-substituted benzene. The conversion process involves removing hydrogen atoms from ethylbenzene to form styrene. This is achieved through dehydrogenation, a reaction that requires a suitable transition metal catalyst to facilitate the process (160, 161).

The commonly used transition metals include: platinum, chromium and iron. Platinum based catalysts being highly expensive are used only if high conversion rates are needed and the process requires very good quality ethylbenzene, as feed (161, 162).

Chromium-based catalysts, such as chromium oxide ($Cr_2O_3$) on an alumina support, are known for their effectiveness in dehydrogenation reactions. This catalyst activates the C-H bonds in ethylbenzene, facilitating the release of hydrogen (163, 164, 165).

$$C_6H_5C_2H_5 \longrightarrow C_6H_5CH=CH_2 + 2H_2 \qquad (xix)$$

The iron catalysts, often used in the form of iron oxide ($Fe_2O_3$), are also employed for dehydrogenation however this catalyst has lower conversion rates and selectivity hence this is used only for the manufacture aniline of low purity (166, 167).

The reaction typically occurs at high temperatures, ranging from 500°C to 600°C. High temperatures are necessary to overcome the activation energy barrier of the endothermic dehydrogenation reaction.Lower pressures favor the removal of hydrogen gas from the reaction mixture, driving the reaction forward, thus the reaction is generally conducted under atmospheric pressure or slightly reduced pressure.After the dehydrogenation reaction, the mixture typically contains styrene, unreacted ethylbenzene, hydrogen gas, and possibly some by-products. Separation and purification processes include distillation or adsorption to isolate and purify styrene. Hydrogen is often recovered for reuse in other processes (160, 161, 162).

## 5.21    Benzoic acid hydrogenation to caprolactum

Caprolactam is a key intermediate in the production of nylon-6, an important synthetic polymer. The conversion of benzoic acid to caprolactam involves several steps: decarboxylation, hydrogenation, oxidation, oximation, and Beckmann rearrangement (168, 168).

### 1.Decarboxylation of Benzoic Acid

The initial step in the conversion is the decarboxylation of benzoic acid ($C_7H_6O_2$) to benzene ($C_6H_6$). This reaction is catalyzed by a transition metal catalyst, typically a metal oxide such as copper(I) oxide ($Cu_2O$) or a base metal catalyst in the presence of a strong base like sodium hydroxide (NaOH) (170).

$$C_7H_6O_2 \longrightarrow C_6H_6 + CO_2 \qquad (xx)$$

Here, benzoic acid loses a carboxyl group (COOH) as carbon dioxide ($CO_2$). The transition metal catalyst helps activate the benzoic acid and accelerate the removal of $CO_2$ under high temperature conditions (around 200-300°C). This step is crucial as it generates benzene, the next intermediate in the process (171).

### 2. Hydrogenation of Benzene to Cyclohexane

Benzene is then subjected to hydrogenation to form cyclohexane ($C_6H_{12}$). This reaction uses a transition metal catalyst such as palladium (Pd), platinum (Pt), or nickel (Ni) (171, 172).

$$C_6H_6 + 3H_2 \longrightarrow C_6H_{12} \qquad (xxi)$$

The hydrogenation occurs under moderate conditions (typically 50-100°C) and atmospheric pressure. The transition metal catalyst facilitates the addition of hydrogen to benzene, converting it to cyclohexane. This step is necessary as cyclohexane serves as a precursor for further transformations (173).

### 3. Oxidation of cyclohexane to cyclohexanone

This reaction is carried out using catalysts such as chromium(VI) oxide ($CrO_3$) or manganese dioxide ($MnO_2$) in the presence of an oxidizing agent. The reaction is given as follows (174, 175, 176):

$$C_6H_{12}+O_2 \longrightarrow C_6H_{10}O+H_2O \qquad \text{(xxii)}$$

Transition metal catalysts help activate oxygen to facilitate the oxidation of cyclohexane to cyclohexanone. This reaction typically occurs at elevated temperatures (400°C to 600°C) and pressures (174).

### 4. Formation of cyclohexanoneoxime

Cyclohexanone is then converted to cyclohexanoneoxime ($C_6H_{10}NO$) through a reaction with hydroxylamine ($NH_2OH$). This step can be catalyzed by a transition metal-based catalyst or proceed under acidic or neutral conditions (177, 178).

$$C_6H_{10}O + NH_2OH \longrightarrow C_6H_{10}NO + H_2O \qquad \text{(xxiii)}$$

Cyclohexanoneoxime is an intermediate used in the subsequent Beckmann rearrangement.

### 5. Beckmann Rearrangement to Caprolactam

The final step is the Beckmann rearrangement of cyclohexanoneoxime to caprolactam ($C_6H_{11}NO$). This reaction requires an acid catalyst, but transition metals such as iron (Fe) or zinc (Zn) can be used to facilitate the process (179, 180).

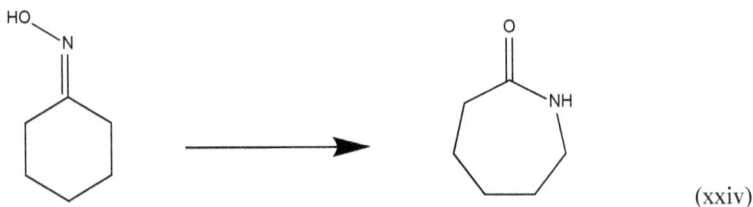

$$\text{(xxiv)}$$

During this rearrangement, cyclohexanoneoxime undergoes a rearrangement to form caprolactam, a six-membered ring lactam. This step is carried out at elevated temperatures (150-200°C) under acidic conditions (179, 180).

### 5.22 Terephthalic acid from p-xylene

The catalytic conversion of para-xylene to terephthalic acid is a critical industrial process used to produce terephthalic acid (TPA), which is a key raw material in the production of polyethylene terephthalate (PET) plastics and fibers. This process involves several key steps and utilizes specific catalysts to ensure high efficiency and selectivity (181).

The conversion process begins with the oxidation of p-xylene to p-toluic acid. This reaction typically occurs in the presence of a catalyst. The most commonly used catalysts for this step are combinations of cobalt and manganese, often with bromine as a promoter. Cobalt and manganese help in the activation of oxygen, while bromine enhances the efficiency and selectivity by reducing unwanted side reactions.The general reaction is as follows (181,182):

$$2C_8H_{10} + 3O_2 \longrightarrow 2C_8H_9COOH + 2H_2O \qquad (xxv)$$

Here, p-xylene ($C_8H_{10}$) is oxidized to form p-toluic acid ($C_8H_9COOH$). The process typically occurs at high temperatures (around 150-200°C) and pressures (up to 20 atm) to achieve effective oxidation. The catalyst helps to control the reaction and improve the yield by promoting the activation of molecular oxygen (182,183).

The p-toluic acid produced in the first step undergoes further oxidation to form TPA. This step is carried out under more stringent conditions using air or pure oxygen andoften with a different set of catalysts, including cobalt-manganese and sometimes iron-based catalysts (184).

$$C_8H_9COOH + O_2 \longrightarrow C_8H_6(COOH)_2 + H_2O \qquad (xxvi)$$

In this reaction, p-toluic acid ($C_8H_9COOH$) is further oxidized to produce TPA ($C_8H_6(COOH)_2$). This step also requires high temperatures (~ 200-250°C) and pressures (up to 40 atm) to ensure complete oxidation. The choice of catalyst and reaction conditions is crucial for maximizing the yield of TPA while minimizing by-products (184).

After the oxidation reactions, the TPA is separated from the reaction mixture by filtering out the solid product, followed by purification using recrystallization or solvent extraction to obtain high-purity TPA (184).

The conventional catalysts used in petroleum and petrochemical industry are given in table 4.

*Table 4: Overview of catalysts used in petrochemistry*

| Name of process | Catalyst used | Reaction conditions | Reactant (feed) | Products | Reference |
|---|---|---|---|---|---|
| **Petroleum refinery** | | | | | |
| Hydrocracking | Zeolite or aluminosilicate supported transition metals viz Ni, Co, W, Mo | 300-400°C 80-200 bar | Residue of atmospheric and vacuum distillation units | Lighter saturated hydrocarbons (fuel range hydrocarbons) | 14-19 |
| Catalytic cracking | ZSM-5, aluminated, transition metals ( Cr, Ce, Zr, Ti and alumina) complex on alumina support | 480-550°C 0.7-1.4 bar | Light naphtha, gas oil and atmospheric residue | Gasoline, diesel and other fuel range hydrocarbons | 20-22 |
| Hydrotreating | Mo or W sulfide active sites with Ni or Co as promoter supported on γ-alumina | 350- 450°C 0.7-5 MPa | All cuts from atmospheric and | Low heteroatom, and low sulphur containing | 23-25 |

| | | | vacuum distillation units | hydrocarbons Demetallized hydrocarbon cuts | |
|---|---|---|---|---|---|
| Catalytic reforming | Traditional catalytic: Pt and Rh Low cost catalyst: Fe, Ni, Co mixed oxide catalyst supported on alumina | 450-550°C 10-30 atm | Light naphtha | High-octane reformate | 26-34 |
| **Petrochemical industries** | | | | | |
| Methanol from syn gas | Cu, Fe and Co oxide supported on alumina | 200-300°C 50 to 100 atm | $CO$, $H_2$, $H_2O$ | $CH_3OH$, $CO_2$, $H_2$ | 35-38 |
| Catalytic production of ethylene | Fe, Ni, Cu oxides supported on alumina foam | 300-500°C Moderate pressure | Light naphtha, ethane | $C_2H_4$ | 39-41 |
| Catalytic naphtha cracking | Fe and Ni on $\gamma$-alumina and Co on silica | 500-800°C, atmospheric pressure | Heavy naphtha | Olefins (ethylene, propylene, butylenes, etc.) | 42-46 |
| Synthesis of vinyl chloride monomer by direct chlorination of ethylene | Fe and Al oxides | 100-200°C Atmospheric pressure | $C_2H_4$, $Cl_2$ | $C_2H_3Cl$ | 47-50 |
| Vinyl acetate monomer synthesis | Metallic or oxide of Cu on alumina | 200-350°C 1-10 atm | $C_2H_4$, $CH_3COOH$, $O_2$ | $CH_2CH(OCOC H_3)$ | 51-56 |
| Acrylonitrile synthesis | Fe-Mo or V-Sb supported on alumina or silica | 300-500°C 1-5 atm | $C_3H_6$, $NH_3$, $O_2$ | $C_3H_3N$ | 57-64 |
| Acetaldehyde synthesis from acetylene and ammonia | Ni or Cu in the form of oxide or chloride supported on alumina | 200-350°C Higher partial pressure of acetylene and ammonia | $C_2H_2$, $NH_3$, $O_2$ | $CH_3CHO$ | 65-71 |
| Catalytic dehydrogenataion of propane and higher paraffins | FeCr mixed oxide catalyst | 500-800°C Moderate pressure | $C_3H_8$, $C_4H_{10}$ | $C_3H_6$, $C_4H_8$ | 72-78 |
| Direct oxidation process for methacrylate synthesis | $V_2O_5$ based bulk catalyst | 200-400°C | $C_4H_8$, $CH_3OH$, $O_2$ | $C_5H_6O_2$ | 79-82 |
| Propylene glycol | Titanium dioxide and | 80-150°C | $C_3H_6O$, | $C_3H_8O_2$ | 83-92 |

| | | | | | |
|---|---|---|---|---|---|
| form propylene oxide | zirconium oxide based catalyst or aluminium chloride based catalyst | Atmospheric pressure | $H_2O$ | | |
| n-butylene from iso and n-butanol | Platinum or gallium supported on zeolite Cr, Zr, W or Mo oxide supported alumina catalyst | 150-300°C Atmospheric pressure | $C_4H_9OH$, $C_4H_{10}O$ | $C_4H_8O$ | 93-103 |
| Oxo route for n-butanol synthesis | Rh or Co complex catalysed hydroformylation followed by reduction using metallic Pd catalyst | 100- 200°C 30-70 bar (for the first step) | Alkenes | Aldehydes | 104-113 |
| Methacrylic acid from isobutylene | Mo or W oxide based catalyst for oxidative cleavage reaction followed by transformation to methacrylic acid catalysed by Pt or Pt impregnated on alumina | 200-400°C Atmospheric pressure (for the first step) 100-200°C Atmospheric pressure (for the second step) | $C_4H_8$ | $C_4H_4O_2$ | 114-121 |
| MTBE from isobutene and methanol | Zeolite-Y or a mixture of W and Mo oxide | 100-200°C Pressure slightly above 1atm | $C_4H_8$, $CH_3OH$ | $C_5H_{12}O$ | 122-127 |
| Butadiene from butane | Pt, Pd Raney Ni or Ni supported on alumina | 500-700°C Moderate pressure | $C_4H_{10}$ | $C_4H_6$ | 128-132 |
| Maleic anhydride synthesis | $V_2O_5$ supported on γ-alumina | 300-500°C (for n-butane) 400-600°C (for benzene) Moderate pressure in both case | $C_4H_{10}$and $O_2$or $C_6H_6$and$O_2$ | $C_4H_2O_3$ | 133-137 |
| Isoprene from isopentane | Pd or Ni in metallic form as catalyst for isopentane dehydrogenation followed by Zr, Cr and Mo mixed oxide catalyst for dehydrocyclization | 300-500°C, atmospheric pressure (for the first step) 400-600°C, reduced pressure (for the | $C_5H_{12}$ | $C_5H_8$ | 138-144 |

| | | second step) | | | |
|---|---|---|---|---|---|
| Polypropylene synthesis | Ziegler-Natta catalysts ($TiCl_4$ or $TiCl_3$ mixed with an organoaluminum viz. TEA or TMA) and metallocene catalysts ($Cp_2TiMe_2$) | 50-80°C 1-10 atm | $C_3H_6$ | $(C_3H_6)_n$ | 145-152 |
| Aniline from nitrobenzene | Pt, Pd or Ni in their metallic form | 50-150°C 30-50 atm | $C_6H_5NO_2$, $H_2$ | $C_6H_5NH_2$ | 153-159 |
| Styrene from ethyl benzene | $Cr_2O_3$ or $Fe_2O_3$ on alumina support | 500-600°C Pressure below atmospheric pressure | $C_6H_5C_2H_5$ | $C_6H_5CH=CH_2$ | 160-162 |
| Benzoic acid hydrogenation to caprolactum | $Cu_2O$ is used for decarboxylation of benzoic acid followed by Pt, Pd or Ni metallic catalyst for the formation of cyclohexane. $CrO_3$ or $MnO_2$ is used in the next step for the oxidation of cyclohexane to cyclohexanone which is converted to cyclohexanoneoxime in acidic or neutral medium ensued by Beckmann Rearrangement to caprolactam in the presence of Fe or Zn oxide catalyst | 200-300°C, moderate pressure (first step) 50-100°C, atmospheric pressure (second step) 400-600°C, high pressure (third step) 90°C, atmospheric pressure Beckmann Rearrangement | $C_7H_6O_2$ | $C_6H_{11}NO$ | 168-180 |
| Terephthalic acid from p-xylene | Co and Mn oxide bulk catalyst is used for the oxidation of p-xylene and similar catalyst with variation of Co and Mn ratio is used in the second step for the formation of TPA | 150-200°C up to 20 atm (for the first step) 200-250°C up to 40 atm (for the second step) | $C_8H_{10}$, $O_2$, | $C_8H_6 (COOH)_2$ | 181-184 |

## Conclusion

The conversion of crude oil to valuable products and fuels requires complex chemical reactions. These processes are also energy intensive. The energy demand can be mitigated and process selectivity increased by the incorporation of various catalysts. Transition metal plays a vital role as catalysts in these processes due to the presence of special characteristic of existing in multiple oxidation states. In petrochemistry, transition metals act as catalysts to enhance the efficiency of key processes like cracking, reforming, hydrocracking and also in petrochemical production. They help convert hydrocarbon molecules into valuable products. For instance, platinum and palladium are used in catalytic reforming to produce high-octane fuels, while nickel and molybdenum are used in hydrocracking to convert heavier fractions into lighter, more valuable products. Their ability to expedite these reactions efficiently makes them essential for optimizing petrochemical production.

## References

[1] Orege, J.I., Oderinde, O., Kifle, G.A., Ibikunle, A.A., Raheem, S.A., Ejeromedoghene, O., Okeke, E.S., Olukowi, O.M., Orege, O.B., Fagbohun, E.O., Ogundipe, T.O., Avor, E.P., Ajayi, O.O., Daramola, M.O., 2022. Recent advances in heterogeneous catalysis for green biodiesel production by transesterification. Energy Conversion and Management. https://doi.org/10.1016/j.enconman.2022.115406

[2] Kumar, S., Kumar, G., Saroha, B., Gulati, K., 2023. Metal oxide heterostructures as catalysts in organic reactions. Metal Oxide-Based heterostructures. https://doi.org/10.1016/b978-0-323-85241-8.00003-7

[3] Melián-Cabrera, I., 2021. Catalytic Materials: Concepts to Understand the Pathway to Implementation. Ind. Eng. Chem. Res. https://doi.org/10.1021/acs.iecr.1c02681

[4] Shin, S., Kim, M.-J., 2024. Hydrothermal synthesis of zeolites from residual waste generated via indirect carbonation of coal fly ash. Sustain Environ Res. https://doi.org/10.1186/s42834-023-00206-6

[5] Basrur, A., Sabde, D., 2016. Catalyst Synthesis and Characterization. Industrial Catalytic Processes for Fine and Specialty Chemicals. https://doi.org/10.1016/b978-0-12-801457-8.00004-5

[6] Mehrabadi, B.A.T., Eskandari, S., Khan, U., White, R.D., Regalbuto, J.R., 2017. A Review of Preparation Methods for Supported Metal Catalysts. Advances in Catalysis. https://doi.org/10.1016/bs.acat.2017.10.001

[7] Munnik, P., de Jongh, P.E., de Jong, K.P., 2015. Recent Developments in the Synthesis of Supported Catalysts. Chem. Rev. https://doi.org/10.1021/cr500486u

[8] GS, A., Raphel M, S., 2024. Cracking Upgrading Process of Biomass. Encyclopedia of Renewable Energy, Sustainability and the Environment. https://doi.org/10.1016/b978-0-323-93940-9.00057-8

[9] GS, A., Raphel M, S., 2024. Cracking Upgrading Process of Biomass. Encyclopedia of Renewable Energy, Sustainability and the Environment. https://doi.org/10.1016/b978-0-323-93940-9.00057-8

[10]   Kurian, M., Thankachan, S., 2023. Ceramics as catalyst supports. Ceramic Catalysts. https://doi.org/10.1016/b978-0-323-85746-8.00008-4

[11]   Xu, G., Yang, P., Yang, S., Wang, H., Fang, B., 2022. Non-natural catalysts for catalytic tar conversion in biomass gasification technology. International Journal of Hydrogen Energy. https://doi.org/10.1016/j.ijhydene.2021.12.094

[12]   Yeetsorn, R., Tungkamani, S., Maiket, Y., 2022. Fabrication of a Ceramic Foam Catalyst Using Polymer Foam Scrap via the Replica Technique for Dry Reforming. ACS Omega. https://doi.org/10.1021/acsomega.1c05841

[13]   Richardson, J.T., Twigg, M.V., 1994. Ceramic Foam Catalyst Supports Preparation and Properties. MRS Proc. https://doi.org/10.1557/proc-368-315

[14]   Aitani, A.M., 2004. Oil Refining and Products. Encyclopedia of Energy. https://doi.org/10.1016/b0-12-176480-x/00259-x

[15]   Catherin, N., Blanco, E., Laurenti, D., Piccolo, L., Simonet, F., Lorentz, C., Leclerc, E., Calemma, V., Geantet, C., 2021. Transition metal sulfides on zeolite catalysts for selective ring opening. Catalysis Today. https://doi.org/10.1016/j.cattod.2020.10.012

[16]   Rigutto, M.S., van Veen, R., Huve, L., 2007. Zeolites in Hydrocarbon Processing. Studies in Surface Science and Catalysis. https://doi.org/10.1016/s0167-2991(07)80812-3

[17]   Weitkamp, J., 2012. Catalytic Hydrocracking—Mechanisms and Versatility of the Process. ChemCatChem. https://doi.org/10.1002/cctc.201100315

[18]   Kapustin, V., Chernysheva, E., Khakimov, R., 2021. Comparison of Moving-Bed Catalytic Tar Hydrocracking Processes. Processes. https://doi.org/10.3390/pr9030500

[19]   M. A.,Halabi, J. Beshara, H. Oabazard, A. Stanislaus, 1995. Catalysts in petroleum refining and petrochemical industries

[20]   Xu, Y., Zuo, Y., Yang, W., Shu, X., Chen, W., Zheng, A., 2023. Targeted Catalytic Cracking to Olefins (TCO): Reaction Mechanism, Production Scheme, and Process Perspectives. Engineering. https://doi.org/10.1016/j.eng.2023.02.018

[21]   Liu, Y., Chen, R., Liu, J., Zhang, X., 2022. Research Progress of Catalysts and Initiators for Promoting the Cracking of Endothermic Hydrocarbon Fuels. Trans. Tianjin Univ. https://doi.org/10.1007/s12209-022-00315-0

[22]   Bai, P., Etim, U.J., Yan, Z., Mintova, S., Zhang, Z., Zhong, Z., Gao, X., 2018. Fluid catalytic cracking technology: current status and recent discoveries on catalyst contamination. Catalysis Reviews. https://doi.org/10.1080/01614940.2018.1549011

[23]   Polischuk, C., Eleeza, J., Vedachalam, S., Dalai, A.K., Adjaye, J., 2023. A review of foulant sources, operational issues, and remedies during the processing of oil sand derived bitumen fractions. Fuel. https://doi.org/10.1016/j.fuel.2023.127516

[24]   Xiaojie, Z., Mukherjee, K., Manna, S., Das, M.K., Kim, J.K., Sinha, T.K., 2022. Efficient management of oil waste: chemical and physicochemical approaches. Advances in Oil-Water Separation. https://doi.org/10.1016/b978-0-323-89978-9.00027-6

Materials Research Forum LLC
https://doi.org/21741/9781644903711-13

[25]  Tomášek, J., Matějovský, L., Lamblová, M., Blažek, J., 2020. Properties and Composition of Products from Hydrotreating of Straight-Run Gas Oil and Its Mixtures with Light Cycle Oil over Sulfidic Ni-Mo/Al2O3 Catalyst. ACS Omega. https://doi.org/10.1021/acsomega.0c03259

[26]  Foong, S.Y., Chan, Y.H., Cheah, W.Y., Kamaludin, N.H., Tengku Ibrahim, T.N.B., Sonne, C., Peng, W., Show, P.-L., Lam, S.S., 2021. Progress in waste valorization using advanced pyrolysis techniques for hydrogen and gaseous fuel production. Bioresource Technology. https://doi.org/10.1016/j.biortech.2020.124299

[27]  Cheremisinoff, N.P., Rosenfeld, P., 2009. The petroleum industry. Handbook of Pollution Prevention and Cleaner Production - Best Practices in The Petroleum Industry. https://doi.org/10.1016/b978-0-8155-2035-1.10001-6

[28]  Fink, J. K., Guide to the Practical Use of Chemicals in Refineries and Pipelines Chapter 11-Processes

[29]  Viswanathan, B., 2017. Petroleum. Energy Sources. https://doi.org/10.1016/b978-0-444-56353-8.00002-2

[30]  Seman, M.H.A., Othman, N.H., Osman, N., Jani, A.M.M., 2023. Nickel based catalysts supported on porous support for methane steam reforming: potential and short review. IOP Conf. Ser.: Earth Environ. Sci. https://doi.org/10.1088/1755-1315/1151/1/012061

[31]  Morlanés, N., 2013. Reaction mechanism of naphtha steam reforming on nickel-based catalysts, and FTIR spectroscopy with CO adsorption to elucidate real active sites. International Journal of Hydrogen Energy. https://doi.org/10.1016/j.ijhydene.2012.12.128

[32]  U.T.Turaga, R. Ramanathan, 2003. Catalytic Naphtha Reforming: Revisiting its Importance in the Modern Refinery. Journal of Scientific & Industrial Research

[33]  Grzybek, G., Greluk, M., Indyka, P., Góra-Marek, K., Legutko, P., Słowik, G., Turczyniak-Surdacka, S., Rotko, M., Sojka, Z., Kotarba, A., 2020. Cobalt catalyst for steam reforming of ethanol–Insights into the promotional role of potassium. International Journal of Hydrogen Energy. https://doi.org/10.1016/j.ijhydene.2020.06.037

[34]  Sun, Y., Zhang, Y., Yin, X., Zhang, C., Li, Y., Bai, J., 2024. Recent advances in the design of high-performance cobalt-based catalysts for dry reforming of methane. Green Chem. https://doi.org/10.1039/d3gc05136f

[35]  Yerga, R.M.N., 2021. Catalysts for Production and Conversion of Syngas. Catalysts. https://doi.org/10.3390/catal11060752

[36]  Mierczynski, P., Stępińska, N., Mosinska, M., Chalupka, K., Albinska, J., Maniukiewicz, W., Rogowski, J., Nowosielska, M., Szynkowska, M.I., 2020. Hydrogen Production via the Oxy-Steam Reforming of LNG or Methane on Ni Catalysts. Catalysts. https://doi.org/10.3390/catal10030346

[37]  Karemore, A.L., Sinha, R., Chugh, P., Vaidya, P.D., 2021. Parametric and Reaction Kinetic Study of Syngas Production from Dry Methane Reforming over Improved Nickel Catalysts. Energy Fuels. https://doi.org/10.1021/acs.energyfuels.0c04037

[38]    Abasaeed, A.E., Lanre, M.S., Kasim, S.O., Ibrahim, A.A., Osman, A.I., Fakeeha, A.H., Alkhalifa, A., Arasheed, R., Albaqi, F., Kumar, N.S., Khan, W.U., Kumar, R., Frusteri, F., Al-Fatesh, A.S., Bagabas, A.A., 2023. Syngas production from methane dry reforming via optimization of tungsten trioxide-promoted mesoporous γ-alumina supported nickel catalyst. International Journal of Hydrogen Energy. https://doi.org/10.1016/j.ijhydene.2022.09.313

[39]    Zhao, B.-H., Chen, F., Wang, M., Cheng, C., Wu, Y., Liu, C., Yu, Y., Zhang, B., 2023. Economically viable electrocatalytic ethylene production with high yield and selectivity. Nat Sustain. https://doi.org/10.1038/s41893-023-01084-x

[40]    Gaffney, A.M., Mason, O.M., 2017. Ethylene production via Oxidative Dehydrogenation of Ethane using M1 catalyst. Catalysis Today. https://doi.org/10.1016/j.cattod.2017.01.020

[41]    Gao, Y., Neal, L., Ding, D., Wu, W., Baroi, C., Gaffney, A.M., Li, F., 2019. Recent Advances in Intensified Ethylene Production—A Review. ACS Catal. https://doi.org/10.1021/acscatal.9b02922

[42]    Yoshimura, Y., Kijima, N., Hayakawa, T., Murata, K., Suzuki, K., Mizukami, F., Matano, K., Konishi, T., Oikawa, T., Saito, M., Shiojima, T., Shiozawa, K., Wakui, K., Sawada, G., Sato, K., Matsuo, S., Yamaoka, N., 2001. . Catalysis Surveys from Japan. https://doi.org/10.1023/a:1011463606189

[43]    Longstaff, D.C., 2019. Naphtha Cracking Kinetics and Process Chemistry on Y and ZSM5 Type Catalysts. Energy Fuels. https://doi.org/10.1021/acs.energyfuels.8b04128

[44]    Tian, Y., He, X., Chen, X., Qiao, C., Wang, H., Diao, Z., Liu, G., 2024. Transition metal modified hierarchical ZSM-5 nanosheet for catalytic cracking of n-pentane to light olefins. Fuel. https://doi.org/10.1016/j.fuel.2024.130902

[45]    Al-Marshed, A., Hart, A., Leeke, G., Greaves, M., Wood, J., 2015. Effectiveness of Different Transition Metal Dispersed Catalysts for In Situ Heavy Oil Upgrading. Ind. Eng. Chem. Res. https://doi.org/10.1021/acs.iecr.5b02953

[46]    Mohiuddin, E., Mdleleni, M.M., Key, D., 2018. Catalytic cracking of naphtha: The effect of Fe and Cr impregnated ZSM-5 on olefin selectivity. Appl Petrochem Res. https://doi.org/10.1007/s13203-018-0200-2

[47]    Ma, H., Wang, Y., Qi, Y., Rout, K.R., Chen, D., 2020. Critical Review of Catalysis for Ethylene Oxychlorination. ACS Catal. https://doi.org/10.1021/acscatal.0c01698

[48]    Bruzzi, V., Colaianni, M., Zanderighi, L., 1998. Energy savings in chemical plants: a vinyl chloride case history. Energy Conversion and Management. https://doi.org/10.1016/s0196-8904(98)00058-2

[49]    L, Andliyani, I, Alfand, N. Inda, Z, Muchtar, 2023.Production of Vinyl Chloride from Ethylene: Technology Review and Process Selection. International Engineering Student Conference 2023

[50]    Scharfe, M., Paunović, V., Mitchell, S., Hauert, R., Xi, S., Borgna, A., Pérez-Ramírez, J., 2020. Dual catalyst system for selective vinyl chloride production via ethane oxychlorination. Catal. Sci. Technol. https://doi.org/10.1039/c9cy01801h

Applications for Earth-Abundant Transition Metals
Materials Research Foundations 179 (2025) 356-399

Materials Research Forum LLC
https://doi.org/21741/9781644903711-13

[51]    Gonzalez Caranton, A.R., Schmal, M., CheccaHuaman, N.R., da Silva Pinto, J.C., 2023. Synthesis of Vinyl Acetate Monomer OverPd Cu Alloys: The Role of Surface Oxygenation in the Reaction Path. Macro Reaction Engineering. https://doi.org/10.1002/mren.202300016

[52]    Gonzalez Caranton, A.R., da Silva Pinto, J.C.C., Stavale, F., Barreto, J., Schmal, M., 2020. Statistical analysis of the catalytic synthesis of Vinyl acetate over Pd-Cu/ZrO2 nanostructured based catalysts. Catalysis Today. https://doi.org/10.1016/j.cattod.2018.10.034

[53]    Thuening, T., Tysoe, W.T., 2017. Kinetics and Mechanism of Vinyl Acetate Monomer Synthesis on Pd(100) Model Catalysts. Catal Lett. https://doi.org/10.1007/s10562-017-2109-2

[54]    Buronov, F., Normurod, F., 2021. ACTIVE CATALYSTS FOR PRODUCING VINYL ACETATE MONOMERS. UniTech. https://doi.org/10.32743/unitech.2021.86.5.11684

[55]    Stacchiola, D., Calaza, F., Burkholder, L., Tysoe, W.T., 2004. Vinyl Acetate Formation by the Reaction of Ethylene with Acetate Species on Oxygen-Covered Pd(111). J. Am. Chem. Soc. https://doi.org/10.1021/ja044641w

[56]    K.M. Vapoyev, B.F. Muhiddinov, J.R. UmarovaL.I. ,Tilavova, 2019. Synthesis of vinyl acetate in a liquid phase. International Journal of Advanced Research in Science, Engineering and Technology.

[57]    Zhang, J., Hu, W., Li, Y., Savoy, A., Sun, J., Chi, T.Y., Wang, Y., 2024. Advances in the catalytic    production    of    acrylonitrile.    Chem    Catalysis. https://doi.org/10.1016/j.checat.2023.100825

[58]    Zhang, J., Hu, W., Li, Y., Savoy, A., Sun, J., Chi, T.Y., Wang, Y., 2024. Advances in the catalytic    production    of    acrylonitrile.    Chem    Catalysis. https://doi.org/10.1016/j.checat.2023.100825

[59]    Brazdil, J.F., 2017. A critical perspective on the design and development of metal oxide catalysts for selective propylene ammoxidation and oxidation. Applied Catalysis A: General. https://doi.org/10.1016/j.apcata.2017.06.022

[60]    Guerrero-Pérez, M.O., Peña, M.A., Fierro, J.L.G., Bañares, M.A., 2006. A Study about the Propane Ammoxidation to Acrylonitrile with an Alumina-Supported Sb−V−O Catalyst. Ind. Eng. Chem. Res. https://doi.org/10.1021/ie051000g

[61]    Bell, A.T., 2022. Insights into the mechanism and kinetics of propene oxidation and ammoxidation over bismuth molybdate catalysts derived from experiments and theory. Journal of Catalysis. https://doi.org/10.1016/j.jcat.2021.05.009

[62]    Singh, B., Gawande, M.B., Kute, A.D., Varma, R.S., Fornasiero, P., McNeice, P., Jagadeesh, R.V., Beller, M., Zbořil, R., 2021. Single-Atom (Iron-Based) Catalysts: Synthesis and Applications. Chem. Rev. https://doi.org/10.1021/acs.chemrev.1c00158

[63]    Catani, R., Centi, G., Trifiro, F., Grasselli, R.K., 1992. Kinetics and reaction network in propane ammoxidation to acrylonitrile on vanadium-antimony-aluminum based mixed oxides. Ind. Eng. Chem. Res. https://doi.org/10.1021/ie00001a016

[64]    Alexopoulos, K., Reyniers, M.-F., Marin, G.B., 2012. Reaction path analysis of propene selective    oxidation    over    V2O5    and    V2O5/TiO2.    Journal    of    Catalysis. https://doi.org/10.1016/j.jcat.2012.08.010

[65] Gebers, J.C., Abu Kasim, A.F.B., Fulham, G.J., Kwong, K.Y., Marek, E.J., 2023. Production of Acetaldehyde via Oxidative Dehydrogenation of Ethanol in a Chemical Looping Setup. ACS Eng. Au. https://doi.org/10.1021/acsengineeringau.2c00052

[66] Ob-eye, J., Praserthdam, P., Jongsomjit, B., 2019. Dehydrogenation of Ethanol to Acetaldehyde over Different Metals Supported on Carbon Catalysts. Catalysts. https://doi.org/10.3390/catal9010066

[67] Pekmezci Karaman, B., Cakiryilmaz, N., Arbag, H., Oktar, N., Dogu, G., Dogu, T., 2017. Performance comparison of mesoporous alumina supported Cu & Ni based catalysts in acetic acid reforming. International Journal of Hydrogen Energy. https://doi.org/10.1016/j.ijhydene.2017.08.155

[68] Akmalaev, K.A., Fayzullaev, N.I., Karjavov, A., 2020. Joint synthesis of acetone and acetaldehyde from acetylene. Asia. Journ. of Multidimensi. Resear. (AJMR). https://doi.org/10.5958/2278-4853.2020.00246.3

[69] Trotuş, I.-T., Zimmermann, T., Schüth, F., 2013. Catalytic Reactions of Acetylene: A Feedstock for the Chemical Industry Revisited. Chem. Rev. https://doi.org/10.1021/cr400357r

[70] Voronin, V.V., Ledovskaya, M.S., Bogachenkov, A.S., Rodygin, K.S., Ananikov, V.P., 2018. Acetylene in Organic Synthesis: Recent Progress and New Uses. Molecules. https://doi.org/10.3390/molecules23102442

[71] HETEROGENEOUS CATALYTIC CONVERSIONS OF ACETYLENE, 2021. CHEMISTRY AND CHEMICAL ENGINEERING. https://doi.org/10.51348/gttb5269

[72] Trotuş, I.-T., Zimmermann, T., Schüth, F., 2013. Catalytic Reactions of Acetylene: A Feedstock for the Chemical Industry Revisited. Chem. Rev. https://doi.org/10.1021/cr400357r

[73] Pitkethly, R.C., Steiner, H., 1939. A note on the mechanism of catalytic dehydrogenation and cyclisation. Trans. Faraday Soc. https://doi.org/10.1039/tf9393500979

[74] Vafajoo, L., Khorasheh, F., Nakhjavani, M.H., Fattahi, M., 2014. Kinetic Parameters Optimization and Modeling of Catalytic Dehydrogenation of Heavy Paraffins to Olefins. Petroleum Science and Technology. https://doi.org/10.1080/10916466.2011.604061

[75] Wang, G., Zhu, X., Li, C., 2019. Recent Progress in Commercial and Novel Catalysts for Catalytic Dehydrogenation of Light Alkanes. The Chemical Record. https://doi.org/10.1002/tcr.201900090

[76] Tan, S., Hu, B., Kim, W.-G., Pang, S.H., Moore, J.S., Liu, Y., Dixit, R.S., Pendergast, J.G., Sholl, D.S., Nair, S., Jones, C.W., 2016. Propane Dehydrogenation over Alumina-Supported Iron/Phosphorus Catalysts: Structural Evolution of Iron Species Leading to High Activity and Propylene Selectivity. ACS Catal. https://doi.org/10.1021/acscatal.6b01286

[77] Zhang, X., Li, J., Zheng, Y., Xin, W., An, J., Zhu, X., Li, X., 2023. Structural reconstruction of iron oxide induces stable catalytic performance in the oxidative dehydrogenation of n-butane to 1,3-butadiene. Chemical Engineering Journal. https://doi.org/10.1016/j.cej.2023.145370

[78]   Li, P.-P., Lang, W.-Z., Xia, K., Luan, L., Yan, X., Guo, Y.-J., 2016. The promotion effects of Ni on the properties of Cr/Al catalysts for propane dehydrogenation reaction. Applied Catalysis A: General. https://doi.org/10.1016/j.apcata.2016.05.007

[79]   Elbashir, N.O., Al-Zahrani, S.M., Abasaeed, A.E., Abdulwahed, M., 2003. Alumina-supported chromium-based mixed-oxide catalysts in oxidative dehydrogenation of isobutane to isobutene. Chemical Engineering and Processing: Process Intensification. https://doi.org/10.1016/s0255-2701(02)00108-3

[80]   Védrine, J., 2016. Heterogeneous Partial (amm) Oxidation and Oxidative Dehydrogenation Catalysis on Mixed Metal Oxides. Catalysts. https://doi.org/10.3390/catal6020022

[81]   Darabi Mahboub, M.J., Dubois, J.-L., Cavani, F., Rostamizadeh, M., Patience, G.S., 2018. Catalysis for the synthesis of methacrylic acid and methyl methacrylate. Chem. Soc. Rev. https://doi.org/10.1039/c8cs00117k

[82]   Di Capua, A., Dubois, J.-L., Fournier, M., 2007. Fine analysis of by-products of the selective oxidation of isobutane into methacrolein and methacrylic acid over Mo–V–P catalyst. Journal of Molecular Catalysis A: Chemical. https://doi.org/10.1016/j.molcata.2006.07.004

[83]   AI, M., 1986. Oxidation of propane to acrylic acid on V2O5&$z.sbnd;P2O5-based catalysts. Journal of Catalysis. https://doi.org/10.1016/0021-9517(86)90266-6

[84]   Horbatenko, Y., Pérez, J.P., Hernández, P., Swart, M., Solà, M., 2014. Reaction Mechanisms for the Formation of Mono- And Dipropylene Glycol from the Propylene Oxide Hydrolysis over ZSM-5 Zeolite. J. Phys. Chem. C. https://doi.org/10.1021/jp504432a

[85]   Alvear, M., Eränen, K., Murzin, D.Yu., Salmi, T., 2021. Study of the Product Distribution in the Epoxidation of Propylene over TS-1 Catalyst in a Trickle-Bed Reactor. Ind. Eng. Chem. Res. https://doi.org/10.1021/acs.iecr.0c06150

[86]   Pandey, D.K., Biswas, P., 2019. Production of propylene glycol (1,2-propanediol) by the hydrogenolysis of glycerol in a fixed-bed downflow tubular reactor over a highly effective Cu–Zn bifunctional catalyst: effect of an acidic/basic support. New J. Chem. https://doi.org/10.1039/c9nj01180c

[87]   Okolie, J.A., 2022. Insights on production mechanism and industrial applications of renewable propylene glycol. Science. https://doi.org/10.1016/j.isci.2022.104903

[88]   Huang, Z., Lin, Y., Li, L., Ye, C., Qiu, T., 2017. Preparation and shaping of solid acid SO42−/TiO2 and its application for esterification of propylene glycol monomethyl ether and acetic acid. Chinese Journal of Chemical Engineering. https://doi.org/10.1016/j.cjche.2016.11.006

[89]   Santhosh, S., Tamizhdurai, P., Kavitha, C., Mangesh, V.L., Kumar, N.S., Basivi, P.K., Al-Fatesh, A.S., Kumaran, R., 2023. ZrO2/SO4/Cu nanoparticles supported on reduced graphene oxide for selective oxidation of propylene glycol in continuous reactor. International Journal of Hydrogen Energy. https://doi.org/10.1016/j.ijhydene.2023.05.027

[90]   Liu, Z., Zhao, W., Xiao, F., Wei, W., Sun, Y., 2010. One-pot synthesis of propylene glycol and dipropylene glycol over strong basic catalyst. Catalysis Communications. https://doi.org/10.1016/j.catcom.2010.01.004

[91]   TAKAHASHI, R., ONISHI, A., SATO, F., KURAMOTO, M., 2017. Preparation of bimodal porous alumina using propylene glycol oligomers. J. Ceram. Soc. Japan. https://doi.org/10.2109/jcersj2.17062

[92]   Doherty, S., Errington, R.J., Housley, N., Clegg, W., 2004. Dimeric Aluminum Chloride Complexes of N-Alkoxyalkyl-β-ketoimines: Activation with Propylene Oxide To Form Efficient      Lactide      Polymerization      Catalysts.      Organometallics. https://doi.org/10.1021/om0343770

[93]   Timofeeva, M.N., Panchenko, V.N., Gil, A., Chesalov, Y.A., Sorokina, T.P., Likholobov, V.A., 2011. Synthesis of propylene glycol methyl ether from methanol and propylene oxide over      alumina-pillared      clays.      Applied      Catalysis      B:      Environmental. https://doi.org/10.1016/j.apcatb.2010.12.020

[94]   Wingad, R.L., Birch, L., Farndon, J., Lee, J., Pellow, K.J., Wass, D.F., 2023. Control in Advanced Biofuels Synthesis via Alcohol Upgrading: Catalyst Selectivity to n-Butanol, sec-Butanol or Isobutanol. ChemCatChem. https://doi.org/10.1002/cctc.202201410

[95]   BRYANT, D., 1967. Dehydration of alcohols over zeolite catalysts. Journal of Catalysis. https://doi.org/10.1016/0021-9517(67)90275-8

[96]   John, M., Alexopoulos, K., Reyniers, M.-F., Marin, G.B., 2015. Reaction path analysis for 1-butanol dehydration in H-ZSM-5 zeolite: Ab initio and microkinetic modeling. Journal of Catalysis. https://doi.org/10.1016/j.jcat.2015.07.005

[97]   Van Daele, S., Minoux, D., Nesterenko, N., Maury, S., Coupard, V., Valtchev, V., Travert, A., Gilson, J.-P., 2021. A highly selective FER-based catalyst to produce n-butenes from      isobutanol.      Applied      Catalysis      B:      Environmental. https://doi.org/10.1016/j.apcatb.2020.119699

[98]   Makarova, M.A., Paukshtis, E.A., Thomas, J.M., Williams, C., Zamaraev, K.I., 1994. Dehydration of n-Butanol on Zeolite H-ZSM-5 and Amorphous Aluminosilicate: Detailed Mechanistic Study and the Effect of Pore Confinement. Journal of Catalysis. https://doi.org/10.1006/jcat.1994.1270

[99]   Zikrata, O.V., Larina, O.V., Balakin, D.Yu.,Nychiporuk, Y.M., Khalakhan, I., Švegovec, M., Volavšek, J., Yaremov, P.S., Soloviev, S.O., Orlyk, S.M., 2024. Influence of Acid-Base Characteristics of Different Structural-Type Zeolites (FER, MFI, FAU, BEA) on Their Activity   and   Selectivity   in   Isobutanol   Dehydration.   ChemCatChem. https://doi.org/10.1002/cctc.202400068

[100]  Potter, M.E., Amsler, J., Spiske, L., Plessow, P.N., Asare, T., Carravetta, M., Raja, R., Cox, P.A., Studt, F., Armstrong, L.-M., 2023. Combining Theoretical and Experimental Methods to Probe Confinement within Microporous Solid Acid Catalysts for Alcohol Dehydration. ACS Catal. https://doi.org/10.1021/acscatal.3c00352

[101] Tian, K., Li, Q., Jiang, W., Wang, X., Liu, S., Zhao, Y., Zhou, G., 2021. Effect of the pore structure of an active alumina catalyst on isobutene production by dehydration of isobutanol. RSC Adv. https://doi.org/10.1039/d1ra00136a

[102] Kella, T., Vennathan, A.A., Dutta, S., Mal, S.S., Shee, D., 2021. Selective dehydration of 1-butanol to butenes over silica supported heteropolyacid catalysts: Mechanistic aspect. Molecular Catalysis. https://doi.org/10.1016/j.mcat.2021.111975

[103] Buniazet, Z., Lorentz, C., Cabiac, A., Maury, S., Loridant, S., 2018. Supported oxides catalysts for the dehydration of isobutanol into butenes: Relationships between acidic and catalytic properties. Molecular Catalysis. https://doi.org/10.1016/j.mcat.2017.12.007

[104] Wu, J., Liu, H.-J., Yan, X., Zhou, Y.-J., Lin, Z.-N., Mi, S., Cheng, K.-K., Zhang, J.-A., 2019. Efficient Catalytic Dehydration of High-Concentration 1-Butanol with Zn-Mn-Co Modified γ-Al2O3 in Jet Fuel Production. Catalysts. https://doi.org/10.3390/catal9010093

[105] Jiang, Y., Liu, J., Jiang, W., Yang, Y., Yang, S., 2015. Current status and prospects of industrial bio-production of n-butanol in China. Biotechnology Advances. https://doi.org/10.1016/j.biotechadv.2014.10.007

[106] Sakai, N., Mano, S., Nozaki, K., Takaya, H., 1993. Highly enantioselective hydroformylation of olefins catalyzed by new phosphine phosphite-rhodium(I) complexes. J. Am. Chem. Soc. https://doi.org/10.1021/ja00068a095

[107] Billig, E., Bryant, D.R., 2000. Oxo Process. Kirk-Othmer Encyclopedia of Chemical Technology. https://doi.org/10.1002/0471238961.15241502091212.a01

[108] Zhang, B., Kubis, C., Franke, R., 2022. Hydroformylation catalyzed by unmodified cobalt carbonyl under mild conditions. Science. https://doi.org/10.1126/science.abm4465

[109] Xu, W., Ma, Y., Wei, X., Gong, H., Zhao, X., Qin, Y., Peng, Q., Hou, Z., 2022. Core–shell Co@CoO catalysts for the hydroformylation of olefins. New J. Chem. https://doi.org/10.1039/d2nj02797f

[110] Zhang, S., Chen, J., Wei, B., Zhou, H., Hua, K., Liu, X., Wang, H., Sun, Y., 2024. Efficient Alkene Hydroformylation by Co–C Symmetry-Breaking Sites. J. Am. Chem. Soc. https://doi.org/10.1021/jacs.3c13092

[111] Wang, A., Yang, Z., Liu, J., Gui, Q., Chen, X., Tan, Z., Shi, J.-C., 2013. Pd-Catalyzed Reduction of Aldehydes to Alcohols Using Formic Acid as the Hydrogen Donor. Synthetic Communications. https://doi.org/10.1080/00397911.2013.804575

[112] Ren, H., Long, P., Zhao, Y., Zhang, K., Fan, P., Wang, B., 2019. Highly selective hydrogenation of aldehydes promoted by a palladium-based catalyst and its application in equilibrium displacement in a one-enzyme procedure using ω-transaminase. Org. Chem. Front. https://doi.org/10.1039/c9qo00018f

[113] Kanth, J.V.B., Periasamy, M., 1991. Selective reduction of carboxylic acids into alcohols using sodium borohydride and iodine. J. Org. Chem. https://doi.org/10.1021/jo00020a052

[114] Chaikin, S.W., Brown, W.G., 1949. Reduction of Aldehydes, Ketones and Acid Chlorides by Sodium Borohydride. J. Am. Chem. Soc. https://doi.org/10.1021/ja01169a033

[115] Gogin, L.L., Zhizhina, E.G., Pai, Z.P., 2021. Production of Methacrylic Acid and Metacrylates. Catal. Ind. https://doi.org/10.1134/s2070050421020057

[116] DarabiMahboub, M.J., Dubois, J.-L., Cavani, F., Rostamizadeh, M., Patience, G.S., 2018. Catalysis for the synthesis of methacrylic acid and methyl methacrylate. Chem. Soc. Rev. https://doi.org/10.1039/c8cs00117k

[117] Wang, B., Dong, H., Lu, L., Liu, H., Zhang, Z., Zhu, J., 2021. Study on the Development of High-Performance P-Mo-V Catalyst and the Influence of Aldehyde Impurities on Catalytic Performance in Selective Oxidation of Methacrolein to Methacrylic Acid. Catalysts. https://doi.org/10.3390/catal11030394

[118] Jing, F., Katryniok, B., Dumeignil, F., Bordes-Richard, E., Paul, S., 2014. Catalytic selective oxidation of isobutane to methacrylic acid on supported (NH4)3HPMo11VO40 catalysts. Journal of Catalysis. https://doi.org/10.1016/j.jcat.2013.09.014

[119] Zhang, L., Paul, S., Dumeignil, F., Katryniok, B., 2021. Selective Oxidation of Isobutane to Methacrylic Acid and Methacrolein: A Critical Review. Catalysts. https://doi.org/10.3390/catal11070769

[120] Bohre, A., Avasthi, K., Novak, U., Likozar, B., 2021. Single-Step Production of Bio-Based Methyl Methacrylate from Biomass-Derived Organic Acids Using Solid Catalyst Material for Cascade Decarboxylation–Esterification Reactions. ACS Sustainable Chem. Eng. https://doi.org/10.1021/acssuschemeng.0c08914

[121] Zhang, L., Paul, S., Dumeignil, F., Katryniok, B., 2021. Selective Oxidation of Isobutane to Methacrylic Acid and Methacrolein: A Critical Review. Catalysts. https://doi.org/10.3390/catal11070769

[122] DarabiMahboub, M.J., Dubois, J.-L., Cavani, F., Rostamizadeh, M., Patience, G.S., 2018. Catalysis for the synthesis of methacrylic acid and methyl methacrylate. Chem. Soc. Rev. https://doi.org/10.1039/c8cs00117k

[123] Van Grieken, R., Ovejero, G., Serrano, D.P., Uguina, M.A., Melero, J.A., 1996. Synthesis of MTBE from isobutane using a single catalytic system based on titanium-containing ZSM-5 zeolite. Chem. Commun. https://doi.org/10.1039/cc9960001145

[124] Ahmed, S., El-Faer, M.Z., Abdillahi, M.M., Shirokoff, J., Siddiqui, M.A.B., Barri, S.A.I., 1997. Production of methyl tert-butyl ether (MTBE) over MFI-type zeolites synthesized by the rapid crystallization method and modified by varying Si/Ai ratio and steaming. Applied Catalysis A: General. https://doi.org/10.1016/s0926-860x(97)00108-7

[125] Xu, X., Zheng, Y., Zheng, G., 1995. Kinetics and Effectiveness of Catalyst for Synthesis of Methyl tert-Butyl Ether in Catalytic Distillation. Ind. Eng. Chem. Res. https://doi.org/10.1021/ie00046a004

[126] Collignon, F., Loenders, R., Martens, J.A., Jacobs, P.A., Poncelet, G., 1999. Liquid Phase Synthesis of MTBE from Methanol and Isobutene over Acid Zeolites and Amberlyst-15. Journal of Catalysis. https://doi.org/10.1006/jcat.1998.2366

[127] Matouq, M., Tagawa, T., Goto, S., 1993. Liquid-phase synthesis of methyl tert-butyl ether on heterogeneous heteropoly acid catalyst. J. Chem. Eng. Japan / JCEJ. https://doi.org/10.1252/jcej.26.254

Applications for Earth-Abundant Transition Metals                    Materials Research Forum LLC
Materials Research Foundations 179 (2025) 356-399              https://doi.org/21741/9781644903711-13

[128] Nogueira, L.S., Neves, P., Gomes, A.C., Amarante, T.A., Paz, F.A.A., Valente, A.A., Gonçalves, I.S., Pillinger, M., 2018. A Comparative Study of Molybdenum Carbonyl and Oxomolybdenum Derivatives Bearing 1,2,3-Triazole or 1,2,4-Triazole in Catalytic Olefin Epoxidation. Molecules. https://doi.org/10.3390/molecules24010105

[129] López Nieto, J.M., Concepción, P., Dejoz, A., Knözinger, H., Melo, F., Vázquez, M.I., 2000. Selective Oxidation of n-Butane and Butenes over Vanadium-Containing Catalysts. Journal of Catalysis. https://doi.org/10.1006/jcat.1999.2689

[130] Jermy, B.R., Ajayi, B.P., Abussaud, B.A., Asaoka, S., Al-Khattaf, S., 2015. Oxidative dehydrogenation of n-butane to butadiene over Bi–Ni–O/γ-alumina catalyst. Journal of Molecular Catalysis A: Chemical. https://doi.org/10.1016/j.molcata.2015.01.016

[131] Kopač, D., Jurković, D.L., Likozar, B., Huš, M., 2020. First-Principles-Based Multiscale Modelling of Nonoxidative Butane Dehydrogenation on Cr2O3(0001). ACS Catal. https://doi.org/10.1021/acscatal.0c03197

[132] Tanimu, G., Elmutasim, O., Alasiri, H., Polychronopoulou, K., 2023. Unravelling the pathway for the dehydrogenation of n-butane to 1,3-butadiene using thermodynamics and DFT studies. Chemical Engineering Science. https://doi.org/10.1016/j.ces.2023.119059

[133] olpe, M., Tonetto, G., de Lasa, H., 2004. Butane dehydrogenation on vanadium supported catalysts under oxygen free atmosphere. Applied Catalysis A: General. https://doi.org/10.1016/j.apcata.2004.05.017

[134] AITLACHGAR, K., TUEL, A., BRUN, M., HERRMANN, J., KRAFFT, J., MARTIN, J., VOLTA, J., ABON, M., 1998. Selective oxidation of n-butane to maleic anhydride on vanadylpyrophosphateII. Characterization of the oxygen-treated catalyst by electrical conductivity, Raman, XPS, and NMR spectroscopic techniques. Journal of Catalysis. https://doi.org/10.1006/jcat.1998.2097

[135] BIELASKI, A., 1988. Catalytic activity of vanadium oxides in the oxidation of benzene. Journal of Catalysis. https://doi.org/10.1016/0021-9517(88)90262-x

[136] BIELASKI, A., 1988. Catalytic activity of vanadium oxides in the oxidation of benzene. Journal of Catalysis. https://doi.org/10.1016/0021-9517(88)90262-x

[137] CENTI, G., 1984. On the mechanism of n-butane oxidation to maleic anhydride: Oxidation in oxygen-stoichiometry-controlled conditions. Journal of Catalysis. https://doi.org/10.1016/0021-9517(84)90278-

[138] Müller, M., Kutscherauer, M., Böcklein, S., Mestl, G., Turek, T., 2020. On the importance of by-products in the kinetics of n-butane oxidation to maleic anhydride. Chemical Engineering Journal. https://doi.org/10.1016/j.cej.2020.126016

[139] Wang, Q., Zhang, C., Zhu, Z., Arslan, M.T., Yang, L., Wei, F., 2016. Comparison study for the oxidative dehydrogenation of isopentenes to isoprene in fixed and fluidized beds. Catalysis Today. https://doi.org/10.1016/j.cattod.2016.05.032

[140] Aminova, E.K., Islamutdinova, A.A., 2021. Selection of an effective catalyst for the stage isoamyl alcohol dehydration during synthesis isoprene from isopentane. IOP Conf. Ser.: Mater. Sci. Eng. https://doi.org/10.1088/1757-899x/1181/1/012031

[141] Valenzuela, R.X., Muñoz Asperilla, J.M., Corberán, V.C., 2008. Isoprene and C5 Olefins Production by Oxidative Dehydrogenation of Isopentane. Ind. Eng. Chem. Res. https://doi.org/10.1021/ie800756p

[142] Sushkevich, V.L., Ordomsky, V.V., Ivanova, I.I., 2016. Isoprene synthesis from formaldehyde and isobutene over Keggin-type heteropolyacids supported on silica. Catal. Sci. Technol. https://doi.org/10.1039/c6cy00761a

[143] Fridman, V.Z., Xing, R., Severance, M., 2016. Investigating the CrOx/Al2O3 dehydrogenation catalyst model: I. identification and stability evaluation of the Cr species on the fresh and equilibrated catalysts. Applied Catalysis A: General. https://doi.org/10.1016/j.apcata.2016.05.008

[144] Lipinski, B.M., Walker, K.L., Clayman, N.E., Morris, L.S., Jugovic, T.M.E., Roessler, A.G., Getzler, Y.D.Y.L., MacMillan, S.N., Zare, R.N., Zimmerman, P.M., Waymouth, R.M., Coates, G.W., 2020. Mechanistic Study of Isotactic Poly(propylene oxide) Synthesis using a Tethered Bimetallic Chromium Salen Catalyst. ACS Catal. https://doi.org/10.1021/acscatal.0c02135

[145] Ezinkwo, G.O., Tretjakov, V.F., Talyshinky, R.M., Ilolov, A.M., Mutombo, T.A., 2013. Overview of the Catalytic Production of Isoprene from different raw materials; Prospects of Isoprene production from bio-ethanol. Catalysis for Sustainable Energy. https://doi.org/10.2478/cse-2013-0006

[146] Govindaswamy, P., Wada, E., Kono, H., Uozumi, T., Funabashi, H., 2022. Propylene Polymerization Performance with Ziegler-Natta Catalysts Combined with U-Donor and T01 Donor as External Donor. Catalysts. https://doi.org/10.3390/catal12080864

[147] Barabanov, A.A., Vereykina, V.V., Matsko, M.A., Zakharov, V.A., 2021. Propylene polymerization over titanium–magnesium catalysts: The effect of internal and external stereoregulating donors on the number of active centers with different stereospecificity and their reactivity in propagation reaction. Journal of Catalysis. https://doi.org/10.1016/j.jcat.2021.09.023

[148] Huang, J., 1995. Ziegler-Natta catalysts for olefin polymerization: Mechanistic insights from metallocene systems. Progress in Polymer Science. https://doi.org/10.1016/0079-6700(94)00039-5

[149] ntinucci, G., Cannavacciuolo, F.D., Ehm, C., Budzelaar, P.H.M., Cipullo, R., Busico, V., 2024. MgCl2-Supported Ziegler–Natta Catalysts for Propene Polymerization: Before Activation. Macromolecules. https://doi.org/10.1021/acs.macromol.4c00932

[150] Kissin, Y.V., Liu, X., Pollick, D.J., Brungard, N.L., Chang, M., 2008. Ziegler-Natta catalysts for propylene polymerization: Chemistry of reactions leading to the formation of active centers. Journal of Molecular Catalysis A: Chemical. https://doi.org/10.1016/j.molcata.2008.02.026

[151] Kaminsky, W., 2016. Production of Polyolefins by Metallocene Catalysts and Their Recycling by Pyrolysis. Macromolecular Symposia. https://doi.org/10.1002/masy.201500127

Materials Research Forum LLC
https://doi.org/21741/9781644903711-13

[152] Shamiri, A., Chakrabarti, M., Jahan, S., Hussain, M., Kaminsky, W., Aravind, P., Yehye, W., 2014. The Influence of Ziegler-Natta and Metallocene Catalysts on Polyolefin Structure, Properties, and Processing Ability. Materials. https://doi.org/10.3390/ma7075069

[153] Kaminsky, W., Laban, A., 2001. Metallocene catalysis. Applied Catalysis A: General. https://doi.org/10.1016/s0926-860x(01)00829-8

[154] Sheng, T., Qi, Y.-J., Lin, X., Hu, P., Sun, S.-G., Lin, W.-F., 2016. Insights into the mechanism of nitrobenzene reduction to aniline over Pt catalyst and the significance of the adsorption of phenyl group on kinetics. Chemical Engineering Journal. https://doi.org/10.1016/j.cej.2016.02.066

[155] Morisse, C.G.A., McCullagh, A.M., Campbell, J.W., Mitchell, C., Carr, R.H., Lennon, D., 2022. Mechanistic Insight Into the Application of Alumina-Supported Pd Catalysts for the Hydrogenation of Nitrobenzene to Aniline. Ind. Eng. Chem. Res. https://doi.org/10.1021/acs.iecr.2c01134

[156] Morisse, C.G.A., McCullagh, A.M., Campbell, J.W., Mitchell, C., Carr, R.H., Lennon, D., 2022. Mechanistic Insight Into the Application of Alumina-Supported Pd Catalysts for the Hydrogenation of Nitrobenzene to Aniline. Ind. Eng. Chem. Res. https://doi.org/10.1021/acs.iecr.2c01134

[157] Sangeetha, P., Shanthi, K., Rao, K.S.R., Viswanathan, B., Selvam, P., 2009. Hydrogenation of nitrobenzene over palladium-supported catalysts—Effect of support. Applied Catalysis A: General. https://doi.org/10.1016/j.apcata.2008.10.044

[158] Daems, N., Wouters, J., Van Goethem, C., Baert, K., Poleunis, C., Delcorte, A., Hubin, A., Vankelecom, I.F.J., Pescarmona, P.P., 2018. Selective reduction of nitrobenzene to aniline over electrocatalysts based on nitrogen-doped carbons containing non-noble metals. Applied Catalysis B: Environmental. https://doi.org/10.1016/j.apcatb.2017.12.079

[159] Sheng, T., Qi, Y.-J., Lin, X., Hu, P., Sun, S.-G., Lin, W.-F., 2016. Insights into the mechanism of nitrobenzene reduction to aniline over Pt catalyst and the significance of the adsorption of phenyl group on kinetics. Chemical Engineering Journal. https://doi.org/10.1016/j.cej.2016.02.066

[160] Sheng, T., Qi, Y.-J., Lin, X., Hu, P., Sun, S.-G., Lin, W.-F., 2016. Insights into the mechanism of nitrobenzene reduction to aniline over Pt catalyst and the significance of the adsorption of phenyl group on kinetics. Chemical Engineering Journal. https://doi.org/10.1016/j.cej.2016.02.066

[161] Tang, R., Zhou, Y., Xie, L., 2024. Experimental, Kinetics, and Reactor Modeling Studies of the Direct Dehydrogenation of Ethylbenzene to Styrene in the Fixed-Bed Reactor. Ind. Eng. Chem. Res. https://doi.org/10.1021/acs.iecr.4c01175

[162] Venugopal, A.K., Venugopalan, A.T., Kaliyappan, P., Raja, T., 2013. Oxidative dehydrogenation of ethyl benzene to styrene over hydrotalcite derived cerium containing mixed metal oxides. Green Chem. https://doi.org/10.1039/c3gc41321g

[163] Zhu, X.M., Schön, M., Bartmann, U., van Veen, A.C., Muhler, M., 2004. The dehydrogenation of ethylbenzene to styrene over a potassium-promoted iron oxide-based

catalyst: a transient kinetic study. Applied Catalysis A: General. https://doi.org/10.1016/j.apcata.2004.02.002

[164] Ye, X., Hua, W., Yue, Y., Dai, W., Miao, C., Xie, Z., Gao, Z., 2004. Ethylbenzene dehydrogenation to styrene in the presence of carbon dioxide over chromia-based catalysts. New J. Chem. https://doi.org/10.1039/b308450g

[165] Węgrzyniak, A., Jarczewski, S., Węgrzynowicz, A., Michorczyk, B., Kuśtrowski, P., Michorczyk, P., 2017. Catalytic Behavior of Chromium Oxide Supported on Nanocasting-Prepared Mesoporous Alumina in Dehydrogenation of Propane. Nanomaterials. https://doi.org/10.3390/nano7090249

[166] Zhang, K., Miao, P., Zhang, H., Wang, Y., Wang, G., Zhu, X., Li, C., 2021. Research on ethylbenzene dehydrogenation over the Fe-Al-based catalysts in a circulating fluidized-bed unit. Journal of the Taiwan Institute of Chemical Engineers. https://doi.org/10.1016/j.jtice.2021.08.049

[167] Castro, A.J.R., Soares, J.M., Filho, J.M., Oliveira, A.C., Campos, A., Milet, É.R.C., 2013. Oxidative dehydrogenation of ethylbenzene with CO2 for styrene production over porous iron-based catalysts. Fuel. https://doi.org/10.1016/j.fuel.2013.02.019

[168] Kumar, R., Shah, S., Paramita Das, P., Bhagavanbhai, G.G.K., Al Fatesh, A., Chowdhury, B., 2019. An overview of caprolactam synthesis. Catalysis Reviews. https://doi.org/10.1080/01614940.2019.165087

[169] Du, Y., Chen, X., Shen, W., Liu, H., Fang, M., Liu, J., Liang, C., 2023. Electrocatalysis as an efficient alternative to thermal catalysis over PtRu bimetallic catalysts for hydrogenation of benzoic acid derivatives. Green Chem. https://doi.org/10.1039/d3gc01540h

[170] Tang, M., Mao, S., Li, X., Chen, C., Li, M., Wang, Y., 2017. Highly effective Ir-based catalysts for benzoic acid hydrogenation: experiment- and theory-guided catalyst rational design. Green Chem. https://doi.org/10.1039/c7gc00387k

[171] Guo, M., Kong, X., Li, C., Yang, Q., 2021. Hydrogenation of benzoic acid derivatives over Pt/TiO2 under mild conditions. Commun Chem. https://doi.org/10.1038/s42004-021-00489-z

[172] alukdar, A.K., Bhattacharyya, K.G., Sivasanker, S., 1993. Hydrogenation of phenol over supported platinum and palladium catalysts. Applied Catalysis A: General. https://doi.org/10.1016/0926-860x(90)80012-4

[173] He, H., Meyer, R.J., Rioux, R.M., Janik, M.J., 2021. Catalyst Design for Selective Hydrogenation of Benzene to Cyclohexene through Density Functional Theory and Microkinetic Modeling. ACS Catal. https://doi.org/10.1021/acscatal.1c02630

[174] Mohammadian, Z., Parsafard, N., 2022. Optimization of Kinetic Study by Response Surface Methodology on the Various Types Ni Supported Catalysts in Competitive Benzene Hydrogenation. Theor Found Chem Eng. https://doi.org/10.1134/s0040579522330053

[175] Fang, X., Yin, Z., Wang, H., Li, J., Liang, X., Kang, J., He, B., 2015. Controllable oxidation of cyclohexane to cyclohexanol and cyclohexanone by a nano-MnOx/Tielectrocatalytic membrane reactor. Journal of Catalysis. https://doi.org/10.1016/j.jcat.2015.05.004

[176] Kumar, R., Sithambaram, S., Suib, S.L., 2009. Cyclohexane oxidation catalyzed by manganese oxide octahedral molecular sieves—Effect of acidity of the catalyst. Journal of Catalysis. https://doi.org/10.1016/j.jcat.2009.01.007

[177] Yuan, H.-X., Xia, Q.-H., Zhan, H.-J., Lu, X.-H., Su, K.-X., 2006. Catalytic oxidation of cyclohexane to cyclohexanone and cyclohexanol by oxygen in a solvent-free system over metal-containing ZSM-5 catalysts. Applied Catalysis A: General. https://doi.org/10.1016/j.apcata.2006.02.037

[178] Wang, K., Wang, F., Zhai, Y., Wang, J., Zhang, Xu, Li, M., Jiang, L., Fan, X., Bing, C., Zhang, J., Zhang, Xubin, 2023. Application of zeolite in Beckmann rearrangement of cyclohexanoneoxime. Molecular Catalysis. https://doi.org/10.1016/j.mcat.2022.112881

[179] Andrade, M.A., Martins, L.M.D.R.S., 2019. Sustainability in Catalytic Cyclohexane Oxidation: The Contribution of Porous Support Materials. Catalysts. https://doi.org/10.3390/catal10010002

[180] Marziano, N.C., Ronchin, L., Tortato, C., Vavasori, A., Badetti, C., 2007. Catalyzed Beckmann rearrangement of cyclohexanoneoxime in heterogeneous liquid/solid system. Journal of Molecular Catalysis A: Chemical. https://doi.org/10.1016/j.molcata.2007.07.046

[181] Kaur, K., Srivastava, S., 2020. Beckmann rearrangement catalysis: a review of recent advances. New J. Chem. https://doi.org/10.1039/d0nj02034f

[182] Tomás, R.A.F., Bordado, J.C.M., Gomes, J.F.P., 2013. p-Xylene Oxidation to Terephthalic Acid: A Literature Review Oriented toward Process Optimization and Development. Chem. Rev. https://doi.org/10.1021/cr300298j

[183] Li, K.-T., Li, S.-W., 2008. CoBr2-MnBr2 containing catalysts for catalytic oxidation of p-xylene to terephthalic acid. Applied Catalysis A: General. https://doi.org/10.1016/j.apcata.2008.02.025

[184] Lapa, H.M., Martins, L.M.D.R.S., 2023. p-Xylene Oxidation to Terephthalic Acid: New Trends. Molecules. https://doi.org/10.3390/molecules28041922

# Keyword Index

# About the Editors

**Dr. Inamuddin** is working as an Assistant Professor at the Department of Applied Chemistry, Aligarh Muslim University, Aligarh, India. He obtained a Master of Science degree in Organic Chemistry from Chaudhary Charan Singh (CCS) University, Meerut, India, in 2002. He received his Master of Philosophy and Doctor of Philosophy degrees in Applied Chemistry from Aligarh Muslim University (AMU), India, in 2004 and 2007, respectively. He has extensive research experience in multidisciplinary fields of Analytical Chemistry, Materials Chemistry, and Electrochemistry and, more specifically, Renewable Energy and Environment. He has worked on different research projects as a project fellow and senior research fellow funded by the University Grants Commission (UGC), Government of India, and the Council of Scientific and Industrial Research (CSIR), Government of India. He has received the Fast Track Young Scientist Award from the Department of Science and Technology, India, to work in the area of bending actuators and artificial muscles. He has also received the Sir Syed Young Researcher of the Year Award 2020 from Aligarh Muslim University. He has completed four major research projects sanctioned by the University Grant Commission, Department of Science and Technology, Council of Scientific and Industrial Research, and Council of Science and Technology, India. He has published 216 research articles in international journals of repute and nineteen book chapters in knowledge-based book editions published by renowned international publishers. He has published 200 edited books with Springer (U.K.), Elsevier, Nova Science Publishers, Inc. (U.S.A.), CRC Press Taylor & Francis Asia Pacific, Trans Tech Publications Ltd. (Switzerland), IntechOpen Limited (U.K.), Wiley-Scrivener, (U.S.A.) and Materials Research Forum LLC (U.S.A). He is a member of various journals' editorial boards. He has served as Associate Editor for journals (Environmental Chemistry Letter, Applied Water Science and Euro-Mediterranean Journal for Environmental Integration, Springer-Nature), Frontiers Section Editor (Current Analytical Chemistry, Bentham Science Publishers), Editorial Board Member (Scientific Reports-Nature)and Review Editor (Frontiers in Chemistry, Frontiers, U.K.) He has also guest-edited various special thematic issues for the journals of Elsevier, Bentham Science Publishers, and John Wiley & Sons, Inc. He has attended as well as chaired sessions at various international and national conferences. He has worked as a Postdoctoral Fellow, leading a research team at the Creative Research Initiative Center for Bio-Artificial Muscle, Hanyang University, South Korea, in the field of renewable energy, especially biofuel cells. He has also worked as a Postdoctoral Fellow at the Center of Research Excellence in Renewable Energy, King Fahd University of Petroleum and Minerals, Saudi Arabia, in the field of polymer electrolyte membrane fuel cells and computational fluid dynamics of polymer electrolyte membrane fuel cells. He is a life member of the Journal of the Indian Chemical Society. His research interest includes ion exchange materials, a sensor for heavy metal ions, biofuel cells, supercapacitors and bending actuators.

**Dr. Arwa Alrooqi** is an Assistant Professor in the Department of Chemistry, Faculty of Science, at Al Baha University, Saudi Arabia. She has published several research articles in internationally recognized journals. Dr. Alrooqi is actively engaged in multidisciplinary research, focusing on the synthesis and engineering of organic semiconductors and nanomaterials, their characterization, and their application in smart windows.

**Dr. Hind Alluqmani** is an Assistant Professor of Analytical Chemistry in the Department of Chemistry, Faculty of Science, at Al Baha University, Saudi Arabia. She earned her Ph.D. with a focus on the "Analytical Study of Physical Behavior in Clay Hydrogel Media." Her research expertise includes surface analyses, such as scanning electron microscopy (SEM) and associated techniques, including surface area (BET) measurements, energy-dispersive X-ray spectroscopy (EDX), and powder X-ray diffraction (PXRD). Additionally, she specializes in thermal analysis, including thermogravimetric analysis (TGA) and differential scanning calorimetry (DSC), as well as rheological analysis.

**Dr. Mohammad A. Jafar Mazumder** has been serving as a Professor of Chemistry at King Fahd University of Petroleum & Minerals (KFUPM), Saudi Arabia. He has extensive experience in designing, synthesizing, and characterizing various organic compounds, ionic and thermo-responsive polymers for corrosion, water treatment, and biomedical applications. Dr. Jafar Mazumder obtained his B.Sc (Hons.), M.Sc (Chemistry) from Aligarh Muslim University, India, MS (Chemistry) from KFUPM, Saudi Arabia, and Ph.D. in Chemistry (2009) from McMaster University, Canada.

In more than 20 years of academic research, Dr. Jafar Mazumder has had the opportunity to work with several international collaborative research groups and has exposed himself to a broad range of research areas. Dr. Jafar Mazumder secured 8 US patents, published more than 85 articles in peer-reviewed journals, 37 conference proceedings, 9 book chapters, and co-edited 4 books with Springers and Trans Tech publications. He is awarded as a Fellow of the Royal Society of Chemistry and Chartered Chemist, Association of Chemical Profession of Ontario, Canada. Besides, Dr. Jafar Mazumder is a member of the American Chemical Society (ACS), Canadian Society for Chemistry (CSC), Canadian Biomaterial Society (CBS), and a life member of the Bangladesh Chemical Society (BCS). In his academic career, he was awarded numerous national and international scholarships and awards including the prestigious Indian Council for Cultural Relations (ICCR) Scholarship from Govt. of India for undergraduate studies in India, Aligarh Muslim University undergraduate & graduate Gold medal, and certificate of excellence from the Ministry of Human Resource Development, Govt. of India, and MITACS postdoctoral fellowship (Canada) for pursuing postdoctoral research in Chemical and Biomedical Engineering.

Currently, Dr. Jafar Mazumder is actively involved in several ongoing university (KFUPM), government (KACST, NSTIP), and client (Saudi Aramco) funded projects in the capacity of principal and co-investigators. His current research interest includes the design, synthesis, and characterization of various modified monomers and polymers for potential use in the inhibition of mild steel corrosion in oil and gas industries and the preparation of multilayered polyelectrolyte coated membranes for the removal of heavy metals and organic contaminants from aqueous water samples. The long-term scientific goal of Dr. Jafar Mazumder is not merely to make science fun and entertaining for people. It is to engage them with a multidisciplinary scientific mission at a deeper level to create a space through which they can interact with scientific ideas, develop connections between science, engineering, and biology, and thoughts of their own to contribute to society. He feels this goal and engaging personality make him a pleasant person to work with and help inspire his co-workers in any professional setting.

www.ingramcontent.com/pod-product-compliance
Lightning Source LLC
Chambersburg PA
CBHW071316210326
41597CB00015B/1246